QUANTUM THEORY AT THE CROSSROADS

The 1927 Solvay conference was perhaps the most important meeting in the history of quantum theory. Contrary to popular belief, the interpretation of quantum theory was not settled at this conference, and no consensus was reached. Instead, a range of sharply conflicting views were presented and extensively discussed, including de Broglie's pilot-wave theory, Born and Heisenberg's quantum mechanics, and Schrödinger's wave mechanics. Today, there is no longer an established or dominant interpretation of quantum theory, so it is important to re-evaluate the historical sources and keep the interpretation debate open.

This book contains a complete translation of the original proceedings, with background essays on the three main interpretations of quantum theory presented at the conference, and an extensive analysis of the lectures and discussions in the light of current research in the foundations of quantum theory. The proceedings contain much unexpected material, including extensive discussions of de Broglie's pilot-wave theory (which de Broglie presented for a many-body system), and a theory of 'quantum mechanics' apparently lacking in wave function collapse or fundamental time evolution. This book will be of interest to graduate students and researchers in physics and in the history and philosophy of quantum theory.

GUIDO BACCIAGALUPPI is a Reader in the Department of Philosophy, University of Aberdeen. His research interests lie mainly in the philosophy of physics. He has contributed significantly to the development and critique of modal interpretations of quantum mechanics. He has since worked widely in various approaches to the foundations of quantum theory, as well as in the philosophy of probability and of time and in the history of quantum mechanics.

ANTONY VALENTINI is Professor of Theoretical Quantum Physics in the Department of Physics and Astronomy, Clemson University. He has proposed that the universe began with a non-quantum distribution of hidden variables, which later relaxed to the quantum equilibrium state we see today. He has pioneered the development of the new physics of quantum non-equilibrium, in de Broglie-Bohm theory and in hidden-variable theories generally, and has explored its possible role in cosmology, black holes and information theory. He also works in the history and philosophy of modern physics.

QUANTUM THEORY AT THE CROSSROADS

Reconsidering the 1927 Solvay Conference

GUIDO BACCIAGALUPPI
ANTONY VALENTINI

CAMBRIDGE
UNIVERSITY PRESS

University Printing House, Cambridge CB2 8BS, United Kingdom

Published in the United States of America by Cambridge University Press, New York

Cambridge University Press is part of the University of Cambridge.

It furthers the University's mission by disseminating knowledge in the pursuit of education, learning and research at the highest international levels of excellence.

www.cambridge.org
Information on this title: www.cambridge.org/9781107698314

First published 2009
First paperback edition 2013

A catalogue record for this publication is available from the British Library

Library of Congress Cataloguing in Publication data
Bacciagaluppi, Guido.
Quantum theory at the crossroads : reconsidering the 1927 Solvay conference /
Guido Bacciagaluppi,
Antony Valentini.
p. cm.
Includes bibliographical references and index.
1. Quantum theory–Congresses. I. Valentini, Antony. II. Title.
QC173.96.B33 2008
530.12–dc22
2008019585

ISBN 978-0-521-81421-8 Hardback
ISBN 978-1-107-69831-4 Paperback

To the memory of James T. Cushing

Contents

List of illustrations

Parts I and II

Chapter 6

Chapter 7

Chapter 10

Part III

W. L. Bragg's report

A. H. Compton's report

Discussion of A. H. Compton's report

Discussion of L. de Broglie's report

General discussion

Preface

> And they said one to another: Go to, let us build us a tower, whose top may reach unto heaven; and let us make us a name. And the Lord said: Go to, let us go down, and there confound their language, that they may not understand one another's speech.
>
> *(Genesis 11: 3–7)*

Anyone who has taken part in a debate on the interpretation of quantum theory will recognise how fitting is the above quotation from the book of Genesis, according to which the builders of the Tower of Babel found that they could no longer understand one another's speech. For when it comes to the interpretation of quantum theory, even the most clear-thinking and capable physicists are often unable to understand each other.

This state of affairs dates back to the genesis of quantum theory itself. In October 1927, during the 'general discussion' that took place in Brussels at the end of the fifth Solvay conference, Paul Ehrenfest wrote the above lines on the blackboard. As Langevin later remarked, the Solvay meeting in 1927 was the conference where 'the confusion of ideas reached its peak'.

Ehrenfest's perceptive gesture captured the essence of a situation that has persisted for three-quarters of a century. According to widespread historical folklore, the deep differences of opinion among the leading physicists of the day led to intense debates, which were satisfactorily resolved by Bohr and Heisenberg around the time of the 1927 Solvay meeting. But in fact, at the end of 1927, a significant number of the main participants (in particular de Broglie, Einstein and Schrödinger) remained unconvinced, and the deep differences of opinion were never really resolved.

The interpretation of quantum theory seems as highly controversial today as it was in 1927. There has also been criticism – on the part of historians as well as physicists – of the tactics used by Bohr and others to propagate their views in the late 1920s, and a realisation that alternative ideas may have been dismissed or unfairly disparaged. For many physicists, a sense of unease lingers over the whole subject. Might it be that things are not as clear-cut as Bohr and Heisenberg would have us believe? Might it be that their opponents had something important to say after all? Because today there is no longer an established interpretation of quantum mechanics, we feel it is important to go back to the sources and re-evaluate them.

In this spirit, we offer the reader a return to a time just before the Copenhagen interpretation was widely accepted, when the best physicists of the day gathered to discuss a range of views, concerning many topics of interest today (measurement, determinism, non-locality, subjectivity, interference and so on), and when three distinct theories – de Broglie's pilot-wave theory, Born and Heisenberg's quantum mechanics and Schrödinger's wave mechanics – were presented and discussed on an equal footing.

*

Since the 1930s, and especially since the Second World War, it has been common to dismiss questions about the interpretation of quantum theory as 'metaphysical' or 'just philosophical'. It will be clear from the lively and wide-ranging discussions of 1927 that at that time, for the most distinguished physicists of the day, the issues were decidedly *physical*: Is the electron a point particle with a continuous trajectory (de Broglie), or a wave packet (Schrödinger), or neither (Born and Heisenberg)? Do quantum outcomes occur when nature makes a choice (Dirac), or when an observer decides to record them (Heisenberg)? Is the non-locality of quantum theory compatible with relativity (Einstein)? Can a theory with trajectories account for the recoil of a single photon on a mirror (Kramers, de Broglie)? Is indeterminism a fundamental limitation, or merely the outcome of coarse-graining over something deeper and deterministic (Lorentz)?

After 1927, the Copenhagen interpretation became firmly established. Rival views were marginalised, in particular those represented by de Broglie, Schrödinger and Einstein, even though these scientists were responsible for many of the major developments in quantum physics itself. (This marginalisation is apparent in most historical accounts written throughout the twentieth century.) From the very beginning, however, there were some notes of caution: for example, when Bohr's landmark paper of 1928 (the English version of his famous Como lecture) was published in *Nature*, an editorial preface expressed dissatisfaction with the 'somewhat vague statistical description' and ended with the hope that this would not be the 'last word on the subject'. And there were a few outstanding alarm bells, in particular the famous paper by Einstein, Podolsky and Rosen in 1935, and the important papers by Schrödinger (in the same year) on the cat paradox and on entanglement. But on the whole, the questioning ceased in all but a few corners. A general opinion arose that the questions had been essentially settled, and that a satisfactory point of view had been arrived at, principally through the work of Bohr and Heisenberg. For subsequent generations of physicists, 'shut up and calculate' emerged as the working rule among the vast majority.

Despite this atmosphere, the questioning never completely died out, and some very significant work was published, for example by Bohm in 1952, Everett in 1957 and Bell in 1964 and 1966. But attitudes changed very slowly. Younger physicists were strongly discouraged from pursuing such questions. Those who persisted generally had difficult careers, and much of the careful thinking about quantum foundations was relegated to departments of philosophy.

Nevertheless, the closing decade of the twentieth century saw a resurgence of interest in the foundations of quantum theory. At the time of writing, a range of alternatives (such as hidden variables, many worlds, collapse models, among others) are being actively pursued, and the Copenhagen interpretation can no longer claim to be the dominant or 'orthodox' interpretation.

The modern reader familiar with current debates and positions in quantum foundations will recognise many of the standard points of view in the discussions reproduced here, although expressed with a remarkable concision and clarity. This provides a welcome contrast with the generally poor level of debate today: as the distinguished cosmologist Dennis Sciama was fond of pointing out, when it comes to the interpretation of quantum theory 'the standard of argument suddenly drops to zero'. We hope that the publication of this book will contribute to a revival of sharp and informed debate about the meaning of quantum theory.

*

Remarkably, the proceedings of the fifth Solvay conference have not received the attention they deserve, neither from physicists nor from historians, and the literature contains numerous major misunderstandings about what took place there.

The fifth Solvay conference is usually remembered for the clash that took place between Einstein and Bohr over the uncertainty relations. It is remarkable, then, to find that not a word of these discussions appears in the published proceedings. It is known that Einstein and Bohr engaged in vigorous informal discussions, but in the formal debates recorded in the proceedings they were relatively silent. Bohr did contribute to the general discussion, but this material was not published. Instead, at Bohr's request, it was replaced by a translation of the German version of his Como lecture, which appeared in *Naturwissenschaften* in 1928. (We do not reproduce this well-known paper here.) The appending of this translation to the published proceedings may be the cause of the common misunderstanding that Bohr gave a report at the conference: in fact, he did not.

Born and Heisenberg present a number of unfamiliar viewpoints concerning, among other things, the nature of the wave function and the role of time and of probability in quantum theory. Particularly surprising is the seeming absence of a collapse postulate in their formulation, and the apparently phenomenological (or effective) status of the time-dependent Schrödinger equation. Born and Heisenberg's 'quantum mechanics' seems remarkably different from quantum mechanics (in the Dirac–von Neumann formulation) as we know it today.

De Broglie's pilot-wave theory was the subject of extensive and varied discussions. This is rather startling in view of the claim – in Max Jammer's classic historical study *The Philosophy of Quantum Mechanics* – that de Broglie's theory 'was hardly discussed at all' and that 'the only serious reaction came from Pauli' (Jammer 1974, pp. 110–11). Jammer's view is typical even today. But in the published proceedings, at the end of de Broglie's report there are 9 pages of discussion devoted to de Broglie's theory; and of the 42 pages of general discussion, 15 contain discussion of de Broglie's theory, with serious reactions and comments coming not only from Pauli but also from Born, Brillouin, Einstein, Kramers, Lorentz, Schrödinger and others. Even the well-known exchange between Pauli and de Broglie has been widely misunderstood.

Finally, another surprise is that in his report de Broglie proposed the many-body pilot-wave dynamics for a system of particles, with the total configuration guided by a wave in configuration space, and not just (as is generally believed) the one-body theory in 3-space. De Broglie's theory is essentially the same as that developed by Bohm in 1952, the only difference being that de Broglie's dynamics (like the form of pilot-wave theory popularised by Bell) is formulated in terms of velocity rather than acceleration.

*

This work is a translation of and commentary on the proceedings of the fifth Solvay conference of 1927, which were published in French in 1928 under the title *Électrons et Photons*.

We have not attempted to give an exhaustive historical analysis of the fifth Solvay conference. Rather, our main aims have been to present the material in a manner accessible to the general physicist, and to situate the proceedings in the context of current research in quantum foundations. We hope that the book will contribute to stimulating and reviving serious debate about quantum foundations in the wider physics community, and that making the proceedings available in English will encourage historians and philosophers to reconsider their significance.

Part I begins with a historical introduction and provides essays on the three main theories presented at the conference (pilot-wave theory, quantum mechanics, wave mechanics). The lectures and discussions that took place at the fifth Solvay conference contain an extensive range of material that is relevant to current research in the foundations of quantum theory. In Part II, after a brief review of the status of quantum foundations today, we summarise what seem to us to be the highlights of the conference, from the point of view of current debates about the meaning of quantum theory. Part III of the book consists of translations of the reports, of the discussions following them, and of the general discussion. Wherever possible, the original (in particular English or German) texts have been used. We have tacitly corrected minor mistakes in punctuation and spelling, and we have uniformised the style of equations, references and footnotes. (Unless otherwise specified, all translations of quotations are ours.)

Part I (except for Chapter 2), the reports by Compton, by Born and Heisenberg and by Schrödinger, and the Appendix to Part III are principally the work of Guido Bacciagaluppi. Chapter 2, all of Part II, the reports by Bragg and by de Broglie, and the general discussion in Part III are principally the work of Antony Valentini.

Chapters 2, 10 and 11 are based on a seminar, 'The early history of Louis de Broglie's pilot-wave dynamics', given by Antony Valentini at the University of Notre Dame in September 1997, at a conference in honour of the sixtieth birthday of the late James T. Cushing.

*

To James T. Cushing, physicist, philosopher, historian and gentleman, we both owe a special and heartfelt thanks. It was he who brought us together on this project, and to him we are indebted for his encouragement and, above all, his example. This book is dedicated to his memory.

Guido Bacciagaluppi wishes to express his thanks to the Humboldt Foundation, which supported the bulk of his work in the form of an Alexander-von-Humboldt Forschungsstipendium, and to his hosts in Germany, Carsten Held and the Philosophisches Seminar I, University of Freiburg, and Harald Atmanspacher and the Institut für Grenzgebiete der Psychologie und Psychohygiene, Freiburg, as well as to Jacques Dubucs and the Institut d'Histoire et de Philosophie des Sciences et des Techniques (CNRS, Paris 1, ENS), for support during the final phase. He also wishes to thank Didier Devriese of the Université Libre de Bruxelles, who is in charge of the archives of the Instituts Internationaux de Physique et de Chimie

Solvay, Université Libre de Bruxelles, for his kindness and availability, and Brigitte Parakenings (formerly Uhlemann) and her staff at the Philosophisches Archiv of the University of Konstanz, for the continuous assistance with the Archive for the History of Quantum Physics. Finally, he wishes to thank Jeff Barrett for suggesting this project to him in Utrecht one day back in 1996, as well as Mark van Atten, Jennifer Bailey, Olivier Darrigol, Felicity Pors, Gregor Schiemann and many others for discussions, suggestions, correspondence, references and other help.

Antony Valentini began studying these fascinating proceedings while holding a postdoctoral position at the University of Rome 'La Sapienza' (1994–6), and is grateful to Marcello Cini, Bruno Bertotti and Dennis Sciama for their support and encouragement during that period. For support in recent years, he is grateful to Perimeter Institute, and wishes to express a special thanks to Howard Burton, Lucien Hardy and Lee Smolin.

We are both grateful in particular to Tamsin van Essen at Cambridge University Press for her support and encouragement during most of the gestation of this book, and to Augustus College for support during the final stages of this work.

Abbreviations

AEA: Albert Einstein Archives, Jewish National and University Library, Hebrew University of Jerusalem.

AHQP: Archive for the History of Quantum Physics.

AHQP-BSC: Bohr Scientific Correspondence, microfilmed from the Niels Bohr Arkiv, Copenhagen.

AHQP-BMSS: Bohr Scientific Manuscripts, microfilmed from the Niels Bohr Arkiv, Copenhagen.

AHQP-EHR: Ehrenfest collection, microfilmed from the Rijksmuseum voor de Geschiedenis van de Natuurwetenschappen en van de Geneeskunde 'Museum Boerhaave', Leiden.

AHQP-LTZ: Lorentz collection, microfilmed from the Algemeen Rijksarchief, Den Haag (now in the Noord-Hollands Archief, Haarlem).

AHQP-RDN: Richardson Collection, microfilmed from the Harry Ransom Humanities Research Center, University of Texas at Austin.

AHQP-OHI: Oral history interview transcripts.

IIPCS: Archives of the Instituts Internationaux de Physique et de Chimie Solvay, Université Libre de Bruxelles.

Ann. d. Phys. or *Ann. der Phys.*: Annalen der Physik.

Bayr. Akad. d. Wiss. Math. phys. Kl.: Sitzungsberichte der Mathematisch-Physikalischen Klasse der Königlich-Bayerischen Akademie der Wissenschaften (München).

Berl. Ber.: Sitzungsberichte der Preussischen Akademie der Wissenschaften (Berlin).

Acad. Roy. Belg. or *Bull. Ac. R. Belg.* or *Bull. Ac. roy. de Belgique* or *Bull. Ac. roy. Belgique* or *Bull. Ac. roy. Belg.* or *Bull. Ac. R. Belg., Cl. des Sciences*: Bulletin de l'Académie Royale des Sciences, des Lettres et des Beaux-arts de Belgique. Classe des Sciences.

Bull. Natl. Res. Coun.: Bulletin of the National Research Council (U.S.).

Comm. Fenn.: Commentationes Physico-mathematicae, Societas Scientiarum Fennica.

C. R. or *C. R. Acad. Sc.* or *Comptes Rendus Acad. Sci. Paris*: Comptes Rendus Hebdomadaires des Séances de l'Académie des Sciences (Paris).

Gött. Nachr.: Nachrichten der Akademie der Wissenschaften in Göttingen. II., Mathematisch-Physikalische Klasse.

J. de Phys. or *Jour. de Phys.* or *Journ. Physique* or *Journ. d. Phys.*: Journal de Physique (until 1919), then Journal de Physique et le Radium.

Jour. Frank. Inst.: Journal of the Franklin Institute.

Lincei Rend.: Rendiconti Lincei.

Manchester Memoirs: Manchester Literary and Philosophical Society, Memoirs and Proceedings.

Math. Ann. or *Mathem. Ann.*: Mathematische Annalen.

Naturw. or *Naturwiss.* or *Naturwissensch.* or *Naturwissenschaften*: Die Naturwissenschaften.

Nat. Acad. Sci. Proc. or *Proc. Nat. Acad. Sci.* or *Proc. Nat. Acad.*: Proceedings of the National Academy of Sciences (U.S.).

Phil. Mag.: Philosophical Magazine.

Phil. Trans. or *Phil. Trans. Roy. Soc.*: Philosophical Transactions of the Royal Society of London.

Phys. Rev.: Physical Review.

Phys. Zeits. or *Phys. Zeitsch.* or *Physik. Zts.*: Physikalische Zeitschrift.

Proc. Camb. Phil. Soc. or *Proc. Cambr. Phil. Soc.* or *Proc. Cambridge Phil. Soc.*: Proceedings of the Cambridge Philosophical Society.

Proc. Phys. Soc.: Proceedings of the Physical Society of London.

Proc. Roy. Soc. or *Roy. Soc. Proc.*: Proceedings of the Royal Society of London.

Upsala Univ. Årsskr.: Uppsala Universitets Årsskrift.

Z. f. Phys. or *Zts. f. Phys.* or *Zeit. f. Phys.* or *Zeits. f. Phys.* or *Zeitsch. f. Phys.* or *Zeitschr. f. Phys.*: Zeitschrift für Physik.

Typographic conventions

The following conventions have been used.

Square brackets [ ] denote editorial amendments or (in the translations) original wordings; [?] denotes an uncertain reading.

Curly brackets { } denote additions (in original typescripts or manuscripts).

Angular brackets < > denote deletions (in original typescripts or manuscripts).

Note on the bibliography and the index

The references cited in Parts I and II, and in the endnotes and editorial footnotes to Part III, are listed in our bibliography. The references cited in the original Solvay volume are found in the translation of the proceedings in Part III.

In the index, under entries for the conference participants, italic page numbers indicate spoken contributions.

Permissions and copyright notices

The material in the original published proceedings of the fifth Solvay conference has been translated and used courtesy of the Solvay institutes. We wish to record our personal thanks to the director, Prof. Marc Henneaux.

The frontispiece of the original publication is reproduced courtesy of Dunod Éditeur, Paris.

We have further prepared our versions of the conference reports with the following permissions.

The report by W. L. Bragg: courtesy of the Royal Institution of Great Britain.

The report by A. H. Compton: courtesy of the *The Journal of the Franklin Institute*, for the use of the version of Compton's paper in volume **205** (1928), pp. 155–78.

The report by L. de Broglie: courtesy of Prof. Georges Lochak and the Fondation Louis de Broglie.

The report by M. Born and W. Heisenberg: courtesy of Prof. Gustav Born and of the Werner Heisenberg Archive, Munich.

The report by E. Schrödinger: courtesy of Mrs Ruth Braunizer.

Parts of the general discussion have employed material with the following permissions.

Richardson's notes from the general discussion: courtesy of the Harry Ransom Humanities Research Center, The University of Texas at Austin, and courtesy of Dr Peter J. Richardson.

Einstein's manuscript of the beginning of his discussion contribution: ©Albert Einstein Archives, the Hebrew University of Jerusalem, Israel. Einstein's other discussion contributions: courtesy of the Albert Einstein Archives, The Hebrew University of Jerusalem.

Dirac's manuscript of part of his discussion contributions: courtesy of the University Libraries, Florida State University, Tallahassee, FL.

The published excerpts of Bohr's discussion contributions, from N. Bohr, *Collected Works*, Vol. 6, ed. J. Kalckar (1985), pp. 103 and 105: ©Elsevier and The Niels Bohr Archive, Copenhagen.

The Appendix is reproduced courtesy of The Niels Bohr Archive and (for Kramers' notes) Ms. S. M. Kramers. We thank the Niels Bohr Archive also for permission to quote from other materials in their collections, and thank in particular Dr Felicity Pors for her assistance.

Unpublished archival material in Parts I and II is quoted in part courtesy of the Archives of the Instituts Internationaux de Physique et de Chimie Solvay, Université Libre de Bruxelles, where we thank in particular Prof. Didier Devriese, and with the following permissions.

The letters by Born: courtesy of Prof. Gustav Born.

The letters by W. H. Bragg and W. L. Bragg: courtesy of the Royal Institution of Great Britain.

The interviews with de Broglie: courtesy of Prof. Georges Lochak and the Fondation Louis de Broglie.

The remark by Ehrenfest on pp. 3 and 8: courtesy of Museum Boerhaave, Leiden.

The letters by Heisenberg: courtesy of the Werner Heisenberg Archive, Munich, where we thank in particular Prof. Helmut Rechenberg.

The materials by Lorentz: courtesy of the Noord-Hollands Archief, Haarlem.

The letter from Schrödinger to Lorentz, 23 June 1927, AHQP-LTZ-13 and AHQP-41: courtesy of Mrs Ruth Braunizer.

The cover photographs are reproduced with the following permissions.

Portrait of Einstein: AIP Emilio Segrè Visual Archives, W. F. Meggers Gallery of Nobel Laureates.

Portrait of de Broglie: AIP Emilio Segrè Visual Archives, Physics Today Collection.

Portrait of Heisenberg: AIP Emilio Segrè Visual Archives, Segrè Collection.

Portrait of Schrödinger: AIP Emilio Segrè Visual Archives.

Portrait of Born: 'Voltiana', Como, Italy – 10 September 1927 issue, courtesy AIP Emilio Segrè Visual Archives.

Portrait of Bohr: National Archives and Records Administration, courtesy AIP Emilio Segrè Visual Archives.

Group photograph of the fifth Solvay conference: photograph by Benjamin Couprie, Institut International de Physique Solvay, courtesy AIP Emilio Segrè Visual Archives.

Part I

Perspectives on the 1927 Solvay conference

1
Historical introduction

Quantum reconciliation very [added, deleted] unpleasant [deleted] tendency [deleted] retrograde [deleted] questionable [added, deleted] idea [deleted] flippant [deleted] title leads to misunderstanding.

(Ehrenfest, on the conference plans[1])

The conference was surely the most interesting scientific conference I have taken part in so far.

(Heisenberg, upon receipt of the conference photograph[2])

The early Solvay conferences were remarkable occasions, made possible by the generosity of Belgian industrialist Ernest Solvay and, with the exception of the first conference in 1911, planned and organised by the indefatigable Hendrik Antoon Lorentz. In this chapter, we shall first sketch the beginnings of the Solvay conferences, Lorentz's involvement and the situation in the years leading up to 1927 (Sections 1.1 and 1.2). Then we shall describe specifically the planning of the fifth Solvay conference, both in its scientific aspects (Section 1.3) and in its more practical aspects (Section 1.4). Section 1.5 presents the day-by-day progress of the conference as far as it can be reconstructed from the sources, while Section 1.6 follows the making of the volume of proceedings, which is the main source of original material from the fifth Solvay conference and forms Part III of this book.

1.1 Ernest Solvay and the Institute of Physics

Ernest Solvay had an extensive record of supporting scientific, educational and social initiatives, as Lorentz emphasises in a two-page document written in September 1914, during the first months of the First World War:[3]

I feel bound to say some words in these days about one of Belgium's noblest citizens, one of the men whom I admire and honour most highly. Mr Ernest Solvay ... is the founder of one of the most flourishing industries of the world, the soda manufacture based on the process invented by him and now spread over Belgium, France, England, Germany, Russia and the United States. . . . The fortune won by an activity of half a century has been largely used by Mr Solvay for the public benefit. In the firm conviction that a better understanding of the laws of nature and of human society will prove one of the most powerful means for promoting the happiness of mankind, he has in many ways and on a large scale encouraged and supported scientific research and teaching.

Part of this activity was centred around the project of the Cité Scientifique, a series of institutes in Brussels founded and endowed by Ernest Solvay and by his brother Alfred Solvay, which culminated in the founding of the Institutes of Physics and of Chemistry in 1912 and 1913.[a]

This project had originally developed through the chance encounter between Ernest Solvay and Paul Héger, physician and professor of physiology at the Université Libre de Bruxelles (ULB), and involved a collaboration between Solvay, the ULB and the city of Brussels. In June 1892, it was agreed that Solvay would construct and equip two Institutes of Physiology on land owned by the city in the Parc Léopold in Brussels.[b] There soon followed in 1893–4 an Institute for Hygiene, Bacteriology and Therapy, funded mainly by Alfred Solvay, and a School of Political and Social Sciences, founded by Ernest Solvay in 1894, which moved to the Cité Scientifique in 1901, and to which a School of Commerce was added in 1904.

The idea for what became known as the first Solvay conference in physics goes back to Wilhelm Nernst and Max Planck,[c] who around 1910 considered that the current problems in the theory of radiation and in the theory of specific heats had become so serious that an international meeting (indeed a 'council') should be convened in order to attempt to resolve the situation. The further encounter between Nernst and Solvay provided the material opportunity for the meeting, and by July 1910, Nernst was sending Solvay the detailed proposals. He had also secured the collaboration of Lorentz (who was eventually asked to preside), of Knudsen and naturally of Planck, who wrote:

. . . anything that may happen in this direction will excite my greatest interest and . . . I promise already my participation in any such endeavour. For I can say without exaggeration that in fact for the past 10 years nothing in physics has so continuously stimulated, excited and irritated me as much as these quanta of action.[d]

[a] The following material on the Cité Scientifique is drawn mainly from Despy-Meyer and Devriese (1997).

[b] One was to become property of the city and given in use to the ULB, while the other was to be leased for 30 years to and run by Solvay himself.

[c] In the rest of this and in part of the following sections, we draw on an unpublished compendium of the contents of the IIPCS archives by J. Pelseneer.[4]

[d] Exquisite ending in the original: '. . . dass mich seit 10 Jahren im Grunde nichts in der Physik so ununterbrochen an-, er-, und aufregt wie diese Wirkungsquanten'.[5]

The first Solvay conference took place between 30 October and 3 November 1911. Lorentz set up a committee to consider questions relating to the new experimental research that had been deemed necessary during the conference. This committee included Marie Curie, Brillouin, Warburg, Kamerlingh Onnes, Nernst, Rutherford and Knudsen. Lorentz in turn was asked to be the president. Further, at the end of the conference, Solvay proposed to Lorentz the idea of a scientific foundation. Lorentz's reply to Solvay's proposals, of 4 January 1912, includes extremely detailed suggestions on the functions and structure of the foundation, all of which were put into practice and which can be summarised as follows.[6]

The foundation would be devoted principally to physics and physical chemistry, as well as to questions relating to physics from other sciences. It would provide international support to researchers ('a Rutherford, a Lenard, a Weiss') in the form of money or loan of scientific instruments, and it would provide scholarships for young Belgian scientists (both men and women) to work in the best laboratories or universities, mostly abroad. The question of a link between the foundation and the 'Conseil de physique' was left open, but Lorentz suggested to provide meeting facilities if Solvay wished to link the two. Lorentz suggested instituting an administrative board (consisting of a Solvay family member or appointee, an appointee of the King, and a member of the Belgian scientific establishment) and a scientific committee (which could initially be the one he had formed during the first Solvay conference). Finally, Lorentz suggested housing the foundation in an annex of one of the existing institutes in the Cité Scientifique.

During January, Solvay sent Paul Héger to Leiden to work with Lorentz on the statutes of the foundation, which Lorentz sent to Solvay on 2 February. Solvay approved them with hardly any modifications (only such as were required by the Belgian legislation of the time). The foundation, or rather the 'Solvay International Institute of Physics', was officially established on 1 May 1912, thus predating by several years the establishment of the comparable Belgian state institutions (Fondation Universitaire, 1920; Fonds National de la Recherche Scientifique, 1928). In this connection, Lorentz hoped 'that governments would understand more and more the importance of scientific research and that in the long run one will arrive at a satisfactory organisation, independent of the individual efforts of private persons',[7] a sentiment echoed by Solvay himself.[8]

The institute, which Solvay had endowed for 30 years, could soon boast of remarkable activity in supporting scientific research. The numerous recipients of subsidies granted during the first two years until the First World War included Lebedew's laboratory, von Laue, Sommerfeld, Franck and Hertz, W. L. Bragg (who was later to become president of the scientific committee), Stark, and Wien. In 1913, an Institute of Chemistry followed suit, organised along similar lines to the Institute of Physics.

1.2 War and international relations

The first meeting of the scientific committee, for the planning of the second Solvay conference, took place on 30 September and 1 October 1912. The conference was held the following year, but the activities of the institute were soon disrupted by the start of the First World War, in particular the German invasion of Belgium.

Immediate practical disruption included the fear of requisitions, the difficulty of communication between the international membership of the scientific committee and, with regard to the publication of the proceedings of the second Solvay conference, the impossibility of sending Lorentz the proofs for correction and the eventual prospect of German censorship.[a]

The war, however, had longer-term negative implications for international intellectual cooperation. In October 1914, a group of 93 representatives of German science and culture signed the manifesto 'An die Kulturwelt!', denying German responsibilities in the war.[b] Among the signatories were both Nernst and Planck. This manifesto was partly responsible for the very strong hostility of French and Belgian scientists and institutions towards renewal of scientific relations with Germany after the war.

No Germans or Austrians were invited to the third Solvay conference of 1921. The only exception (which remained problematic until the last minute) was Ehrenfest, who was Austrian, but who had remained in Leiden throughout the war as Lorentz's successor. Similarly, no Germans participated in the fourth Solvay conference of 1924. French and Belgian armies had occupied the Ruhr in January 1923, and the international situation was particularly tense. Einstein had (temporarily) resigned from the League of Nations' Committee on Intellectual Cooperation, and wrote to Lorentz that he would not participate in the Solvay conference because of the exclusion of the German scientists, and if he could please make sure that no invitation was sent.[9] Bohr also declined to participate in the conference apparently because of the continued exclusion of German scientists (Moore 1989, p. 157). Schrödinger, however, who was Austrian and working in Switzerland, was invited.[c]

[a] The proceedings of the first Solvay conference had had both a French and a German edition. Those of the second Solvay conference were printed in three languages in 1915, but never published in this form and later mostly destroyed. Only under the changed conditions after the war, in 1921, were the proceedings published in a French translation (carried out, as on later occasions, by J.-É. Verschaffelt).

[b] The main claims of the manifesto were: '... *It is not true* that Germany is the cause of this war.... *It is not true* that we have wantonly [freventlich] infringed the neutrality of Belgium. ... *It is not true* that the life and property of a single Belgian citizen has been touched by our soldiers, except when utter self-defence required it.... *It is not true* that our troops have raged brutally against Leuven.... *It is not true* that our conduct of war disregards the laws of international right. ... *It is not true* that the struggle against our so-called militarism is not a struggle against our culture....' (translated from Böhme 1975, pp. 47–9).

[c] Van Aubel (a member of the scientific committee) objected strongly in 1923 to the possibility of Einstein being invited to the fourth Solvay conference, and resigned when it was decided to invite him. It appears he was convinced to remain on the committee.[10]

Einstein had distinguished himself by assuming a pacifist position during the war.[a] Lorentz was pointing out Einstein's exceptional case to Solvay already in January 1919:

However, in talking about the Germans, we must not lose sight of the fact that they come in all kinds of nuances. A man like Einstein, the great and profound physicist, is not 'German' in the sense one often attaches to the word today; his judgement on the events of the past years will not differ at all from yours or mine.[11]

In the meantime, after the treaty of Locarno of 1925, Germany was going to join the League of Nations, but the details of the negotiations were problematic.[b] As early as February 1926, one finds mention of the prospect of renewed inclusion of German scientists at the Solvay conferences.[13] In the same month, Kamerlingh Onnes died, and at the next meeting of the scientific committee, in early April (at which the fifth Solvay conference was planned), it was decided to propose both to invite Einstein to replace Onnes and to include again the German scientists.

On 1 April, Charles Lefébure, then secretary of the administrative commission, wrote to commission members Armand Solvay and Jules Bordet,[c] enquiring about the admissibility of 'moderate' figures like Einstein, Planck[d] and others.[16] On 2 April, Lorentz himself had a long interview with the King, who gave his approval.

Thus, finally, Lorentz wrote to Einstein on 6 April, informing him of the unanimous decision by the members of the committee present at the meeting,[e] as well as of the whole administrative commission, to invite him to succeed Kamerlingh Onnes. The Solvay conferences were to readmit Germans, and if Einstein were a member of the committee, Lorentz hoped this would encourage the German scientists to accept the invitation.[17] Einstein was favourably impressed by the positive Belgian attitude and glad to accept the invitation under the altered conditions.[18] Lorentz proceeded to invite the German scientists, 'not because there

[a] For instance, Einstein was one of only four signatories of the counter-manifesto 'Aufruf an die Europäer' (Nicolai 1917). Note also that Einstein had renounced his German citizenship and had become a Swiss citizen in 1901, although there was some uncertainty about his citizenship when he was awarded the Nobel prize (Pais 1982, pp. 45, 503–4).

[b] Lorentz to Einstein on 14 March 1926: 'Things are bad with the League of Nations; if only one could yet find a way out until the day after tomorrow'.[12] Negotiations provisionally broke down on 17 March, but Germany eventually joined the League in September 1926.

[c] Lefébure was the appointee of the Solvay family to the administrative commission, and as such succeeded Eugène Tassel, who had been a long-standing collaborator of Ernest Solvay since 1886, and had died in October 1922. Armand Solvay was the son of Ernest Solvay, who had died on 26 May 1922. Bordet was the royal appointee to the commission, and had just been appointed in February 1926, following the death of Paul Héger.[14]

[d] According to Lorentz, Planck had always been helpful to him when he had tried to intervene with the German authorities during the war. Further, Planck had somewhat qualified his position with regard to the Kulturwelt manifesto in an open letter, which he asked Lorentz to publish in the Dutch newspapers in 1916. On the other hand, he explicitly ruled out a public disavowal of the manifesto in December 1923.[15]

[e] Listed as Marie Curie, Langevin, Richardson, Guye and Knudsen (with two members absent, W. H. Bragg and Van Aubel).

should be such a great haste in the thing, rather to show the Germans as soon as possible our good will',[19] and sent the informal invitations to Born, Heisenberg and Planck (as well as to Bohr) in or around June 1926.[20]

As late as October 1927, however, the issue was still sensitive. Van Aubel (who had not been present at the April 1926 meeting of the scientific committee) declined the official invitation to the conference, apparently because of the presence of the German scientists.[21] Furthermore, it was proposed to release the list of participants to the press only after the conference to avoid public demonstrations. Lorentz travelled in person to Brussels on 17 October to discuss the matter.[22]

Lorentz's own position during and immediately after the war, as a physicist from one of the neutral countries, had possibly been rather delicate. In the text on Ernest Solvay from which we have quoted at the beginning of this chapter, for instance, he appears to be defending the impartiality of the policies of the Institute of Physics in the years leading up to the war. Lorentz started working for some form of reconciliation as soon as the war was over, writing as follows to Solvay in January 1919:

> All things considered, I think I must propose to you not to exclude formally the Germans, that is, not to close the door on them forever. I hope that it may be open for a new generation, and even that maybe, in the course of the years, one may admit those of today's scholars who one can believe regret sincerely and honestly the events that have taken place. Thus German science will be able to regain the place that, despite everything, it deserves for its past.[23]

It should be noted that Lorentz was not only the scientific organiser of the Solvay institute and the Solvay conferences, but also a prime mover behind efforts towards international intellectual cooperation, through his heavy involvement with the Conseil International de Recherches, as well as with the League of Nations' Committee on Intellectual Cooperation, of which he was a member from 1923 and president from 1925.[a]

Lorentz's figure and contributions to the Solvay conferences are movingly recalled by Marie Curie in her obituary of Lorentz in the proceedings of the fifth Solvay conference (which opens Part III of this volume).

1.3 Scientific planning and background

What was at issue in the remark that heads this chapter,[b] scribbled by Ehrenfest in the margin of a letter from Lorentz, was the proposed topic for the fifth

[a] The Conseil International de Recherches (founded in 1919) has today become the International Council for Science (ICSU). The Committee on Intellectual Cooperation (founded in 1922) and the related International Institute of Intellectual Cooperation (inaugurated in Paris in 1926) were the forerunners of UNESCO.[24]

[b] In the original: 'Quantenverzoening <{zeer} antipathik<e>> <tendentie> <retrograde> <{bedenkelijk<e>}> <idee> <loszinnige> [?] titel wekt misverstand'. Many thanks to Mark van Atten for help with this passage.

Solvay conference, namely 'the conflict and the possible reconciliation between the classical theories and the theory of quanta'.[25] Ehrenfest found the phrasing objectionable in that it encouraged one to 'swindle away the fruitful and suggestive harshness of the conflict by most slimy unclear thinking, quite in analogy with what happened also even after 1900 with the mechanical ether theories of the Maxwell equations', pointing out that 'Bohr feels even more strongly than me against this slogan [Schlagwort], precisely because he takes it so particularly to heart to find the foundations of the future theory'.[26]

Lorentz took Ehrenfest's suggestion into account, and dropped the reference to reconciliation both from the title and from later descriptions of the focus of the meeting.[a]

The meeting of the scientific committee for the planning of the fifth Solvay conference took place in Brussels on 1 and 2 April 1926. Lorentz reported a few days later to Einstein:

> As the topic for 1927 we have chosen 'The quantum theory and the classical theories of radiation', and we hope to have the following reports or lectures:
> *1* W. L. Bragg. New tests of the classical theory.
> *2* A. H. Compton. Compton effect and its consequences.
> *3* C. T. R. Wilson. Observations on photoelectrons and collision electrons by the condensation method.
> *4* L. de Broglie. Interference and light quanta.
> *5* (short note): Kramers. Theory of Slater–Bohr–Kramers and analogous theories.
> *6* Einstein. New derivations of Planck's law and applications of statistics to quanta.
> *7* Heisenberg. Adaptation of the foundations of dynamics to the quantum theory.[28]

Another report, by the committee's secretary Verschaffelt,[29] qualifies point *5*, making it conditional on Kramers judging that it is still useful; it further lists a few alternative speakers: Compton or Debye for *2*, Einstein or Ehrenfest for *6*, and Heisenberg or Schrödinger for *7*.[b]

Thus, the fifth Solvay conference, as originally planned, was to focus mainly on the theory of radiation and on light quanta, including only one report on the new quantum theory of matter. The shift in focus between 1926 and 1927 was clearly due to major theoretical advances (for example by Schrödinger and Dirac) and new experimental results (such as the Davisson–Germer experiments), and it can be partly followed as the planning of the conference progressed.

[a] To Bohr in June 1926: '... the conflict between the classical theories and the quantum theory ....'; to Schrödinger in January 1927: '... the contrast between the current and the earlier conceptions [Auffassungen] and the attempts at development of a new mechanics'.[27]

[b] For details of the other participants, see the next section.

Schrödinger's wave mechanics was one of the major theoretical developments of the year 1926. Einstein, who had been alerted to Schrödinger's first paper by Planck (cf. Przibram 1967, p. 23), suggested to Lorentz that Schrödinger should talk at the conference instead of himself, on the basis of his new 'theory of quantum states', which he described as a development of genius of de Broglie's ideas.[30] While it is unclear whether Lorentz knew of Schrödinger's papers by the time of the April meeting,[a] Schrödinger was listed a week later as a possible substitute for Heisenberg, and Lorentz himself was assuring Einstein at the end of April that Schrödinger was already being considered, specifially as a substitute for the report on the new foundations of dynamics rather than for the report on quantum statistics.[b]

Lorentz closely followed the development of wave mechanics, indeed contributing some essential critique in his correspondence with Schrödinger from this period, for the most part translated in Przibram (1967) (see Chapter 4, especially Sections 4.3 and 4.4, for some more details on this correspondence). Lorentz also gave a number of colloquia and lectures on wave mechanics (and on matrix mechanics) in the period leading up to the Solvay conference, in Leiden, Ithaca and Pasadena.[31] In Pasadena he also had the opportunity of discussing with Schrödinger the possibility that Schrödinger may also give a report at the conference, as in fact he did.[c] Schrödinger's wave mechanics had also made a great impression on Einstein, although he repeatedly expressed his unease to Lorentz at the use of wave functions on configuration space (calling it obscure,[33] harsh,[34] a mystery[35]), and again during the general discussion (p. 442).

One sees Lorentz's involvement with the recent developments also in his correspondence with Ehrenfest. In particular, Lorentz appears to have been struck by Dirac's contributions to quantum mechanics.[d] In June 1927, Lorentz invited Dirac to spend the following academic year in Leiden,[37] and asked Born and Heisenberg to include a discussion of Dirac's work in their report.[38] Finally, in late August, Lorentz decided that Dirac, and also Pauli, ought to be invited to the conference, for indeed:

Since last year, quantum mechanics, which will be our topic, has developed with an unexpected rapidity, and some physicists who were formerly in the second tier have made extremely notable contributions. For this reason I would be very keen to invite also Mr Dirac of Cambridge and Mr Pauli of Copenhagen. ... Their collaboration would be very

[a] Cf. Section 4.1.

[b] Note that Schrödinger (1926a) had written on 'Einstein's gas theory' in a paper that is an immediate precursor to his series of papers on quantisation.

[c] Lorentz was at Cornell from September to December 1926, then in Pasadena until March 1927.[32] On Schrödinger's American voyage, see Moore (1989, pp. 230–3).

[d] This correspondence includes for instance a 15-page commentary by Lorentz on Dirac (1927a).[36]

useful to us . . . I need not consult the scientific committee because Mr Dirac and Mr Pauli were both on a list that we had drawn up last year . . . .[39]

Lorentz invited Pauli on 5 September 1927 (Pauli 1979, pp. 408–9) and Dirac sometime before 13 September 1927.[40]

On the experimental side, some of the main achievements of 1927 were the experiments on matter waves. While originally de Broglie was listed to give a report on light quanta, the work he presented was about both light quanta and material particles (indeed, electrons and photons!), and Lorentz asked him explicitly to include some discussion of the recent experiments speaking in favour of the notion of matter waves, specifically discussing Elsasser's (1925) proposals, and the experimental work of Dymond (1927) and of Davisson and Germer (1927).[41] Thus, in the final programme of the conference, we find three reports on the foundations of a new mechanics, by de Broglie, Heisenberg (together with Born) and Schrödinger.

The talks given by Bragg and Compton, instead, reflect at least in part the initial orientation of the conference. Here is how Compton presents the division of labour (p. 301):

Professor W. L. Bragg has just discussed a whole series of radiation phenomena in which the electromagnetic theory is confirmed. . . . I have been left the task of pleading the opposing cause to that of the electromagnetic theory of radiation, seen from the experimental viewpoint.

Bragg focusses in particular on the technique of X-ray analysis, as the 'most direct way of analysing atomic and molecular structure' (p. 260), in the development of which, as he had mentioned to Lorentz, he had been especially interested.[42] This includes in particular the investigation of the electronic charge distribution. At Lorentz's request, he had also included a discussion of the refraction of X-rays (Section 8 of his report), which is directly relevant to the discussion after Compton's report.[43] As described by Lorentz in June 1927, Bragg was to report 'on phenomena that still somehow allow a classical description'.[44] A few more aspects of Bragg's report are of immediate relevance for the rest of the conference, especially to the discussion of Schrödinger's interpretation of the wave function in terms of an electric charge density (pp. 280, 285, Section 4.4), and so are some of the issues taken up further in the discussion (Hartree approximation, problems with waves in three dimensions), but it is fair to say that the report provides a rather distant background for what followed it.

Compton's report covers the topics of points *2* and *3* listed above. The explicit focus of his report is the three-way comparison between the photon hypothesis, the Bohr–Kramers–Slater (BKS) theory of radiation, and the classical theory of radiation. Note, however, that Compton introduces many of the topics of later discussions. For instance, he discusses the problem of how to explain atomic radiation (section on 'The emission of radiation', p. 304), which is inexplicable from the point of view of the classical theory, given that the 'orbital frequencies' in the atom do not correspond to the emission frequencies. This problem was one of Schrödinger's main concerns and one of the main points of conflict between Schrödinger and, for instance, Heisenberg (see in particular the discussion after Schrödinger's report and Sections 4.4 and 4.6). Compton's discussion of the photon hypothesis relates to the question of 'guiding fields' (pp. 302 and 323) and of the localisation of particles or energy quanta within a wave (pp. 309 and 317). These in turn are closely connected with some of de Broglie's and Einstein's ideas (see Chapter 7, especially Section 7.2, and Chapter 9), and with de Broglie's report on pilot-wave theory and Einstein's remark about locality in the general discussion (p. 441).

Bohr had been a noted sceptic of the photon hypothesis, and in 1924 Bohr, Kramers and Slater had developed a theory that was able to maintain a wave picture of radiation, by introducing a description of the atom based on 'virtual oscillators' with frequencies equal to the frequencies of emission (Bohr, Kramers and Slater 1924a,b).[a] A stationary state of an atom, say the nth, is associated with a state of excitation of the oscillators with frequencies corresponding to transitions from the energy E_n. Such oscillators produce a classical radiation field (a 'virtual' one), which in turn determines the *probabilities* for spontaneous emission in the atom, that is, for the emission of energy from the atom and the jump to a stationary state of lower energy. The virtual field of one atom also interacts with the virtual oscillators in other atoms (which in turn produce secondary virtual radiation) and influences the probabilities for induced emission and absorption in the other atoms. While the theory provides a mechanism for radiation consistent with the picture of stationary states (cf. Compton's remarks, p. 306), it violates energy and momentum conservation for single events, in that an emission in one atom is not connected directly to an absorption in another atom, but only indirectly through the virtual radiation field. Energy and momentum conservation hold only at a statistical level. The BKS proposal was short-lived, because the Bothe–Geiger and Compton–Simon experiments established the conservation laws for individual processes (as explained in detail by Compton in his report, pp. 319ff.). Thus, at the time of the planning of the fifth Solvay conference, the experimental evidence had ruled out the BKS theory (hence the above remark about whether a talk by Kramers

[a] As Darrigol (1992, p. 257) emphasises, while the free virtual fields obey the Maxwell equations, that is, can be considered to be classical, the virtual oscillators and the interaction between the fields and the oscillators are non-classical in several respects.

would still be useful).[a] The short note 5, indeed, dropped out of the programme altogether.[b]

The description of the interaction between matter and radiation, in particular the Compton effect, continued to be a problem for Bohr, and contributed to the development of his views on wave-particle dualism and complementarity. In his contribution to the discussion after Compton's report (p. 328, the longest of his published contributions in the Solvay volume[c]), Bohr sketches the motivations behind the BKS theory, the conclusions to be drawn from the Bothe–Geiger and Compton–Simon experiments, and the further development of his views.

Lorentz, in his report of the meeting to Einstein, had mentioned 'Slater–Bohr–Kramers and analogous theories'. This may refer to the further developments (independent of the validity of the BKS theory) that led in particular to Kramer's (1924) dispersion theory (and from there towards matrix mechanics), or to Slater's original ideas, which were roughly along the lines of guiding fields for the photons (even though the photons were dropped from the final BKS proposal).[d] Note that Einstein at this time was also thinking about guiding fields (in three dimensions). Pais (1982, pp. 440–1) writes that, according to Wigner, Einstein did not publish these ideas because they also led to problems with energy-momentum conservation.[e]

Einstein was asked by Lorentz to contribute a report on 'New derivations of Planck's law and applications of statistics to quanta' (point *6*), clearly referring to the work by Bose (1924) on Planck's law, championed by Einstein and applied by him to the theory of the ideal gas (Einstein 1924, 1925a,b). The second of these papers is also where Einstein famously endorses de Broglie's idea of matter waves. Einstein thought that his work on the subject was already too well-known, but he accepted after Lorentz repeated his invitation.[46] On 17 June 1927, however, at about the time when Lorentz was sending detailed requests to the speakers, Einstein informed him in the following terms that he would not, after all, present a report:

I recall having committed myself to you to give a report on quantum statistics at the Solvay conference. After much reflection back and forth, I come to the conviction that I am not

[a] Bothe and Geiger had been working on their experiments since June 1924 (Bothe and Geiger 1924), and provisional results were being debated by the turn of the year. For two differing views on the significance of these results, see for instance Einstein to Lorentz, 16 December 1924 (the same letter in which he wrote to Lorentz about de Broglie's results)[45] and the exchange of letters between Born and Bohr in January 1925 (Bohr 1984, pp. 302–6). By April 1925, Bothe and Geiger had clear-cut results against the BKS theory (Bothe and Geiger 1925a,b; see also the letters between Geiger and Bohr in Bohr (1984, pp. 352–4).

[b] On the BKS theory and related matters, see also Chapter 3 (especially Sections 3.3.1 and 3.4.2), Chapter 9, Darrigol (1992, chapter 9), the excellent introduction by Stolzenburg to Part I of Bohr (1984), and Mehra and Rechenberg (1982a, section V.2).

[c] See below for the fate of his contributions to the general discussion.

[d] Cf. Slater (1924) and Mehra and Rechenberg (1982a, pp. 543–6). See also Pauli's remark during the discussion of de Broglie's report (p. 365).

[e] Cf. Einstein's contribution to the general discussion (p. 440) and the discussion in Chapter 9.

competent [to give] such a report in a way that really corresponds to the state of things. The reason is that I have not been able to participate as intensively in the modern development of quantum theory as would be necessary for this purpose. This is in part because I have on the whole too little receptive talent for fully following the stormy developments, in part also because I do not approve of the purely statistical way of thinking on which the new theories are founded ... Up until now, I kept hoping to be able to contribute something of value in Brussels; I have now given up that hope. I beg you not to be angry with me because of that; I did not take this lightly but tried with all my strength ... (Quoted in Pais 1982, pp. 431–2)

Einstein's withdrawal may be related to the following circumstances. On 5 May 1927, during a meeting of the Prussian Academy of Sciences in Berlin, Einstein had read a paper on the question: 'Does Schrödinger's wave mechanics determine the motion of a system completely or only in the sense of statistics?'[47] As discussed in detail by Belousek (1996), the paper attempts to define deterministic particle motions from Schrödinger's wave functions, but was suddenly withdrawn on 21 May.[a]

The plans for the talks were finalised by Lorentz around June 1927. An extract from his letter to Schrödinger on the subject reads as follows:

[W]e hope to have the following reports [Referate] (I give them in the order in which we might discuss them):

1. From Mr W. L. Bragg on phenomena that still somehow allow a classical description (reflexion of X-rays by crystals, diffraction and total reflection of X-rays).
2. From Mr Compton on the effect discovered by him and what relates to it.
3. From Mr de Broglie on his theory. I am asking him also to take into account the application of his ideas to free electrons (Elsasser, quantum mechanics of free electrons; Dymond, Davisson and Germer, scattering of electrons).
4. From Dr Heisenberg *or* Prof. Born (the choice is left to them) on matrix mechanics, including Dirac's theory.
5. Your report [on wave mechanics].

Maybe another one or two short communications [Berichte] on special topics will be added.[49]

This was, indeed, the final programme of the conference, with Born and Heisenberg deciding to contribute a joint report.[b]

[a] The news of Einstein's communication prompted an exchange of letters between Heisenberg and Einstein, of which Heisenberg's letters, of 19 May and 10 June, survive.[48] The second of these is particularly interesting, because Heisenberg presents in some detail his view of theories that include particle trajectories. Both Einstein's hidden-variables proposal and Heisenberg's reaction will be described in Section 11.3.

[b] See Section 3.2. Note that, as we shall see below, while Bohr contributed significantly to the general discussion and reported the views he had developed in Como (Bohr 1949, p. 216; 1985, pp. 35–7), he was unable to prepare an edited version of his comments in time and therefore suggested that a translation of his Como lecture, in the version for *Naturwissenschaften* (Bohr 1928), be included in the volume instead. This has given rise to a common belief that Bohr gave a report on a par with the other reports, and that the general discussion at the conference was the discussion following it. See for instance Mehra (1975, p. 152), and Mehra and Rechenberg (2000, pp. 246, 249), who appear further to believe that Bohr did not participate in the official discussion.

1.4 Further details of planning

In 1926–7 the scientific committee and the administrative commission of the Solvay institute were composed as follows.

Scientific committee: Lorentz (Leiden) as president, Knudsen (Copenhagen) as secretary, W. H. Bragg (until May 1927, London),[a] Marie Curie (Paris), Einstein (since April 1926, Berlin), Charles-Eugène Guye (Geneva),[b] Langevin (Paris), Richardson (London), Edmond van Aubel (Gent).

Administrative commission: Armand Solvay, Jules Bordet (ULB), Maurice Bourquin (ULB), Émile Henriot (ULB); the administrative secretary since 1922, and thus main correspondent of Lorentz and others, was Charles Lefébure.

The secretary of the meeting was Jules-Émile Verschaffelt (Gent), who had acted as secretary since the third Solvay conference.[51]

The first provisional list of possible participants (in addition to Ehrenfest) appears in Lorentz's letter to Ehrenfest of 29 March 1926:

> Einstein, Bohr, Kramers, Born, Heisenberg (Jordan surely more mathematician), Pauli, Ladenburg (?), Slater, the young Bragg (because of the 'correspondence' to the classical theory that his work has often resulted in), J. J. Thomson, another one or two Englishmen (Darwin? Fowler?), Léon Brillouin (do not know whether he has worked on this, he has also already been there a number of times), *Louis* de Broglie (light quanta), one or two who have concerned themselves with diffraction of X-rays (Bergen Davis?, Compton, Debye, Dirac (?)).[52]

Lorentz asked for further suggestions and comments, which Ehrenfest sent in a letter dated 'Leiden 30 March 1926. Late at night':

> Langevin, Fowler, Dirac, J. Fran[c]k (already for the experiments he devised by Hanle on the destruction of resonance polarisation through Larmor rotation and for the work he proposed by Hund on the Ramsauer effect[c] – undisturbed passage of slow electrons through atoms and so on), Fermi (for interesting continuation of the experiments by Hanle), Oseen (possibly a wrong attempt at explanation of needle radiation and as sharpwitted critic), Schrödinger (was perhaps the first to give quantum interpretation of the Doppler effect, thus close to Compton effect), Bothe (for Bothe–Geiger experiment on correlation of Compton quantum and electron, which destroys Bohr–Slater theory, altogether a fine brain!) (Bothe should be considered perhaps *before* Schrödinger), Darwin, Smekal (is indeed a very deserving connoisseur of quantum finesses, only he writes so *frightfully* much).
>
> Léon Brillouin has published something recently on matrix physics, but I have not read it yet.[53]

[a] W. H. Bragg resigned due to overcommitment and was later replaced by Cabrera (Madrid).[50]

[b] In 1909 the university of Geneva conferred on Einstein his first honorary degree. According to Pais (1982, p. 152), this was probably due to Guye. Coincidentally, Ernest Solvay was honoured at the same time.

[c] For more on the special interest of the Ramsauer effect, see Section 3.4.2.

At the April meeting (as listed in the report by Verschaffelt[54]) it was then decided to invite: Bohr, Kramers, Ehrenfest, two among Born, Heisenberg and Pauli, Planck, Fowler, W. L. Bragg, C. T. R. Wilson, L. de Broglie, L. Brillouin, Deslandres, Compton, Schrödinger and Debye. Possible substitutes were listed as: M. de Broglie or Thibaud for Bragg, Dirac for Brillouin, Fabry for Deslandres, Kapitza for Wilson, Darwin or Dirac for Fowler, Bergen Davis for Compton, and Thirring for Schrödinger.[a] The members of the scientific commitee would all take part *ex officio*, and invitations would be sent to the professors of physics at ULB, that is, to Piccard, Henriot and De Donder[56] (the latter apparently somewhat to Lefébure's chagrin, who, just before the beginning of the conference, felt obliged to warn Lorentz, reminding him that De Donder was a 'paradoxical' mind, always ready to seize the word[57]).

Both the number of actual participants and of observers was to be kept limited,[58] which partly explains why it was thought that one should invite only two among Born, Heisenberg and Pauli. The choice initially fell on Born and Heisenberg (although Franck was also considered as an alternative).[59] Eventually, as noted above, Pauli was also included, as was Dirac.[60] Lorentz was also keen to invite Millikan – and possibly Hall – when he heard that Millikan would be in Europe anyway for the Como meeting (Einstein and Richardson agreed).[61] However, nothing came of this plan.

When Einstein eventually withdrew as a speaker, he suggested Fermi or Langevin as possible substitutes (Pais 1982, p. 432). For a while it was not clear whether Langevin (who was anyway a member of the scientific committee) would be able to come, since he was in Argentina over the (Northern) summer and due to go on to Pasadena from there. Ehrenfest suggested F. Perrin instead, in rather admiring tones. Langevin was needed in Paris in October, however, and was able to come to the conference.[62] At Brillouin's suggestion, finally, a few days before the beginning of the conference Lorentz extended the invitation to Irving Langmuir,[63] who happened to be in Europe at the time.[b]

Lorentz sent most of the informal invitations around January 1927.[64] In May 1927, he sent to Lefébure the list of all the people he had 'provisionally invited',[65] including all the members of the scientific committee and the prospective invitees as listed above by Verschaffelt (that is, as yet without Pauli and Dirac). All had already replied and accepted, except Deslandres (who eventually replied much later declining the invitation[66]). Around early July, Lorentz invited the physicists from

[a] A few days later, Guye suggested also Auger as a possible substitute for Wilson.[55]

[b] Relevant correspondence by Langmuir is to be found in the Library of Congress (our thanks to Patrick Coffey for details). To Langmuir we owe some fascinating footage of the conference, available (at the time of writing) at http://www.maxborn.net/index.php?page=filmnews; see the report in the *AIP Bulletin of Physics News*, number 724 (2005).

the university,[67] and presumably sent a new invitation to W. H. Bragg, who thanked him but declined.[68] Formal letters of confirmation were sent out by Lefébure shortly before the conference.[69]

Around June 1927, Lorentz wrote to the planned speakers inviting them in the name of the scientific committee to contribute written reports, to reach him preferably by 1 September. The general guidelines were: to focus on one's own work, without mathematical details, but rather so that 'the principles are highlighted as clearly as possible, and the open questions as well as the connections [Zusammenhänge] and contrasts are clarified'. The material in the reports did not have to be unpublished, and a bibliography would be welcome.[70] Compton wrote that he would aim to deliver his manuscipt by 20 August, de Broglie easily before the end of August, Bragg, as well as Born and Heisenberg, by 1 September, and Schrödinger presumably only in the second half of September.[71] (For further details of the correspondence between some of the authors and Lorentz, see the relevant chapters.)

The written reports were to be sent to all participants in advance of the conference.[a] De Broglie's, which had been written directly in French, was sent by Lorentz to the publishers, Gauthier-Villars in Paris, before he left for the Como meeting. They hoped to send 35 proofs to Lorentz by the end of September. In the meantime, Verschaffelt and Lorentz's son, Rudolf, had the remaining reports mimeographed by the 'Holland Typing Office' in Amsterdam, and Verschaffelt with the help of a student added in the formulas by hand, managing to mail on time to the participants at least Compton's and Born and Heisenberg's reports, if not all of them.[73] Lorentz had further written to all speakers (except Compton) to ask them to bring reprints of their papers.[74]

Late during planning, a slight problem emerged, namely an unfortunate overlap of the Brussels conference with the festivities for the centenary of Fresnel in Paris, to be officially opened Thursday 27 October. Lorentz informed Lefébure of the clash writing from Naples after the Como meeting: neither the date of the conference could be changed nor that of the Fresnel celebrations, which had been fixed by the French President. The problem was compounded by the fact that de Broglie had accepted to give a lecture to the Société de Physique on the occasion.[b] Lorentz suggested the compromise solution of a general invitation to attend the celebrations. Those who wished to participate could travel to Paris on

[a] Mimeographed copies of Bragg's, Born and Heisenberg's and Schrödinger's reports are to be found in the Richardson collection, Harry Ransom Humanities Research Center, University of Texas at Austin.[72]

[b] In Lorentz's letter, the date of de Broglie's lecture is mentioned as 28 October, but the official invitations state that it was Zeeman who lectured then, and de Broglie the next evening, after the end of the Solvay conference. A report on de Broglie's lecture, which was entitled 'Fresnel's œuvre and the current development of physics', was published by Guye in the *Journal de Genève* of 16 and 18 April 1928.[75]

27 October, returning to Brussels the next day, when sessions would be resumed in the afternoon. This was the solution that was indeed adopted.[76]

1.5 The Solvay meeting

The fifth Solvay conference took place from 24 to 29 October 1927 in Brussels. As on previous occasions, the participants stayed at the Hôtel Britannique, where a dinner invitation from Armand Solvay awaited them.[77] Other meals were taken at the institute, which was housed in the building of the Institute of Physiology in the Parc Léopold; catering for 50–55 people had been arranged.[78] The participants were guests of the administrative commission and all travel expenses within Europe were met.[79] From the evening of 23 October onwards, three seats were reserved in a box at the Théatre de la Monnaie.[80]

The first session of the conference started at 10:00 on Monday 24 October. A tentative reconstruction of the schedule of the conference is as follows.[81] We assume that the talks were given in the order they were described in the plans and printed in the volume.

- Monday 24 October: W. L. Bragg's report, followed by discussion (possibly also Compton's report in the afternoon).
- Tuesday 25 October, starting 9:00 a.m.: reception offered by the ULB, followed by Compton's report and discussion (or possibly only the discussion).
- Tuesday 25 October, afternoon: L. de Broglie's report, followed by discussion.
- Wednesday 26 October, morning: M. Born and W. Heisenberg's report, followed by discussion.
- Wednesday 26 October, afternoon: E. Schrödinger's report, followed by discussion.
- Thursday 27 October, morning: general discussion (Bohr's main contribution and intervening discussion).
- Thursday 27 October, afternoon and evening: travel to Paris and centenary of Fresnel.[a]
- Friday 28 October, morning: return to Brussels.
- Friday 28 October, afternoon: general discussion (up to and including Dirac's remarks and the ensuing discussion).
- Saturday 29 October: general discussion,[b] lunch with the King and Queen of the Belgians, and dinner offered by Armand Solvay.

It appears the languages used were English, German and French. Schrödinger had volunteered to give his talk in English,[c] while Born had suggested that he

[a] Most of the participants at the Solvay conference, with the exception of Knudsen, Dirac, Ehrenfest, Planck, Schrödinger, Henriot, Piccard and Herzen, travelled to Paris to attend the inauguration of the celebrations, at 8:30 p.m. in the *grand amphithéatre* of the Sorbonne.[82]

[b] The final session of the conference also included a homage to Ernest Solvay's widow.[83]

[c] Note that Schrödinger was fluent in English from childhood, his mother and aunts being half-English (Moore 1989, chapter 1).

and Heisenberg could provide additional explanations in English (while he thought that neither of them knew French).[84] The phrasing used by Born referred to who should 'explain orally the contents of the report', suggesting that the speakers did not present the exact or full text of the reports as printed.

Multiplicity of languages had long been a characteristic of the Solvay conferences. A well-known letter by Ehrenfest[85] tells us of '[p]oor Lorentz as interpreter between the British and the French who were absolutely unable to understand each other. Summarising Bohr. And Bohr responding with polite despair' (as quoted in Bohr 1985, p. 38).[a]

On the last day of the conference, as reported by Bohr (1963) and in more detail by Pelseneer, Ehrenfest went to the blackboard and evoked the image of the tower of Babel (presumably in a more metaphorical sense than the mere multiplicity of spoken languages):[b]

And they said one to another: ... Go to, let us build us ... a tower, whose top may reach unto heaven; and *let us make us a name*[c] ... And the Lord said: ... Go to, let us go down, and there confound their language, that they may not understand one another's speech. (Genesis 11: 3–7)

Informal discussions at the conference must have been plentiful, but information about them has to be gathered from other sources. Famously, Einstein and Bohr engaged in discussions that were described in detail in later recollections by Bohr (1949), and vividly recalled by Ehrenfest within days of the conference in the well-known letter quoted above (see also Chapter 12).

Little known, if at all, is another reference by Ehrenfest to the discussions between Bohr and Einstein, which appears to relate more directly to the issues raised by Einstein in the general discussion:[d]

Bohr had given a very pretty argument in a conversation with Einstein, that one could not hope ever to master many-particle problems with three-dimensional Schrödinger machinery. He said something like the following (more or less!!!!!!): a wave packet can never simultaneously determine EXACTLY the position and the velocity of a particle. Thus if one has for instance TWO particles, then they cannot possibly be represented in *three-dimensional* space such that one can simultaneously represent exactly their kinetic energy and the potential energy OF THEIR INTERACTION. Therefore .... (What

[a] Both W. H. Bragg and Planck deplored in letters to Lorentz that they were very poor linguists. Indeed, in a letter explaining in more detail why he would not participate in the conference, W. H. Bragg wrote: 'I find it impossible to follow the discussions even though you so often try to make it easy for us', and Planck was in doubt about coming, particularly because of the language difficulties.[86]

[b] In Bohr's words: 'In the very lively discussions ...., ambiguities of terminology presented great difficulties for agreement regarding the epistemological problems. This situation was humorously expressed by Ehrenfest, who wrote on the blackboard the sentence from the Bible describing the confusion of languages that disturbed the building of the Babel tower' (Bohr 1963, p. 92).

[c] This may have been Ehrenfest's own emphasis. Note that, if not necessarily present at the sessions, Pelseneer had some connection with the conference, having taken Lorentz's photograph reproduced in the proceedings.[87]

[d] See also the notes on the Saturday session of the general discussion, given in the appendix to Part III.

comes after this therefore I already cannot reproduce properly.) In the multidimensional representation instead the potential energy of the interaction appears totally sharp in the relevant coefficients of the wave equation and one does [?] not get to see the kinetic energy at all.[88]

1.6 The editing of the proceedings

The editing of the proceedings of the fifth Solvay conference was largely carried out by Verschaffelt, who reported regularly to Lorentz and Lefébure. During the last months of 1927 Lorentz was busy writing up the lecture he had given at the Como meeting in September.[89] He then died suddenly on 4 February, before the editing work was complete.

The translation of the reports into French was carried out after the conference, except for de Broglie's report which, as mentioned, was written directly in French. From Verschaffelt's letters we gather that by 6 January 1928 all the reports had been translated, Bragg's and Compton's had been sent to the publishers, and Born and Heisenberg's and Schrödinger's were to be sent on that day or the next. Several proofs were back by the beginning of March.[90]

Lorentz had envisaged preparing with Verschaffelt an edited version of the discussions from notes taken during the conference, and sending the edited version to the speakers at proof stage.[a] In fact, stenographed notes appear to have been taken, typed up and sent to the speakers, who for the most part used them to prepare drafts of their contributions. From these, Verschaffelt then edited the final version, with some help from Kramers (who specifically completed two of Lorentz's contributions).[91] The Bohr archive in Copenhagen contains some related material, which is not microfilmed, in particular on Bohr's contributions and on the general discussion, most of which is reproduced in the appendix to Part III. A copy of the galley proofs of the general discussion, dated 1 June 1928, also survives in the Bohr archive,[92] and includes a few contributions that appear to have been still largely unedited at that time.

By January, the editing of the discussions was proceeding well, and at the beginning of March it was almost completed. Some contributions, however, were still missing, most notably Bohr's. The notes sent by Verschaffelt had many gaps; Bohr wanted Kramers's advice and help with the discussion contributions, and travelled to Utrecht for this purpose at the beginning of March.[93] At the end of March, Kramers sent Verschaffelt the edited version of Bohr's contributions to the discussion after Compton's report (pp. 328, 337) and after Born and Heisenberg's

[a] According to D. Devriese, curator of the IIPCS archives, the original notes have not survived.

report (p. 403), remarking that these were all of Bohr's contributions to the discussions during the first three days of the conference.[a]

In contrast, material on Bohr's contributions to the general discussion survives only in the notes in the Bohr archive.[b] Some notes by Richardson also relate to Bohr's contributions.[95] A translation of the version of the Como lecture for *Naturwissenschaften* (Bohr 1928) was included instead, reprinted on a par with the other reports, and accompanied by the following footnote (p. 215 of the published proceedings):

This article, which is the translation of a note published very recently in *Naturwissenschaften*, vol. 16, 1928, p. 245, has been added at the author's request to replace the exposition of his ideas that he gave in the course of the following general discussion. It is essentially the reproduction of a talk on the current state of quantum theory that was given in Como on 16 September 1927, on the occasion of the jubilee festivities in honour of Volta.

The last remaining material was sent to Gauthier-Villars sometime in September 1928, and the volume was finally published in early December of that year.[96]

1.7 Conclusion

The fifth Solvay conference was by any standards an important and memorable event. On this point all participants presumably agreed, as shown by numerous letters, such as Ehrenfest's letter quoted above (reproduced in Bohr 1985), Heisenberg's letter to Lefébure at the head of this chapter, or various other letters of thanks addressed to the organisers after the conference:[97]

I would like to take this opportunity of thanking you for your kind hospitality, and telling you how much I enjoyed this particular Conference. I think it has been the most memorable one which I have attended for the subject which was discussed was of such vital interest and I learned so much. (W. L. Bragg to Armand Solvay, 3 November 1927)

It was the most stimulating scientific meeting I have ever taken part in. (Max Born to Charles Lefébure, 8 November 1927)

Perceptions of the significance of the conference differed from each other, however. In the official history, the fifth Solvay conference went down (perhaps together with the Como meeting) as the occasion on which the interpretational issues were finally clarified. This was presumably a genuine sentiment on the part of Bohr, Heisenberg

[a] Note that Bohr also asked some brief questions after Schrödinger's report. Kramers further writes that Bohr suggested to omit the end of the discussion after Born and Heisenberg's report, as well as a remark by Fowler from the general discussion (both of which are indeed absent from the published proceedings). Again, thanks to Mark van Atten for help with this letter.[94]

[b] See also Bohr (1985, pp. 35–7, 100, 478–9).

and the other physicists of the Copenhagen–Göttingen school. We find it explicitly as early as 1929:

> In relating the development of the quantum theory, one must in particular not forget the discussions at the Solvay conference in Brussels in 1927, chaired by Lorentz. Through the possibility of exchange [Aussprache] between the representatives of different lines of research, this conference has contributed extraordinarily to the clarification of the physical foundations of the quantum theory; it forms so to speak the outward completion of the quantum theory . . .
>
> *(Heisenberg 1929, p. 495)*

On the other hand, the conference was also described (by Langevin) as the one where 'the confusion of ideas reached its peak'.[98] From a distance of some 80 years, the beginnings of a more dispassionate evaluation should be possible. In the following chapters, we shall revisit the fifth Solvay conference, focussing in particular on the background and contributions relating to the three main 'lines of research' into quantum theory represented there: de Broglie's pilot-wave theory, Born and Heisenberg's quantum mechanics and Schrödinger's wave mechanics.

Archival notes

1 Handwritten remark (by Ehrenfest) in the margin of Lorentz to Ehrenfest, 29 March 1926, AHQP-EHR-23 (in Dutch).
2 Heisenberg to Lefébure, 19 December [1927], IIPCS 2685 (in German).
3 Noord-Hollands Archief (Hoarlem, NL), papers of Prof. Dr H. A. Lorentz (1858–1928), 1866–1930, inventory number 157, AHQP-LTZ-12, talk X 23, 'Ernest Solvay', dated 28 September 1914 (in English). Cf. also the French version of the same, X 10. (The two pages of the latter are in separate places on the microfilm.)
4 Pelseneer, J., *Historique des Instituts Internationaux de Physique et de Chimie Solvay depuis leur fondation jusqu'à la deuxième guerre mondiale* [1962], 103pp., AHQP-58, section 1 (hereafter simply 'Pelseneer').
5 Planck to Nernst, 11 June 1910, Pelseneer, p. 7 (in German).
6 Lorentz to Solvay, 4 January 1912, Pelseneer, pp. 20–6 (in French). See also the reply, Solvay to Lorentz, 10 January 1912, AHQP-LTZ-12 (in French).
7 Lorentz to Solvay, 6 March 1912, Pelseneer, p. 27 (in French).
8 Héger to Lorentz, 16 February 1912, AHQP-LTZ-11 (in French).
9 Einstein to Lorentz, 16 August 1923, AHQP-LTZ-7 (in German).
10 Van Aubel to Lorentz, 16 April, 16 May and 19 July 1923, AHQP-LTZ-11 (in French).
11 Lorentz to Solvay, 10 January 1919, Pelseneer, p. 37 (in French).
12 Lorentz to Einstein, 14 March 1926, AHQP-86 (in German).
13 Lefébure to Lorentz, 12 February 1926, AHQP-LTZ-12 (in French).
14 Cf. Lefébure to Lorentz, 12 February 1926, AHQP-LTZ-12 (in French). Obituary of Tassel, *L'Éventail*, 15 October 1922, AHQP-LTZ-13 (in French).
15 Letter by Lorentz, 7 January 1919, Pelseneer, pp. 35–6 (in French), two letters from Planck to Lorentz, 1915, AHQP-LTZ-12 (in German), Planck to Lorentz, March 1916, Pelseneer, pp. 34–5 (in German), and Planck to Lorentz, 5 December 1923, AHQP-LTZ-9 (in German).
16 From Lefébure [possibly a copy for Lorentz], 1 April 1926, AHQP-LTZ-12; cf. also Lefébure to Lorentz, 6 April 1926, AHQP-LTZ-12 (both in French).
17 Lorentz to Einstein, 6 April 1926, AHQP-86 (in German).
18 Einstein to Lorentz, 14 April 1926 and 1 May 1927, AHQP-LTZ-11 (in German).
19 Lorentz to Einstein, 28 April 1926, AHQP-86 (in German).
20 Compare the invitation by Lorentz to Bohr, 7 June 1926, AHQP-BSC-13 (in English), and the replies by Bohr to Lorentz, 24 June 1926, AHQP-LTZ-11 (in English), Planck to Lorentz, 13 June 1926, AHQP-LTZ-8 (in German), Born to Lorentz, 19 June 1926, AHQP-LTZ-11 (in German), and Heisenberg to Lorentz, 4 July 1926, AHQP-LTZ-12 (in German).
21 Van Aubel to Lefébure, 6 October 1927, IIPCS 2545 (in French), with a handwritten comment by Lefébure.
22 Lefébure to Lorentz, 14 October 1927, IIPCS 2534, and 15 October 1927, IIPCS 2536, telegramme Lorentz to Lefébure, 15 October 1927, IIPCS 2535, and Lefébure to the King, 19 October 1927, IIPCS 2622 (all in French).
23 Lorentz to Solvay, 10 January 1919, Pelseneer, p. 37 (in French).
24 There is a large amount of relevant correspondence in AHQP-LTZ.
25 Lorentz to Ehrenfest, 29 March 1926, AHQP-EHR-23 (in Dutch), with Ehrenfest's handwritten comments.
26 Ehrenfest to Lorentz, 30 March 1926, Noord-Hollands Archief (Haarlem, NL), papers of Prof. Dr H. A. Lorentz (1858–1928), 1866–1930, inventory number 20, AHQP-LTZ-11 (in German).
27 Lorentz to Bohr, 7 June 1926, AHQP-BSC-13, section 3 (in English), Lorentz to Schrödinger, 21 January 1927, AHQP-41, section 9 (in German).
28 Lorentz to Einstein, 6 April 1926, AHQP-86 (in German and French).
29 Verschaffelt to Lefébure, 8 April 1926, IIPCS 2573 (in French).
30 Einstein to Lorentz, 12 April 1926, AHQP-LTZ-11 (in German).
31 See for instance the catalogue of Lorentz's manuscripts in AHQP-LTZ-11.
32 See also Lorentz to Schrödinger, 21 January and 17 June 1927, AHQP-41, section 9 (in German).

33 Einstein to Lorentz, 1 May 1926, AHQP-LTZ-11 (in German).
34 Einstein to Lorentz, 22 June 1926, AHQP-LTZ-8 (in German).
35 Einstein to Lorentz, 16 February 1927, AHQP-LTZ-11 (in German).
36 Enclosed with Lorentz to Ehrenfest, 3 June [1927, erroneously amended to 1925], AHQP-EHR-23 (in Dutch).
37 Lorentz to Dirac, 9 June 1927, AHQP-LTZ-8 (in English).
38 Cf. Born to Lorentz, 23 June 1927, AHQP-LTZ-11 (in German).
39 Lorentz to Lefébure, 27 August 1927, IIPCS 2532A/B (in French). [There appears to be a different item also numbered 2532 in the archive].
40 Dirac to Lorentz, 13 September 1927, AHQP-LTZ-11 (in English).
41 Cf. de Broglie to Lorentz, 22 June 1927, AHQP-LTZ-11 (in French), and Ehrenfest to Lorentz, 14 June 1927, AHQP-EHR-23 (in Dutch and German).
42 W. L. Bragg to Lorentz, 7 February 1927, AHQP-LTZ-11 (in English).
43 Cf. also W. L. Bragg to Lorentz, 27 June 1927, AHQP-LTZ-11 (in English).
44 Lorentz to Schrödinger, 17 June 1927, AHQP-41, section 9 (in German).
45 Einstein to Lorentz, 16 December 1924, AHQP-LTZ-7 (in German).
46 Einstein to Lorentz, 12 April 1926, AHQP-LTZ-11, Lorentz to Einstein, 28 April 1926, AHQP-86, and Einstein to Lorentz, 1 May 1926, AHQP-LTZ-11 (all in German).
47 'Bestimmt Schrödingers Wellenmechanik die Bewegung eines Systems vollständig oder nur im Sinne der Statistik?', AEA 2-100.00 (in German), available on-line at http://www.alberteinstein.info/db/ViewDetails.do?DocumentID=34338.
48 Heisenberg to Einstein, 19 May 1927, AEA 12-173.00, and 10 June 1927, AEA 12-174.00 (both in German).
49 Lorentz to Schrödinger, 17 June 1927, AHQP-41, section 9 (in German).
50 W. H. Bragg to Lorentz, 11 May 1927, AHQP-LTZ-11 (in English).
51 Tassel to Verschaffelt, 24 February 1921, AHQP-LTZ-13 (in French).
52 Lorentz to Ehrenfest, 29 March 1926, AHQP-EHR-23 (in Dutch).
53 Ehrenfest to Lorentz, 30 March 1926, Noord-Hollands Archief, H A. Lorentz papers, inventory number 20, AHQP-LTZ-11 (in German).
54 Verschaffelt to Lefébure, 8 April 1926, IIPCS 2573 (in French).
55 Guye to Lorentz, 14 April 1926, AHQP-LTZ-8 (in French).
56 Cf. for instance Lorentz to Schrödinger, 21 January 1927, AHQP-41, section 9 (in German).
57 Lefébure to Lorentz, 22 October 1927, AHQP-LTZ-12 (in French).
58 Lorentz to Einstein, 28 April 1926, AHQP-86 (in German), Lorentz to Lefébure, 9 October 1927, IIPCS 2530A/B (in French).
59 Lorentz to Einstein, 28 April 1926, AHQP-86 (in German), Einstein to Lorentz, 1 May 1926, AHQP-LTZ-11 (in German).
60 For the latter, cf. also Brillouin to Lorentz, 20 August 1927, AHQP-LTZ-11 (in French).
61 Lorentz to Einstein, 30 January 1927, AHQP-86 (in German), Einstein to Lorentz, 16 February 1927, AHQP-LTZ-11 (in German), Richardson to Lorentz, 19 February 1927, AHQP-LTZ-12 (in English).
62 Brillouin to Lorentz, 20 August 1927, AHQP-LTZ-11 (in French), Ehrenfest to Lorentz, 18 August 1927, AHQP-EHR-23 (in German).
63 Telegramme Lorentz to Lefébure, 19 October 1927, IIPCS 2541 (in French), with Lefébure's note: 'Oui'.
64 Lorentz to Brillouin, 15 December 1926, AHQP-LTZ-12 (in French), Lorentz to Ehrenfest, 18 January 1927, AHQP-EHR-23 (in Dutch), Lorentz to Schrödinger, 21 January 1927, AHQP-41, section 9 (in German); compare various other replies to Lorentz's invitation, in AHQP-LTZ-11, AHQP-LTZ-12 and AHQP-LTZ-13: Brillouin, 8 January 1927 (in French), de Broglie, 8 January 1927 (in French), W. L. Bragg, 7 February 1927 (in English), Wilson, 11 February 1927 (in English), Kramers, 14 February 1927 (in German), Debye, 24 February 1927 (in Dutch), Compton, 3 April 1927 (in English) [late because he had been away for two months 'in the Orient'].
65 Lorentz to Lefébure, 21 May 1927, IIPCS 2521A (in French).
66 Deslandres to Lorentz, 19 July 1927, AHQP-LTZ-11 (in French).

67 See the letters of acceptance to Lorentz: De Donder, 8 July 1927, AHQP-LTZ-11, Henriot, 10 July 1927, AHQP-LTZ-12, Piccard, 2 October 1927, AHQP-LTZ-12 (all in French).
68 W. H. Bragg to Lorentz, 12 and 17 July 1927, AHQP-LTZ-11 (in English).
69 Copies in AHQP-LTZ-12 and IIPCS 2543. Various replies: IIPCS 2544–51, 2553–6, 2558, 2560–3.
70 Cf. Lorentz to Schrödinger, 17 June 1927, AHQP-41, section 9 (in German).
71 Compare the answers by W. L. Bragg, 27 June 1927, and by Compton, 7 July 1927, both AHQP-LTZ-11 (in English), and the detailed ones by de Broglie, 27 June 1927, AHQP-LTZ-11 (in French) and by Born, 23 June 1927, AHQP-LTZ-11 (in German); also Schrödinger to Lorentz, 23 June 1927, AHQP-LTZ-13 (original with Schrödinger's corrections), and AHQP-41, section 9 (carbon copy) (in German).
72 Microfilmed in AHQP-RDN, documents M-0059 (Bragg, catalogued as 'unidentified author'), M-0309 (Born and Heisenberg, with seven pages of notes by Richardson) and M-1354 (Schrödinger).
73 See in particular the already quoted Lorentz to Lefébure, 27 August 1927, IIPCS 2532A/B (in French), as well as Lorentz to Lefébure, 23 September 1927, IIPCS 2523A/B, and 4 October 1927, IIPCS 2528 (both in French), Gauthier-Villars to Lefébure, 16 September 1927, IIPCS 2755 (in French), Verschaffelt to Lorentz, 6 October 1927, AHQP-LTZ-13 (in Dutch), de Broglie to Lorentz, 29 August 1927, AHQP-LTZ-8, and 11 October 1927, AHQP-LTZ-11 (both in French), and Verschaffelt to Lefébure, 15 October 1927, IIPCS 2756 (in French).
74 Lorentz to Ehrenfest, 13 October 1927, AHQP-EHR-23 (in Dutch). See also Born to Lorentz, 11 October 1927, AHQP-LTZ-11 (in German), de Broglie to Lorentz, 11 October 1927, AHQP-LTZ-11 (in French), Planck to Lefébure, 17 October 1927, IIPCS 2558, and Richardson to Lefébure, IIPCS 2561.
75 'Une crise dans la physique moderne, I & II', IIPCS 2750–1 (in French).
76 Lorentz to Lefébure, 29 September 1927, IIPCS 2523A/B, Brillouin to Lorentz, 11 October 1927, AHQP-LTZ-11, Fondation Solvay to Lefébure, IIPCS 2582. Invitation: IIPCS 2615 and 2619, AHPQ-LTZ-8. Replies: IIPCS 2617 and 2618. (All documents in French).
77 IIPCS 2530A/B, 2537.
78 IIPCS 2533, 2586A/B/C/D/E (the proposed menus from the Taverne Royale), 2587A/B.
79 Cf. Lorentz to Schrödinger, 21 January 1927, AHQP-41, section 9 (in German).
80 IIPCS 2340.
81 For the dates and times of Compton's report and of the general discussion, see the appendix to Part III. For the time and place of the first session, see IIPCS 2523A/B. For the reception at ULB, see IIPCS 2540 and 2629. There is a seating plan for the lunch with the royal couple, IIPCS 2627. For the dinner with Armand Solvay, see IIPCS 2533, 2624 and 2625. See also Pelseneer, pp. 49–50.
82 IIPCS 2621.
83 See AHQP-LTZ-12 (draft), and presumably IIPCS 2667.
84 Schrödinger to Lorentz, 23 June 1927, AHQP-LTZ-13 and AHQP-41, section 9 (in German), Lorentz to Schrödinger, 8 July 1927, AHPQ-41, section 9 (in German), Born to Lorentz, 23 June 1927, AHQP-LTZ-11 (in German).
85 Ehrenfest to Goudsmit, Uhlenbeck and Dieke, 3 November 1927, AHQP-61 (in German).
86 W. H. Bragg to Lorentz, 17 July [1927], AHQP-LTZ-11 (in English), Planck to Lorentz, 2 February 1927, AHQP-LTZ-8 (in German).
87 See Pelseneer, pp. 50–1.
88 Ehrenfest to Kramers, 6 November 1927, AHQP-9, section 10 (in German). The words 'bekommt man' [?] are very faint.
89 Lorentz to Lefébure, 30 December 1927, IIPCS 2670A/B (in French).
90 Verschaffelt to Lefébure, 6 January 1928, IIPCS 2609, 2 March 1928, IIPCS 2610 (both in French).
91 Lefébure to Lorentz, [after 27 August 1927], IIPCS 2524 (in French), Bohr to Kramers, 17 February 1928, AHQP-BSC-13 (in Danish), Brillouin to Lorentz, 31 December 1927,

AHQP-LTZ-8 (in French), and Kramers to Verschaffelt, 28 March 1928, AHQP-28, section 4 (in Dutch).
92 Microfilmed in AHQP-BMSS-11, section 5.
93 Bohr to Kramers, 17 and 27 February 1928, AHQP-BSC-13 (both in Danish).
94 Kramers to Verschaffelt, 28 March 1928, AHQP-28, section 4 (in Dutch).
95 Included with the copy of Born and Heisenberg's report in AHQP-RDN, document M-0309.
96 Gauthier-Villars to Verschaffelt, 6 December 1928, IIPCS 2762 (in French), and Verschaffelt to Lefébure, 11 [December] 1928, IIPCS 2761 (in French).
97 IIPCS 2671 (in English), IIPCS 2672 (in German).
98 Cf. Pelseneer, p. 50.

2

De Broglie's pilot-wave theory

2.1 Background

At a time when no single known fact supported this theory, Louis de Broglie asserted that a stream of electrons which passed through a very small hole in an opaque screen must exhibit the same phenomena as a light ray under the same conditions.

(Prof. C. W. Oseen, Chairman of the Nobel Committee for Physics, presentation speech, 12 December 1929 (Oseen 1999))

In September 1923, Prince Louis de Broglie[a] made one of the most astonishing predictions in the history of theoretical physics: that material bodies would exhibit the wave-like phenomena of diffraction and interference upon passing through sufficiently narrow slits. Like Einstein's prediction of the deflection of light by the sun, which was based on a reinterpretation of gravitational force in terms of geometry, de Broglie's prediction of the deflection of electron paths by narrow slits was made on the basis of a fundamental reappraisal of the nature of forces and of dynamics. De Broglie had proposed that Newton's first law of motion be abandoned, and replaced by a new postulate, according to which a freely moving body follows a trajectory that is orthogonal to the surfaces of equal phase of an associated guiding wave. The resulting 'de Broglian dynamics' – or pilot-wave theory as de Broglie later called it – was a new approach to the theory of motion, as radical as Einstein's interpretation of the trajectories of falling bodies as geodesics of a curved spacetime, and as far-reaching in its implications. In 1929 de Broglie received the Nobel Prize 'for his discovery of the wave nature of electrons'.

[a] The de Broglies had come to France from Italy in the seventeenth century, the original name 'Broglia' eventually being changed to de Broglie. On his father's side, de Broglie's ancestors included dukes, princes, ambassadors, and marshals of France. Nye (1997) considers the conflict between de Broglie's pursuit of science and the expectations of his aristocratic family. For a biography of de Broglie, see Lochak (1992).

Strangely enough, however, even though de Broglie's prediction was confirmed experimentally a few years later, for most of the twentieth century single-particle diffraction and interference were routinely cited as evidence *against* de Broglie's ideas: even today, some textbooks on quantum mechanics assert that such interference demonstrates that particle trajectories cannot exist in the quantum domain (see Section 6.1). It is as if the deflection of light by the sun had come to be widely regarded as evidence against Einstein's general theory of relativity. This remarkable misunderstanding illustrates the extent to which de Broglie's work in the 1920s has been underestimated, misrepresented, and indeed largely ignored, not only by physicists but also by historians.

De Broglie's PhD thesis of 1924 is of course recognised as a landmark in the history of quantum theory. But what is usually remembered about that thesis is the proposed extension of Einstein's wave-particle duality from light to matter, with the formulas $E = h\nu$ and $p = h/\lambda$ (relating energy and momentum to frequency and wavelength) being applied to electrons or 'matter waves'. Usually, little attention is paid to the fact that a central theme of de Broglie's thesis was the construction of a new form of dynamics, in which classical (Newtonian or Einsteinian) laws are abandoned and replaced by new laws according to which particle *velocities* are determined by guiding waves, in a specific manner that unifies the variational principles of Maupertuis and Fermat. Nor, indeed, have historians paid much attention to de Broglie's later and more complete form of pilot-wave dynamics, which he arrived at in a paper published in May 1927 in *Journal de Physique*, and which he then presented in October 1927 at the fifth Solvay conference.

Unlike the other main contributors to quantum theory, de Broglie worked in relative isolation, having little contact with the principal research centres in Berlin, Copenhagen, Göttingen, Cambridge and Munich. While Bohr, Heisenberg, Born, Schrödinger, Pauli and others visited each other frequently and corresponded regularly, de Broglie worked essentially alone in Paris.[a] In France at the time, while pure mathematics was well represented, there was very little activity in theoretical physics. In addition, after the First World War, scientific relations with Germany and Austria were interrupted.[b] All this seems to have suited de Broglie's rather solitary temperament. De Broglie's isolation, and the fact that France was outside the mainstream of theoretical physics, may account in part for why so much of de Broglie's work went relatively unnoticed at the time, and has remained largely ignored even to the present day.

[a] In his typed 'Replies to Mr Kuhn's questions' in the Archive for the History of Quantum Physics,[1] de Broglie writes (p. 7): 'Between 1919 and 1928, I worked *very much in isolation*' (emphasis in the original). Regarding his PhD thesis de Broglie recalled (p. 9): 'I worked very much alone and almost without any exchange of ideas'.

[b] For more details on de Broglie's situation in France at the time, see Mehra and Rechenberg (1982a, pp. 578–84).

For some 70 years, the physics community tended to believe either that 'hidden-variables' theories like de Broglie's were impossible, or that such theories had been disproven experimentally. The situation changed considerably in the 1990s, with the publication of textbooks presenting quantum mechanics in the pilot-wave formulation (Bohm and Hiley 1993; Holland 1993). Pilot-wave theory – as originated by de Broglie in 1927 and elaborated by Bohm 25 years later (Bohm 1952a,b) – is now accepted as an alternative (if little used) formulation of quantum theory.

Focussing for simplicity on the non-relativistic quantum theory of a system of N (spinless) particles with 3-vector positions $\mathbf{x}_i$ ($i = 1, 2, \ldots, N$), it is now generally agreed that, with appropriate initial conditions, quantum physics may be accounted for by the deterministic dynamics defined by two differential equations: the Schrödinger equation

$$i\hbar\frac{\partial\Psi}{\partial t} = \sum_{i=1}^{N} -\frac{\hbar^2}{2m_i}\nabla_i^2\Psi + V\Psi \tag{2.1}$$

for a 'pilot wave' $\Psi(\mathbf{x}_1, \mathbf{x}_2, \ldots, \mathbf{x}_N, t)$ in configuration space and the de Broglie guidance equation

$$m_i\frac{d\mathbf{x}_i}{dt} = \nabla_i S \tag{2.2}$$

for particle trajectories $\mathbf{x}_i(t)$, where the phase $S(\mathbf{x}_1, \mathbf{x}_2, \ldots, \mathbf{x}_N, t)$ is locally defined by $S = \hbar \operatorname{Im} \ln \Psi$ (so that $\Psi = |\Psi|\, e^{(i/\hbar)S}$).

This, as we shall see, is how de Broglie presented his dynamics in 1927. Bohm's presentation of 1952 was rather different. If one takes the time derivative of (2.2), then using (2.1) one obtains Newton's law of motion for acceleration

$$m_i\ddot{\mathbf{x}}_i = -\nabla_i(V + Q), \tag{2.3}$$

where

$$Q \equiv -\sum_i \frac{\hbar^2}{2m_i}\frac{\nabla_i^2|\Psi|}{|\Psi|} \tag{2.4}$$

is the 'quantum potential'. Bohm regarded (2.3) as the law of motion, with (2.2) added as a constraint on the initial momenta, a constraint that Bohm thought could be relaxed (see Section 11.1). For de Broglie, in contrast, the law of motion (2.2) for velocity had a fundamental status, and for him represented the unification of the principles of Maupertuis and Fermat. One should then distinguish between de Broglie's first-order (velocity-based) dynamics of 1927, and Bohm's second-order (acceleration-based) dynamics of 1952.

In this chapter, we shall be concerned with the historical origins of de Broglie's 1927 dynamics defined by (2.1) and (2.2). Some authors have referred to this dynamics as 'Bohmian mechanics'. Such terminology is misleading: it disregards de Broglie's priority, and misses de Broglie's physical motivations for recasting dynamics in terms of velocity; it also misrepresents Bohm's 1952 formulation, which was based on (2.1) and (2.3). These and other historical misconceptions concerning de Broglie–Bohm theory will be addressed in Section 11.1.

The two equations (2.1) and (2.2) define a deterministic (de Broglian or pilot-wave) dynamics for a single multiparticle system: given an initial wave function $\Psi(\mathbf{x}_1, \mathbf{x}_2, \ldots, \mathbf{x}_N, 0)$ at $t = 0$, (2.1) determines $\Psi(\mathbf{x}_1, \mathbf{x}_2, \ldots, \mathbf{x}_N, t)$ at all times t; and given an initial configuration $(\mathbf{x}_1(0), \mathbf{x}_2(0), \ldots, \mathbf{x}_N(0))$, (2.2) then determines the trajectory $(\mathbf{x}_1(t), \mathbf{x}_2(t), \ldots, \mathbf{x}_N(t))$. For an ensemble of systems with the same initial wave function $\Psi(\mathbf{x}_1, \mathbf{x}_2, \ldots, \mathbf{x}_N, 0)$, and with initial configurations $(\mathbf{x}_1(0), \mathbf{x}_2(0), \ldots, \mathbf{x}_N(0))$ distributed according to the Born rule

$$P(\mathbf{x}_1, \mathbf{x}_2, \ldots, \mathbf{x}_N, 0) = |\Psi(\mathbf{x}_1, \mathbf{x}_2, \ldots, \mathbf{x}_N, 0)|^2 , \tag{2.5}$$

the statistical distribution of outcomes of quantum measurements will agree with the predictions of standard quantum theory. This is shown by treating the measuring apparatus, together with the system being measured, as a single multiparticle system obeying de Broglian dynamics, so that $(\mathbf{x}_1, \mathbf{x}_2, \ldots, \mathbf{x}_N)$ defines the 'pointer position' of the apparatus as well as the configuration of the measured system. Given the initial condition (2.5) for any multiparticle system, the statistical distribution of particle positions at later times will also agree with the Born rule $P = |\Psi|^2$. Thus, the statistical distribution of pointer positions in any experiment will agree with the predictions of quantum theory, yielding the correct statistical distribution of outcomes for standard quantum measurements.

In his 1927 Solvay report, de Broglie gave some simple applications of pilot-wave theory, with the assumed initial condition (2.5). He applied the theory to single-photon interference, to atomic transitions, and to the scattering (or diffraction) of electrons by a crystal lattice. But a detailed demonstration of equivalence to quantum theory, and in particular a pilot-wave account of the general quantum theory of measurement, was not provided until the work of Bohm in 1952.

How did de Broglie come to propose this theory in 1927? In this chapter, we trace de Broglie's work in this direction, from his early work leading to his doctoral thesis of 1924 (de Broglie 1924e, 1925), to his crucial paper of 1927 published in *Journal de Physique* (de Broglie 1927b), and culminating in his presentation of pilot-wave theory at the fifth Solvay conference. We examine in detail how de Broglie arrived at this new form of particle dynamics, and what his attitude towards it was. Later, in Chapter 10, we shall consider some of the discussions of de Broglie's theory that took place at the conference, in particular the famous (and widely misunderstood) clash between de Broglie and Pauli.

De Broglie's dynamics has the striking feature that electrons and photons are regarded as both particles and waves. Like many scientific ideas, this mingling of particle-like and wave-like aspects had precursors. In Newton's *Opticks* (first published in 1704), both wave-like and particle-like properties are attributed to light. Newton's so-called 'corpuscular' theory was formulated on the basis of extensive and detailed experiments (carried out by Grimaldi, Hooke and Newton himself) involving what we would now call interference and diffraction. According to Newton, light corpuscles – or light 'Rays' as he called them[a] – generate 'Waves of Vibrations' in an 'Aethereal Medium', much as a stone thrown into water generates water waves (Newton 1730, reprint, pp. 347–9). In addition, Newton supposed that the waves in turn affect the motion of the corpuscles, which 'may be alternately accelerated and retarded by the Vibrations' (p. 348). In particular, Newton thought that the effect of the medium on the motion of the corpuscles was responsible for the phenomena of interference and diffraction. He writes, for example (p. 350):

> And doth not the gradual condensation of this Medium extend to some distance from the Bodies, and thereby cause the Inflexions of the Rays of Light, which pass by the edges of dense Bodies, at some distance from the Bodies?

Newton understood that, for diffraction to occur, the motion of the light corpuscles would have to be affected at a distance by the diffracting body – 'Do not Bodies act upon Light at a distance, and by their action bend its Rays ... ?' (p. 339) – and his proposed mechanism involved waves in an inhomogeneous ether. Further, according to Newton, to account for the coloured fringes that had been observed by Grimaldi in the diffraction of white light by opaque bodies, the corpuscles would have to execute an oscillatory motion 'like that of an Eel' (p. 339):

> Are not the Rays of Light in passing by the edges and sides of Bodies, bent several times backwards and forwards, with a motion like that of an Eel? And do not the three Fringes of colour'd Light above-mention'd arise from three such bendings?

For Newton, of course, such non-rectilinear motion could be caused only by a force emanating from the diffracting body.

It is interesting to note that, in the general discussion at the fifth Solvay conference (p. 461), de Broglie commented on this very point, with reference to the 'emission' (or corpuscular) theory, and pointed out that if pilot-wave dynamics were written in terms of acceleration (as done later by Bohm) then just such forces appeared:

> In the corpuscular conception of light, the existence of diffraction phenomena occuring at the edge of a screen requires us to assume that, in this case, the trajectory of the photons is curved. The supporters of the emission theory said that the edge of the screen exerts a force

[a] The opening definition of the *Opticks* defines 'Rays' of light as 'its least Parts'.

> on the corpuscle. Now, if in the new mechanics as I develop it, one writes the Lagrange equations for the photon, one sees appear on the right-hand side of these equations a term ... [that] ... represents a sort of force of a new kind, which exists only ... where there is interference. It is this force that will curve the trajectory of the photon when its wave ψ is diffracted by the edge of a screen.

The striking similarity between Newton's qualitative ideas and pilot-wave theory has also been noted by Berry, who remarks that during interference or diffraction the de Broglie–Bohm trajectories indeed 'wriggle like an eel' (Berry 1997, p. 42), in some sense vindicating Newton.

A mathematical precursor to de Broglian dynamics is found in the early nineteenth century, in Hamilton's formulation of geometrical optics and particle mechanics. As de Broglie points out in his Solvay report (pp. 344, 349), Hamilton's theory is in fact the short-wavelength limit of pilot-wave dynamics: for in that limit, the phase of the wave function obeys the Hamilton–Jacobi equation, and de Broglie's trajectories reduce to those of classical mechanics.

A physical theory of light as both particles and waves – in effect a revival of Newton's views – emerged again with Einstein in 1905. It is less well known that, after 1905, Einstein tried to construct theories of localised light quanta coupled to vector fields in 3-space. As we shall see in Chapter 9, Einstein's ideas in this vein show some resemblance to de Broglie's but also differ from them.

It should also be mentioned that, in the autumn of 1923 (the same year in which de Broglie first elaborated his ideas), Slater tried to develop a theory in which the motion of photons was guided by the electromagnetic field. It appears that Slater first attempted to construct a deterministic theory, but had trouble defining an appropriate velocity vector; he then came to the conclusion that photons and the electromagnetic field were related only statistically, with the photon probability density being given by the intensity of the field. After discussing his ideas with Bohr and Kramers in 1924, the photons were removed from the theory, apparently against Slater's wishes (Mehra and Rechenberg 1982a, pp. 542–6). Note that, while de Broglie applied his theory to photons, he made it clear (for example in the general discussion, p. 460) that in his theory the guiding 'ψ-wave' was distinct from the electromagnetic field.

In the case of light, then, the idea of combining both particle-like and wave-like aspects was an old one, going back indeed to Newton. In the case of ordinary matter, however, de Broglie seems to have been the first to develop a physical theory of this form.

It is sometimes claimed that, for the case of electrons, ideas similar to de Broglie's were put forward by Madelung in 1926. What Madelung proposed, however, was to regard an electron with mass m and wave function ψ not as a point-like particle within the wave, but as a continuous fluid spread over space with mass density $m\,|\psi|^2$ (Madelung 1926a,b). In this 'hydrodynamical' interpretation, mathematically the fluid velocity coincides with de Broglie's velocity field; but physically, Madelung's theory seems more akin to Schrödinger's theory than to de Broglie's.

Finally, before we examine de Broglie's work, we note what appears to be a recurring historical opposition to dualistic physical theories containing both waves and particles. In 1801–3, Thomas Young, who by his own account regarded his theory as a development of Newton's ideas,[a] removed the corpuscles from Newton's theory and produced a purely undulatory account of light. In 1905, Einstein's dualist view of light was not taken seriously, and did not win widespread support until the discovery of the Compton effect in 1923. In 1924, Bohr and Kramers, who regarded the Bohr–Kramers–Slater theory as a development of Slater's original idea, insisted on removing the photons from Slater's theory of radiation.[b] And in 1926, Schrödinger, who regarded his work as a development of de Broglie's ideas, removed the trajectories from de Broglie's theory and produced a purely undulatory 'wave mechanics'.[c]

2.2 A new approach to particle dynamics: 1923–1924

In this section we show how de Broglie took his first steps towards a new form of dynamics.[d] His aim was to explain the quantum phenomena known at the time – in particular the Bohr–Sommerfeld quantisation of atomic energy levels, and the apparently dual nature of radiation – by unifying the physics of particles with the physics of waves. To accomplish this, de Broglie began by extending Einstein's wave-particle duality for light to all material bodies, by introducing a 'phase wave' accompanying every material particle. Then, inspired by the optical-mechanical analogy,[e] de Broglie proposed that Newton's first law of motion should be abandoned, and replaced by a new principle that unified Maupertuis' variational principle for mechanics with Fermat's variational principle for optics. The result was a new form of dynamics in which the velocity $\mathbf{v}$ of a particle is determined by the gradient of the phase ϕ of an accompanying wave – in contrast with classical mechanics, where accelerations are determined by forces. (Note that de Broglie's phase ϕ has a sign opposite to the phase S as we would normally define it now.)

[a] See, for example, Bernard Cohen's preface to Newton's *Opticks* (1730, reprint).

[b] In 1925, Born and Jordan attempted to restore the photons, proposing a stochastic theory reminiscent of Slater's original ideas; it appears that they were dissuaded from publication by Bohr. See Darrigol (1992, p. 253) and Section 3.4.2.

[c] Cf. Section 4.5.

[d] An insightful and general account of de Broglie's early work, up to 1924, has been given by Darrigol (1993).

[e] Possibly, de Broglie was also influenced by the philosopher Henri Bergson's writings concerning time, continuity and motion (Feuer 1974, pp. 206–14); though this is denied by Lochak (1992).

This new approach to dynamics enabled de Broglie to obtain a wave-like explanation for the quantisation of atomic energy levels, to explain the observed interference of single photons, and to predict for the first time the new and unexpected phenomenon of the diffraction and interference of electrons.

As we shall see, the theory proposed by de Broglie in 1923–4 was, in fact, a simple form of pilot-wave dynamics, for the special case of independent particles guided by waves in 3-space, and without a specific wave equation.

2.2.1 First papers on pilot-wave theory (1923)

De Broglie's earliest experience of physics was closely tied to experiment. During the First World War he worked on wireless telegraphy, and after the war his first papers concerned X-ray spectroscopy. In 1922 he published a paper on treating blackbody radiation as a gas of light quanta (de Broglie 1922). In this paper, de Broglie made the unusual assumption that photons had a very small but non-zero rest mass m_0. He was therefore now applying Einstein's relations $E = h\nu$ and $p = h/\lambda$ (relating energy and momentum to frequency and wavelength) to *massive* particles, even if these were still only photons. It seems that de Broglie made the assumption $m_0 \neq 0$ so that light quanta could be treated in the same way as ordinary material particles. It appears that this paper was the seed from which de Broglie's subsequent work grew.[a]

According to de Broglie's later recollections (L. de Broglie, AHQP interview, 7 January 1963, p. 1),[2] his first ideas concerning a pilot-wave theory of massive particles arose as follows. During conversations on the subject of X-rays with his older brother Maurice de Broglie,[b] he became convinced that X-rays were both particles and waves. Then, in the summer of 1923, de Broglie had the idea of extending this duality to ordinary matter, in particular to electrons. He was drawn in this direction by consideration of the optical-mechanical analogy; further, the presence of whole numbers in quantisation conditions suggested to him that waves must be involved.

This last motivation was recalled by de Broglie (1999) in his Nobel lecture of 1929:

> ... the determination of the stable motions of the electrons in the atom involves whole numbers, and so far the only phenomena in which whole numbers were involved in physics were those of interference and of eigenvibrations. That suggested the idea to me

[a] In a collection of papers by de Broglie and Brillouin, published in 1928, a footnote added to de Broglie's 1922 paper on blackbody radiation and light quanta remarks: 'This paper ... was the origin of the ideas of the author on wave mechanics' (de Broglie and Brillouin 1928, p. 1).

[b] Maurice, the sixth duc de Broglie, was a distinguished experimental physicist, having done important work on the photoelectric effect with X-rays – experiments that were carried out in his private laboratory in Paris.

that electrons themselves could not be represented as simple corpuscles either, but that a periodicity had also to be assigned to them too.

De Broglie first presented his new ideas in three notes published (in French) in the *Comptes Rendus* of the Academy of Sciences in Paris (de Broglie 1923a,b,c), and also in two papers published in English – one in *Nature* (de Broglie 1923d), the other in the *Philosophical Magazine* (de Broglie 1924a).[a]

The ideas in these papers formed the basis for de Broglie's doctoral thesis. The paper in the *Philosophical Magazine* reads, in fact, like a summary of much of the material in the thesis. Since the thesis provides a more systematic presentation, we shall give a detailed summary of it in the next subsection; here, we give only a brief account of the earlier papers, except for the crucial second paper, whose conceptual content warrants more detailed commentary.[b]

The first communication (de Broglie 1923a), entitled 'Waves and quanta', proposes that an 'internal periodic phenomenon' should be associated with *any* massive particle (including light quanta). In the rest frame of a particle with rest mass m_0, the periodic phenomenon is assumed to have a frequency $\nu_0 = m_0c^2/h$. In a frame where the particle has uniform velocity v, de Broglie considers the two frequencies ν and ν_1, where $\nu = \nu_0/\sqrt{1-v^2/c^2}$ is the frequency $\nu = mc^2/h$ associated with the relativistic mass increase $m = m_0/\sqrt{1-v^2/c^2}$ and $\nu_1 = \nu_0\sqrt{1-v^2/c^2}$ is the time-dilated frequency. De Broglie shows that, because $\nu_1 = \nu(1-v^2/c^2)$, a 'fictitious' wave of frequency ν and phase velocity $v_{\text{ph}} = c^2/v$ (propagating in the same direction as the particle) will remain in phase with the internal oscillation of frequency ν_1. De Broglie then considers an atomic electron moving uniformly on a circular orbit. He proposes that orbits are stable only if the fictitious wave remains in phase with the internal oscillation of the electron. From this condition, de Broglie derives the Bohr–Sommerfeld quantisation condition.

The second communication (de Broglie 1923b), entitled 'Light quanta, diffraction and interference', has a more conceptual tone. De Broglie begins by recalling his previous result, that a moving body must be associated with 'a *non-material* sinusoidal wave'. He adds that the particle velocity v is equal to the group velocity of the wave, which de Broglie here calls 'the phase wave' because its phase at the location of the particle is equal to the phase of the internal oscillation of the particle. De Broglie then goes on to make some very significant observations about diffraction and the nature of the new dynamics that he is proposing.

De Broglie asserts that diffraction phenomena prove that light quanta cannot always propagate in a straight line, even in what would normally be called empty

[a] It seems possible that *Comptes Rendus* was not widely read by physicists outside France, but this certainly was not true of *Nature* or the *Philosophical Magazine*.

[b] We do not always keep to de Broglie's original notation.

space. He draws the bold conclusion that Newton's first law of motion (the 'principle of inertia') must be abandoned (p. 549):

The light quanta [atomes de lumière] whose existence we assume do not always propagate in a straight line, as proved by the phenomena of diffraction. It then seems *necessary* to modify the principle of inertia.

De Broglie then suggests replacing Newton's first law with a new postulate (p. 549):

We propose to adopt the following postulate as the basis of the dynamics of the free material point: 'At each point of its trajectory, a free moving body follows in a uniform motion the *ray* of its phase wave, that is (in an isotropic medium), the normal to the surfaces of equal phase'.

The diffraction of light quanta is then explained since, as de Broglie notes, 'if the moving body must pass through an opening whose dimensions are small compared to the wavelength of the phase wave, in general its trajectory will curve like the ray of the diffracted wave'.

In retrospect, de Broglie's postulate for free particles may be seen as a simplified form of the law of motion of what we now know as pilot-wave dynamics – except for the statement that the motion along a ray be 'uniform' (that is, have constant speed), which in pilot-wave theory is true only in special cases.[a] De Broglie notes that his postulate respects conservation of energy but not of momentum. And indeed, in pilot-wave theory the momentum of a 'free' particle is generally not conserved: in effect (from the standpoint of Bohm's Newtonian formulation), the pilot wave or quantum potential acts like an 'external source' of momentum (and in general of energy too).[b] The abandonment of something as elementary as momentum conservation is certainly a radical step by any standards. On the other hand, if one is willing – as de Broglie was – to propose a fundamentally new approach to the theory of motion, then the loss of classical conservation laws is not surprising, as these are really properties of classical equations of motion.

De Broglie then makes a remarkable prediction, that *any* moving body (not just light quanta) can undergo diffraction:

[a] From Bohm's second-order equation (2.3) applied to a single particle, for time-independent V it follows that $d(\frac{1}{2}mv^2 + V + Q)/dt = \partial Q/\partial t$ (the usual energy conservation formula with a time-dependent contribution Q to the potential). In free space ($V = 0$), the speed v is constant if and only if $dQ/dt = \partial Q/\partial t$ or $\mathbf{v} \cdot \nabla Q = 0$ (so that the 'quantum force' does no work), which is true only in special cases.

[b] Again from (2.3), in free space the rate of change of momentum $\mathbf{p} = m\mathbf{v}$ (where $\mathbf{v} = \nabla S/m$) is $d\mathbf{p}/dt = -\nabla Q$, which is generally non-zero. Further, in general (2.3) implies $d(\frac{1}{2}mv^2 + V)/dt = -\mathbf{v} \cdot \nabla Q$, so that the standard (classical) expression for energy is conserved if and only if $\mathbf{v} \cdot \nabla Q = 0$. If, on the other hand, one defines $\frac{1}{2}mv^2 + V + Q$ as the 'energy', it will be conserved if and only if $\partial Q/\partial t = 0$, which is true only for special cases (in particular for stationary states, since for these $|\Psi|$ is time-independent).

... any moving body could in certain cases be diffracted. A stream of electrons passing through a small enough opening will show diffraction phenomena. It is in this direction that one should perhaps look for experimental confirmation of our ideas.

Next, de Broglie puts his proposals in a general conceptual and historical perspective. Concerning the role of the phase wave, he writes (p. 549):

We therefore conceive of the phase wave as guiding the movements of energy, and this is what can allow the synthesis of waves and quanta.

Here, for the first time, de Broglie characterises the phase wave as a 'guiding' wave. De Broglie then remarks that, historically speaking, the theory of waves 'went too far' by denying the discontinuous structure of radiation and 'not far enough' by not playing a role in dynamics. For de Broglie, his proposal has a clear historical significance (p. 549, italics in the original):

The new dynamics of the free material point is to the old dynamics (including that of Einstein) what wave optics is to geometrical optics. Upon reflection one will see that the proposed synthesis appears as the logical culmination of the comparative development of dynamics and of optics since the seventeenth century.

In the second part of this note, de Broglie considers the explanation of optical interference fringes. He assumes that the probability for an atom to absorb or emit a light quantum is determined by 'the resultant of one of the vectors of the phase waves crossing each other there [se croisant sur lui]' (pp. 549–50). In Young's interference experiment, the light quanta passing through the two holes are diffracted, and the probability of them being detected behind the screen will vary from point to point, depending on the 'state of interference' of the phase waves. De Broglie concludes that there will be bright and dark fringes as predicted by the wave theories, no matter how feeble the incident light is.

This approach to optical interference – in which interfering phase waves determine the probability for interaction between photons and the atoms in the detection apparatus – is elaborated in de Broglie's thesis (see Section 2.2.2). Soon after completing his thesis (apparently), de Broglie abandoned this idea in favour of a simpler approach, in which the interfering phase waves determine the number density of photon trajectories (see Section 2.2.3).

In de Broglie's third communication (de Broglie 1923c), entitled 'Quanta, the kinetic theory of gases and Fermat's principle', part 1 considers the statistical treatment of a gas of particles accompanied by phase waves. De Broglie makes the following assumption:

The state of the gas will then be stable only if the waves corresponding to all of the atoms form a system of stationary waves.

In other words, de Broglie considers the stationary modes, or standing waves, associated with a given spatial volume. He assumes that each mode 'can transport zero, one, two or several atoms', with probabilities determined by the Boltzmann factor.[a] According to de Broglie, for a gas of non-relativistic atoms his method yields the Maxwell distribution, while for a gas of photons it yields the Planck distribution.[b]

In part 2 of the same note, de Broglie shows how his new dynamical postulate amounts to a unification of Maupertuis' principle of least action with Fermat's principle of least time in optics. Let us recall that, in the mechanical principle of Maupertuis for particle trajectories,

$$\delta \int_a^b m\mathbf{v} \cdot d\mathbf{x} = 0, \tag{2.6}$$

the condition of stationarity determines the particle paths. (In (2.6) the energy is fixed on the varied paths; at the end points, $\Delta \mathbf{x} = 0$ but Δt need not be zero.) While in the optical principle of Fermat for light rays,

$$\delta \int_a^b d\phi = 0, \tag{2.7}$$

the stationary line integral for the phase change – the stationary 'optical path length' – provides a condition that determines the path of a ray connecting two points, in space (for the time-independent case) or in spacetime. Now, according to de Broglie's basic postulate: 'The rays of the phase waves coincide with the dynamically possible trajectories' (p. 632). The rays are described by Fermat's principle (for the case of a dispersive medium), which de Broglie shows coincides with Maupertuis' principle, as follows: writing the element of phase change as $d\phi = 2\pi \nu dl / v_{\mathrm{ph}}$, where dl is an element of path and $v_{\mathrm{ph}} = c^2/v$ is the phase velocity, and using the relation $\nu = E/h = mc^2/h$, the element of phase change may be rewritten as $(2\pi/h)mv dl$, so that (2.7) coincides with (2.6). As de Broglie puts it (p. 632):

> In this way the fundamental link that unites the two great principles of geometrical optics and of dynamics is brought fully to light.

De Broglie remarks that some of the dynamically possible trajectories will be 'in resonance with the phase wave', and that these correspond to Bohr's stable orbits, for which $\int \nu dl / v_{\mathrm{ph}}$ is a whole number.

[a] As remarked by Pais (1982, pp. 435–6), in this paper de Broglie 'evaluated independently of Bose (and published before him) the density of radiation states in terms of particle (photon) language'.

[b] As shown by Darrigol (1993), de Broglie made some errors in his application of the methods of statistical mechanics.

Soon afterwards, de Broglie introduces a covariant 4-vector formulation of his basic dynamical postulate (de Broglie 1924a,b). He defines a 4-vector $w_\mu = (\nu/c, -(\nu/v_{ph})\hat{\mathbf{n}})$, where $\hat{\mathbf{n}}$ is a unit vector in the direction of a ray of the phase wave, and assumes it to be related to the energy-momentum 4-vector $p_\mu = (E/c, -\mathbf{p})$ by $p_\mu = hw_\mu$. De Broglie notes that the identity of the principles of Maupertuis and Fermat then follows immediately. We shall discuss this in more detail in the next subsection.

2.2.2 *Thesis (1924)*

He has lifted a corner of the great veil.

(Einstein, commenting on de Broglie's thesis[a]*)*

De Broglie's doctoral thesis (de Broglie 1924e) was mostly based on the above papers. It seems to have been completed in the summer of 1924, and was defended at the Sorbonne in November. The thesis was published early in 1925 in the *Annales de Physique* (de Broglie 1925).[b]

When writing his thesis, de Broglie was well aware that his theory had gaps. As he put it (p. 30):[c]

> . . . the main aim of the present thesis is to present a more complete account of the new ideas that we have proposed, of the successes to which they have led, and also of the many gaps they contain.

De Broglie begins his thesis with a historical introduction. Newtonian mechanics, he notes, was eventually formulated in terms of the principle of least action, which was first given by Maupertuis and then later in another form by Hamilton. As for the science of light and optics, the laws of geometrical optics were eventually summarised by Fermat in terms of a principle whose form is reminiscent of the principle of least action. Newton tried to explain some of the phenomena of wave optics in terms of his corpuscular theory, but the work of Young and Fresnel led to the rise of the wave theory of light, in particular the successful wave explanation of the rectilinear propagation of light (which had been so clear in the corpuscular or 'emission' theory). On this, de Broglie comments (p. 25):

> When two theories, based on ideas that seem entirely different, account for the same experimental fact with equal elegance, one can always wonder if the opposition between

[a] Letter to Langevin, 16 December 1924 (quoted in Darrigol 1993, p. 355).

[b] An English translation of extracts from de Broglie's thesis appears in Ludwig (1968). A complete translation has been done by A. F. Kracklauer (currently online at http://www.ensmp.fr/aflb/LDB-oeuvres/De_Broglie_Kracklauer.htm). All translations here are ours.

[c] Here and below, page references for de Broglie's thesis correspond to the published version in *Annales de Physique* (de Broglie 1925).

the two points of view is truly real and is not due solely to an inadequacy of our efforts at synthesis.

This remark is, of course, a hint that the aim of the thesis is to effect just such a synthesis. De Broglie then turns to the rise of electrodynamics, relativity and the theory of energy quanta. He notes that Einstein's theory of the photoelectric effect amounts to a revival of Newton's corpuscular theory. De Broglie then sketches Bohr's 1913 theory of the atom, and goes on to point out that observations of the photoelectric effect for X- and γ-rays seem to confirm the corpuscular character of radiation. At the same time, the wave aspect continues to be confirmed by the observed interference and diffraction of X-rays. Finally, de Broglie notes the very recent corpuscular interpretation of Compton scattering. De Broglie concludes his historical introduction with a mention of his own recent work (p. 30):

... the moment seemed to have arrived to make an effort towards unifying the corpuscular and wave points of view and to go a bit more deeply into the true meaning of the quanta. That is what we have done recently ...

De Broglie clearly regarded his own work as a synthesis of earlier theories of dynamics and optics, a synthesis increasingly forced upon us by accumulating experimental evidence.

Chapter 1 of the thesis is entitled 'The phase wave'. De Broglie begins by recalling the equivalence of mass and energy implied by the theory of relativity. Turning to the problem of quanta, he remarks (pp. 32–3):

It seems to us that the fundamental idea of the quantum theory is the impossibility of considering an isolated quantity of energy without associating a certain frequency with it. This connection is expressed by what I shall call the quantum relation:

$$\text{energy} = h \times \text{frequency}$$

where h is Planck's constant.

To make sense of the quantum relation, de Broglie proposes that (p. 33)

... to each energy fragment of proper mass m_0 there is attached a periodic phenomenon of frequency ν_0 such that one has:

$$h\nu_0 = m_0c^2$$

ν_0 being measured, of course, in the system tied to the energy fragment.

De Broglie asks if the periodic phenomenon must be assumed to be localised inside the energy fragment. He asserts that this is not at all necessary, and that it will be seen to be 'without doubt spread over an extensive region of space' (p. 34).

De Broglie goes on to consider the apparent contradiction between the frequency $\nu = mc^2/h = \nu_0/\sqrt{1 - v^2/c^2}$ and the time-dilated frequency $\nu_1 = \nu_0\sqrt{1 - v^2/c^2}$. He proposes that the contradiction is resolved by the following 'theorem of phase harmony' (p. 35): in a frame where the moving body has velocity v, the periodic phenomenon tied to the moving body and with frequency ν_1 is always in phase with a wave of frequency ν propagating in the same direction as the moving body with phase velocity $v_{\rm ph} = c^2/v$. This is shown by applying the Lorentz transformation to a rest-frame wave sin $(\nu_0 t_0)$, yielding a wave

$$\sin\left[\nu_0\left(t - vx/c^2\right)/\sqrt{1-v^2/c^2}\right] \quad (2.8)$$

of frequency $\nu = \nu_0/\sqrt{1-v^2/c^2}$ and phase velocity c^2/v. Regarding the nature of this wave de Broglie says that, because its velocity is greater than c, it cannot be a wave transporting energy: rather, 'it represents the spatial distribution of the *phases* of a phenomenon; it is a "phase wave"' (p. 36). De Broglie shows that the group velocity of the phase wave is equal to the velocity of the particle. In the final section of chapter 1 ('The phase wave in spacetime'), he discusses the appearance of surfaces of constant phase for differently moving observers, from a spacetime perspective.

Chapter 2 is entitled 'Maupertuis' principle and Fermat's principle'. The aim is to generalise the results of the first chapter to non-uniform, non-rectilinear motion. In the introduction to chapter 2 de Broglie writes (p. 45):

Guided by the idea of a deep unity between the principle of least action and that of Fermat, from the beginning of my investigations on this subject I was led to *assume* that, for a given value of the total energy of the moving body and therefore of the frequency of its phase wave, the dynamically possible trajectories of the one coincided with the possible rays of the other.

De Broglie discusses the principle of least action, in the different forms given by Hamilton and by Maupertuis, and also for relativistic particles in an external electromagnetic field. He writes Hamilton's principle as

$$\delta \int_P^Q p_\mu dx^\mu = 0 \quad (2.9)$$

($\mu = 0, 1, 2, 3$, with $dx^0 = cdt$), where P, Q are points in spacetime and p_μ is the canonical energy-momentum 4-vector, and notes that if p_0 is constant the principle becomes

$$\delta \int_A^B p_i dx^i = 0 \quad (2.10)$$

($i = 1, 2, 3$), where A, B are the corresponding points in space – that is, Hamilton's principle reduces to Maupertuis' principle.

De Broglie then discusses wave propagation and Fermat's principle from a spacetime perspective. He considers a sinusoidal function sin ϕ, where the phase ϕ

has a spacetime-dependent differential $d\phi$, and writes the variational principle for the ray in spacetime in the Hamiltonian form

$$\delta \int_P^Q d\phi = 0. \tag{2.11}$$

De Broglie then introduces a 4-vector field w_μ on spacetime, defined by

$$d\phi = 2\pi w_\mu dx^\mu, \tag{2.12}$$

where the w_μ are generally functions on spacetime. (Of course, this implies that $2\pi w_\mu = \partial_\mu \phi$, though de Broglie does not write this explicitly.) De Broglie also notes that $d\phi = 2\pi(\nu dt - (\nu/v_{\rm ph})dl)$ and $w_\mu = (\nu/c, -(\nu/v_{\rm ph})\hat{\mathbf{n}})$, where $\hat{\mathbf{n}}$ is a unit vector in the direction of propagation; and that if ν is constant, the principle in the Hamiltonian form

$$\delta \int_P^Q w_\mu dx^\mu = 0 \tag{2.13}$$

reduces to the principle in the Maupertuisian form

$$\delta \int_A^B w_i dx^i = 0, \tag{2.14}$$

or

$$\delta \int_A^B \frac{\nu}{v_{\rm ph}} dl = 0, \tag{2.15}$$

which is Fermat's principle.

De Broglie then discusses an 'extension of the quantum relation' (that is, an extension of $E = h\nu$). He states that the two 4-vectors p_μ and w_μ play perfectly symmetrical roles in the motion of a particle and in the propagation of a wave. Writing the 'quantum relation' $E = h\nu$ as $w_0 = \frac{1}{h} p_0$, de Broglie proposes the generalisation

$$w_\mu = \frac{1}{h} p_\mu, \tag{2.16}$$

so that

$$d\phi = 2\pi w_\mu dx^\mu = \frac{2\pi}{h} p_\mu dx^\mu. \tag{2.17}$$

Fermat's principle then becomes

$$\delta \int_A^B p_i dx^i = 0, \tag{2.18}$$

which is the same as Maupertuis' principle. Thus, de Broglie arrives at the following statement (p. 56):

Fermat's principle applied to the phase wave is identical to Maupertuis' principle applied to the moving body; the dynamically possible trajectories of the moving body are identical to the possible rays of the wave.

He adds that (p. 56):

We think that this idea of a deep relationship between the two great principles of Geometrical Optics and Dynamics could be a valuable guide in realising the synthesis of waves and quanta.

De Broglie then discusses some particular cases: the free particle, a particle in an electrostatic field, and a particle in a general electromagnetic field. He calculates the phase velocity, which depends on the electromagnetic potentials. He notes that the propagation of a phase wave in an external field depends on the charge and mass of the moving body. And he shows that the group velocity along a ray is still equal to the velocity of the moving body.

For the case of an electron of charge e and velocity v in an electrostatic potential φ, de Broglie writes down the following expressions for the frequency ν and phase velocity v_{ph} of the phase wave (p. 57):

$$\nu = (mc^2 + e\varphi)/h, \quad v_{\mathrm{ph}} = (mc^2 + e\varphi)/mv \tag{2.19}$$

(where again $m = m_0/\sqrt{1 - v^2/c^2}$). He shows that v_{ph} may be rewritten as the free value c^2/v multiplied by a factor $h\nu/(h\nu - e\varphi)$ that depends on the potential φ. The expressions (2.19) formed the starting point for Schrödinger's work on the wave equation for de Broglie's phase waves (as reconstructed by Mehra and Rechenberg (1987, pp. 423–5), see Section 2.3).

While de Broglie does not explicitly say so in his thesis, note that from the definition (2.12) of w_μ, the generalised quantum relation (2.16) may be written in the form

$$p_\mu = \hbar \partial_\mu \phi. \tag{2.20}$$

This is what we would now call a relativistic guidance equation, giving the velocity of a particle in terms of the gradient of the phase of a pilot wave (where here de Broglie defines the phase ϕ so as to be dimensionless). In other words, the extended quantum relation is a first-order equation of motion. In the presence of an electromagnetic field, the canonical momentum p_μ contains the 4-vector potential. For a free particle, with $p_\mu = (E/c, -\mathbf{p})$, the guidance equation has the components

$$E = \hbar\dot{\phi}, \quad \mathbf{p} = -\hbar\nabla\phi, \tag{2.21}$$

where the spatial components may also be written as

$$m\mathbf{v} = \frac{m_0\mathbf{v}}{\sqrt{1 - v^2/c^2}} = -\hbar\nabla\phi. \quad (2.22)$$

For a plane wave of phase $\phi = \omega t - \mathbf{k}\cdot\mathbf{x}$, we have

$$E = \hbar\omega, \quad \mathbf{p} = \hbar\mathbf{k}. \quad (2.23)$$

Thus, de Broglie's unification of the principles of Maupertuis and Fermat amounts to a new dynamical law, (2.16) or (2.20), in which the phase of a guiding wave determines the particle velocity. This new law of motion is the essence of de Broglie's new, first-order dynamics.

Chapter 3 of de Broglie's thesis is entitled 'The quantum conditions for the stability of orbits'. De Broglie reviews Bohr's condition for circular orbits, according to which the angular momentum of the electron must be a multiple of $\hbar$, or equivalently $\int_0^{2\pi} p_\theta d\theta = nh$ (p_θ conjugate to θ). He also reviews Sommerfeld's generalisation, $\oint p_i dq_i = n_i h$ (integral n_i) and Einstein's invariant formulation $\oint \sum_{i=1}^{3} p_i dq_i = nh$ (integral n). De Broglie then provides an explanation for Einstein's condition. The trajectory of the moving body coincides with one of the rays of its phase wave, and the phase wave moves along the trajectory with a constant frequency (because the total energy is constant) and with a variable speed whose value has been calculated. To have a stable orbit, claims de Broglie, the length l of the orbit must be in 'resonance' with the wave: thus $l = n\lambda$ in the case of constant wavelength, and $\oint(\nu/v_{\text{ph}})dl = n$ (n integral) generally. De Broglie notes that this is precisely the integral appearing in Fermat's principle, which has been shown to be equal to the integral giving the Maupertuisian action divided by h. The resonance condition is then identical to the required stability condition. For the simple case of circular orbits in the Bohr atom de Broglie shows, using $v_{\text{ph}} = \nu\lambda$ and $h/\lambda = m_0 v$, that the resonance condition becomes $\oint m_0 v dl = nh$ or $m_0\omega R^2 = n\hbar$ (with $v = \omega R$), as originally given by Bohr.[a] (Note that the simple argument commonly found in textbooks, about the fitting of whole numbers of wavelengths along a Bohr orbit, originates in this work of de Broglie's.)

De Broglie thought that his explanation of the stability or quantisation conditions constituted important evidence for his ideas. As he puts it (p. 65):

> This beautiful result, whose demonstration is so immediate when one has accepted the ideas of the preceding chapter, is the best justification we can give for our way of attacking the problem of quanta.

[a] De Broglie also claims to generalise his results from closed orbits to quasi-periodic (or multi-periodic) motion; however, as shown by Darrigol (1993), de Broglie's derivation is faulty.

Certainly, de Broglie had achieved a concrete realisation of his initial intuition that quantisation conditions for atomic energy levels could arise from the properties of waves.

In his chapter 4, de Broglie considers the two-body problem, in particular the hydrogen atom. He expresses concern over how to define the proper masses, taking into account the interaction energy. He discusses the quantisation conditions for hydrogen from a two-body point of view: he has two phase waves, one for the electron and one for the nucleus.

The subject of chapter 5 is light quanta. De Broglie suggests that the classical (electromagnetic) wave distribution in space is some sort of time average over the true distribution of phase waves. His light quantum is assigned a very small proper mass: the velocity v of the quantum and the phase velocity c^2/v of the accompanying phase wave are then both very close to c.

De Broglie points out that radiation is sometimes observed to violate rectilinear propagation: a light wave striking the edge of a screen diffracts into the geometrical shadow, and rays passing close to the screen deviate from a straight line. De Broglie notes the two historical explanations for this phenomenon – on the one hand the explanation for diffraction given by the wave theory, and on the other the explanation given by Newton in his emission theory: 'Newton assumed [the existence of] a force exerted by the edge of the screen on the corpuscle' (p. 80). De Broglie asserts that he can now give a unified explanation for diffraction, by abandoning Newton's first law of motion (p. 80):

> ... the ray of the wave would curve as predicted by the theory of waves, and the moving body, for which the principle of inertia would no longer be valid, would suffer the same deviation as the ray with which its motion is bound up [solidaire] ...

De Broglie's words here deserve emphasis. As is also very clear in his second paper of the preceding year (see Section 2.2.1), de Broglie regards his explanation of particle diffraction as based on a new form of dynamics in which Newton's first law – the principle that a free body will always move uniformly in a straight line – is abandoned. At the same time, de Broglie recognises that one can always adopt a classical-mechanical viewpoint if one wishes (pp. 80–1):

> ... perhaps one could say that the wall exerts a force on it [the moving body] if one takes the curvature of the trajectory as a criterion for the existence of a force.

Here, de Broglie recognises that one may still think in Newtonian terms, if one continues to identify acceleration as indicative of the presence of a force. Similarly, as we shall see, in 1927 de Broglie notes that his pilot-wave dynamics may if one wishes be written in Newtonian form with a quantum potential. But de Broglie's preferred approach, throughout his work in the period 1923–7, is to abandon Newton's first law and base his dynamics on velocity rather than acceleration.

After considering the Doppler effect, reflection by a moving mirror, and radiation pressure, all from a photon viewpoint, de Broglie turns to the phenomena of wave optics, noting that (p. 86):

> The stumbling block of the theory of light quanta is the explanation of the phenomena that constitute wave optics.

Here it becomes apparent that, despite his understanding of how non-rectilinear particle trajectories arise during diffraction and interference, de Broglie is not sure of the details of how to explain the observed bright and dark fringes in diffraction and interference experiments with light. In particular, de Broglie did not have a precise theory of the assumed statistical relationship between his phase waves and the electromagnetic field. Even so, he went on to make what he called 'vague suggestions' (p. 87) towards a detailed theory of optical interference. De Broglie's idea was that the phase waves would determine the probability for the light quanta to interact with the atoms constituting the equipment used to observe the radiation, in such a way as to account for the observed fringes (p. 88):

> ... the probability of reactions between atoms of matter and atoms of light is at each point tied to the resultant (or rather to the mean value of this) of one of the vectors characterising the phase wave; where this resultant vanishes the light is undetectable; there is interference. One then conceives that an atom of light traversing a region where the phase waves interfere will be able to be absorbed by matter at certain points and not at others. This is the still very qualitative principle of an explanation of interference ....

As we shall see in the next section, after completing his thesis de Broglie arrived at a simpler explanation of optical interference fringes.

The final section of chapter 5 considers the explanation of Bohr's frequency condition $h\nu = E_1 - E_2$ for the light emitted by an atomic transition from energy state E_1 to energy state E_2. De Broglie derives this from the assumption that each transition involves the emission of a single light quantum of energy $E = h\nu$ (together with the assumption of energy conservation).

De Broglie's chapter 6 discusses the scattering of X- and γ-rays.

In his chapter 7, de Broglie turns to statistical mechanics, and shows how the concept of statistical equilibrium is to be modified in the presence of phase waves. If each particle or atom in a gas is accompanied by a phase wave, then a box of gas will be 'criss-crossed in all directions' (p. 110) by the waves. De Broglie finds it natural to assume that the only stable phase waves in the box will be those that form stationary or standing waves, and that only these will be relevant to thermodynamic equilibrium. He illustrates his idea with a simple example of molecules moving in one dimension, confined to an interval of length l. In the non-relativistic limit, each phase wave has a wavelength $\lambda = h/m_0 v$ and the 'resonance condition' is $l = n\lambda$ with n integral. Writing $v_0 = h/m_0 l$, one then has $v = nv_0$. As de Broglie notes (p. 112): 'The speed will then be able to take only values equal to integer multiples of v_0'. (This is, of course, the well-known quantisation of momentum for particles confined to a box.) De Broglie then argues that a velocity element δv corresponds to a number $\delta n = (m_0 l/h)\delta v$ of states of a molecule (compatible with the existence of stationary phase waves), so that an element $m_0 \delta x \delta v$ of phase space volume corresponds to a number $m_0 \delta x \delta v / h$ of possible states. Generalising to three dimensions, de Broglie is led to take the element of phase space volume divided by h^3 as the measure of the number of possible states of a molecule, as assumed by Planck.

De Broglie then turns to the photon gas, for which he obtains Wien's law. He claims that, in order to get Planck's law, the following further hypothesis is required (p. 116):

> If two or several atoms [of light] have phase waves that are exactly superposed, of which one can therefore say that they are transported by the same wave, their motions can no longer be considered as entirely independent and these atoms can no longer be treated as separate units in calculating the probabilities.

In de Broglie's approach, the stationary phase waves play the role of the elementary objects of statistical mechanics. De Broglie defines stationary waves as a superposition of two waves of the form

$$\begin{matrix}\sin \\ \cos\end{matrix} \left[2\pi \left(\nu t - \frac{x}{\lambda} + \phi_0 \right) \right] \quad \text{and} \quad \begin{matrix}\sin \\ \cos\end{matrix} \left[2\pi \left(\nu t + \frac{x}{\lambda} + \phi_0 \right) \right], \tag{2.24}$$

where ϕ_0 can take any value from 0 to 1 and ν takes one of the allowed values. Each elementary wave can carry any number $0, 1, 2, \ldots$ of atoms, and the probability of carrying n atoms is given by the Boltzmann factor $e^{-nh\nu/kT}$. Applying this method to a gas of light quanta, de Broglie claims to derive the Planck distribution.[a]

De Broglie's thesis ends with a summary and conclusions (pp. 125–8). The seeds of the problem of quanta have been shown, he claims, to be contained in the historical 'parallelism of the corpuscular and wave-like conceptions of radiation'. He has postulated a periodic phenomenon associated with each energy fragment, and shown how relativity requires us to associate a phase wave with every uniformly moving body. For the case of non-uniform motion, Maupertuis' principle and Fermat's principle 'could well be two aspects of a single law', and this new approach to dynamics led to an extension of the quantum relation, giving the speed of a phase wave in an electromagnetic field. The most important consequence is the interpretation of the quantum conditions for atomic orbits in terms of a resonance of the phase wave along the trajectories: 'this is the first physically plausible

[a] Again, as shown by Darrigol (1993), de Broglie's application of statistical mechanics contains some errors.

explanation proposed for the Bohr–Sommerfeld stability conditions'. A 'qualitative theory of interference' has been suggested. The phase wave has been introduced into statistical mechanics, yielding a derivation of Planck's phase volume element, and of the blackbody spectrum. De Broglie has, he claims, perhaps contributed to a unification of the opposing conceptions of waves and particles, in which the dynamics of the material point is understood in terms of wave propagation. He adds that the ideas need further development: first of all, a new electromagnetic theory is required that takes into account the discontinuous structure of radiation and the physical nature of phase waves, with Maxwell's theory emerging as a statistical approximation. The final paragraph of de Broglie's thesis emphasises the incompleteness of his theory at the time:

> I have deliberately left rather vague the definition of the phase wave, and of the periodic phenomenon of which it must in some sense be the translation, as well as that of the light quantum. The present theory should therefore be considered as one whose physical content is not entirely specified, rather than as a consistent and definitively constituted doctrine.

As de Broglie's concluding paragraph makes clear, his theory of 1924 was rather abstract. There was no specified basis for the phase waves (they were certainly not regarded as 'material' waves); nor was any particular wave equation suggested. It should also be noted that at this time de Broglie's waves were real-valued functions of space and time, of the form $\propto \sin(\omega t - \mathbf{k} \cdot \mathbf{x})$, with a real oscillating amplitude. They were *not* complex waves $\propto e^{i(\mathbf{k} \cdot \mathbf{x} - \omega t)}$ of uniform amplitude. Thus, de Broglie's 'phase waves' had an oscillating amplitude as well as a phase. (De Broglie seems to have called them 'phase waves' only because of his theorem of phase harmony.) Note also that, in his treatment of particles in a box, de Broglie superposes waves propagating in opposite directions, yielding stationary waves whose amplitudes oscillate in time.

In his thesis de Broglie does not explicitly discuss diffraction or interference experiments with electrons, even though in his second communication of 1923 (de Broglie 1923b) he had suggested electron diffraction as an experimental test. According to de Broglie's later recollections (L. de Broglie, AHQP Interview, 7 January 1963, p. 6),[3] at his thesis defence on 25 November 1924:

> Mr Jean Perrin, who chaired the committee, asked me if my ideas could lead to experimental confirmation. I replied that yes they could, and I mentioned the diffraction of electrons by crystals. Soon afterwards, I advised Mr Dauvillier ... to try the experiment, but, absorbed by other research, he did not do it. I do not know if he believed, or if he said to himself that it was perhaps very uncertain, that he was going to go to a lot of trouble for nothing – it's possible ... But the following year it was discovered in America by Davisson and Germer.

2.2.3 Optical interference fringes: November 1924

On 17 November 1924, just a few days before de Broglie defended his thesis, a further communication of de Broglie's was presented to the Academy of Sciences: entitled 'On the dynamics of the light quantum and interference' and published in the *Comptes Rendus*, this short note gave a new and improved account of optical interference in terms of light quanta (de Broglie 1924d).

De Broglie began his note by recalling his unsatisfactory discussion of optical interference in his recent work on the quantum theory (p. 1039):

> ... I had not reached a truly satisfying explanation for the phenomena of wave optics which, in principle, all come down to interference. I limited myself to putting forward a certain connection between the state of interference of the waves and the probability for the absorption of light quanta by matter. This viewpoint now seems to me a bit artificial and I tend towards adopting another, more in harmony with the broad outlines of my theory itself.

As we have seen, in his thesis de Broglie was unsure about how to account for the bright and dark fringes observed in optical interference experiments. He did not have a theory of the electromagnetic field, which he assumed emerged as some sort of average over his phase waves. To account for optical fringes, he had suggested that the phase waves somehow determined the probability for interactions between photons and the atoms in the apparatus. Now, after completing his thesis, he felt he had a better explanation that was based purely on the spatial distribution of the photon trajectories.

De Broglie's note continues by outlining his 'new dynamics', in which the energy-momentum 4-vector of every material point is proportional to the 'characteristic' 4-vector of an associated wave, even when the wave undergoes interference or diffraction. He then gives his new view of interference fringes (p. 1040):

> The rays predicted by the wave theories would then be in every case the possible trajectories of the quantum. In the phenomena of interference, the rays become concentrated in those regions called 'bright fringes' and become diluted in those regions called 'dark fringes'. In my first explanation of interference, the dark fringes were dark because the action of fragments of light on matter was zero there; in my current explanation, these fringes are dark because the number of quanta passing through them is small or zero.

Here, then, de Broglie explains bright and dark fringes simply in terms of a high or low density of photon trajectories in the corresponding regions.

When de Broglie speaks of the trajectories being concentrated and diluted in regions of bright and dark fringes respectively, he presumably had in mind that the number density of particles in an interference zone should be proportional to the classical wave intensity, though he does not say this explicitly. We can discern the essence of the more precise and complete explanation of optical interference given by de Broglie three years later in his Solvay report: there, de Broglie has the same photon trajectories, with a number density specified as proportional to the amplitude-squared of the guiding wave (see pp. 350–1, and our discussion in Section 6.1.1).

In his note of November 1924, de Broglie goes on to illustrate his proposal for the case of Young's interference experiment with two pinholes acting as point sources. De Broglie cites the well-known facts that in this case the surfaces of equal phase are ellipsoids with the pinholes as foci, and that the rays (which are normal to the ellipsoidal surfaces) are concentrated on hyperboloids of constructive interference where the classical intensity has maxima. He then notes (p. 1040):

> Let r_1 and r_2 be the distances from a point in space to the two holes and let ψ be the function $\frac{1}{2}(r_1 + r_2)$, which is constant on each surface of equal phase. One easily shows that the phase velocity of the waves along a ray is equal to the value it would have in the case of free propagation divided by the derivative of ψ taken along the ray; as for the speed of the quantum, it will be equal to the speed of free motion multiplied by the same derivative.

De Broglie gives no further details, but these are easily reconstructed. For an incident beam of wavelength $\lambda = 2\pi/k$, each pinhole acts as a source of a spherical wave of wavelength λ, yielding a resultant wave proportional to the real part of

$$\frac{e^{i(kr_1-\omega t)}}{r_1} + \frac{e^{i(kr_2-\omega t)}}{r_2}. \tag{2.25}$$

If the pinholes have a separation d, then at large distances ($r_1, r_2 >> d$) from the screen the resultant wave may be approximated as

$$2\frac{e^{i(kr-\omega t)}}{r} \cos\left(\frac{kd}{2}\theta\right) \tag{2.26}$$

where θ is the (small) angular deviation from the normal to the screen. The amplitude shows the well-known interference pattern. As for the phase $\phi = kr - \omega t$, with $r = \frac{1}{2}(r_1+r_2)$, the surfaces of equal phase are indeed the well-known ellipsoids. Further, the phase velocity is given by

$$v_{\text{ph}} = \frac{|\partial\phi/\partial t|}{|\nabla\phi|} = \frac{\omega}{k}\frac{1}{|\nabla r|} = \frac{c}{|\nabla r|}, \tag{2.27}$$

while de Broglie's particle velocity – given by (2.22) – has magnitude

$$v = \frac{\hbar\,|\nabla\phi|}{m} = \frac{\hbar\,|\nabla\phi|}{(\hbar\omega/c^2)} = c^2\frac{k}{\omega}\,|\nabla r| = c\,|\nabla r| \tag{2.28}$$

(where m is the relativistic photon mass), in agreement with de Broglie's assertions.

At the end of his note, de Broglie comments that this method may be applied to the study of scattering.

In November 1924, then, de Broglie understood how interfering phase waves would affect photon trajectories, causing them to bunch together in regions coinciding with the observed bright fringes.

Note that in this paper de Broglie treats his phase waves as if they were a direct representation of the electromagnetic field. In his discussion of Young's interference experiment, he has phase waves emerging from the two holes and interfering, and he identifies the interference fringes of his phase waves with optical interference fringes. However, he seems quite aware that this is a simplification,[a] remarking that 'the whole theory will become truly clear only if one manages to define the structure of the light wave'. De Broglie is still not sure about the precise relationship between his phase waves and light waves, a situation that persists even until the fifth Solvay conference: there, while he gives (pp. 350–1) a precise account of optical interference in his report, he points out (p. 460) that the connection between his guiding wave and the electromagnetic field is still unknown.

2.3 Towards a complete pilot-wave dynamics: 1925–1927

On 16 December 1924, Einstein wrote to Lorentz:[b]

> A younger brother of the de Broglie known to us [Maurice de Broglie] has made a very interesting attempt to interpret the Bohr–Sommerfeld quantization rules (Paris Dissertation, 1924). I believe that it is the first feeble ray of light to illuminate this, the worst of our physical riddles. I have also discovered something that supports his construction.

What Einstein had discovered, in support of de Broglie's ideas, appeared in the second of his famous papers on the quantum theory of the ideal gas (Einstein 1925a). Einstein showed that the fluctuations associated with the new Bose–Einstein statistics contained two distinct terms that could be interpreted as particle-like and wave-like contributions – just as Einstein had shown, many years earlier, for blackbody radiation. Einstein argued that the wave-like contribution should be interpreted in terms of de Broglie's matter waves, and he cited de Broglie's thesis. It was largely through this paper by Einstein that de Broglie's work became known outside France.

In the same paper, Einstein suggested that a molecular beam would undergo diffraction through a sufficiently small aperture. De Broglie had already made

[a] De Broglie may have thought of this as analogous to scalar wave optics, which predicts the correct optical interference fringes by treating the electromagnetic field simply as a scalar wave.

[b] Quoted in Mehra and Rechenberg (1982a, p. 604).

a similar suggestion for electrons, in his second communication to the *Comptes Rendus* (de Broglie 1923b). Even so, in their report at the fifth Solvay conference, Born and Heisenberg state that in his gas theory paper Einstein 'deduced from de Broglie's daring theory the possibility of "diffraction" of material particles' (p. 388), giving the incorrect impression that Einstein had been the first to see this consequence of de Broglie's theory. It seems likely that Born and Heisenberg did not notice de Broglie's early papers in the *Comptes Rendus*.

In 1925 Elsasser – a student of Born's in Göttingen – read de Broglie's thesis. Like most others outside France, Elsasser had heard about de Broglie's thesis through Einstein's gas theory papers. Elsasser suspected that two observed experimental anomalies could be explained by de Broglie's new dynamics. First, the Ramsauer effect – the surprisingly large mean free path of low-velocity electrons in gases – which Elsasser thought could be explained by electron interference. Second, the intensity maxima observed by Davisson and Kunsman at certain angles of reflection of electrons from metal surfaces, which had been assumed to be caused by atomic shell structure, and which Elsasser thought were caused by electron diffraction. Elsasser published a short note sketching these ideas in *Die Naturwissenschaften* (Elsasser 1925). Elsasser then tried to design an experiment to test the ideas further, with low-velocity electrons, but never carried it out. According to Heisenberg's later recollection, Elsasser's supervisor Born was sceptical about the reality of matter waves, because they seemed in conflict with the observed particle tracks in cloud chambers.[a]

On 3 November 1925, Schrödinger wrote to Einstein: 'A few days ago I read with the greatest interest the ingenious thesis of Louis de Broglie ...'.[b] Schrödinger too had become interested in de Broglie's thesis by reading Einstein's gas theory papers, and he set about trying to find the wave equation for de Broglie's phase waves. As we have seen (Section 2.2.2), in his thesis de Broglie had shown that, in an electrostatic potential φ, the phase wave of an electron of charge e and velocity v would have (see equation (2.19)) frequency $\nu = (mc^2 + V)/h$ and phase velocity $v_{\mathrm{ph}} = (mc^2 + V)/mv$, where $m = m_0/\sqrt{1 - v^2/c^2}$ and $V = e\varphi$ is the potential energy. These expressions for ν and v_{ph}, given by de Broglie, formed the starting point for Schrödinger's work on the wave equation.

Schrödinger took de Broglie's formulas for ν and v_{ph} and applied them to the hydrogen atom, with a Coulomb field $\varphi = -e/r$.[c] Using the formula for ν to

[a] For this and further details concerning Elsasser, see Mehra and Rechenberg (1982a, pp. 624–7). See also the discussion in Section 3.4.2.

[b] Quoted in Mehra and Rechenberg (1987, p. 412).

[c] Here we follow the analysis by Mehra and Rechenberg (1987, pp. 423–5) of what they call Schrödinger's 'earliest preserved [unpublished] manuscript on wave mechanics'. Similar reasoning is found in a letter from Pauli to Jordan of 12 April 1926 (Pauli 1979, pp. 315–20).

eliminate v, Schrödinger rewrote the expression for the phase velocity v_{ph} purely in terms of the frequency ν and the electron–proton distance r:

$$v_{ph} = \frac{h\nu}{m_0 c} \frac{1}{\sqrt{\left(h\nu/m_0c^2 - V/m_0c^2\right)^2 - 1}} \tag{2.29}$$

(where $V = -e^2/r$). Then, writing de Broglie's phase wave as $\psi = \psi(\mathbf{x}, t)$, he took the equation for ψ to be the usual wave equation

$$\nabla^2 \psi = \frac{1}{v_{ph}^2} \frac{\partial^2 \psi}{\partial t^2} \tag{2.30}$$

with phase velocity v_{ph}. Assuming ψ to have a time dependence $\propto e^{-2\pi i \nu t}$, Schrödinger then obtained the time-independent equation

$$\nabla^2 \psi = -\frac{4\pi^2 \nu^2}{v_{ph}^2} \psi \tag{2.31}$$

with v_{ph} given by (2.29). This was Schrödinger's original (relativistic) equation for the energy states of the hydrogen atom.

As is well known, Schrödinger found that the energy levels predicted by (2.31) – that is, the eigenvalues $h\nu$ – disagreed with experiment. He then adopted a non-relativistic approximation, and found that this yielded the correct energy levels for the low-energy limit.

It is in fact straightforward to obtain the correct non-relativistic limit. Writing $E = h\nu - m_0c^2$, in the low-energy limit $|E|/m_0c^2 << 1$ and $|V|/m_0c^2 << 1$, we have

$$\frac{\nu^2}{v_{ph}^2} = \frac{2m_0}{h^2}(E - V), \tag{2.32}$$

so that (2.31) reduces to what we now know as the non-relativistic time-independent Schrödinger equation for a single particle in a potential V:

$$-\frac{\hbar^2}{2m_0}\nabla^2 \psi + V\psi = E\psi. \tag{2.33}$$

Historically speaking, then, Schrödinger's original equation (2.31) for stationary states amounted to a mathematical transcription – into the language of the standard wave equation (2.30) – of de Broglie's expressions for the frequency and phase velocity of an electron wave. By studying the eigenvalue problem of this equation in its non-relativistic form (2.33), Schrödinger was able to show that the eigenvalues agreed remarkably well with the observed features of the hydrogen spectrum. Thus, by adopting the formalism of a wave equation, Schrödinger transformed de Broglie's elementary derivation of the quantisation of energy levels into a rigorous and powerful technique.

The time-dependent Schrödinger equation

$$i\hbar\frac{\partial\psi}{\partial t} = -\frac{\hbar^2}{2m_0}\nabla^2\psi + V\psi, \tag{2.34}$$

with a *single* time derivative, was eventually obtained by Schrödinger in his fourth paper on wave mechanics, completed in June 1926 (Schrödinger 1926g). The path taken by Schrödinger in these four published papers was rather tortuous. In the fourth paper, he actually began by considering a wave equation that was of second order in time and of fourth order in the spatial derivatives. However, he eventually settled on (2.34), deriving it by the following argument: for a wave with time dependence $\propto e^{-(i/\hbar)Et}$, one may write the term $E\psi$ in (2.33) as $i\hbar\partial\psi/\partial t$ and thus obtain (2.34), which must be valid for any E and therefore for any ψ that can be expanded as a Fourier time series.[a] The price paid for having an equation that was of only first order in time was that ψ had to be complex.

In retrospect, the time-dependent Schrödinger equation for a free particle ((2.34) with $V = 0$) may be immediately derived as the simplest wave equation obeyed by a complex plane wave $e^{i(\mathbf{k}\cdot\mathbf{x}-\omega t)}$ with the non-relativistic dispersion relation $\hbar\omega = (\hbar k)^2/2m$ (that is, $E = p^2/2m$ combined with de Broglie's relations $E = \hbar\omega$ and $p = \hbar k$). This derivation is, in fact, often found in textbooks.

Not only did Schrödinger adopt de Broglie's idea that quantised energy levels could be explained in terms of waves, he also took up de Broglie's conviction that classical mechanics was merely the short-wavelength limit of a broader theory. Thus in his second paper on wave mechanics Schrödinger (1926c, p. 497) wrote:

> Maybe our classical mechanics is the *complete* analogue of geometrical optics and as such is wrong, not in agreement with reality; it fails as soon as the radii of curvature and the dimensions of the trajectory are no longer large compared to a certain wavelength, which has a real meaning in the q-space. Then one has to look for an 'undulatory mechanics' –[b] and the most natural path towards this is surely the wave-theoretical elaboration of the Hamiltonian picture.

Schrödinger used the optical-mechanical analogy as a guide in the construction of wave equations for atomic systems, particularly in his second paper. Schrödinger tended to think of the optical-mechanical analogy in terms of the Hamilton–Jacobi equation and the equation of geometrical optics, whereas de Broglie tended to think

[a] Actually, Schrödinger considered the time dependence $\propto e^{\pm(i/\hbar)Et}$, so that $E\psi = \pm i\hbar\partial\psi/\partial t$, leading to *two* possible wave equations differing by the sign of i. He wrote: '*We shall require that the complex wave function ψ satisfy one of these two equations.* Since at the same time the complex conjugate function $\bar{\psi}$ satisfies the *other* equation, one may consider the real part of ψ as a real wave function (if one needs it)' (Schrödinger 1926g, p. 112, original italics).

[b] Here Schrödinger adds a footnote: 'Cf. also A. Einstein, *Berl. Ber.*, pp. 9ff. (1925)'.

of it in terms of the principles of Maupertuis and Fermat. These are of course two different ways of drawing the same analogy.

While Schrödinger took up and developed many of de Broglie's ideas, he did *not* accept de Broglie's view that the particle was localised within an extended wave. In effect, Schrödinger removed the particle trajectories from de Broglie's theory, and worked only with the extended (non-singular) waves. (For a detailed discussion of Schrödinger's work, see Chapter 4.)

In the period 1925–6, then, many of the ideas in de Broglie's thesis were taken up and developed by other workers, especially Schrödinger. But what of de Broglie himself? Unlike in his earlier work, during this period de Broglie considered specific equations for his waves. Like Schrödinger, he took standard relativistic wave equations as his starting point. Unlike Schrödinger, however, de Broglie was guided by the following two ideas. First, that particles are really small singular regions, of very large wave amplitude, within an extended wave. Second, that the motion of the particles – or singularities – must in some sense satisfy the condition

$$\text{Maupertuis} \equiv \text{Fermat}, \tag{2.35}$$

so as to bring about a synthesis of the dynamics of particles with the theory of waves, along the lines already sketched in de Broglie's thesis. This work came to a head in a remarkable paper published in May 1927, to which we now turn.

2.3.1 *'Structure':* Journal de Physique, *May 1927*

In the last number of the Journal de Physique, a paper by de Broglie has appeared ... de Broglie attempts here to reconcile the full determinism of physical processes with the dualism between waves and corpuscles ... even if this paper by de Broglie is off the mark (and I hope that actually), still it is very rich in ideas and very sharp, and on a much higher level than the childish papers by Schrödinger, who even today still thinks he may ... abolish material points.

(Pauli, letter to Bohr, 6 August 1927 (Pauli 1979, pp. 404–5))

What we now know as pilot-wave theory first appears in a paper by de Broglie (1927b) entitled 'Wave mechanics and the atomic structure of matter and of radiation', which was published in *Journal de Physique* in May 1927, and which we shall discuss in detail in this section.[a] We shall refer to this crucial paper as 'Structure' for short.

In 'Structure', de Broglie presents a theory of particles as moving singularities. It is argued, on the basis of certain assumptions, that the equations of what we now

[a] Again, we do not always follow de Broglie's notation. Note also that de Broglie moves back and forth between solutions of the form $f\cos\phi$ and solutions 'written in complex form' $fe^{i\phi}$.

call pilot-wave dynamics will emerge from this theory. At the end of the paper de Broglie proposes, as a provisional theory, simply taking the equations of pilot-wave dynamics as given, without trying to derive them from something deeper. It is this last, provisional theory that de Broglie presents a few months later at the fifth Solvay conference.

For historical completeness we should point out that, according to a footnote to the introduction, 'Structure' was 'the development of two notes' published earlier in *Comptes Rendus* (de Broglie 1926, 1927a). The first, of 28 August 1926, considers a model of photons as moving singularities: arguments are given (similar to those found in 'Structure') leading to the usual velocity formula, and to the conclusion that the probability density of photons is proportional to the classical wave intensity. The second note, of 31 January 1927, sketches analogous ideas for material bodies in an external potential, with the probability density now proportional to the amplitude-squared of the wave function. Unlike 'Structure', neither of these notes contains the suggestion that pilot-wave theory may be adopted as a provisional view.

We now turn to a detailed analysis of de Broglie's 'Structure' paper. De Broglie first considers a 'material point of proper mass m_0' moving in free space with a constant velocity $\mathbf{v}$ and represented by a wave $u(\mathbf{x}, t)$ satisfying what we would now call the Klein–Gordon equation

$$\nabla^2 u - \frac{1}{c^2}\frac{\partial^2 u}{\partial t^2} = \frac{4\pi^2\nu_0^2}{c^2}u, \tag{2.36}$$

with $\nu_0 = m_0c^2/h$. De Broglie considers solutions of the form

$$u(\mathbf{x}, t) = f(\mathbf{x} - \mathbf{v}t)\cos\frac{2\pi}{h}\phi(\mathbf{x}, t), \tag{2.37}$$

where the amplitude f is *singular* at the location $\mathbf{x} = \mathbf{v}t$ of the moving body and the phase[a]

$$\phi(\mathbf{x}, t) = \frac{h\nu_0}{\sqrt{1 - v^2/c^2}}\left(t - \frac{\mathbf{v}\cdot\mathbf{x}}{c^2}\right) \tag{2.38}$$

is equal to the classical Hamiltonian action of the particle. A single particle is represented by a wave (2.37), where the moving singularity has velocity $\mathbf{v}$ and the phase $\phi(\mathbf{x}, t)$ is extended over all space.

De Broglie then considers an ensemble (or 'cloud') of similar free particles with no mutual interaction, all having the same velocity $\mathbf{v}$, and with singular amplitudes f centred at different points in space. According to de Broglie, this ensemble of

[a] For simplicity we ignore an arbitrary constant in the phase, which was included by de Broglie.

moving singularities may be represented by a *continuous* solution of the same wave equation (2.36), of the form

$$\Psi(\mathbf{x}, t) = a \cos \frac{2\pi}{h} \phi(\mathbf{x}, t), \tag{2.39}$$

where a is a constant and ϕ is given by (2.38). This continuous solution is said to 'correspond to' the singular solution (2.37). (De Broglie points out that the continuous solutions Ψ are the same as those considered by Schrödinger.)

The number density of particles in the ensemble is taken to have a constant value ρ, which may be written as

$$\rho = Ka^2 \tag{2.40}$$

for some constant K. Thus, de Broglie notes, (2.39) gives the 'distribution of phases in the cloud of points' as well as the 'density of the cloud'; and for a single particle with known velocity and unknown position, $a^2 d^3\mathbf{x}$ will measure the probability for the particle to be in a volume element $d^3\mathbf{x}$.

Having discussed the free particle, de Broglie moves on to the case of a single particle in a static external potential $V(\mathbf{x})$. The wave u 'written in complex form' now satisfies what we would call the Klein–Gordon equation in an external potential[a]

$$\nabla^2 u - \frac{1}{c^2}\frac{\partial^2 u}{\partial t^2} + \frac{2i}{\hbar c^2} V \frac{\partial u}{\partial t} - \frac{1}{\hbar^2}\left(m_0^2 c^2 - \frac{V^2}{c^2}\right) u = 0. \tag{2.41}$$

De Broglie assumes that the particle begins in free space, represented by a singular wave (2.37), and then enters a region where $V \neq 0$, into which the free solution (2.37) must be extended. De Broglie writes (2.37) 'in complex form', substitutes into (2.41), and takes the real and imaginary parts, yielding two coupled partial differential equations for f and ϕ. Writing

$$\phi(\mathbf{x}, t) = Et - \phi_1(\mathbf{x}) \tag{2.42}$$

(where $E = h\nu$ is constant), one of the said partial differential equations becomes

$$\hbar^2 \frac{\Box f}{f} = (\nabla\phi_1)^2 - \frac{1}{c^2}(E - V)^2 + m_0^2 c^2 \tag{2.43}$$

(where $\Box \equiv \nabla^2 - (1/c^2)\partial^2/\partial t^2$). De Broglie notes that if the left-hand side is negligible, (2.43) reduces to the 'Jacobi equation' for relativistic dynamics in a static potential, so that ϕ_1 reduces to the 'Jacobi function'. Deviations from classical mechanics occur, de Broglie notes, when $\Box f$ is non-zero.

[a] This follows from (2.36) by the substitution $i\hbar\partial/\partial t \to i\hbar\partial/\partial t + V$.

De Broglie now remarks that, in the classical limit, the velocity of the particle has the same direction as the vector $\nabla\phi_1$; he then explicitly *assumes* that this is still true in the general case. Here, the identity (2.35) of the principles of Maupertuis and Fermat is being invoked. With this assumption, de Broglie shows that the singularity must have velocity

$$\mathbf{v} = \frac{c^2 \nabla\phi_1}{E - V}, \tag{2.44}$$

and he adds that in the non-relativistic approximation, $E - V \approx m_0 c^2$, so that

$$\mathbf{v} = \frac{1}{m_0} \nabla\phi_1. \tag{2.45}$$

De Broglie remarks that, in the classical approximation, ϕ_1 becomes the 'Jacobi function' and that (2.44) agrees with the relativistic relation between velocity and momentum. He adds (p. 230):

The aim of the preceding arguments is to make it plausible that the relation [(2.44)] is strictly valid in the new mechanics.

The importance of the relation (2.44) for de Broglie is, of course, that it embodies the identity of the principles of Maupertuis and Fermat.

As in the free case, de Broglie then goes on to consider an ensemble of such particles, with each particle represented by a moving singularity. He assumes that the phase function ϕ_1 is the same for all the particles, whose velocities are then given by (2.44). The (time-independent) particle density $\rho(\mathbf{x})$ must then satisfy the continuity equation

$$\nabla \cdot (\rho \mathbf{v}) = 0. \tag{2.46}$$

Once again, de Broglie introduces a representation of the ensemble by a continuous solution of (2.41). In the presence of the potential V, the solution is taken to have the form

$$\Psi(\mathbf{x}, t) = a(\mathbf{x}) \cos \frac{2\pi}{h} \phi'(\mathbf{x}, t) = a(\mathbf{x}) \cos 2\pi \left(\nu t - \frac{1}{h} \phi'_1(\mathbf{x}) \right). \tag{2.47}$$

Again, writing Ψ 'in complex form', substituting into (2.41) and taking real and imaginary parts, de Broglie obtains two coupled partial differential equations, now in a and ϕ'_1. One of these reads

$$\hbar^2 \frac{\nabla^2 a}{a} = \left(\nabla\phi'_1\right)^2 - \frac{1}{c^2}(E - V)^2 + m_0^2 c^2. \tag{2.48}$$

In this case de Broglie notes that if the left-hand side is negligible, (2.48) reduces to the equation of geometrical optics associated with the wave equation (2.41).

Comparing (2.43) and (2.48), de Broglie notes that when the left-hand sides are negligible, one recovers both classical mechanics (for a moving singularity) and geometrical optics (for a continuous wave). In this limit, the functions ϕ_1 and ϕ_1' are identical, being both equal to the 'Jacobi function'.

At this point de Broglie makes a crucial assumption. He proposes to assume, as a hypothesis, that ϕ_1 and ϕ_1' are *always* equal, regardless of any approximation (pp. 231–2):

We now make the essential hypothesis that ϕ_1 and ϕ_1' are still identical when [the left-hand sides of (2.43) and (2.48)] can no longer be neglected. Obviously, this requires that one have:

$$\frac{\nabla^2 a}{a} = \frac{\Box f}{f}.$$

We shall refer to this postulate by the name of 'principle of the double solution', because it implies the existence of two sinusoidal solutions of equation [(2.41)] having the same phase factor, the one consisting of a point-like singularity and the other having, on the contrary, a continuous amplitude.

This hypothesis expresses, in a more concrete form, de Broglie's earlier idea of 'phase harmony' between the internal oscillation of a particle and the oscillation of an accompanying extended wave: the condition that ϕ_1 and ϕ_1' should coincide amounts to a phase harmony between the singular u-wave representing the point-like particle and the continuous Ψ-wave.

Given this condition, de Broglie deduces that the ratio $\rho/a^2(E - V)$ is constant along the particle trajectories. Since each particle is assumed to begin in free space, where $V = 0$ and (according to (2.40)) $\rho = Ka^2$, de Broglie deduces that in general

$$\rho(\mathbf{x}) = Ka^2(\mathbf{x})\left(1 - \frac{V(\mathbf{x})}{E}\right), \tag{2.49}$$

where K is a constant. The continuous wave then gives the ensemble density at each point. In the non-relativistic approximation, this becomes

$$\rho(\mathbf{x}) = Ka^2(\mathbf{x}). \tag{2.50}$$

De Broglie notes that each possible initial position for the particle gives rise to a possible trajectory, and that an ensemble of initial positions gives rise to an ensemble of motions. Again, de Broglie takes $\rho(\mathbf{x})d^3\mathbf{x}$ as the probability for a single particle to be in a volume element $d^3\mathbf{x}$. This probability is, as he notes, given in terms of Ψ by (2.49) or (2.50). De Broglie adds that Ψ also determines the trajectories (p. 232):

The form of the trajectories is moreover equally determined by knowledge of the continuous wave, since these trajectories are orthogonal to the surfaces of equal phase.

Here we see the first suggestion that Ψ itself may be regarded as determining the trajectories, an idea that de Broglie proposes more fully at the end of the paper.

After sketching how the above could be used to calculate probabilities for electron scattering off a fixed potential, de Broglie goes on to generalise his results to the case of a particle of charge e in a time-dependent electromagnetic field $(\mathcal{V}(\mathbf{x}, t), \mathbf{A}(\mathbf{x}, t))$. He writes down the corresponding Klein–Gordon equation, and once again considers solutions of the form (2.37) with a moving singularity, following the same procedure as before. Because of the time-dependence of the potentials, ϕ no longer takes the form (2.42). Instead of (2.44), the velocity of the singularity is now found to be

$$\mathbf{v} = -c^2 \frac{\nabla\phi + \frac{e}{c}\mathbf{A}}{\dot{\phi} - e\mathcal{V}}, \tag{2.51}$$

or, in the non-relativistic approximation,

$$\mathbf{v} = -\frac{1}{m_0}\left(\nabla\phi + \frac{e}{c}\mathbf{A}\right). \tag{2.52}$$

Once again, de Broglie then considers an ensemble of such moving singularities, with the same phase function $\phi(\mathbf{x}, t)$. The density $\rho(\mathbf{x}, t)$ now obeys a continuity equation

$$\frac{\partial\rho}{\partial t} + \nabla\cdot(\rho\mathbf{v}) = 0, \tag{2.53}$$

with velocity field $\mathbf{v}$ given by (2.51). Again, de Broglie proceeds to represent the motion of the ensemble by means of a continuous solution Ψ of the wave equation, this time of the form

$$\Psi(\mathbf{x}, t) = a(\mathbf{x}, t)\cos\frac{2\pi}{h}\phi'(\mathbf{x}, t), \tag{2.54}$$

with a time-dependent amplitude and a phase no longer linear in t. And again, de Broglie assumes the principle of the double solution, that the phase functions $\phi(\mathbf{x}, t)$ and $\phi'(\mathbf{x}, t)$ are the same. From this, de Broglie deduces that the ratio $\rho/a^2(\dot{\phi} - e\mathcal{V})$ is constant along particle trajectories. Using once more the expression $\rho = Ka^2$ for free space, where $\mathcal{V} = 0$ and $\dot{\phi} = E_0$, de Broglie argues that in general

$$\rho(\mathbf{x}, t) = \frac{K}{E_0}a^2(\mathbf{x}, t)\left(\dot{\phi} - e\mathcal{V}\right) = K'a^2\left(\dot{\phi} - e\mathcal{V}\right), \tag{2.55}$$

and that in the non-relativistic limit ρ is still proportional to a^2. As before, the ensemble may be regarded as composed of all the possible positions of a single particle, of which only the initial velocity is known, and the probability that the particle is in the volume element $d^3\mathbf{x}$ at time t is equal to $\rho(\mathbf{x},t)d^3\mathbf{x}$ and is given in terms of Ψ by (2.55).

De Broglie now shows how the above results for the motion of a particle in a potential $V = e\mathcal{V}$ (ignoring for simplicity the vector potential $\mathbf{A}$) may be obtained from *classical* mechanics, with a Lagrangian of the standard form

$$L = -M_0c^2\sqrt{1 - v^2/c^2} - V, \tag{2.56}$$

by assuming that the particle has a variable proper mass

$$M_0(\mathbf{x},t) = \sqrt{m_0^2 - \frac{\hbar^2}{c^2}\frac{\Box a}{a}}, \tag{2.57}$$

where a is the amplitude of the continuous Ψ wave. Further, he considers this point of view in the non-relativistic limit. Writing $M_0(\mathbf{x},t) = m_0 + \varepsilon(\mathbf{x},t)$ with ε small, the Lagrangian takes the approximate form

$$L = -m_0c^2 + \frac{1}{2}m_0v^2 - \varepsilon c^2 - V. \tag{2.58}$$

As de Broglie remarks (p. 237):

Everything then takes place as if there existed, in addition to V, a potential energy term εc^2.

This extra term coincides, of course, with Bohm's quantum potential Q (equation (2.4)), since to lowest order

$$\varepsilon c^2 = -\frac{\hbar^2}{2m_0}\frac{\Box a}{a} = -\frac{\hbar^2}{2m_0}\frac{\nabla^2 a}{a} + \frac{\hbar^2}{2m_0c^2}\frac{\ddot{a}}{a} \tag{2.59}$$

and the second term is negligible in the non-relativistic limit ($c \to \infty$).

As we have already repeatedly remarked, for de Broglie the guidance equation $\mathbf{v} \propto \nabla\phi$ for *velocity* is the fundamental equation of motion, expressing as it does the identity of the principles of Maupertuis and Fermat. Even so, de Broglie points out that one *can* write the dynamics of the particle in classical (Newtonian or Einsteinian) terms, provided one includes a variable proper mass or, in the non-relativistic limit, an additional quantum potential. It is this latter, Newtonian formulation that Bohm proposes in 1952.

De Broglie now turns to the case of a many-body system, consisting of N particles with proper masses $m_1, m_2, \ldots, m_N$. In the non-relativistic approximation, if the system has total (Newtonian) energy E and potential energy $V(\mathbf{x}_1, \ldots, \mathbf{x}_N)$, then following Schrödinger one may consider the propagation of a wave u in the $3N$-dimensional configuration space, satisfying the wave equation

$$\sum_{i=1}^{N} -\frac{\hbar^2}{2m_i}\nabla_i^2 u + Vu = Eu \tag{2.60}$$

(the time-independent Schrödinger equation). De Broglie remarks that this seems natural, because (2.60) is the obvious generalisation of the one-body Schrödinger equation, which follows from the non-relativistic limit of de Broglie's wave equation (2.41). However, de Broglie criticises Schrödinger's interpretation, according to which a particle is identified with an extended, non-singular wave packet, having no precise position or trajectory. De Broglie objects that, without well-defined particle positions, the coordinates $\mathbf{x}_1, \ldots, \mathbf{x}_N$ used to construct the configuration space would have no meaning. Further, de Broglie asserts that configuration space is 'purely abstract', and that a wave propagating in this space cannot be a physical wave: instead, the physical picture of the system must involve N waves propagating in 3-space.

'What then', asks de Broglie (p. 238), 'is the true meaning of the Schrödinger equation?' To answer this question, de Broglie considers, for simplicity, the case of two particles, which for him are singularities of a wave-like phenomenon in 3-space. Neglecting the vector potential, the two singular waves $u_1(\mathbf{x}, t)$ and $u_2(\mathbf{x}, t)$ satisfy the coupled equations

$$\begin{aligned} \Box u_1 + \frac{2i}{\hbar c^2} V_1 \frac{\partial u_1}{\partial t} - \frac{1}{\hbar^2}\left(m_1^2 c^2 - \frac{V_1^2}{c^2}\right) u_1 &= 0, \\ \Box u_2 + \frac{2i}{\hbar c^2} V_2 \frac{\partial u_2}{\partial t} - \frac{1}{\hbar^2}\left(m_2^2 c^2 - \frac{V_2^2}{c^2}\right) u_2 &= 0, \end{aligned} \tag{2.61}$$

with potentials

$$V_1 = V(|\mathbf{x} - \mathbf{x}_2|), \ V_2 = V(|\mathbf{x} - \mathbf{x}_1|), \tag{2.62}$$

where $\mathbf{x}_1$, $\mathbf{x}_2$ are the positions of the two particles (or singularities). The propagation of each wave then depends, through V, on the position of the singularity in the other wave.

De Broglie is now faced with the formidable problem of solving the simultaneous partial differential equations (2.61), in order to obtain the motions of the two singularities. Not surprisingly, de Broglie does not attempt to carry through such a solution. Instead he notes that, in classical mechanics in the non-relativistic limit, it is possible to find a 'Jacobi function' $\phi(\mathbf{x}_1, \mathbf{x}_2)$ for the two-particle system, such that the particle momenta are given by

$$m_1 \mathbf{v}_1 = -\nabla_1 \phi, \quad m_2 \mathbf{v}_2 = -\nabla_2 \phi. \tag{2.63}$$

De Broglie then asks (p. 238): 'Can the new mechanics ... define such a function ϕ?'

De Broglie is asking whether, if one could solve the coupled equations (2.61), the resulting motions of the singularities would (in the non-relativistic limit) satisfy the guidance equations (2.63) for some function $\phi(\mathbf{x}_1, \mathbf{x}_2)$. He then asserts, on the basis of an incorrect argument (to which we shall return in a moment), that this will indeed be the case. Having concluded that a function $\phi(\mathbf{x}_1, \mathbf{x}_2)$ generating the motions of the singularities will in fact exist, he goes on to identify $\phi(\mathbf{x}_1, \mathbf{x}_2)$ with the phase of a continuous solution $\Psi(\mathbf{x}_1, \mathbf{x}_2, t)$ of the Schrödinger equation (2.60) for the two-particle system. While de Broglie does not say so explicitly, in effect he *assumes* that there exists a continuous solution of (2.60) whose phase coincides with the function $\phi(\mathbf{x}_1, \mathbf{x}_2)$ that generates the motions of the singularities via (2.63). This seems to be de Broglie's answer to the question he raises, as to the 'true meaning' of the Schrödinger equation: solving the Schrödinger equation (2.60) in configuration space provides an effective means of obtaining the motions of the singularities without having to solve the coupled partial differential equations (2.61) in 3-space.

De Broglie's incorrect argument for the existence of an appropriate function ϕ, derived from the equations (2.61), proceeds as follows. He first imagines that the motion of singularity 2 is already known, so that the problem of motion for singularity 1 reduces to a case that has already been discussed, of a single particle in a given time-dependent potential. For this case, it has already been shown that the ensemble of possible trajectories may be represented by a continuous wave

$$a_1(\mathbf{x}, t)e^{(i/\hbar)\phi_1(\mathbf{x},t)}. \tag{2.64}$$

Further, as was shown earlier, the equations of motion for the particle may be written in Lagrangian form with an additional potential $\varepsilon_1(\mathbf{x}_1, t)c^2$. Similarly, considering the motion of singularity 1 as known, the motion of singularity 2 can be described by a Lagrangian with an additional potential $\varepsilon_2(\mathbf{x}_2, t)c^2$. The two sets of (classical) equations of motion can be derived from a single Lagrangian, de Broglie argues, only if ε_1 and ε_2 reduce to a *single* function $\varepsilon(|\mathbf{x}_1 - \mathbf{x}_2|)$ of the interparticle separation. 'If we assume this', says de Broglie, the total Lagrangian for the system will be

$$L = \frac{1}{2}m_1 v_1^2 + \frac{1}{2}m_2 v_2^2 - V(|\mathbf{x}_1 - \mathbf{x}_2|) - \varepsilon(|\mathbf{x}_1 - \mathbf{x}_2|)c^2, \tag{2.65}$$

and one will be able to deduce, in the usual fashion, the existence of a function $\phi(\mathbf{x}_1, \mathbf{x}_2)$ satisfying the equations (2.63). It is not entirely clear from the text if de Broglie is really convinced that ε_1 and ε_2 will indeed reduce to a single function $\varepsilon(|\mathbf{x}_1 - \mathbf{x}_2|)$. With hindsight one sees, in fact, that this will *not* usually be the case. For it amounts to requiring that the quantum potential $Q(\mathbf{x}_1, \mathbf{x}_2)$ for two particles should take the form $Q(\mathbf{x}_1, \mathbf{x}_2) = Q(|\mathbf{x}_1 - \mathbf{x}_2|)$, which is generally false.[a]

[a] One might also question the meaning of de Broglie's wave function (2.64) for particle 1, in the context of a system of two interacting particles, which must have an entangled wave function $\Psi(\mathbf{x}_1, \mathbf{x}_2, t)$.

After giving his incorrect argument for the existence of ϕ, de Broglie points out that, since the particles have definite trajectories, it is meaningful to consider their six-dimensional configuration space, and to represent the two particles by a single point in this space, with velocity components given by (2.63). If the initial velocity components are given, but the initial positions of the two particles are not, one may consider an ensemble of representative points $(\mathbf{x}_1, \mathbf{x}_2)$, whose density $\rho(\mathbf{x}_1, \mathbf{x}_2)$ in configuration space satisfies the continuity equation,

$$\nabla \cdot (\rho \mathbf{v}) = 0, \tag{2.66}$$

where $\mathbf{v}$ is the (six-dimensional) velocity field given by (2.63).

De Broglie then considers, as we have said, a continuous solution of the Schrödinger equation (2.60), of the form

$$\Psi(\mathbf{x}_1, \mathbf{x}_2, t) = A(\mathbf{x}_1, \mathbf{x}_2)e^{(i/\hbar)\phi}, \tag{2.67}$$

with the tacit assumption that the phase ϕ is the same function ϕ whose existence has been (apparently) established by the above (incorrect) argument. He then shows, from (2.60), that A^2 satisfies the configuration-space continuity equation (2.66). De Broglie concludes that $A^2 d^3\mathbf{x}_1 d^3\mathbf{x}_2$ is the probability for the representative point $(\mathbf{x}_1, \mathbf{x}_2)$ to be present in a volume element $d^3\mathbf{x}_1 d^3\mathbf{x}_2$ of configuration space.

Here, then, de Broglie has arrived at a higher-dimensional analogue of his previous results: the wave function Ψ in configuration space determines the ensemble probability density through its amplitude, as well as the motion of a single system through its phase.

De Broglie ends this section by remarking that it seems difficult to find an equation playing a role similar to (2.60) outside of the non-relativistic approximation, and that Fermi's calculation concerning the scattering of electrons by a rotator may be regarded as illustrating the above.

From the point of view of the history of pilot-wave theory, the most important section of de Broglie's paper now follows. Entitled 'The pilot wave' ('L'onde pilote'), it begins by recalling the results obtained in the case of a single particle in a time-dependent potential, which de Broglie states may be summarised by 'the two fundamental formulas' (2.51) and (2.55) for the particle velocity and probability density respectively. De Broglie notes that he arrived at the velocity formula (2.51) by invoking the principle of the double solution – a principle that is valid in free space but which 'remains a hypothesis in the general case' (p. 241). At this point, de Broglie makes a remarkable suggestion: that instead of trying to derive the velocity field (2.51) from an underlying theory, one could simply take it as a postulate, and regard Ψ as a physically-real 'pilot wave' guiding the motion of the particle. To quote de Broglie (p. 241):

But if one does not wish to invoke the principle of the double solution, it is acceptable to adopt the following point of view: one will assume the existence, as distinct realities, of the material point and of the continuous wave represented by the function Ψ, and one will take it as a postulate that the motion of the point is determined as a function of the phase of the wave by the equation [(2.51)]. One then conceives the continuous wave as guiding the motion of the particle. It is a pilot wave.

This is the first appearance in the literature of what we now know as pilot-wave or de Broglian dynamics (albeit stated explicitly for only a single particle). The pilot wave Ψ satisfying the Schrödinger equation, and the material point, are regarded as 'distinct realities', with the former guiding the motion of the latter according to the velocity law (2.51).

De Broglie made it clear, however, that he thought such a dynamics could be only provisional (p. 241):

By thus taking [(2.51)] as a postulate, one avoids having to justify it by the principle of the double solution; but this can only be, I believe, a provisional attitude. No doubt one will indeed have to *reincorporate* the particle into the wave-like phenomenon, and one will probably be led back to ideas analogous to those that have been developed above.

As we shall discuss in the next section, de Broglie's proposal of pilot-wave theory as a provisional measure has striking analogues in the early history of Newtonian gravity and of Maxwellian electromagnetic theory.

De Broglie goes on to point out two important applications of the formulas (2.51) and (2.55). First, applying the theory to light, according to (2.55) 'the density of the photons is proportional to the square of the amplitude' of the guiding wave Ψ, yielding agreement with the predictions of wave optics. Second, an ensemble of hydrogen atoms with a definite state Ψ will have a mean electronic charge density proportional to $|\Psi|^2$, as assumed by Schrödinger.

De Broglie's 'Structure' paper ends with some remarks on the pressure exerted by particles on a wall (for example of a box of gas). He notes that according to (2.51), because of the interference between the incident and reflected waves, the particles will not actually strike the wall, raising the question of how the pressure is produced. De Broglie claims that the pressure comes from stresses in the interference zone, as appear in an expression obtained by Schrödinger (1927c) for the stress-energy tensor associated with the wave Ψ.

2.3.2 Significance of de Broglie's 'Structure' paper

The exceptional quality of de Broglie's 'Structure' paper in *Journal de Physique* was noted by Pauli, in a letter to Bohr dated 6 August 1927, already quoted as the epigraph to the last subsection: '... it is very rich in ideas and very sharp, and on a much higher level than the childish papers by Schrödinger ...' (Pauli

1979, pp. 404–5). In the same letter, Pauli suggested that Bohr would have to refer to de Broglie's paper in his Como lecture (in which Bohr developed the idea of complementarity between waves and particles); and in fact, in what was to remain an unpublished version of the Como lecture, Bohr did take an explicit stand against de Broglie's ideas (Bohr 1985, pp. 89–98). Bohr characterised de Broglie's work as attempting 'to reconcile the two apparently contradictory sides of the phenomena by regarding the individual particles or light quanta as singularities in the wave field'; further, Bohr suggested that de Broglie's view rested upon 'the concepts of classical physics' and was therefore not 'suited to help us over the fundamental difficulties' (Bohr 1985, p. 92). On the whole, though, this paper by de Broglie has been essentially ignored, by both physicists and historians.

De Broglie's 'Structure' paper may be summed up as follows. De Broglie has a model of particles as singularities of 3-space waves u_i, in which the motion of the individual particles, as well as the ensemble distribution of the particles, are determined by a continuous wave function Ψ. (We emphasise that de Broglie has both the wave function Ψ and the singular u-waves.) The result is a first-order theory of motion, based on the velocity law $m_i \mathbf{v}_i = \nabla_i \phi$, in which the principles of Maupertuis and Fermat are unified. Then, at the end of the paper, de Broglie recognises that his singularity model of particles can be dropped, and that the results he has obtained – the formulas for velocity and probability density in terms of Ψ – can be simply postulated, yielding pilot-wave theory as a provisional measure.

A few months later, in October 1927, de Broglie presented this pilot-wave theory at the fifth Solvay conference, for a non-relativistic system of N particles guided by a wave function Ψ in configuration space. Before discussing de Broglie's Solvay report, however, it is worth pausing to consider the role played by the 'Structure' paper in de Broglie's thinking.

From a historical point of view, the significance of de Broglie's 'Structure' paper inevitably depends on the significance one ascribes to the provisional pilot-wave theory arrived at in that paper. Given what we know today – that pilot-wave theory provides a consistent account of quantum phenomena – de Broglie's 'Structure' paper now seems considerably more significant than it has seemed in the past.

From the point of view of pilot-wave theory as we know it today, the singular u-waves played a similar role for de Broglie as the material ether did for Maxwell in electromagnetic theory. In both cases, there was a conceptual scaffolding that was used to build a new theory, and that could be dropped once the results had been arrived at. De Broglie recognised at the end of his paper that, if one took pilot-wave dynamics as a provisional theory, then the scaffolding he had used to construct this theory could be dropped. At the same time, de Broglie insisted that taking such a step was indeed only provisional, and that an underlying theory was still needed, probably along the lines he had been pursuing.

Similar situations have arisen before in the history of science. Abstracting away the details of a model based on an older theory sometimes results in a new theory in its own right, involving new concepts, where, however, the author regards the new theory as only a provisional measure, and expects that the model based on the older theory will eventually provide a proper basis for the provisional theory. Thus, for example, Newton tried to explain gravitation on the basis of action by contact, involving a material medium filling space – the same 'Aethereal Medium', in fact, as he thought responsible for the interference and diffraction of light (Newton 1730; reprint, pp. 350–3). Newton regarded his theory of gravitation, with action at a distance, as merely a provisional and phenomenological theory, that would later find an explanation in terms of contiguous action. Eventually, however, the concept of 'gravitational action at a distance' became widely accepted in its own right (though it was later to be overthrown, of course, by general relativity). Similarly, Maxwell used mechanical models of an ether to develop his theory of the electromagnetic field. Maxwell himself may or may not have recognised that this scaffolding could be dropped (Hendry 1986). But certainly, many of his immediate followers did not: they regarded working only with Maxwell's equations as provisional and phenomenological, pending the development of an underlying mechanical model of an ether. Again, the concept of 'electromagnetic field' eventually became widely accepted as a physical entity in its own right.

In the case of de Broglie in 1927, in his 'Structure' paper (and subsequently at the fifth Solvay conference), he arrived at the new concept of 'pilot wave in configuration space', an entirely new kind of physical entity that, according to de Broglie's (provisional) theory, guides the motion of material systems.

2.4 1927 Solvay report: the new dynamics of quanta

We now turn to a summary of and commentary on de Broglie's report at the fifth Solvay conference.[a] As we have noted, the theory presented in this report is pilot-wave theory as we know it today, for a non-relativistic many-body system, with a guiding wave in configuration space that determines the particle velocities according to de Broglie's basic law of motion. De Broglie's ideas about particles as singularities of 3-space waves are mentioned only briefly. The theory de Broglie presents in Brussels in October 1927 is, indeed, just the provisional theory he proposed a few months earlier at the end of his 'Structure' paper (though now explicitly applied to many-body systems as well).

[a] Again, our notation sometimes departs from de Broglie's.

De Broglie begins part I of his report by reviewing the results obtained in his doctoral thesis. The energy E and momentum $\mathbf{p}$ of a particle are determined by the phase ϕ of an associated wave:[a]

$$E = \partial\phi/\partial t, \quad \mathbf{p} = -\nabla\phi. \tag{2.68}$$

(These are the relativistic guidance equations of de Broglie's early pilot-wave theory of 1923–4.) It follows that, as de Broglie remarks, 'the principles of least action and of Fermat are identical' (p. 342). Quantisation conditions appear in a natural way, and there are far-reaching implications for statistical mechanics. De Broglie ends the review of his early work with some general remarks. He points out (p. 344) that he 'has always assumed that the material point occupies a well-defined position in space', and he asserts that as a result the wave amplitude must be singular, or take very large values in a small region, somewhere within the extended wave. Significantly, de Broglie adds: 'But, in fact, the form of the amplitude plays no role in the results reviewed above'. This seems to be a first hint that the actual results, such as (2.68), do not depend on the details of any underlying model of the particles as moving singularities.

De Broglie then outlines the work of Schrödinger. He notes that Schrödinger's wave equation is constructed in order that the phase ϕ of the wave function

$$\Psi = a \cos \frac{2\pi}{h}\phi \tag{2.69}$$

be a solution of the Hamilton–Jacobi equation in the geometrical-optics limit. He points out that Schrödinger identifies particles with localised wave packets instead of with a small concentration within an extended wave, and that for a many-body system Schrödinger has a wave Ψ propagating in configuration space. For both the one-body and many-body cases, de Broglie writes down the time-independent Schrödinger equation only, with a static potential energy function. As we shall see, de Broglie in fact considers non-stationary wave functions as well. For the one-body case, de Broglie also writes down a relativistic, and time-dependent, equation – what we now know as the Klein–Gordon equation in an external electromagnetic field.

De Broglie then raises two conceptual difficulties with Schrödinger's work (similar in spirit to the objections he raises in 'Structure'): (1) He questions how one can meaningfully construct a configuration space without a real configuration, asserting that it 'seems a little paradoxical to construct a configuration space with the coordinates of points that do not exist' (p. 346). (2) He claims that the physical meaning of the wave Ψ cannot be compared with that of an ordinary wave in

[a] Again, de Broglie's phase ϕ has a sign opposite to the phase S as we would normally define it now.

3-space, because the number of dimensions of the abstract space on which Ψ is defined is determined by the number of degrees of freedom of the system. (This second point seems indeed a very effective way to make clear that Ψ is quite different from a conventional wave or field on 3-space.)

Part I of the report ends with some remarks on Born's statistical interpretation, which de Broglie asserts is analogous to his own.

In part II of his report, de Broglie presents his own interpretation of the wave function. Pilot-wave theory is clearly formulated, first for a single particle, and then for a system of N particles. Several applications are outlined. It is interesting to see how de Broglie motivates his theory and compares it with the contenders.

De Broglie begins by asking what the relationship is between particles and the wave Ψ. He first considers a single relativistic particle in an external electromagnetic field with potentials $(\mathcal{V}, \mathbf{A})$. De Broglie notes that, in the classical limit, the phase ϕ of Ψ obeys the Hamilton–Jacobi equation, and the velocity of the particle is given by

$$\mathbf{v} = -c^2 \frac{\nabla\phi + \frac{e}{c}\mathbf{A}}{\dot{\phi} - e\mathcal{V}} \tag{2.70}$$

(the same formula (2.51) discussed in 'Structure'). De Broglie then proposes that this velocity formula is valid even outside the classical limit (p. 349, italics in the original):

We propose to assume by induction that this formula is still valid when the old Mechanics is no longer sufficient, that is to say when [ϕ] is no longer a solution of the Jacobi equation. If one accepts this hypothesis, which appears justified by its consequences, the formula [(2.70)] completely determines the motion of the corpuscle *as soon as one is given its position at an initial instant.* In other words, the function [ϕ], just like the Jacobi function of which it is the generalisation, determines a whole class of motions, and to know which of these motions is actually described it suffices to know the initial position.

We emphasise that there is no appeal to singular u-waves anywhere in de Broglie's report. No use is made of the principle of the double solution. As he had suggested at the end of his 'Structure' paper a few months earlier, de Broglie simply postulates the basic equations of pilot-wave dynamics, without trying to derive them from anything else. To motivate the guidance equation, de Broglie simply generalises the classical Hamilton–Jacobi velocity formula to the non-classical domain: he assumes 'by induction' that the formula holds even outside the classical limit. And de Broglie is quite explicit that he is proposing a first-order theory of motion, based on velocities: given the wave function, the initial position alone determines the trajectory.

So far, then, for a single particle de Broglie has a (relativistic) wave function $\Psi = a \cos \frac{2\pi}{h}\phi$ whose phase ϕ determines the particle velocity by the guidance equation (2.70). De Broglie now goes on to point out that, for an ensemble of particles guided by the velocity field (2.70), the distribution

$$Ka^2\left(\dot{\phi} - e\mathcal{V}\right) \tag{2.71}$$

is preserved in time (that is, equivariant). He concludes that, if the initial position of the particle is ignored, then the probability for the particle to be present (at time t) in a spatial volume $d\tau$ is

$$\pi\, d\tau = Ka^2\left(\dot{\phi} - e\mathcal{V}\right) d\tau. \tag{2.72}$$

(This is the same probability formula (2.55) arrived at in 'Structure'.)

De Broglie's expression 'probabilité de présence' makes it clear that we have to do with a probability for the electron *being* somewhere, and not merely with a probability for an experimenter *finding* the electron somewhere – cf. Bell (1990, p. 29). De Broglie simply assumes that the equivariant distribution is the correct probability measure for a particle of unknown position. In fact, this distribution is only an equilibrium distribution, analogous to thermal equilibrium in classical statistical mechanics (see Section 8.2).

De Broglie sums up his results so far (p. 349):

> In brief, in our hypotheses, each wave Ψ determines a 'class of motions', and each one of these motions is governed by equation [(2.70)] when one knows the initial position of the corpuscle. If one ignores this initial position, the formula [(2.72)] gives the probability for the presence of the corpuscle in the element of volume $d\tau$ at the instant t. The wave Ψ then appears as both a *pilot wave* (Führungsfeld of Mr Born) and a *probability wave*.

There are, de Broglie adds, no grounds for abandoning determinism, and in this his theory differs from that of Born.

De Broglie adds further that, in the non-relativistic approximation (where $\dot{\phi} - e\mathcal{V} \approx m_0 c^2$), the guidance and probability formulas (2.70) and (2.72) reduce to

$$\mathbf{v} = -\frac{1}{m_0}\left(\nabla\phi + \frac{e}{c}\mathbf{A}\right) \tag{2.73}$$

and

$$\pi = \mathrm{const} \cdot a^2. \tag{2.74}$$

These are the standard pilot-wave equations for a single particle (in an external electromagnetic field).

De Broglie then remarks on how the above formulas may be applied to the scattering of a single particle by a fixed potential. The ensemble of incident particles may be represented by a plane wave with a uniform probability distribution. Upon entering a region of non-zero field, the behaviour of the wave function may be calculated using Born's perturbation theory (for example, for the

Rutherford scattering of an electron by an atomic nucleus). De Broglie draws an analogy between the scattering of the wave function Ψ and the classical scattering of light by a refracting medium.

As in his 'Structure' paper, de Broglie remarks in passing that, for a relativistic particle governed by the guidance equation (2.70), one may write down the equations of classical dynamics with a variable rest mass M_0 given by equation (2.57). As we have already noted, in the non-relativistic approximation this yields an additional potential energy term, which is precisely Bohm's 'quantum potential'. And, once again, we emphasise that for de Broglie the equation $\mathbf{v} \propto \nabla\phi$ for *velocity* is the fundamental equation of motion, expressing the identity of the principles of Maupertuis and Fermat: de Broglie merely points out that, if one wishes, the dynamics can be written in classical terms (as Bohm did in 1952).

De Broglie then turns to the interpretation of interference and diffraction, for the case of photons. Here, the guiding wave Ψ is similar to but not the same as a light wave.[a] De Broglie considers scattering by fixed obstacles, in which case the guiding wave may be taken to have a constant frequency ν. The relativistic equations (2.70) and (2.72) then become

$$\mathbf{v} = -\frac{c^2}{h\nu}\nabla\phi, \qquad \pi = \text{const} \cdot a^2. \tag{2.75}$$

De Broglie points out that the second equation predicts the well-known interference and diffraction patterns. He argues that the results will be the same, whether the experiment is done with an intense beam over a short time or with a feeble beam over a long time.

De Broglie remarks, as he did at the end of 'Structure', on the stresses appearing in Schrödinger's expression for the stress-energy tensor of the wave Ψ. According to de Broglie, these stresses provide an explanation for the pressure exerted by light reflecting on a mirror – despite the fact that, according to (2.70), the photons never actually strike the surface of the mirror (as shown explicitly by Brillouin in the discussion after de Broglie's report).

De Broglie then turns to the generalisation of the above dynamics to a (non-relativistic) *many-body* system. Remarkably, many historians and commentators have not noticed this proposal by de Broglie of a many-body dynamics in configuration space, a proposal that is usually attributed to Bohm (cf. Section 11.1).

De Broglie begins by pointing out how the two difficulties he has raised against Schrödinger's wave mechanics might be solved. First, if a real configuration exists

[a] De Broglie did not identify the photonic pilot wave with the electromagnetic field. In the general discussion, on p. 460, de Broglie explicitly states that in his theory the wave Ψ for the case of photons is distinct from the electromagnetic field.

at all times, one can meaningfully construct the configuration space. As for the second difficulty, regarding the meaning of a wave in the abstract configuration space, de Broglie makes the following preliminary remark (p. 352):

It appears to us certain that if one wants to *physically* represent the evolution of a system of N corpuscles, one must consider the propagation of N waves in space, each of the N propagations being determined by the action of the $N-1$ corpuscles connected to the other waves.

This seems a clear reference to the theory of interacting singular u-waves discussed a few months earlier in de Broglie's 'Structure' paper.[a] (The propagation of each wave must depend on the motion of the other $N-1$ particles, of course, in order to account for interactions between the particles.) As we saw above (Section 2.3.1), in 'Structure' de Broglie explicitly considered the case of two particles, which were represented by two singular waves $u_1(\mathbf{x}, t)$ and $u_2(\mathbf{x}, t)$ satisfying a pair of coupled partial differential equations (2.61), where the equation for each wave contained a potential depending on the position of the singularity in the other wave. Instead of trying to solve the equations, de Broglie assumed that the resulting velocities of the moving singularities could be written as the gradient of a function $\phi(\mathbf{x}_1, \mathbf{x}_2)$, which could be identified with the phase of a solution $\Psi(\mathbf{x}_1, \mathbf{x}_2, t)$ of the (time-independent) Schrödinger equation. In 'Structure', then, the wave function in configuration space appeared as an effective description of the motions of the particles (or singularities). An echo of this view is discernible in the Solvay report, which continues with the following justification for introducing a guiding wave in configuration space (p. 352):

Nevertheless, if one focusses one's attention only on the corpuscles, one can represent their states by a point in configuration space, and one can try to relate the motion of this representative point to the propagation of a fictitious wave Ψ in configuration space.

De Broglie seems to be saying that, if one is concerned only with a succinct mathematical account of particle motion (as opposed to a physical representation), then this can be obtained by introducing a guiding wave in configuration space. Certainly, the dynamics is much simpler with a single, autonomous wave Ψ in configuration space. As in his 'Structure' paper, this seems to be de Broglie's explanation for Schrödinger's otherwise mysterious configuration-space wave.

[a] It might be thought that here de Broglie has in mind the fact that N moving particles may be associated with N velocity fields in 3-space, which may be associated with N (non-singular) guiding waves in 3-space. From a pilot-wave perspective, such N guiding waves may be identified with the N 'conditional' wave functions $\psi_i(\mathbf{x}_i, t) \equiv \Psi(\mathbf{X}_1, \mathbf{X}_2, \ldots, \mathbf{x}_i, \ldots, \mathbf{X}_N, t)$, where $\mathbf{x}_i$ $(i = 1, 2, \ldots, N)$ ranges over all positions in 3-space and $\mathbf{X}_j$ is the *actual* position of the jth particle. Each $\psi_i(\mathbf{x}_i, t)$ defines a wave in 3-space that determines the velocity of the ith particle (through the gradient of its phase); and each $\psi_i(\mathbf{x}_i, t)$ depends on the positions of the other $N-1$ particles. However, given the context (in particular the recent publication of 'Structure'), it seems clear that here de Broglie's N waves are the singular u-waves of his double-solution theory.

When de Broglie states that the wave Ψ is 'fictitious', he presumably means that it has only mathematical, and not physical, significance.[a] It provides a convenient mathematical account of particle motion, pending a full physical description by a more detailed theory. As we discussed in Section 2.3.2, the provisional introduction of a 'mathematical' description, pending the development of a proper 'physical' model, has distinguished precedents in the history of Newtonian gravity and Maxwellian electrodynamics.

Having motivated the introduction of a guiding wave Ψ in configuration space, de Broglie suggests (p. 352) that Ψ

> ... plays for the representative point of the system in configuration space the same role of pilot wave and of probability wave that the wave Ψ plays in ordinary space in the case of a single material point.

Thus, in the non-relativistic approximation, de Broglie considers N particles with positions $\mathbf{x}_1, \mathbf{x}_2, \ldots, \mathbf{x}_N$. He states that the wave Ψ determines the velocity of the representative point in configuration space by the formula

$$\mathbf{v}_k = -\frac{1}{m_k}\nabla_k\phi \tag{2.76}$$

(where m_k is the mass of the kth particle). As in the case of a single particle, notes de Broglie, the probability for the system to be present in a volume element $d\tau$ of configuration space is

$$\pi\, d\tau = \text{const} \cdot a^2 d\tau. \tag{2.77}$$

These are the standard pilot-wave equations for a many-body system, replacing the single-particle formulas (2.73) and (2.74). De Broglie remarks that (2.77) seems to agree with Born's results for electron scattering by an atom and with Fermi's for scattering by a rotator.

Finally, at the end of his section on the many-body case, de Broglie notes that it seems difficult to construct a wave Ψ that can generate the motion of a relativistic many-body system (in contrast with the relativistic one-body case), a point that de Broglie had already noted in 'Structure'. From the very beginning, then, it was recognised that it would be difficult to formulate a fundamentally Lorentz-invariant pilot-wave theory for a many-body system, a situation that persists to this day.

Thus, in his report at the fifth Solvay conference, de Broglie arrived at pilot-wave theory for a many-body system, with a guiding wave in configuration space.

[a] This is somewhat in contrast with de Broglie's introduction of the pilot wave at the end of 'Structure', where he refers to the particle and the guiding wave as 'distinct realities'. However, there de Broglie explicitly proposed pilot-wave theory for a single particle only, and it is likely that while he was comfortable with the idea of a physically real pilot wave in 3-space (for the one-body case), he could not regard a pilot wave in configuration space as having more than mathematical significance.

Judging by de Broglie's comments about a physical representation requiring N waves in 3-space, and his characterisation of Ψ as a 'fictitious' wave, it seems clear that he still regarded his new dynamics as an effective, mathematical theory only (just as he had a few months earlier in his 'Structure' paper).

De Broglie's report then moves on to sketch some applications to atomic theory. For stationary states of hydrogen, de Broglie notes that the electron motion is circular, except in the case of magnetic quantum number $m = 0$ for which the electron is at rest. (Note that de Broglie expresses no concern that the electron is predicted to be motionless in the ground state of hydrogen. This result may seem puzzling classically, just as Bohr's quantised atomic orbits seem puzzling classically: but both are natural consequences of de Broglie's non-classical dynamics.)

De Broglie also points out that one can calculate the electron velocity during an atomic transition $i \rightarrow j$, so that such transitions *can* be visualised. In this example, de Broglie's guiding wave is a solution of the time-*dependent* Schrödinger equation, with a time-dependent amplitude as well as phase. (The atomic wave function is taken to have the form $\Psi = c_i\Psi_i + c_j\Psi_j$, where Ψ_i, Ψ_j are eigenfunctions corresponding to the atomic states i, j, and c_i, c_j are functions of time.) Clearly, then, de Broglie applied his pilot-wave dynamics not just to stationary states, but to quite general wave functions – as he had also done (in the case of a single particle) in 'Structure'.

De Broglie then outlines how, in his theory, one can obtain expressions for the mean charge and current density for an ensemble of atoms. These expressions are the same as those used by Schrödinger and others. De Broglie remarks that, 'denoting by Ψ the wave written in *complex* form, and by [Ψ^*] the conjugate function', in the non-relativistic limit the charge density is proportional to $|\Psi|^2$: the electric dipole moment then contains the correct transition frequencies. By using these expressions as sources in Maxwell's equations, says de Broglie, one can correctly predict the mean energy radiated by an atom. This is just semiclassical radiation theory, which is still widely used today in quantum optics.[a]

Part II of de Broglie's report ends with some general remarks. First and foremost, de Broglie makes it clear that he regards pilot-wave theory as only provisional (p. 355):

> So far we have considered the corpuscles as 'exterior' to the wave Ψ, their motion being only determined by the propagation of the wave. This is, no doubt, only a provisional point of view: a true theory of the atomic structure of matter and radiation should, it seems to us, *incorporate* the corpuscles in the wave phenomenon by considering singular solutions of the wave equations.

[a] Cf. Section 4.4.

De Broglie goes on to suggest that in a deeper theory one could 'show that there exists a correspondence between the singular waves and the waves Ψ, such that the motion of the singularities is connected to the propagation of the waves Ψ', just as he had suggested in 'Structure'. Part II ends by noting the incomplete state of the theory with respect to the electromagnetic field and electron spin.

De Broglie's final part III contains a lengthy discussion and review of recent experiments involving the diffraction, interference and scattering of electrons (as had been requested by Lorentz[4]). He regards the results as evidence for his 'new Dynamics'. For the case of the diffraction of electrons by a crystal lattice, de Broglie points out that the scattered wave function Ψ has maxima in certain directions, and notes that according to his theory the electrons should be preferentially scattered in these directions (p. 357):

> Because of the role of pilot wave played by the wave Ψ, one must then observe a selective scattering of the electrons in these directions.

What de Broglie is (briefly) describing here is the separation of the incident wave function Ψ into distinct (non-overlapping) emerging beams, with each outgoing electron occupying one beam, and with an ensemble of electrons being distributed among the emerging beams according to the Born rule. This is relevant to a proper understanding of the de Broglie–Pauli encounter, discussed in Section 10.2.

Concerning the scattering maxima observed recently by Davisson and Germer, for electrons incident on a crystal, de Broglie remarks (p. 360):

> There is direct numerical confirmation of the formulas of the new Dynamics . . ..

De Broglie also discusses the inelastic scattering of electrons by atoms. According to Born's calculations, one should observe maxima and minima in the angular dependence of the differential scattering cross section. According to de Broglie, it is premature to speak of an agreement with experiment. The results of Dymond, for the inelastic scattering of electrons by helium atoms, do however show maxima in the cross section, in qualitative agreement with the predictions. De Broglie makes it quite clear that, at the time, theory was far behind experiment.

The discussion following de Broglie's report was extensive, detailed, and varied. A number of participants raised queries, and de Broglie replied to most of them. Some of this discussion will be considered in detail in Chapters 10 and 11. Here, we restrict ourselves to a brief summary of the questions raised.

Lorentz asked how, in the simple pilot-wave theory of 1924, de Broglie derived quantisation conditions for the case of multiperiodic atomic orbits. Born questioned the validity of de Broglie's guidance equation for an elastic collision, while Pauli suggested that the key idea behind de Broglie's theory was the association of particle trajectories with a locally conserved current. Schrödinger raised the question of an alternative velocity field different from that assumed by de Broglie, while Kramers raised the question of how the Maxwell energy-momentum tensor could arise from independently moving photons. Lorentz, Ehrenfest and Schrödinger asked about the properties of electron orbits in hydrogen. Brillouin discussed at length the simple example of a photon colliding with a mirror: in his Fig. 2 (p. 368), the incident and emergent photon trajectories are located inside packets of limited extent (a point relevant for the de Broglie–Pauli encounter in the general discussion). Finally, Lorentz considered, in classical Maxwell theory, the near-field attractive stress between two prisms, and claimed that this 'negative pressure' could not be produced by the motion of corpuscles (photons).

2.5 Significance of de Broglie's work from 1923 to 1927

In his papers and thesis of 1923–4, de Broglie proposed a simple form of a new, non-classical dynamics of particles with velocities determined by the phase gradient of abstract waves in 3-space. De Broglie constructed the dynamics in such a way as to unify the mechanical principle of Maupertuis with the optical principle of Fermat. He showed that it gave an account of some simple quantum phenomena, including single-particle interference and quantised atomic energy levels.

As we have seen, the scope and ambition of de Broglie's doctoral thesis went far beyond a mere extension of the relations $E = h\nu$, $p = h/\lambda$ from photons to other particles. Yet, the thesis is usually remembered solely for this idea. The depth and inner logic of de Broglie's thinking is not usually appreciated, neither by physicists nor by historians. An exception is Darrigol (1993), who on this very point writes (pp. 303–4):

> For one who only knows of de Broglie's relation $\lambda = h/p$, two explanations of his originality offer themselves. The first has him as a lucky dreamer who hit upon a great idea amidst a foolish play with analogies and formulas. The second has him a providential deep thinker, who deduced unsuspected connections by rationally combining distant concepts. The first explanation is the most popular, though rarely expressed in print. ... The second explanation of Louis de Broglie's originality, though also extreme, is certainly closer to the truth. Anyone who has read de Broglie's thesis cannot help admiring the unity and inner consistency of his views, the inspired use of general principles, and a necessary reserve.

In 1926, starting from de Broglie's expressions for the frequency and phase velocity of an electron wave in an external potential, Schrödinger found the (non-relativistic) wave equation for de Broglie's waves. It was de Broglie's work, beginning in 1923, that initiated the notion of a wave function for material particles. And it was de Broglie's view of the significance of the optical-mechanical analogy, and of the role waves could play in bringing about the existence of integer-valued quantum numbers, that formed the basis for Schrödinger's development of the wave equation and the associated eigenvalue problem.

In 1927, de Broglie proposed what we now know as pilot-wave or de Broglian dynamics for a many-body system, with a guiding wave in configuration space. De Broglie regarded this theory as provisional: he thought it should emerge as an effective theory, from a more fundamental theory in which particles are represented by singularities of 3-space waves. Even so, he did propose the theory, in the first-order form most commonly used today. Further, as we have seen, de Broglie's view of the pilot wave as merely phenomenological is strikingly reminiscent of (for example) late-nineteenth-century views of the electromagnetic field. In retrospect, one may regard de Broglie as having unwittingly arrived at a new and fundamental concept, that of a pilot wave in configuration space.

De Broglie was unable to show that his new dynamics accounted for all quantum phenomena. In particular, as we shall discuss in Chapters 10 and 11, de Broglie did not understand how to describe the process of measurement of arbitrary quantum observables in pilot-wave theory: as shown in detail by Bohm in 1952, this requires an application of de Broglie's dynamics to the measuring device itself. De Broglie did, however, possess the fundamental dynamics in complete form. Furthermore, de Broglie did understand how his theory accounted for single-particle interference, for the directed scattering associated with crystal diffraction, and for electron scattering by atoms (for the latter, see Section 10.2). Clearly, in 1927, many applications of pilot-wave theory remained to be developed; just as, indeed, many applications of quantum theory – including the general quantum theory of measurement – remained to be developed and clarified.

In retrospect, de Broglian dynamics seems as radical as – and indeed somewhat reminiscent of – Einstein's theory of gravity. According to Einstein, there is no gravitational force, and a freely falling body follows the straightest path in a curved spacetime. According to de Broglie, a massive body undergoing diffraction and following a curvilinear path is not acted upon by a Newtonian force: it is following the ray of a guiding wave. De Broglie's abandonment of Newton's first law of motion in 1923, and the adoption of a dynamics based on velocity rather than acceleration, amounts to a far-reaching departure from classical mechanics and (arguably) from classical kinematics too – with implications for the structure of spacetime that have perhaps not been understood (Valentini 1997). Certainly, the extent to which de Broglie's dynamics departs from classical ideas was unfortunately obscured by Bohm's presentation of it, in 1952, in terms of acceleration and a pseudo-Newtonian quantum potential, a formulation that today seems artificial and inelegant compared with de Broglie's (much as the rewriting of general relativity as a field theory on flat spacetime seems unnatural and hardly illuminating). The fundamentally second-order nature of classical physics is today embodied in the formalism of Hamiltonian dynamics in phase space. In contrast, de Broglie's first-order approach to the theory of motion seems more naturally cast in terms of a dynamics in configuration space.

Regardless of how one may wish to interpret it, by any standards de Broglie's work from 1923 to 1927 shows a remarkable progression of thought, beginning from early intuitions and simple models of the relationship between particles and waves, and ending (with Schrödinger's help) with a complete and new form of dynamics for non-relativistic systems – a deterministic dynamics that was later shown by Bohm to be empirically equivalent to quantum theory (given a Born-rule distribution of initial particle positions). The inner logic of de Broglie's work in this period, his drive to unite the physics of particles with the physics of waves by unifying the principles of Maupertuis and Fermat, his wave-like explanation for quantised energy levels, his prediction of electron diffraction, his explanation for single-particle interference, his attempts to construct a field-theoretical picture of particles as moving singularities, and his eventual proposal of pilot-wave dynamics as a provisional theory – all this compels admiration, all the more for being largely unknown and unappreciated.

Today, pilot-wave theory is often characterised as simply adding particle trajectories to the Schrödinger equation. An understanding of de Broglie's thought from 1923 to 1927, and of the role it played in Schrödinger's work, shows the gross inaccuracy of this characterisation: after all, it was actually Schrödinger who removed the trajectories from de Broglie's theory (cf. Section 11.1). It is difficult to avoid the conclusion that de Broglie's stature as a major contributor to quantum theory has suffered unduly from the circumstance that, for most of the twentieth century, the theory proposed by him was incorrectly regarded as untenable or inconsistent with experiment. Regardless of whether or not de Broglie's pilot-wave theory is closer to the truth than other interpretations, the fact that it is a consistent and viable approach to quantum physics – which has no measurement problem, which shows that objectivity and determinism are not incompatible with quantum physics, and which stimulated Bell to develop his famous inequalities – necessarily entails a reappraisal of de Broglie's place in the history of twentieth-century physics.

Archival notes

1 AHQP-OHI, Louis de Broglie, 'Replies to Mr Kuhn's questions'.
2 AHQP-OHI, Louis de Broglie, session 1 (tape 43a), 7 January 1963, Paris; 0.75 h, in French, 13 pp.; by T. S. Kuhn, T. Kahan and A. George.
3 Ibid.
4 Cf. de Broglie to Lorentz, 27 June 1927, AHQP-LTZ-11 (in French).

3

From matrix mechanics to quantum mechanics

The report by Born and Heisenberg on 'quantum mechanics' may seem surprisingly difficult to the modern reader. This is partly because Born and Heisenberg are describing various stages of development of the theory that are quite different from today's quantum mechanics. Among these, it should be noted in particular that the theory developed by Heisenberg, Born and Jordan in the years 1925–6 and known today as matrix mechanics (Heisenberg 1925b [1], Born and Jordan 1925 [2], Born, Heisenberg and Jordan 1926 [4])[a] differs from standard quantum mechanics in several important respects. At the same time, the interpretation of the theory (the topic of Section II of the report) also appears to have undergone important modifications, in particular regarding the notion of the state of a system. Initially, Born and Heisenberg insist on the notion that a system is always in a stationary state (performing quantum jumps between different stationary states). Then the notion of the wave function is introduced and related to probabilities for the stationary states. At a later stage, probabilistic notions (in particular, what one now calls transition probabilities) are extended to arbitrary observables, but it remains somewhat unclear whether the wave function itself should be regarded as a fundamental entity or merely as an effective one. This may reflect the different routes followed by Born and by Heisenberg in the development of their ideas. The common position presented by Born and Heisenberg emphasises the probabilistic aspect of the theory as fundamental, and the conclusion of the report expresses strong confidence in the resulting picture.

The two main sections of this chapter, Section 3.3 and Section 3.4, will be devoted, respectively, to providing more details on the various stages of development of the theory, and to disentangling various threads of interpretation that appear to be present in Born and Heisenberg's report. Before that, we provide

[a] Throughout this chapter, numbers in square brackets refer to entries in the original bibliography at the end of Born and Heisenberg's report.

a summary (Section 3.1) and a few remarks on the authorship and writing of the various sections of the report (Section 3.2).

3.1 Summary of Born and Heisenberg's report

Born and Heisenberg's report has four sections (together with an introduction and conclusion): I on formalism, II on interpretation, III on axiomatic formulations and on uncertainty, and IV on applications. The formalism that is described is initially that of matrix mechanics, which is then extended beyond the original framework (among other things, in order to make the connection with Schrödinger's wave mechanics). Then, further developments of matrix mechanics are sketched. These allow one to incorporate a 'statistical' interpretation. After a brief discussion of Jordan's (1927b,c [39]) axiomatic formulation, the uncertainty relations are used to justify the statistical element of the interpretation. A few applications of special interest and some brief final remarks conclude the report. As we discuss below, Born drafted Sections I and II, and Heisenberg drafted the introduction, Sections III and IV and the conclusion.

Born and Heisenberg's introduction stresses the continuity of quantum mechanics with the old quantum theory of Planck, Einstein and Bohr, and touches briefly on such themes as discontinuity, observability in principle, 'Anschaulichkeit' (for which see Section 4.6) and the statistical element in quantum mechanics.

Section I, 'The mathematical methods of quantum mechanics', first sketches matrix mechanics roughly as developed in the 'three-man paper' by Born, Heisenberg and Jordan (1926 [4]): the basic framework of position and momentum matrices and of the canonical equations, the perturbation theory, and the connection with the theory of quadratic forms (the latter leading to both discrete and continuous spectra).[a] Next, Born and Heisenberg describe two generalisations of matrix mechanics: Dirac's (1926a [7]) q-number theory and what they refer to as Born and Wiener's (1926a [21], 1926b) operator calculus. As a matter of fact, they already sketch von Neumann's representation of physical quantities by operators in Hilbert space. This is identified as a mathematically rigorous version of the transformation theory of Dirac (1927a [38]) and Jordan (1927b,c [39]).[b] Born and Heisenberg note that, if one takes as the Hilbert space the

[a] Note that Born and Heisenberg use the term 'quantum mechanics' to refer also to matrix mechanics, in keeping with the terminology of the original papers. See also Mehra and Rechenberg (1982c, fn. 72 on pp. 61–2). Heisenberg expresses his dislike for the term 'matrix physics' in Heisenberg to Pauli, 16 November 1925 (Pauli 1979, pp. 255–6).

[b] See Sections 3.3.3 and 3.4.1 for some discussion of Born and Wiener's work, and Section 3.4.5 for the transformation theory (mainly Dirac's). Note that von Neumann had published a series of papers on the Hilbert-space formulation of quantum mechanics before his well-known treatise of 1932. It is to two of these that Born and Heisenberg refer: Hilbert, von Neumann and Nordheim (1928 [41], submitted April 1927) and von Neumann (1927 [42]).

appropriate function space, the problem of diagonalising the Hamiltonian operator leads to the time-independent Schrödinger equation. Thus, in this (formal) sense, the Schrödinger theory is a special case of the operator version of quantum mechanics. The eigenfunctions of the Hamiltonian can be associated with the unitary transformation that diagonalises it. Matrix mechanics, too, is a special case of this more general formalism, if one takes the (discrete) energy eigenstates as the basis for the matrix representation.

Section II on 'Physical interpretation' is probably the most striking. It begins by stating that matrix mechanics describes neither the actual state of a system, nor when changes in the actual state take place, suggesting that this is related to the idea that matrix mechanics describes only closed systems. In order to obtain some description of change, one must consider open systems. Two methods for doing this are introduced. First of all, following Heisenberg (1926b [35]) and Jordan (1927a [36]), one can consider the matrix mechanical description of two coupled systems that are in resonance. One can show that the resulting description can be interpreted in terms of quantum jumps between the energy levels of the two systems, with an explicit expression for the transition probabilities. The second method uses the generalised formalism of Section I, in which time dependence can be introduced explicitly via the time-dependent Schrödinger equation. Again, expressions are found that can be interpreted as transition probabilities, and similarly the squared modulus of the wave function's coefficients in the energy basis is interpreted as the probability for the occurrence of the corresponding stationary state. Born and Heisenberg introduce the notion of interference of probabilities. Only at this stage are experiments mentioned as playing a conceptually crucial role, namely in an argument why such interference 'does not represent a contradiction with the rules of the probability calculus' (p. 387). Interference is then related to the wave theory of de Broglie and Schrödinger, and the interpretation of the squared modulus of the wave function as a probability density for position is introduced. Born and Heisenberg also define the 'relative' (that is, conditional) position density given an energy value, as the squared modulus of the amplitude of a stationary state. Finally, in the context of Dirac's and Jordan's transformation theory, the notions of transition probability (for a single quantity) and of conditional probability (for a pair of quantities) are generalised to arbitrary physical quantities.

Section III opens with a concise exposition of Jordan's (1927b,c [39]) axioms for quantum mechanics, which take the notion of probability amplitude as primary. Born and Heisenberg point out some formal drawbacks, such as the use of δ-functions and the presence of unobservable phases in the probability amplitudes. They note that such drawbacks have been overcome by the formulation of von Neumann (which they do not go on to discuss further). Born and Heisenberg then proceed to discuss in particular whether the statistical element in the theory can be reconciled with macroscopic determinism. To this effect, they first justify the necessity of using probabilistic notions by appeal to the notion of uncertainty (discussed using the example of diffraction of light by a single slit); then they point out that, while for instance in cases of diffraction the laws of propagation of the probabilities in quantum mechanics are very different from the classical evolution of a probability density, there are cases where the two coincide to a very good approximation; they write that this justifies the classical treatment of α- and β-particles in a cloud chamber.

Section IV discusses briefly some applications, chosen for their 'close relation to questions of principle' (p. 395). Mostly, these applications highlight the importance of the generalised formalism (introduced in Section I), by going beyond matrix mechanics proper or wave mechanics proper. The first example is that of spin, which, requiring finite matrices, is taken to be a problematic concept for wave mechanics (but not for matrix mechanics). The main example is given by the discussion of identical particles, the Pauli principle and quantum statistics. Born and Heisenberg note that the choice of whether the wave function should be symmetric or antisymmetric appears to be arbitrary. They note, moreover, that the appropriate choice arises naturally if one quantises the normal modes of a black body, leading to Bose–Einstein statistics, or, if one adopts Jordan's (1927d [54]) quantisation procedure, to Fermi–Dirac statistics.[a] Finally, Born and Heisenberg comment on Dirac's quantum electrodynamics (Dirac 1927b,c [51,52]), noting in particular that it yields the transition probabilities for spontaneous emission.

The conclusion discusses quantum mechanics as a 'closed theory' (see Section 3.4.7) and addresses the question of whether indeterminism in quantum mechanics is fundamental. In particular, Born and Heisenberg state that the assumption of indeterminism agrees with experience and that the treatment of electrodynamics will not modify this state of affairs. They conclude by noting that the existing problems concern rather the development of a fully relativistic theory of quantum electrodynamics, but that there is progress also in this direction.

The discussion is comparatively brief. Dirac describes the analogy between the matrix method and the Hamilton–Jacobi theory. Then Lorentz makes a few remarks, in particular emphasising that the freedom to choose the phases in the matrices q and p, (3.2) and (3.5) below, is not limited to a different choice of the time origin, a possibility already noted by Heisenberg (1925b [1]), but extends to arbitrary phase factors of the form $e^{i(\delta_m-\delta_n)}$, as noted by Born and Jordan (1925 [2]).[b] According to Lorentz, this fact suggests that the 'true oscillators'

[a] The general discussion returns to these issues in more detail; see for instance Dirac's critical remarks on Jordan's procedure, p. 454.

[b] This corresponds of course to the choice of a phase factor for each energy state.

are associated with the different stationary states rather than with the spectral frequencies.[1] The last few and brief contributions address the question of the phases.[a]

3.2 Writing of the report

The respective contributions by Born and by Heisenberg to the report become clear from their correspondence with Lorentz. As mentioned in Chapter 1, Lorentz had originally planned to have a report on quantum mechanics from either Born or Heisenberg, leaving to them the choice of who was to write it. In reply, Born and Heisenberg suggested that they would provide a joint report.[2] Heisenberg was going to visit Born in Göttingen in early July 1927, on which occasion they would decide on the structure of the report and divide up the work, planning to meet again in August to finish it. The report (as requested by Lorentz) would include the results by Dirac, who was in Göttingen at the time.[b] As Born explains, the report would 'above all emphasise the viewpoints that are presumably taken less into account by Schrödinger, namely the statistical conception of quantum mechanics'. Either author, or both, could orally present the report at the conference (Heisenberg, in a parallel letter, suggested that Born should be the appropriate choice, since he was the senior scientist[3]); Born and Heisenberg could also give additional explanations in English. Born did not speak French, nor (Born thought) did Heisenberg.

In a later letter,[4] Born reports that Heisenberg had sent him at the end of July the draft of introduction, conclusion and Sections III and IV, upon which Born had written Sections I and II, reworked Heisenberg's draft, and sent everything back to Heisenberg. Heisenberg in turn had made some further small changes. Due to illness and a small operation, Born had been unable to go to Munich in mid August, so that he had not seen the changes, but he assured Lorentz that he was in agreement with Heisenberg on all essential matters.[c] Born also mentions that he and Heisenberg would like to go through the text again, at least at proof stage.[d] The presentation would be split between Born (introduction and Sections I and II) and Heisenberg (Sections III and IV and conclusion). Born concludes with the

[a] As mentioned in footnote on p. 21, Bohr suggested to omit this entire discussion from the published proceedings.

[b] Dirac was in Göttingen from February 1927 to the end of June, after having spent September to February in Copenhagen. Dirac's paper on transformation theory (1927a [38]) and his paper on emission and absorption (Dirac 1927b [51]) were written in Copenhagen, while the paper on dispersion (Dirac 1927c [52]) was written in Göttingen. Compare Kragh (1990, ch. 2, pp. 37ff.).

[c] Heisenberg had sent the final text to Lorentz on 24 August.[5]

[d] The published version and the typescript show only minor discrepancies, and many of these are clearly mistakes in the published version (detailed in our endnotes to the report). This and the fact that Born and Heisenberg appear not to have spoken French suggests that there was in fact no proofreading on their part.

following words: 'It would be particularly important for us to come to an agreement with Schrödinger regarding the physical interpretation of the quantum formalisms'.

3.3 Formalism

3.3.1 Before matrix mechanics

In the old Bohr–Sommerfeld theory, electron orbits are described as classical Kepler orbits subject to additional constraints (the 'quantum conditions'), yielding discrete stationary states. Such a procedure, in Born and Heisenberg's introduction, is criticised as artificial.[a] An atom is assumed to 'jump' between its various stationary states, and energy differences between stationary states are related to the spectral frequencies via Bohr's frequency condition. Spectral frequencies and orbital frequencies, however, appear to be quite unrelated. This is the crucial 'radiation problem' of the old quantum theory (see also Section 4.4).

The BKS theory of radiation[b] includes a rather different picture of the atom, and arguably provides a solution to the radiation problem just mentioned. In the BKS theory, one associates to each atom a set of 'virtual' oscillators, with frequencies equal to the spectral frequencies. Specifically, when the atom is in the stationary state n, the oscillators that are excited are those with frequencies corresponding to transitions from the energy E_n to both lower and higher energy levels. These oscillators produce a virtual radiation field, which propagates (classically) according to Maxwell's equations and interacts (non-classically) with the virtual oscillators, in particular influencing the probabilities for induced emission and absorption in other atoms and determining the probabilities for spontaneous emission in the atom itself. The actual emission or absorption of energy is associated with the corresponding transition between stationary states. Note that an emission in one atom is not connected directly to an absorption in another, so that energy and momentum are conserved only statistically.

Several important results derived by developing correspondence arguments into translation rules from classical to quantum expressions were formulated within the context of the BKS theory, in particular Kramers' (1924) dispersion theory. Other examples were Born's (1924) perturbation formula, Kramers and Heisenberg's (1925) joint paper on dispersion and Heisenberg's (1925a) paper on polarisation of fluorescence radiation.[c] The same arguments also led to a natural reformulation of the quantum conditions independent of the old 'classical models' (Kuhn

[a] The remark is rather brief, but note that Born and Heisenberg seem to consider the introduction of photons into the classical electromagnetic theory (a corpuscular discontinuity) to be as artificial as the introduction of discrete stationary states into classical mechanics.

[b] See also the description in Section 1.3.

[c] See Darrigol (1992, pp. 224–46) and Mehra and Rechenberg (1982b, sections III.4 and III.5).

1925, Thomas 1925). As first argued by Pauli (1925), such results were entirely independent of the presumed mechanism of radiation.[a]

Indeed, after the demise of the BKS theory following the experimental verification of the conservation laws in the Bothe–Geiger experiments, these results and the corresponding techniques of symbolic translation formed the basis for Heisenberg's formulation of matrix mechanics in his famous paper of 1925 (Heisenberg 1925b [1]). Heisenberg's stroke of genius was to give up altogether the kinematical description in terms of spatial coordinates, while retaining the classical form of the equations of motion, to be solved under suitably reformulated quantum conditions (those of Kuhn and Thomas). While Heisenberg had dropped both the mechanism of radiation and the older electronic orbits, there were points of continuity with the BKS theory and with the old quantum theory; in particular, as we shall see, the new variables were at least formally related to the virtual oscillators, and the picture of quantum jumps was retained, at least for the time being.

3.3.2 Matrix mechanics

Heisenberg's original paper is rather different from the more definitive presentations of matrix mechanics: the equations are given in Newtonian (not Hamiltonian) form, energy conservation is not established to all orders, and, as is well known, Heisenberg at first had not recognised that his theory used the mathematical machinery of matrices. Like Born and Heisenberg in their report, we shall accordingly follow in this section mainly the papers by Born and Jordan (1925 [2]) and Born, Heisenberg and Jordan (1926 [4]), as well as the book by Born (1926d,e [58]), supplementing the report with more details when useful.

Kinematically, the description of 'motion' in matrix mechanics generalises that given by the set of the components of a classical Fourier series,

$$q_n e^{2\pi i n \nu t} \tag{3.1}$$

(and thus generalises the idea of a periodic motion). As frequencies, one chooses, instead of $\nu(n) = n\nu$, the *transition frequencies* of the system under consideration, which are required to obey Ritz's combination principle, $\nu_{ij} = T_i - T_j$ (relating the frequencies to the spectroscopic terms $T_i = E_i/h$, or, in Born and Heisenberg's notation, $T_i = W_i/h$). The components thus defined form a doubly infinite array:

$$q = \begin{pmatrix} q_{11}e^{2\pi i \nu_{11} t} & q_{12}e^{2\pi i \nu_{12} t} & \cdots \\ q_{21}e^{2\pi i \nu_{21} t} & q_{22}e^{2\pi i \nu_{22} t} & \cdots \\ \vdots & \vdots & \ddots \end{pmatrix}. \tag{3.2}$$

[a] Cf. Darrigol (1992, pp. 244–5). Cf. also Jordan (1927e [63]), among others.

From the combination principle for the ν_{ij}, it follows that $\nu_{ij} = -\nu_{ji}$. Further, it is required of the (generally complex) amplitudes that

$$q_{ji} = q_{ij}^*, \tag{3.3}$$

by analogy with a classical Fourier series, so that the array is in fact a Hermitian matrix.

In modern notation, the above matrix consists of the elements of the position observable (in Heisenberg picture) in the energy basis. For a closed system with Hamiltonian H, these indeed take the form

$$\begin{aligned} \langle E_i | Q(t) | E_j \rangle &= \langle E_i | e^{(i/\hbar)Ht} Q(0) e^{-(i/\hbar)Ht} | E_j \rangle \\ &= \langle E_i | Q(0) | E_j \rangle e^{(i/\hbar)(E_i - E_j)t}. \end{aligned} \tag{3.4}$$

Along with the position matrix q, one considers also a momentum matrix:

$$p = \begin{pmatrix} p_{11}e^{2\pi i\nu_{11}t} & p_{12}e^{2\pi i\nu_{12}t} & \cdots \\ p_{21}e^{2\pi i\nu_{21}t} & p_{22}e^{2\pi i\nu_{22}t} & \cdots \\ \vdots & \vdots & \ddots \end{pmatrix}, \tag{3.5}$$

as well as all other matrix quantities that can be obtained as functions of q and p, defined as polynomials or as power series (leaving aside questions of convergence). Again from the combination principle, it follows that under multiplication of two such matrices, one obtains another matrix of the same form (that is, with the same time-dependent phases). Thus one justifies taking these two-dimensional arrays as matrices. The most general physical quantity in matrix mechanics is a matrix whose elements (in modern terminology) are the elements of an arbitrary Heisenberg-picture observable in the energy basis.

The values of these matrices (in particular the amplitudes q_{ij} and the frequencies ν_{ij}) will have to be determined by solving the equations of motion under some suitable quantum conditions. Specifically, the equations of motion are postulated to be the analogues of the classical Hamiltonian equations:

$$\frac{dq}{dt} = \frac{\partial H}{\partial p}, \qquad \frac{dp}{dt} = -\frac{\partial H}{\partial q}, \tag{3.6}$$

where H is a suitable matrix function of q and p. The quantum conditions are formulated as

$$pq - qp = \frac{h}{2\pi i} \cdot 1, \tag{3.7}$$

where 1 is the identity matrix. In the limit of large quantum numbers, (3.7) corresponds to the 'old' quantum conditions.

Of course, differentiation with respect to t and with respect to matrix arguments need to be suitably defined. The former is defined elementwise as

$$\dot{f}_{ij} = 2\pi i \nu_{ij} f_{ij}, \tag{3.8}$$

which can be written equivalently as

$$\dot{f} = \frac{2\pi i}{h}(Wf - fW), \tag{3.9}$$

where W is the diagonal matrix with $W_{ij} = W_i \delta_{ij}$. Differentiation with respect to matrix arguments is defined (in slightly modernised form) as

$$\frac{df}{dx} = \lim_{\alpha \to 0} \frac{f(x + \alpha 1) - f(x)}{\alpha}, \tag{3.10}$$

where the limit is also understood elementwise.[a]

Once the preceding scheme has been set up, the next question is how to go about solving the equations of motion. First, for the special case of differentiation with respect to q or p, one shows that

$$fq - qf = \frac{h}{2\pi i}\frac{\partial f}{\partial p} \tag{3.11}$$

and

$$pf - fp = \frac{h}{2\pi i}\frac{\partial f}{\partial q} \tag{3.12}$$

(by induction on the form of f and using (3.7)). From this and from (3.9), one can easily show that the equations of motion are equivalent to

$$Wq - qW = Hq - qH \tag{3.13}$$

and

$$Wp - pW = Hp - pH. \tag{3.14}$$

Thus, $W - H$ commutes with both q and p, and therefore also with H. It follows that

$$WH - HW = 0, \tag{3.15}$$

which by (3.9) means that

$$\dot{H} = 0, \tag{3.16}$$

[a] This definition was agreed upon after some debate. See Mehra and Rechenberg (1982c, pp. 69–71 and 97–100).

that is, energy is conserved. In the non-degenerate case, it follows that H is a diagonal matrix. Denoting its diagonal terms by H_i, (3.13) yields further the Bohr frequency condition,

$$H_i - H_j = W_i - W_j = h\nu_{ij}, \tag{3.17}$$

and fixing an arbitrary constant one can set $H = W$. The above proof shows also that conversely, (3.16), (3.17) and the commutation relations imply the equations of motion. Therefore, the problem of solving the equations of motion essentially reduces to that of diagonalising the energy matrix.

This discussion leads also to the notion of a 'canonical transformation' (a transformation that leaves the equations of motion invariant) as a transformation that preserves the commutation relations. Obviously, any transformation of the form

$$Q = SqS^{-1}, \qquad P = SpS^{-1} \tag{3.18}$$

(with S invertible) will be such a transformation, and it was conjectured that these were the most general canonical transformations. At this stage, S is required only to be invertible; indeed, the notion of unitarity has not been introduced yet. If H is a suitably symmetrised function of q and p, however, it will also be a Hermitian matrix, and the problem of diagonalising it can be solved using the theory of quadratic forms of infinitely many variables (assuming that the theory as known at the time extends also to unbounded quadratic forms). In particular, it will be possible to diagonalise H using a unitary S, which also guarantees that the new coordinates Q and P will be Hermitian.

What does such a solution to the equations of motion yield? In the first place, one obtains the values of the frequencies ν_{ij} and of the energies W_i (along with the values of other conserved quantities). In the second place, based on correspondence arguments, one can identify the diagonal elements of a matrix with 'time averages' in the respective stationary states,[a] and relate the squared amplitudes $|q_{ij}|^2$ to the intensities of spontaneous emission, or equivalently to the corresponding transition probabilities between stationary states.[b] This, however, is not an argument based on first principles. Indeed, as Born and Heisenberg repeatedly stress, matrix mechanics in the above form describes a closed, conservative system, in which no change should take place. As Born and Heisenberg note in Section II, actual transfer of energy is to be expected only when the atom is coupled to some other system, specifically the radiation field. Born and Jordan's paper (1925 [2]) and Born, Heisenberg and Jordan's paper (1926 [4]) include a first attempt at treating the

[a] The concept of 'time averages' can also be thought of as related to the BKS theory, as remarked by Heisenberg himself in a letter to Einstein of February 1926.[6]

[b] The determination of intensities had been a pressing problem since the work by Born (1924) and the 'Utrecht observations' referred to in the report (p. 376). See Darrigol (1992, pp. 234–5).

radiation field quantum mechanically and at justifying the assumed interpretation of the squared amplitudes. A satisfactory treatment of spontaneous radiation was later given by Dirac (1927b [51]), as also mentioned in the report.[a] In the third place, by setting up relations between the amplitudes q_{ij} and the matrix elements of other conserved quantities such as angular momentum, one is able to identify which transitions are possible between states with the corresponding quantum numbers (derivation of 'selection rules'). Expectation values, other than in the form of time averages for the stationary states, are lacking.

It should be clear that matrix mechanics in its historical formulation is not to be identified with today's quantum mechanics in the Heisenberg picture. The matrices allowed as solutions in matrix mechanics are basis-dependent. Moreover, quite apart from the fact that solutions to the equations of motion are hard to find,[b] the results obtained are relatively modest. In particular, there are no general expectation values for physical quantities. In discussing both the mathematical formalism and the physical interpretation, Born and Heisenberg suggest that the original matrix theory is inadequate and in need of extension.

3.3.3 Formal extensions of matrix mechanics

As presented in the report, the chief formal rather than interpretational difficulty for matrix mechanics is the failure to describe *aperiodic* quantities. As we have seen, the matrices were understood as a generalisation of classical Fourier series. Strictly speaking, if one follows this analogy, a periodic quantity is represented by a discrete (if doubly infinite) matrix, and one can envisage representing aperiodic quantities by continuous matrices – analogously to the representation of classical quantities by Fourier integrals. Indeed, already Heisenberg (1925b [1]) points out that his quadratic arrays would have in general both a periodic part (discrete) and an aperiodic part (continuous). The connection between matrices and quadratic forms showed in fact that matrices in general had also a continuous spectrum (if the theory extended to the unbounded case). Still, continuous matrices are unwieldy, and in certain cases the matrix elements (which are always evaluated in the energy basis) will become singular, as when trying to describe a free particle. (This is not at all surprising, considering that also in the classical analogy the Fourier integral of the function $at + b$ fails to converge.)

This problem was addressed by generalising the notion of a matrix to that of a q-number (Dirac 1926a [7]) and to that of an operator (Born and Wiener 1926a

[a] Note that by this time Dirac (1927a [38]) had already introduced a probabilistic interpretation based on the transformation theory.

[b] In particular, one was unable to introduce action-angle variables to solve the Hamiltonian equations as in classical mechanics; cf. Born and Heisenberg's remarks on 'aperiodic motions'.

[21], 1926b). Dirac's q-numbers are abstract objects that are assumed to form a non-commutative algebra, while Born and Wiener's operators are characterised not by their elements in some basis, but by their action on a space of functions. In both approaches, one can deal with aperiodic quantities and quantities with a singular matrix representation.

Born and Wiener's approach is the lesser-known of the two, and shall therefore be briefly sketched.[a] The operators in question are linear operators acting on a space of functions $x(t)$ (which are not given any specific physical interpretation). These functions are understood as generalising the functions having the form

$$x(t) = \sum_{n=1}^{\infty} x_n e^{\frac{2\pi i}{h} W_n t}. \tag{3.19}$$

That is, they generalise the functions that can be written in a Fourier-like series with frequencies equal to the spectral frequencies. Therefore, the operators generalise the matrices that act on infinite-dimensional vectors indexed by the energy values, that is, they generalise the matrices in the energy basis. By using these operators, Born and Wiener free themselves from the need to operate with the matrix elements, and they are able to solve the equations of motion explicitly for systems more general than those treated in matrix mechanics until then. Their main example is the (aperiodic) one-dimensional free particle.

However, instead of presenting this formalism in the report, Born and Heisenberg present directly von Neumann's formalism of operators on Hilbert space (evidently but tacitly considering Born and Wiener's formalism as its natural precursor). They note that in this formalism it is possible to consider matrices in arbitrary, even continuous bases, and they use this fact to make the connection with Schrödinger's theory. In particular, they point out that solving the (time-independent) Schrödinger equation is equivalent to diagonalising the Hamiltonian quadratic form, in the sense that the set of Schrödinger eigenfunctions $\varphi(q, W)$ (wave functions in q indexed by W) yields the transformation matrix S from the position basis to the energy basis. Thus, one can use the familiar methods of partial differential equations to diagonalise H.[b]

Despite providing very useful formal extensions of matrix mechanics, the q-numbers and the operators (at least as presented in Section I of the report) do

[a] Born and Wiener published two very similar papers on their operator theory, one in German, referred to in the report (1926a [21]), and one in English (1926b).

[b] Cf. Section 3.4.5. The connection between matrix and wave mechanics was discovered independently by Schrödinger (1926d), Eckart (1926 [22]) and Pauli (Pauli to Jordan, 12 April 1926, in Pauli 1979, pp. 315–20). For a description of the various contributions, including the work of Lanczos (1926 [23]), see Mehra and Rechenberg (1987, pp. 636–84). Muller (1997) argues that matrix mechanics and wave mechanics, as formulated and understood at the time, were nevertheless inequivalent theories. For concrete examples in which the two theories were indeed considered to yield different predictions, see p. 129 (including the footnote) and the discussion on p. 426.

not provide further insights into the physical interpretation of the theory. In the following Section II, on 'Physical interpretation', Born and Heisenberg discuss the problem of describing actual states and processes in matrix mechanics, suitably extended, and the surprising ramifications of this problem.

3.4 Interpretation

Born and Heisenberg's Section II begins with the following statement (p. 383):

> The most noticeable defect of the original matrix mechanics consists in the fact that at first it appears to give information not about actual phenomena, but rather only about possible states and processes. It allows one to calculate the possible stationary states of a system; further it makes a statement about the nature of the harmonic oscillation that can manifest itself as a light wave in a quantum jump. But it says nothing about when a given state is present, or when a change is to be expected. The reason for this is clear: matrix mechanics deals only with closed periodic systems, and in these there are indeed no changes. In order to have true processes, as long as one remains in the domain of matrix mechanics, one must direct one's attention to a *part* of the system; this is no longer closed and enters into interaction with the rest of the system. The question is what matrix mechanics can tell us about this.

(This is again one of the sections originally drafted by Born.) As raised here, the question to be addressed is how to incorporate into matrix mechanics the (actual) state of a system, and the time development of such a state. The discussion given in the report may give rise to some confusion, because it arguably contains at least two, if not three, disparate approaches to what is a state in quantum mechanics. The first, reflected in the above quotation, is the idea that a state of a system is always a stationary state, which stems from Bohr's quantum theory and which appears to have lived on through the BKS phase until well into the development of matrix mechanics, indeed at least as late as Heisenberg's paper on resonance (Heisenberg 1926b [35]). The second is the idea that the state of a system is given by its wave function, but it is bound up with the question of whether the latter should be seen as a 'spread-out' entity, a 'guiding field', a 'statistical state' or something else.[a] Born's papers on collisions (Born 1926a,b [30]) can be said to contain elements of both these approaches. Yet a third approach may well be present in the report, an approach in which the notion of state would be purely an effective one. Some pronouncements by Heisenberg, in correspondence and in the uncertainty paper (Heisenberg 1927 [46]), may support this further (tentative) suggestion.

It seems to us that Born and Heisenberg's statements become clearer if one is aware of the different backgrounds to their discussion. Accordingly, we shall discuss in turn, briefly, various developments that appear to have fed into their conception of quantum mechanics, in particular previous work by Born and Wiener

[a] Cf. the next chapter.

on generalising matrix mechanics (Section 3.4.1), by Born on guiding fields (Section 3.4.2), by Bohr as well as famously by Born on collision processes (Sections 3.4.2 and 3.4.3), by Heisenberg on atoms in resonance (Section 3.4.4) and by Dirac on the transformation theory (Section 3.4.5). We shall then discuss the treatment of interpretational issues given in the report in Sections 3.4.6 and 3.4.7. The latter includes some brief comments on the notion of a 'closed theory', which is prominent in some of Heisenberg's later writings (for example, Heisenberg 1948), and which appears to be used here for the first time. We do not claim to have settled the interpretational issues, and will return to several of them in Part II. Overall, the 'physical interpretation' of the report requires careful reading and assessment.

3.4.1 Matrix mechanics, Born and Wiener

As we have seen, in the old quantum theory the only states allowed for atomic systems are stationary states, understood in terms of classical orbits subject to the quantum conditions of Bohr and Sommerfeld. The same is true in the BKS theory, where stationary states become more abstract and are represented by the collection of virtual oscillators corresponding to the transitions of the atom from a given energy level. In both theories, discontinuous transitions between the stationary states, so-called quantum jumps, are assumed to occur. Although in today's quantum theory one usually still talks about discontinuous transitions, these are associated with the collapse postulate and with the concept of a measurement. The states performing these transitions are not necessarily stationary states, that is, eigenstates of energy (whether one describes them dynamically in the Schrödinger picture or statically in the Heisenberg picture).

This modern notion of state is strikingly absent also in matrix mechanics, as we have seen it formulated in the original papers. Indeed, the interpretation of the theory is still ostensibly in terms of stationary states and quantum jumps, but the formalism itself contains only matrices, which can at most be seen as a collective representation of all stationary states, in the following sense. The matrix (3.2) incorporates the oscillations corresponding to all possible transitions of the system. Each matrix element is formally analogous to a virtual oscillator with frequency $\nu_{ij} = (E_i - E_j)/h$ (corresponding to an atomic transition $E_i \rightarrow E_j$). Therefore, each row (or column) contains all the frequencies corresponding to the transitions from (or to) a given energy level, much like a stationary state in the BKS theory. As in the above quotation, the matrix can thus be seen as the collective representation of all the stationary states of a closed system.[a]

[a] A somewhat similar idea appears to be expressed by Dirac at the beginning of the discussion after the report (p. 402), where he emphasises the parallel between matrix mechanics and classical Hamilton–Jacobi theory, with the latter also describing not single trajectories but whole families of trajectories.

This analysis is further supported by examining Born and Wiener's work, in which the picture of stationary states as the rows of the position matrix (now the position operator) becomes even more explicit. While the functions $x(t)$ remain uninterpreted, Born and Wiener appear to associate the 'rows' or, rather, the 'columns' of their operators with (stationary) states of motion. This is clear from the following: Born and Wiener introduce a notion of 'column sum', which is a generalisation of the sum of the elements in the column of a matrix. In discussing their main example, the free particle, they then show that, for the position operator, this generalised column sum $q(t, W)$ takes the form

$$q(t, W) = t\sqrt{\frac{2W}{\mu}} + F(W), \tag{3.20}$$

where μ is the mass of the particle and $F(W)$ is a complex-valued expression independent of t. They explicitly draw the conclusion that at least the real part of the generalised column sum represents a classical inertial motion with the energy W.

Note that this indicates not only that Born and Wiener associate the state of a system with a column of the position operator but also suggests that, in their view, at least a limited spatio-temporal picture of particle trajectories in the absence of interactions is possible (analogously to the earlier limited use of spatio-temporal orbits in describing atomic states in the absence of transitions).

3.4.2 Born and Jordan on guiding fields, Bohr on collisions

The early history of the guiding field idea, in connection with Einstein and with the BKS theory (in the case of photons), and in connection with de Broglie (in the case of both photons and material particles), is discussed mainly in Chapters 9 and 2, respectively. Slater's original intention was in fact to have the virtual fields to be guiding fields for the photons, which were to carry energy and momentum, but this aspect was not incorporated in the BKS theory. However, after this theory was rejected precisely because of the results on energy and momentum conservation in individual processes (detailed, for instance, in Compton's report), Slater's original idea was fleetingly revived.

On 24 April 1925, after learning from Franck that Bohr had in fact given up the BKS theory, Born sent Bohr the description of such a proposal (which, as he wrote, he had been working on for some weeks with Jordan). A manuscript by Born and Jordan followed, entitled 'Zur Strahlungstheorie', which is found today in the Bohr archive.[a] The proposal combines the BKS idea of emission of waves

[a] Cf. Darrigol (1992, p. 253). We are especially grateful to Olivier Darrigol for helpful correspondence on this matter.

while the atom is in a stationary state with the emission of a light quantum during an instantaneous quantum jump, in order to give a spacetime picture of radiation. The light quantum thus follows the rear end of the wave. It can be scattered or absorbed by other atoms (both processes depending on the dipole moment of the appropriate virtual oscillators in the atoms), in the latter case leaving a 'dead' wave that has no further physical effects. Born and Jordan had applied this picture with some success to a few simple examples, and were intending to publish the idea in *Naturwissenschaften*, provided Bohr or Kramers did not find fault with it (Bohr 1984, pp. 84–5, 308–10).

Bohr replied on 1 May, after receipt of both the letter and the manuscript, criticising the proposal, on the grounds, first, that the proposed mechanism did not guarantee that the trajectories of the light quanta would coincide with the propagation of the wave, and second, that the cross section for the absorption ought to be a constant in order for the particle number density to be proportional to the intensity of the wave. Bohr reiterated the beliefs he had expressed to Franck: that the coupling between state transitions in different atoms excluded a description that used anschaulich pictures, and that he suspected the same conclusion to be likely in the case of collision phenomena (Bohr 1984, pp. 85, 310–11).[a]

Bohr's work on collisions, to which he alluded here, was an extension of the BKS idea of merely statistical conservation laws (Bohr 1925).[b] The idea, as paraphrased by Born, was 'to regard the field of the particle passing by in the same way as the field of a light wave; thus, it only produces a probability for the absorption of energy, and this [absorption] only occurs when the "collision" lasts sufficiently long (the particle passes slowly)' (Born to Bohr, 15 January 1925, quoted in Bohr 1984, p. 73).[c] However, the Ramsauer effect – the anomalously low cross section of atoms of certain gases for slow electrons – could not be accommodated in Bohr's scheme. Bohr, therefore, was developing doubts about statistical conservation laws (and further, about the feasibility altogether of a spacetime picture of collisions[d]),

[a] See Bohr to Franck, 21 April 1925, in Bohr (1984, pp. 350–1).

[b] Cf. also Darrigol (1992, pp. 249–51) and pp. 89–93 in Stolzenburg's introduction to Part I of Bohr (1984).

[c] Another colourful paraphrase is in Bohr to Franck, 30 March 1925: 'If two atoms have the possibility of settling their mutual account, it is, of course, simplest that they do so. However, when the invoices cannot be submitted simultaneously, they must be satisfied with a running account' (quoted in Bohr 1984, p. 74).

[d] One of the most striking expressions of this is a passage in Bohr to Heisenberg, 18 April 1925: 'Stimulated especially by talks with Pauli, I am forcing myself these days with all my strength to familiarise myself with the mysticism of nature and am attempting to prepare myself for all eventualities, indeed even for the assumption of a coupling of quantum processes in separated atoms. However, the costs of this assumption are so great that they cannot be estimated within the ordinary spacetime description' (Bohr 1984, pp. 360–1). For other qualms about such 'quantum nonlocality', cf. also Jordan's habilitation lecture (1927f [62]), in which Jordan considers the idea of microscopic indeterminism to be comprehensible only if the elementary random events are independent. (This lecture was translated into English by Oppenheimer and published in *Nature* as Jordan (1927g).)

even as he was submitting his paper on collisions,[a] that is, even before the results of the Bothe–Geiger experiments were confirmed in April 1925. Bohr's paper was published only after Bohr included an addendum in July 1925, which draws the consequences from both the Bothe–Geiger experiments and the difficulties with the Ramsauer effect. In the same month of July, an explanation for the Ramsauer effect was suggested by Elsasser (1925) in Göttingen, on the basis of de Broglie's matter waves.

3.4.3 Born's collision papers

The above provides a useful backdrop for discussing Born's own work on collisions (Born 1926a,b [30]),[b] which treats collision problems on the basis of Schrödinger's wave mechanics.[c]

This work also reflects the idea that the states of a system are stationary states undergoing transitions. Indeed, Born presents the problem as that of including in matrix mechanics a description of the transitions between stationary states. The case of collisions, say between an atom and an electron, is chosen as the simplest for treating this problem (while still leading to interesting predictions), because it is natural to expect that in this case the combined system is asymptotically in a stationary state for the atom and a state of uniform translational motion for the electron. (Note that this is connected to the treatment of the free particle by Born and Wiener.) If one can manage to describe the asymptotic behaviour of the combined system mathematically, this will give concrete indications as to the transitions between the initial and final asymptotic states. Born managed to find the solution specifically by wave mechanical methods.[d]

Note that Born considers two conceptually distinct objects, the wave function on the one hand and the stationary states of the atom and the electron on the other, the connection between them being that the wave function defines a probability distribution over the stationary states. (Note also that he reserves the word 'state' only for the stationary states.)

Born solves by perturbation methods the time-independent Schrödinger equation for the combined system of atom and electron under the condition that asymptotically for $z \to \infty$, the solution has the form $\psi_n(q)e^{\frac{2\pi}{\lambda}z}$ (a product of

[a] The paper was received by *Zeitschrift für Physik* on 30 March 1925; on that very day Bohr was expressing his doubts in a letter to Franck (Bohr 1984, pp. 348–50).

[b] For a modern discussion of collisions, cf. Section 10.1.

[c] The Ramsauer effect, however, is excluded from Born's discussion (1926b [30], footnote on p. 824). A few passages might be seen to refer to Born's exchange with Bohr, in particular the remark that, at the price of dropping causality, the usual spacetime picture can be maintained (1926b [30], p. 826).

[d] Cf. Born to Schrödinger, 16 May 1927: 'the simple possibility of treating with it aperiodic processes (collisions) made me first believe that your conception was superior' (quoted in Mehra and Rechenberg 2000, p. 135).

the n-th eigenstate of the atom with a plane wave coming from the z-direction), with energy τ. Born's solution has the form:

$$\sum_m \int_{\alpha x+\beta y+\gamma z>0} \Phi^{\tau}_{nm}(\alpha, \beta, \gamma)\psi_m(q)e^{k^{\tau}_{nm}(\alpha x+\beta y+\gamma z+\delta)}d\omega, \tag{3.21}$$

where the energy corresponding to the wave number k^{τ}_{nm} equals

$$E^{\tau}_{nm} = h\nu_{nm} + \tau, \tag{3.22}$$

the ν_{nm} being the transition frequencies of the atom. The components of the superposition can thus be associated with various, generally inelastic, collisions in which energy is conserved, and Born interprets $|\Phi^{\lambda}_{nm}(\alpha, \beta, \gamma)|^2$ as the probability for the atom to be in the stationary state m and the electron to be scattered in the direction (α, β, γ).[a]

But now, crucially, since the initial wave function corresponds to a fully determined stationary state and inertial motion, this probability is also the probability for a quantum jump from the given initial state to the given final state, that is, a transition probability.

This idea is linked to that of a guiding field. The link is made explicitly at the beginning of Born's second paper (which includes the details of the derivation and some quantitative predictions). While Born judges that in the context of optics one ought to wait until the development of a proper quantum electrodynamics, in the context of the quantum mechanics of material particles the guiding field idea can be applied already, using the de Broglie–Schrödinger waves as guiding fields. The trajectories of material particles, however, are determined by the guiding field merely probabilistically (1926b [30], pp. 803–4). In his conclusion, Born regards the picture of the guiding field as fundamentally indeterministic. A deterministic completion, if possible, would not have any practical use. Born also expresses the hope that the 'laws of motion for light quanta' will find a similar treatment to the one given for electrons and refers to the difficulties 'so far' of pursuing a guiding field approach in optics (pp. 826–7).[b]

[a] A statistical interpretation for the modulus squared of the coefficients of the wave function in the energy basis was introduced also by Dirac (1926c [37]), at the same time as and presumably independently of Born (cf. Darrigol 1992, p. 333). Note further that, even though it may be tempting to assume that each trajectory proceeds from the scattering centre, strictly speaking to each stationary state of the free particle corresponds a whole family of inertial trajectories. (Similarly, in the case treated by Born and Wiener, each generalised column sum associated with a stationary state corresponds to two different inertial trajectories, depending on the sign of the square root in (3.20).)

[b] Born's views on quantum mechanics from this period are also presented in Born (1927), an expanded version (published March 1927) of a talk given by Born in August 1926.

3.4.4 Heisenberg on energy fluctuations

As discussed in the next chapter, Heisenberg in particular among matrix physicists was opposed to Schrödinger's attempt to recast and reinterpret quantum theory on the basis of continuous wave functions. Schrödinger's wave functions were meant from the start as descriptions of individual states of a physical system. Even though in general they are abstract functions (on configuration space), they can provide a picture of the stationary states. Furthermore, the solution of the time-dependent Schrödinger equation appears to provide a generalisation of the state of a system as it evolves in time.

Heisenberg appears to have been disturbed initially also by Born's use of Schrödinger's theory in the treatment of collisions, an attitude reflected in particular in his correspondence with Pauli.[a] In this connection, the fact that Pauli – in his letter of 19 October (Pauli 1979, pp. 340–9) – was in effect able to sketch how one could reinterpret Born's results in terms of matrix elements, must have been of particular significance: 'Your calculations have given me again great hope, because they show that Born's somewhat dogmatic viewpoint of the probability waves is only one of many possible schemes'.[b]

A few days later, Heisenberg sent Pauli the manuscript of his paper on fluctuation phenomena (Heisenberg 1926b [35]), in which he developed considerations similar to Pauli's in the context of a characteristic example, that of two atoms in resonance. By focussing on the subsystems of a closed system, Heisenberg was able to derive expressions for (transition) probabilities within matrix mechanics proper, without having to introduce the wave function as an external aid. A very similar result was derived at the same time by Jordan (1927a [36]), using two systems with a single energy difference in common.

We shall now sketch Heisenberg's reasoning. We adapt the presentation of the argument given by Heisenberg in *The Physical Principles of the Quantum Theory* (Heisenberg 1930b, pp. 142–7),[c] which is more general than the one given in the paper and clearer than the one given in Born and Heisenberg's report.

Take two systems, 1 and 2, that are in resonance. Consider, to begin with, that the frequency of the transition $n_1 \to m_1$ in system 1 corresponds to exactly one transition frequency in system 2, say

$$\nu^{(1)}(n_1 m_1) = \nu^{(2)}(n_2 m_2). \tag{3.23}$$

[a] 'One sentence [of Born's paper] reminded me vividly of a chapter in the Christian creed: "An electron *is* a plane wave ... "' (Heisenberg to Pauli, 28 July 1926, in Pauli 1979, p. 338, original emphasis).
[b] Heisenberg to Pauli, 28 October 1926, in Pauli (1979, p. 350).
[c] This very remarkable book is an expanded English edition of Heisenberg (1930a).

If the systems are uncoupled, the combined system has the degenerate eigenvalue of energy

$$W_{n_1}^{(1)} + W_{m_2}^{(2)} = W_{m_1}^{(1)} + W_{n_2}^{(2)}. \tag{3.24}$$

If we couple weakly the two systems, the degeneracy will be lifted. Let us label the new eigenstates of the combined system by a and b. We can now consider the matrix S that transforms the basis of eigenstates of energy of the coupled system to the (product) basis of eigenstates of energy of the uncoupled systems. In particular, we can consider the submatrix

$$\begin{pmatrix} S_{n_1m_2,a} & S_{n_1m_2,b} \\ S_{n_2m_1,a} & S_{n_2m_1,b} \end{pmatrix}. \tag{3.25}$$

Choose one of the stationary states of the combined system, say a. If the combined system is in the state a, what can one say about the energy of the subsystems, for instance $H^{(1)}$? Heisenberg's answer, in the terminology and notation of his book (1930b), is that in the state a, the time average of $H^{(1)}$ (which is no longer a diagonal matrix, thus no longer time-independent), or of any function $f(H^{(1)})$, is

$$\overline{f(H^{(1)})} = f(W_{n_1}^{(1)})|S_{n_1m_2,a}|^2 + f(W_{m_1}^{(1)})|S_{m_1n_2,a}|^2. \tag{3.26}$$

Since f is arbitrary, Heisenberg concludes that $|S_{n_1m_2,a}|^2$ is the probability that the state n_1 has remained the same (and the state m_2 has remained the same), and $|S_{m_1n_2,a}|^2$ is the probability that the state n_1 has jumped to the state m_1 (and m_2 has jumped to n_2).

The associated transfer of energy between the systems appears to be instantaneous, in that a quantum jump in one system (from a higher to a lower energy level) is accompanied by a corresponding jump in the other system (from a lower to a higher energy level). The paper merely mentions, without elaborating further, that a light quantum (or better a 'sound quantum') is exchanged over and over again between the two systems. The picture thus avoids the non-conservation of energy of the BKS theory, at the price of what appears to be an explicit correlation at a distance.

In modern terms, Heisenberg has calculated the expectation value of the observable $f(H^{(1)}) \otimes 1$ in the state a:

$$\begin{aligned} &\langle a|f(H^{(1)}) \otimes 1|a\rangle \\ &\quad = \langle a\,|n_1m_2\rangle\langle n_1m_2|\,f(H^{(1)}) \otimes 1\,|n_1m_2\rangle\langle n_1m_2|\,a\rangle \\ &\qquad + \langle a\,|m_1n_2\rangle\langle m_1n_2|\,f(H^{(1)}) \otimes 1\,|m_1n_2\rangle\langle m_1n_2|\,a\rangle \\ &\quad = \langle n_1|f(H^{(1)})|n_1\rangle|\langle n_1m_2|a\rangle|^2 \\ &\qquad + \langle m_1|f(H^{(1)})|m_1\rangle|\langle m_1n_2|a\rangle|^2. \end{aligned} \tag{3.27}$$

Note, however, that rather than regarding

$$|S_{n_1m_2,a}|^2 = |\langle n_1m_2|a\rangle|^2 \tag{3.28}$$

as a conditional probability (what we would today call the transition probability between n_1m_2 and a), Heisenberg relates it to transitions between the stationary states of the subsystems, as in Born's work on collisions.

3.4.5 Transformation theory

Less than three weeks after completing his draft on fluctuation phenomena, we find Heisenberg reporting to Pauli about Dirac's transformation theory, which generalises precisely the formal expression of a conditional probability given by Heisenberg in terms of the transformation matrix: 'Here [in Copenhagen] we have also been thinking more about the question of the meaning of the transformation function S and *Dirac* has achieved an extraordinarily broad generalisation of this assumption from my note on fluctuations' (Pauli 1979, p. 357, original emphasis).

Dirac indeed presents his results in his paper, significantly titled 'The physical interpretation of quantum dynamics' (1927a [38]), as a generalisation of Heisenberg's approach. The main goal of the paper is the following.[a]

Take any pair of conjugate matrix quantities ξ and η, and any 'constant of integration' $g(\xi, \eta)$.[b] Given a value of ξ as a c-number, find the fraction of η-space for which g lies between any two numerical values. If η is assumed to be distributed uniformly,[c] this result will yield the frequency of the given values of g in an ensemble of systems. Equivalently, we can state Dirac's goal as that of finding the expectation value (or more precisely, the η-average) of any fixed-time observable g, given a certain value of ξ.

The main part of the paper is devoted to developing a 'transformation theory' that will allow Dirac to write the quantity g not in the usual energy representation but in an arbitrary ξ-representation. Given a c-number value ξ' for ξ, Dirac then suggests taking the η-averaged value of g to be the diagonal element $g_{\xi'\xi'}$ of g in this representation. This is in fact a natural if 'extremely broad' generalisation, to an arbitrary pair of conjugate quantities (ξ, η), of the assumption that the diagonal elements of a matrix in the energy representation (such as Heisenberg's $f(H^{(1)})$ from the previous section) are time averages, although the justification Dirac gives for this assumption is merely that 'the diagonal elements ... certainly would

[a] In our presentation we shall partly follow the analysis by Darrigol (1992, pp. 337–45).
[b] This term is meant to include any value of a dynamical quantity at a specified time $t = t_0$ (Dirac 1927 [38], p. 623, footnote).
[c] Cf. Dirac's remarks on interpretation below.

[determine the average values] in the limiting case of large quantum numbers' (Dirac 1927a [38], p. 637).

Dirac writes the elements of a general transformation matrix between two complete sets of variables ξ and α (whether discrete or continuous) as (ξ'/α') (what we would now write $\langle\xi'|\alpha'\rangle$), so that the matrix elements transform as

$$g_{\xi'\xi''} = \int (\xi'/\alpha') g_{\alpha'\alpha''} (\alpha''/\xi'') d\alpha' d\alpha'' \tag{3.29}$$

(Dirac's notation is meant to include the possibility of discrete sums).

His main analytic tool is the manipulation of δ-functions and their derivatives. Dirac shows in particular that for the quantity ξ itself,

$$\xi_{\xi'\xi''} = \xi'\delta(\xi' - \xi''), \tag{3.30}$$

and that for the quantity η canonically conjugate to ξ,

$$\eta_{\xi'\xi''} = -i\hbar\delta'(\xi' - \xi''). \tag{3.31}$$

In a mixed representation, one has

$$\xi_{\xi'\alpha'} = \xi'(\xi'/\alpha'), \tag{3.32}$$

and

$$\eta_{\xi'\alpha'} = -i\hbar \frac{\partial}{\partial \xi'}(\xi'/\alpha'), \tag{3.33}$$

from which follows

$$g_{\xi'\alpha'} = g\left(\xi, -i\hbar\frac{\partial}{\partial \xi'}\right)(\xi'/\alpha') \tag{3.34}$$

for arbitrary $g = g(\xi, \eta)$.

Choosing α such that g is diagonal in the α-representation, one has

$$g_{\xi'\alpha'} = g_{\alpha'}(\xi'/\alpha'), \tag{3.35}$$

where the $g_{\alpha'}$ are the eigenvalues of g. Therefore, (3.34) becomes a differential equation for the (ξ'/α') (seen as functions of ξ'), which generalises the time-independent Schrödinger equation.[a]

Once this equation is solved, one could obtain the desired $g_{\xi'\xi'}$ from the $g_{\xi'\alpha'}$ by the appropriate transformation. The way Dirac states his final result, however, is by considering the matrix $\delta(g - g')$. The numerical function $\delta(g - g')$, when integrated,

$$\int_a^b \delta(g - g')dg', \tag{3.36}$$

[a] Dirac also gives a generalisation of the time-dependent Schrödinger equation.

yields the characteristic function of the set $a < g < b$. Therefore, in Dirac's proposed interpretation, the diagonal elements of the corresponding matrix in the ξ-representation yield, for each value $\xi = \xi'$, the fraction of the η-space for which $a < g < b$. That is, if η is assumed to be distributed uniformly, these diagonal elements yield the conditional probability for $a < g < b$ given $\xi = \xi'$. Thus, the diagonal elements of the matrix $\delta(g - g')$ yield the corresponding conditional probability density for g given $\xi = \xi'$. But now, for example, since for any function $f(g)$ one has

$$f(g)_{\xi'\xi''} = \int (\xi'/g'')f(g'')(g''/\xi'')dg'', \tag{3.37}$$

one has in particular

$$\delta(g - g')_{\xi'\xi'} = \int (\xi'/g'')\delta(g'' - g')(g''/\xi')dg'' = (\xi'/g')(g'/\xi'). \tag{3.38}$$

Therefore, the conditional probability density for g given $\xi = \xi'$ is equal to $|(g'/\xi')|^2$, a result that Dirac illustrates by discussing Heisenberg's example of transition probabilities in resonant atoms and Born's collision problem.

In parallel with Dirac's development of transformation theory, Jordan (1927b,c [39]) also arrived at a similar theory, following on directly from his paper on quantum jumps (1927a [36]) and from his earlier work on canonical transformations (1926a,b), to which Dirac also makes an explicit connection. Although Born and Heisenberg state in the report (p. 392) that the two methods are equivalent, Darrigol (1992, pp. 343–4) points to some subtle differences, which are also related to Dirac's criticism in the general discussion of Jordan's introduction of anticommuting fields (p. 454). Dirac also notes that his theory generalises the work by Lanczos (1926 [23]). The development of transformation theory from the idea of canonical transformations led further towards the realisation that quantum mechanical operators act on a Hilbert space and that the natural transformations are in fact unitary.[a]

Dirac concludes his paper with an intriguing suggestion of

> . . . a point of view for regarding quantum phenomena rather different from the usual ones. One can suppose that the initial state of a system determines definitely the state of the system at any subsequent time. If, however, one describes the state of the system at an arbitrary time by giving numerical values to the co-ordinates and momenta, then one cannot actually set up a one-one correspondence between the values of these co-ordinates and

[a] See Lacki (2004), who gives details also of London's (1926a,b) contributions to this development. Note also the connection between transformation theory and the work on the 'equivalence' between matrix mechanics and wave mechanics.

momenta initially and their values at a subsequent time. All the same one can obtain a good deal of information (of the nature of averages) about the values at the subsequent time considered as functions of the initial values. The notion of probabilities does not enter into the ultimate description of mechanical processes: only when one is given some information that involves a probability (*e.g.*, that all points in η-space are equally probable for representing the system) can one deduce results that involve probabilities.

(Dirac 1927a [38], p. 641)

Here Dirac does not impute indeterminism to nature itself (the matrix equations after all are deterministic), but instead apparently identifies the source of the statistical element in the choice of probabilistic initial data.[a]

According to Heisenberg, however, and despite the generality of the results and the 'extraordinary progress' obtained (Pauli 1979, p. 358), Dirac's transformation theory did not resolve the question of the meaning of quantum mechanics. As Darrigol (1992, p. 344) emphasises, there is no notion of state vector in Dirac's paper (the well-known bras and kets do not appear yet). C-number values, and probability distributions over c-number values, now refer to arbitrary quantities or pairs of quantities. As Heisenberg wrote: 'there are too many c-numbers in all our utterances used to describe a fact' (Pauli 1979, p. 359). Crucially, however, the energy variable and the stationary states no longer played a privileged role.

3.4.6 Development of the 'statistical view' in the report

In the report, Born and Heisenberg appear to understand Born's collision papers (1926a,b [30]) on the one hand and the papers by Heisenberg (1926b [35]) and Jordan (1927a [36]) on the other broadly in the same way, as seeking to obtain 'information . . . about actual phenomena', by 'direct[ing] one's attention to a *part* of the system' (p. 383 of the report). And indeed, by considering coupled systems all of these papers manage to derive quantitative expressions for the *probabilities* of quantum jumps between energy eigenstates.

Heisenberg's setting is the one chosen in the report, and since Heisenberg's treatment of interacting systems does not use the formalism of wave mechanics, this choice may be intended to make the point that matrix mechanics can indeed account for time-dependent phenomena without the aid of wave mechanics.

[a] This should be compared to the general discussion, in which Dirac (a) talks of 'an irrevocable choice of nature' (p. 447) in relation to the outcomes of an experiment, (b) uses explicitly the notion of the state vector, when he affirms that '[a]ccording to quantum mechanics the state of the world at any time is describable by a wave function ψ, which normally varies according to a causal law, so that its initial value determines its value at any later time' (p. 447), and in which (c) he describes the initial data taken for quantum mechanical calculations as describing 'acts of free will', namely 'the disturbances that an experimenter applies to a system to observe it' (p. 446); see also Section 8.2.

The form of Heisenberg's result (3.26) as given in the report is in terms of the expected deviation of the value of energy from a given initial value, for instance n_1:[a]

$$\begin{aligned} \delta \overline{f_{n_1}} =& \overline{f(H^{(1)})} - f(W_{n_1}^{(1)}) \\ =& \{f(W_{n_1}^{(1)}) - f(W_{n_1}^{(1)})\}|S_{n_1 m_2, a}|^2 \\ &+ \{f(W_{m_1}^{(1)}) - f(W_{n_1}^{(1)})\}|S_{m_1 n_2, a}|^2. \end{aligned} \tag{3.39}$$

If we write $\Phi_{n_1 m_1}$ for the probability of the transition $n_1 \rightarrow m_1$ in system 1, this becomes equation (20) of the report, except that Born and Heisenberg label the matrix elements $S_{n_1 m_2, a}$ and $S_{m_1 n_2, a}$, respectively, by the transitions $n_1 m_2 \rightarrow n_1 m_2$ and $n_1 m_2 \rightarrow m_1 n_2$, that is, as $S_{n_1 m_2, n_1 m_2}$ and $S_{n_1 m_2, m_1 n_2}$, omitting reference to the state a.[b] One further difference between our description above and the one given in the report is that Born and Heisenberg are treating the case in which the transition $n_1 \rightarrow m_1$ in system 1 may resonate with more than one transition in system 2. In this case, the total transition probability $\Phi_{n_1 m_1}$ is no longer equal to $|S_{n_1 m_2, m_1 n_2}|^2$ but to

$$\Phi_{n_1 m_1} = \sum_{m_2 n_2} |S_{n_1 m_2, m_1 n_2}|^2. \tag{3.40}$$

Thus, we also have equation (21) of the report.

It is only after this matrix mechanical discussion that Born and Heisenberg introduce the time-dependent Schrödinger equation, as a more 'convenient' formalism for 'thinking of the system under consideration as coupled to another one and neglecting the reaction on the latter' (p. 385). This suggests that Born and Heisenberg may consider the time-dependent Schrödinger equation only as an effective description. This impression is reinforced by comparing with Heisenberg's book (1930b, pp. 148–50) where, after the above derivation, Heisenberg continues with a more general derivation of time-dependent probabilities, which he then relates to the usual time-dependent Schrödinger equation. Nevertheless, Born and Heisenberg use the wave function throughout the ensuing discussion of probabilities, noting that this formalism 'leads to a further development of the statistical view', by which they mean in particular the idea of interference of probabilities.[c]

First of all, Born and Heisenberg relate the wave function to probabilities. They take the time-dependent transformation matrix $S(t)$ given by the unitary evolution.

[a] Note that $|S_{n_1 m_2, a}|^2 + |S_{m_1 n_2, a}|^2 = 1$, because a is normalised.

[b] In Heisenberg's paper, the matrix (3.25) corresponds to a 45°-rotation, so the probabilities are independent of the choice of a or b.

[c] On these matters, see also Section 8.1.

For the coefficients of the wave function in the energy basis one has (equation (25) in the report):

$$c_n(t) = \sum_m S_{nm}(t)c_m(0). \tag{3.41}$$

If now all $c_m(0)$ except one (say, $c_k(0)$) are zero, from the assumption that a system is always in a stationary state it is natural to conclude that the $|S_{nk}(t)|^2$ are the probabilities for transitions to the respective energy states ('transition probabilities'), and the $|c_n(t)|^2$ are the resulting probabilities for the stationary states ('state probabilities'). In reference to this interpretation, the report quotes Born's paper on the adiabatic theorem (1926c [34]), which contains a more complete version of Born's interpretational views than the collision papers. In particular, the report quotes Born's proof that in the adiabatic case the transition probabilities between different states tend to zero, in accordance with Ehrenfest's (1917) principle.[a]

Then Born and Heisenberg come to discussing interference.[b] This is also the first time in the presentation of the physical interpretation of the theory that *measurements* enter the picture. Born and Heisenberg note that if c_k is not the only non-zero coefficient at $t = 0$, then (3.41) does not imply

$$|c_n(t)|^2 = \sum_m |S_{nm}(t)|^2 |c_m(0)|^2, \tag{3.42}$$

but that instead one has

$$|c_n(t)|^2 = |\sum_m S_{nm}(t)c_m(0)|^2. \tag{3.43}$$

Up to this point, Born and Heisenberg have more or less followed Born's presentation in the adiabatic paper (Born 1926c [34]). But now, the passage immediately following this is both remarkable and, in our opinion, very significant (p. 387):[c]

it should be noted that this 'interference' does not represent a contradiction with the rules of the probability calculus, that is, with the assumption that the $|S_{nk}|^2$ are quite usual probabilities. In fact, the composition rule [(3.42)] follows from the concept of

[a] Ehrenfest had the idea that since quantised variables cannot change by arbitrarily small amounts, they should remain constant under adiabatic perturbations. This led him to formulate the principle stating that the classical variables to be quantised are the adiabatic invariants of the system. Cf. Born (1969, pp. 113ff.).

[b] Born and Heisenberg give credit to Pauli for the notion of interference of probabilities (p. 387). Note that Pauli contributed significantly to the development of the 'statistical view', albeit mainly in correspondence and discussion. Contrary to what is commonly assumed, the idea of a probability density for position is not contained in Born's collision papers, but appears in fact in Pauli's letter to Heisenberg of 19 October 1926 (Pauli 1979, p. 340–9), together with the idea of the corresponding momentum density, and in print in a footnote of Pauli's paper on gas degeneracy and paramagnetism (1927 [44]). (See also Heisenberg to Pauli, 28 October 1926, in Pauli 1979, pp. 340–52.) Jordan (1927b [39]), in his second paper on the transformation theory, even gives credit to Pauli for the introduction of arbitrary transition probabilities and amplitudes.

[c] For further discussion, see Section 6.1.2.

probability for the problem treated here when and only when the relative number, that is, the probability $|c_n|^2$ of the atoms in the state n, has been *established* beforehand *experimentally*. In this case the phases γ_n are unknown in principle, so that [(3.43)] then naturally goes over to [(3.42)] . . ..

(The passage then gives a reference to Heisenberg's uncertainty paper.) How do Born and Heisenberg propose to resolve this apparent contradiction?

It would make sense to say that the $|S_{nk}(t)|^2$ cannot be taken in general as probabilities for quantum jumps, because the derivation of $|S_{nk}(t)|^2$ as a transition probability works only in a special case (that is, presumably, if the energy at $t = 0$ has in fact been measured). On this reading, there might conceivably exist some quite different transition probabilities, which lie outside the scope of quantum mechanics and are presumably of no practical value (like a deterministic completion in the case of collision processes).[a]

However, this reading does not seem to fit what Born and Heisenberg actually say. Their suggestion seems to be that the $|S_{nk}(t)|^2$ are indeed always transition probabilities, but that the $|c_n|^2$ are *not* always state probabilities: the $|c_n|^2$ will be state probabilities if and only if the energies have been measured (non-selectively). This seems analogous to Heisenberg's (1927 [46], p. 197) idea in the uncertainty paper that the 'law of causality' is inapplicable because it is impossible in principle to know the present with sufficient accuracy (that is, the antecedent of the law of causality fails).[b]

Born and Heisenberg swiftly move on to generalising the discussion to the case of arbitrary observables, on the basis of Dirac's and Jordan's transformation theory (Dirac 1927a [38], Jordan 1927b,c [39]). They introduce the interpretation of $|\varphi|^2$ as a position density, and consider in particular the density $|\varphi(q', W')|^2$ defined by the stationary state $\varphi(q', W')$ with the energy W', or in modern notation,

$$|\varphi(q', W')|^2 = |\varphi_{W'}(q')|^2 = |\langle q'|W'\rangle|^2. \tag{3.44}$$

This is immediately generalised to arbitrary pairs of observables q and Q with values q' and Q':

$$|\varphi(q', Q')|^2 = |\langle q'|Q'\rangle|^2. \tag{3.45}$$

Born and Heisenberg call this a 'relative state probability', reserving the term 'transition probability' for the case of a single observable evolving in time (or

[a] That such probabilities can be defined (albeit non-uniquely), leading to well-defined stochastic processes for the quantum jumps, was shown explicitly by Bell (1984).

[b] It appears not to be well known that the last 20 pages of the original typescript of Heisenberg's uncertainty paper are contained in AHQP, miscatalogued as an 'incomplete and unpublished paper (pp. 12–31)' by Kramers.[7] The typescript contains slight textual variants (as compared with the published version) and manuscript corrections in what appears to be Heisenberg's hand, but does not include the famous addendum in proof in response to Bohr's criticism (cf. p. 132); cf. Bacciagaluppi (2008b). [Added in proof: currently (summer of 2008), the original of the typescript is in the Noord-Hollands Archief, Haarlem, among other materials relating to Kramers.] On Heisenberg's treatment of the 'law of causality', see also Beller (1999, pp. 110–13).

depending on some external parameter), always with the proviso that in general one should expect interference of the corresponding 'transition amplitudes'.

Note that the physical interpretation of this generalisation makes sense only if one takes over from the above the idea that probabilities such as $|\varphi|^2$ are well-defined only upon measurement, or more precisely, that actual frequencies upon measurement will be given by the expression $|\varphi|^2$. Born and Heisenberg's terminology, however, is somewhat ambiguous.[a]

A major conceptual shift appears to be taking place, which may be easy to miss. Do quantum jumps still occur whenever two systems interact, or do they now occur only between measurements? Indeed, are systems always in stationary states, as has been explicitly assumed until now, or only when the energy is measured? Heisenberg's uncertainty paper (on pp. 190–1), as well as the correspondence with Pauli (Heisenberg to Pauli, 23 February 1927, in Pauli 1979, pp. 376–82) both mention explicitly the loss of a privileged status for stationary states. It seems that, even though we are not explicitly told so, the picture of quantum jumps (that is, of probabilistic transitions between possessed values of energy) is shifting to that of probabilistic transitions from one measurement to the next.

The idea that frequencies are well-defined only upon measurement appears to play the same role as von Neumann's collapse postulate. As discussed in more detail in Chapter 6, however, it is far from clear whether that is what Born and Heisenberg have in mind. A notion similar to collapse is introduced in Heisenberg's uncertainty paper (1927 [46], p. 186), but in terms that are 'somewhat mystical' (Pauli to Bohr, 17 October 1927, in Pauli 1979, p. 411). The postulate appears explicitly in the proceedings only in the general discussion, in Born's main contribution (p. 437) and in the intriguing exchange between Dirac and Heisenberg (pp. 446ff.). In Chapter 6 and Section 8.1, we shall return to Born and Heisenberg's view of interference and to the question of whether, according to them, the collapse of the wave function and the time-dependent Schrödinger evolution are at all fundamental processes.[b]

3.4.7 *Justification and overall conclusions*

The following Section III (drafted by Heisenberg) presents Jordan's axiomatic formulation of quantum mechanics (Jordan 1927b,c [39]),[c] and justifies the

[a] In one paragraph they refer to 'the probability that for given energy W' the coordinate q' *is* in some given element dq'', whereas in the next they refer to 'the probability, given q', to *find* the value of Q' in dQ'' (p. 389f.; italics added).

[b] For a more radical view of the statistical interpretation as proposed by Born and Heisenberg, see Bacciagaluppi (2008a).

[c] This formulation, which is how Jordan presents his transformation theory, was explicitly intended as a generalisation of the formalisms of matrix mechanics, wave mechanics, q-number theory and of Born and Wiener's original operator formalism.

necessity of a statistical view in the context of Heisenberg's notion of uncertainty (Heisenberg 1927 [46]). Born and Heisenberg argue as follows. Even in classical mechanics, if certain quantities (for instance the phases of the motion) were known only with a certain imprecision, the future evolution of the system would be only statistically constrained. Now, the uncertainty relations prevent one from determining the values of all physical quantities, providing a fundamental limit of precision. In addition, quantum mechanics prescribes different laws for the time evolution of the statistical constraints. Imprecise initial conditions can be described by choosing certain 'probability functions' (this is the closest Born and Heisenberg come to discussing the 'reduction of the wave packet' as presented in the uncertainty paper), and 'the quantum mechanical laws determine the change (wave-like propagation) of these probability functions' (p. 394). Born and Heisenberg claim that discussion of the cases in which these laws coincide to a very good approximation with the classical evolution of a probability density justifies the classical treatment of α- and β-particle trajectories in a cloud chamber.[a] They thus maintain that the statistical element in the theory can be reconciled with macroscopic determinism.[b]

Section III arguably addresses the dual task set in the introduction of ensuring that quantum mechanics is 'consistent in itself' and of showing that quantum mechanics can be taken to 'predict unambiguously the results for all experiments conceivable in its domain' (p. 372f.). This task appears to be related to two conceptual *desiderata*, that the theory be intuitive (anschaulich) and closed (abgeschlossen). These are touched upon briefly in the report, especially in the introduction and conclusion, but are important both in the debate with Schrödinger (see Section 4.6) and in some of Heisenberg's later writings (in particular, Heisenberg 1948). The report appears to be the first instance in which Heisenberg uses the concept of a closed theory.[c]

As defined in the report, a closed theory is one that has achieved a definitive form, and is no longer liable to modification, either in its mathematical formulation or in its physical meaning. This is made more precise in later presentations (for

[a] Born and Heisenberg's remarks about different laws of propagation of the probabilities may refer to the conditions under which (3.43) reduces to (3.42), which would arguably be an early example of decoherence considerations. However, the remark is too brief, and Born and Heisenberg may be merely comparing the spreading of the quantum probabilities with that of a Liouville distribution. For a modern treatment of the latter comparison, see Ballentine (2003).

[b] For further discussion of these issues, see Sections 6.2.1 and 6.4.

[c] Compare also Scheibe (1993) on the concept of closed theories in Heisenberg's thought. The origin of this concept has also been traced to two earlier papers (Heisenberg 1926a [28] and 1926c [60]); see for instance Chevalley (1988). However, in the first paper there is no mention of closed theories, only of closed systems of terms (symmetric and antisymmetric), a point also repeated in the second paper. In the latter, Heisenberg mentions also the need to introduce equations for the matrix variables in order to obtain a 'closed theory', but this does not seem to be the same use of the term as in the report. We wish to thank Gregor Schiemann for correspondence and references on this topic.

example, Heisenberg 1948), in which Heisenberg includes the applicability of the concepts of a theory in the analysis. All closed theories possess a specific domain of application, within which they are and will always remain correct. Indeed, their concepts always remain part of the scientific language and are constitutive of our physical understanding of the world. In the report, quantum mechanics (without the inclusion of electrodynamics) is indeed taken to be a closed theory,[a] so that different assumptions about the physical meaning of quantum mechanics (such as Schrödinger's idea of taking $|\varphi|^2$ to be a charge density[b]), would lead to contradictions with experience. Thus, the report ends on a note of utmost confidence.

[a] Heisenberg (1948) lists Newtonian mechanics, Maxwellian electrodynamics and special-relativistic physics, thermodynamics, and non-relativistic quantum mechanics as the four main examples of such theories.

[b] Described in more detail in Section 4.4.

Archival notes

1 Cf. also Lorentz to Ehrenfest, 4 July 1927, AHQP-EHR-23 (in Dutch).
2 This and the following remarks are based on Born to Lorentz, 23 June 1927, AHQP-LTZ-11 (in German).
3 Heisenberg to Lorentz, 23 June 1927, AHQP-LTZ-12 (in German).
4 Born to Lorentz, 29 August 1927, AHQP-LTZ-11 (in German).
5 Heisenberg to Lorentz, 24 August 1927, AHQP-LTZ-12 (in German).
6 Heisenberg to Einstein, 18 February 1926, AEA 12-172.00 (in German).
7 AHQP-28 (H. A. Kramers, notes and drafts 1926–52), section 6.

4

Schrödinger's wave mechanics

Schrödinger's work on wave mechanics in 1926 appears to have been driven by the idea that one could give a purely wave-theoretical description of matter. Key elements in this picture were the idea of particles as wave packets (Section 4.3) and the possible implications for the problem of radiation (Section 4.4). This pure wave theory, in contrast to de Broglie's work, did away with the idea of particle trajectories altogether (Section 4.5). The main conflict, however, was between Schrödinger and the proponents of quantum mechanics (in particular Heisenberg, Section 4.6), both in its form at the time of Schrödinger's papers and in its further developments as sketched in the previous chapter.

For reference, we provide a brief chronology of Schrödinger's writings relating to wave mechanics up to the Solvay conference:

- Paper on Einstein's gas theory, submitted 15 December 1925, published 1 March 1926 (Schrödinger 1926a).
- First paper on quantisation, submitted 27 January 1926, addendum in proof 28 February 1926, published 13 March 1926 (Schrödinger 1926b).
- Second paper on quantisation, submitted 23 February 1926, published 6 April 1926 (Schrödinger 1926c).
- Paper on the relation between wave and matrix mechanics ('equivalence paper'), submitted 18 March 1926, published 4 May 1926 (Schrödinger 1926d).
- Paper on micro- and macromechanics (coherent states for the harmonic oscillator), published 9 July 1926 (Schrödinger 1926e).
- Third paper on quantisation, submitted 10 May 1926, published 13 July 1926 (Schrödinger 1926f).
- Fourth paper on quantisation, submitted 21 June 1926, published 5 September 1926 (Schrödinger 1926g).
- Review paper in English for the *Physical Review*, submitted 3 September 1926, published December 1926 (Schrödinger 1926h).

- Preface to the first edition of *Abhandlungen zur Wellenmechanik*, dated November 1926 (Schrödinger 1926i).
- Paper on the Compton effect in wave mechanics, submitted 30 November 1926, published 10 January 1927 (Schrödinger 1927a).
- Paper on the energy-momentum tensor, submitted 10 December 1926, published 10 January 1927 (Schrödinger 1927b).
- Paper on energy exchange in wave mechanics, submitted 10 June 1927, published 9 August 1927 (Schrödinger 1927c).

We shall now discuss the above points in turn, after a brief discussion of the planning of Schrödinger's report for the conference (Section 4.1) and a summary of the report itself (Section 4.2).

4.1 Planning of Schrödinger's report

As reported in Chapter 1, the scientific committee of the Solvay institute met in Brussels on 1 and 2 April 1926 to plan the fifth Solvay conference. Lorentz had asked Ehrenfest to suggest some further names of possible participants, and it is in Ehrenfest's letter of 30 March[1] that Schrödinger's name is first mentioned in connection with the conference.[a] On this occasion, Ehrenfest suggested Schrödinger on the basis of a paper in which Schrödinger proposed an expression for the broadening of spectral lines due to the Doppler effect, and which applied the conservation laws to phenomena involving single light quanta (Schrödinger 1922).[b] Evidently, neither Lorentz nor Ehrenfest were yet aware of Schrödinger's work on wave mechanics.

In the meantime, Schrödinger sent to Lorentz the proof sheets of his first two papers on quantisation, also on 30 March,[2] thus initiating his well-known correspondence with Lorentz on wave mechanics.[c] Accordingly, already in a report of 8 April 1926 to the administrative commission,[3] Schrödinger is listed as a possible substitute for Heisenberg for a lecture on the 'adaptation of the foundations of dynamics to the quantum theory'. (It is unlikely, however, that the papers reached

[a] Schrödinger had already been a participant in the fourth Solvay conference, though not a speaker; cf. Moore (1989, pp. 157–8).

[b] As Schrödinger points out, the calculated broadening of the spectral lines is small compared to that expected on the basis of the thermal agitation of the radiating gas, otherwise the effect could be used as a test of the light quantum hypothesis.

[c] Most of this correspondence is translated in Przibram (1967). In the letter of 30 March, not included there, Schrödinger suggests reading the second paper on quantisation before the first, which should be seen rather an as example of an application. Also, he writes that the variational principle of the first paper is given a sensible formulation only in the addendum in proof. Finally, he mentions the paper on Einstein's gas theory in the *Physikalische Zeitschrift* (1926a) as a kind of preparatory work. Note that Lorentz on 27 May thanks Schrödinger for the proof sheets of three papers rather than the two mentioned in Schrödinger's letter. This third paper is clearly the equivalence paper (1926d), and was presumably sent separately (cf. Przibram 1967, pp. 43, 55–6).

Lorentz before the meeting in Brussels.[a]) In January 1927 then, like most of the other participants, Schrödinger was invited to the fifth Solvay conference.[5]

A few weeks later, Schrödinger had the opportunity to discuss personally with Lorentz the plans for the report 'under the beautiful palms of Pasadena', as he recalls in a letter of June 1927. In the same letter, we find a useful sketch of the theme and focus of Schrödinger's report; we also gather that Schrödinger was wary of the potential for a confrontation in Brussels:[6]

> ... I nurtured the quiet hope you would yet return to your first plan and entrust only Messrs [d]e Broglie and Heisenberg with reports on the new mechanics. But now you have decided otherwise and I will of course happily perform my duty.
>
> Yet I fear that the 'matricians' (as Mr Ehrenfest used to say) will feel disadvantaged. Should other thoughts emerge, after all bringing about the wish in the committee to limit the reports to two, you know, dear Professor, that I shall always happily remit my charge into your hands.

According to Schrödinger's sketch, the report is to stress points of principle, rather than the (by then many) applications of the theory. First of all, one has to distinguish clearly between two wave mechanical theories: a theory of waves in space and time (which however runs into difficulties especially with the many-electron problem), and the highly successful theory of waves in configuration space (which however is not relativistic). A difficulty of principle to be discussed in the context of the spacetime theory is the possibility of developing an interacting theory, which seems to require distinguishing between the fields generated by different particles, and whether this can be done in a sensible way, or perhaps be avoided.[b] In the context of the configuration-space theory, the main question is how to interpret the wave function. Schrödinger mentions the widespread view that the wave describes only ensembles, as well as his own 'preferred interpretation as a real description of the individual system, which thereby becomes a kind of "mollusc"'. In the letter (but not in the report), he is explicit about some of his misgivings about the ensemble view (as well as about the difficulties with his own preferred understanding, namely the 'failure of the electrons to stay together and similar'). Indeed, he points out that the Schrödinger equation is time-symmetric (if one includes complex conjugation), while experience teaches us that the statistical behaviour of ensembles cannot be described time-symmetrically. Also, insistence on a statistical interpretation leads to 'mystical' calculations with amplitudes and thus to problems with the laws of probability.[7]

[a] Lorentz had written to Ehrenfest on 29 March from Paris, where he had another meeting, and he appears to have travelled to Brussels directly from there.[4]

[b] Cf. p. 417 of the report.

4.2 Summary of the report

The eventual form of Schrödinger's report follows roughly the sketch given above, with an introduction, followed by three main sections, respectively on the configuration-space theory, on the spacetime theory and on the many-electron problem.

Introduction. Schrödinger draws the distinction between the spacetime theory (four-dimensional) and the configuration-space theory (multi-dimensional). He states that the use of configuration space is a mathematical way for describing what are in fact events in space and time. However, it is the multi-dimensional theory that is the most successful and has proved to be a powerful analytic tool in relation to Heisenberg and Born's matrix mechanics. The multi-dimensional point of view has not been reconciled yet with the four-dimensional one.[a]

I. *Multi-dimensional theory.* Schrödinger sketches a derivation of his time-independent wave equation, noting that it reproduces or improves on the results of Bohr's quantum theory. He also notes that the stationary states allow one to calculate the transition probabilities encountered in matrix mechanics. If one wishes to consistently develop a formalism in which there are only discontinuous transitions, he suggests one should take seriously the idea that the transitions do not occur against a continuous time background; the appearance of a continuous time parameter would be purely statistical, so to speak.[b] As an alternative, he suggests interpreting the time-independent equation as arising from a time-dependent one from which the time variable is eliminated by assuming a stationary solution. He thus arrives at the description of a quantum system in terms of a time-dependent wave function on configuration space. He then asks what the meaning of this wave function is: '*how does the system described by it really look like in three dimensions*?' (p. 411, Schrödinger's emphasis). He briefly mentions the view that the ψ-function describes an ensemble of systems, which Born and Heisenberg are going to discuss. Schrödinger instead finds it useful (if perhaps 'a bit naive') to imagine an individual system as continuously filling the whole of space somehow weighted by $|\psi|^2$ (as he further clarifies in the discussion). Schrödinger then carefully spells out that the spatial density resulting from the configuration space density is not a classical charge density, in the sense that the action on the particles by external fields and the interaction between the particles are already described by the potentials in the wave equation, and that it is inconsistent to assume that

[a] As becomes clear in the discussion, Schrödinger thinks that the multi-dimensional theory may prove indispensable, so that one should accept the notion of a ψ-function on configuration space and try instead to understand its physical meaning in terms of its manifestation in space and time. (See the discussion in Section 4.4.)

[b] Cf. the discussion in Chapter 8.1.

this spatial density is also acted upon in the manner of a classical charge density. Instead, it is possible to interpret it as a charge density (with some qualifications, some of which are spelled out only in the discussion) for the purpose of calculating the (classical) radiation field, thus yielding a partial vindication of the idea of spatial densities. This, however, must be an approximation, since the observation of such emitted radiation is itself an interaction between the emitting atom and some other absorbing atom or molecule, to be described again by the appropriate potentials in the wave equation.

II. *Four-dimensional theory.* Schrödinger shows that the time-dependent wave equation for a single particle is a non-relativistic approximation (with subtraction of a rest frequency) to the wave equation for the de Broglie phase wave of the particle. The latter can be made manifestly relativistic by including vector potentials. (In modern terminology, this is the Klein–Gordon equation.) If one couples the Maxwell field to it, the same spatial densities discussed in Section I appear as charge (and current) densities. However, in the application to the electron in the hydrogen atom, it becomes apparent that adding the self-field of the electron to the (external) field of the nucleus yields the wrong results.[a] Thus Schrödinger argues that if one hopes to develop a spacetime theory of interacting particles, it will be necessary to consider not just the overall field generated by the particles, but to distinguish between the (spatially overlapping) fields generated by each individual particle, each field acting only on the *other* particles of the system. Finally, he notes that the Klein–Gordon equation needs to be modified in order to describe spin effects, and that it may be possible to do so by considering a vectorial instead of a scalar ψ.

III. *The many-electron problem.* Schrödinger returns to the multi-dimensional theory and its treatment, by approximation, of the many-electron atom. His interest in this specific example relates to the question of whether this multi-dimensional system can be understood in spacetime terms. The treatment first neglects the interaction potentials between the electrons, and as a first approximation takes products $\psi_{k_1} \ldots \psi_{k_n}$ of the single-electron wave functions as solutions. One then expands the solution of the full equation in terms of the product wave functions. The time-dependent coefficients $a_{k_1 \ldots k_n}(t)$ in this expansion can then be calculated approximately if the interaction between the electrons is small. Schrödinger shows that – before any approximation – the coefficients in the equations for the $a_{k_1 \ldots k_n}(t)$ depend only on potentials calculated from the spatial charge densities associated with the ψ_{k_i}. Thus, although the solution to the full equation is a function on configuration space, it is determined by purely spatial charge densities. According

[a] Cf. Schrödinger (1927b), as mentioned in Section 4.6.

to Schrödinger, this reinforces the hope of providing a spatial interpretation of the wave function. A sketch of the approximation method concludes this section and the report.

4.3 Particles as wave packets

The idea of particles as wave packets is crucial to the development of Schrödinger's ideas and appears to provide one of the main motivations, at least initially, behind the idea of a description of matter purely in terms of waves.

Schrödinger's earliest speculation about wave packets (for both material particles and light quanta) is found in his paper on Einstein's gas statistics (1926a), in section 5, 'On the possibility of representing molecules or light quanta through interference of phase waves'. As he explains, Schrödinger finds it uncanny that in de Broglie's theory one should consider the phase waves of the corpuscles to be plane waves, since it is clear that by appropriate superposition of different plane waves one can construct a 'signal', which following Debye (1909) and von Laue (1914, section 2) can be constrained to a small spatial volume. He then continues:

> On the other hand, it is of course *not* to be achieved by the classical wave laws, that the constructed 'model of a light quantum' – which by the way extends indeed for *many* wavelengths in every direction – also *permanently* stays together. Rather, it spreads itself out [zerstreut sich] over ever larger volumes after passing through a focal point.
>
> If one could avoid this last conclusion by a quantum theoretical modification of the classical wave laws, then a way to deliverance from the light quantum dilemma would appear to be paved [angebahnt]. (1926a, p. 101)

Wave packets are first discussed at length in the second paper on quantisation (1926c): after describing the optical-mechanical analogy, Schrödinger discusses how one can construct wave packets (using the analogues of the optical constructions by Debye and von Laue), and then shows that the centroid of such a wave packet follows the classical equations of motion. Schrödinger conjectures that material points are in fact described by wave packets of small dimensions. He notes also that, for systems moving along very small orbits, the packet will be spread out, so that the idea of the trajectory or of the position of the electron inside the atom loses its meaning. In fact, the main problem that Schrödinger was to face with regard to wave packets turned out to be that spreading of wave packets is a much more generic feature than he imagined at first.

As mentioned, Schrödinger sent to Lorentz the proofs of his first two papers on quantisation (1926b,c) on 30 March 1926. In his reply, among many other things, Lorentz discussed explicitly the idea of wave packets, indeed doubting that they would stay together (Przibram 1967, pp. 47–8).[a] Schrödinger commented on this both in his reply of 6 June (Przibram 1967, pp. 55–66) and in an earlier letter of 31 May to Planck (Przibram 1967, pp. 8–11). In the latter, he admits that there will always exist spread-out states, because of linearity, but still hopes it will be possible to construct packets that stay together for hydrogen orbits of high quantum number. As he notes in the reply to Lorentz, this would imply that there is no general identification between hydrogen eigenstates and Bohr orbits, since a Bohr orbit of high quantum number would be represented by a wave packet rather than a stationary wave. (For an electron in a hydrogen orbit of low quantum number, Schrödinger did not envisage an orbiting packet but indeed a spread-out electron.) With the reply to Lorentz, Schrödinger further sent his paper on micro- and macromechanics (1926e), in which he showed that for the harmonic oscillator, wave packets do stay together. He hoped that the result would generalise to all quasi-periodic motions (admitting that maybe there would be 'dissolution' for a free electron). However, on 19 June Lorentz sent Schrödinger a calculation showing that wave packets on a high hydrogen orbit would indeed spread out (cf. Przibram 1967, pp. 69–71, where the details of the calculation, however, are omitted).

The relevance of high hydrogen orbits is, of course, the role they play in correspondence arguments. The fact that wave packets along such orbits do not stay together was thus a blow for any hopes Schrödinger might have had of explaining the transition from micro- to macromechanics along the lines of correspondence arguments.

A further blow to the idea of wave packets must have come with Born's papers on collision theory during the summer of 1926 (Born 1926a,b; cf. Section 3.4.3). In the equivalence paper, Schrödinger had included a remark about scattering, for which, he wrote, it is 'indeed necessary to understand clearly the continuous transition between the macroscopic anschaulich mechanics and the micromechanics of the atom' (1926d, p. 753; see Section 4.6 for the notion of Anschaulichkeit). Given that at the time Schrödinger thought that electrons on high quantum orbits should be described by wave packets, this remark may indicate that Schrödinger also thought that scattering should involve a deflection of wave packets, which would move asymptotically in straight lines.[b]

Born's work on collisions in the summer of 1926 made essential use of wave mechanics, but suggested a very different picture of scattering. Born explicitly

[a] As Lorentz remarks, the alternative would be 'to dissolve the electron completely ... and to replace it by a system of waves' (p. 48), which, however, would make it difficult to understand phenomena such as the photoelectric effect. The latter was in fact one of the criticisms levelled at Schrödinger's theory by Heisenberg (see Section 4.6).

[b] If thus was indeed Schrödinger's intuition, it may seem quite remarkable. On the other hand, so is the fact that Schrödinger does not seem to discuss diffraction of material particles, which seems equally problematic for the idea of wave packets.

understood his work as providing an alternative interpretation both to his earlier views on matrix mechanics and to Schrödinger's views. In turn, Schrödinger appeared to be sceptical of Born's suggestions, writing on 25 August to Wilhelm Wien:

From an offprint of Born's last work in the *Zeitsch. f. Phys.* I know more or less how he thinks of things: the *waves* must be strictly causally determined through field laws, the *wavefunctions* on the other hand have only the meaning of probabilities for the *actual* motions of light- or material-particles. I believe that Born thereby overlooks that ... it would depend on the taste of the observer which he now wishes to regard as real, the particle or the guiding field.

(Quoted in Moore 1989, p. 225)

A few days later, Schrödinger submitted a review paper on 'undulatory mechanics' to the *Physical Review* (1926h). In it, he qualified rather strongly the idea that 'material points consist of, or are nothing but, wave systems' (p. 1049). Indeed, he continued (pp. 1049–50):

This extreme conception may be wrong, indeed it does not offer as yet the slightest explanation of why such wave-systems seem to be realized in nature as correspond to mass-points of definite mass and charge. ... a thorough correlation of all features of physical phenomena can probably be afforded only by a harmonious union of these two extremes.

During the general discussion at the Solvay conference, Schrödinger summarised the situation with the following words (this volume, p. 469):

The original picture was this, that what moves is in reality not a point but a domain of excitation of finite dimensions .... One has since found that the naive identification of an electron, moving on a macroscopic orbit, with a wave packet encounters difficulties and so cannot be accepted to the letter. The main difficulty is this, that with certainty the wave packet spreads in all directions when it strikes an obstacle, an atom for example. We know today, from the interference experiments with cathode rays by Davisson and Germer, that this is part of the truth, while on the other hand the Wilson cloud chamber experiments have shown that there must be something that continues to describe a well-defined trajectory after the collision with the obstacle. I regard the compromise proposed from different sides, which consists of assuming a combination of waves and point electrons, as simply a provisional manner of resolving the difficulty.

The problem of the relation between micro- and macrophysics is connected of course to the linearity of the wave equation, which appears to lead directly to highly non-classical states (witness Schrödinger's famous 'cat' example, Schrödinger 1935).[a] One might ask whether Schrödinger himself considered the idea of a non-linear wave equation. In this connection, a few remarks by Schrödinger may

[a] Note, on the other hand, that Schulman (1997) shows there are 'classical' solutions to the linear equation in quite realistic models of coupling between micro- and macrosystems (measurements), provided one requires them to satisfy appropriate boundary conditions at *both* the initial and the final time.

be worth investigating further. One explicit, if early, reference to non-linearity is contained in the letter of 31 May to Planck, where, after noticing that linearity forces the existence of non-classical states, Schrödinger indeed speculates that the equations might be only approximately linear (Przibram 1967, p. 10). In the correspondence with Lorentz, the question of non-linearity arises in other contexts. In the context of the problem of radiation, it appears at first to be necessary for combination tones to arise (pp. 49–50). In the context of radiation reaction, Schrödinger writes that the exchange with Lorentz has convinced him of the necessity of non-linear terms (p. 62). In print, Schrödinger mentioned the possibility of a non-linear term in order to include radiation reaction in his fourth paper on quantisation (1926g, p. 130).[a] Finally, a few years later, in his second paper on entanglement (1936, pp. 451–2), Schrödinger mentioned the possibility of spontaneous decay of entanglement at spatial separation, which would have meant a yet untested violation of the Schrödinger equation.

Modern collapse theories, such as those by Ghirardi, Rimini and Weber (1986) or by Pearle (1976, 1979, 1989), modify the Schrödinger equation stochastically, in a way that successfully counteracts spreading with increasing scale of the system. Schrödinger's strategy based on wave packets thus appears to be viable at least if one accepts stochastic modifications to Schrödinger's equation (which Schrödinger was not necessarily likely to do).

Note that while crucial, the failure of the straightforward idea of wave packets for representing the macroscopic, or classical, regime of the theory is distinct from the question of whether Schrödinger's wave picture could adequately describe what appeared to be other examples or clear indications of particulate or 'discontinuous' behaviour, such as encountered in the photoelectric effect. This question was particularly important in the dialogue with Heisenberg and Bohr (Section 4.6). We shall also see that Schrödinger continued to explore how continuous waves might provide descriptions of apparently particulate or discontinuous quantum phenomena, such as the Compton effect, quantum jumps, blackbody radiation and even the photoelectric effect (Section 4.6.3).

[a] An explicit proposal for including radiation reaction through a non-linear term is due to Fermi (1927). Classically, if one includes radiation reaction, one has the third-order non-relativistic Abraham–Lorentz equation

$$\ddot{\mathbf{x}} - \tau \dddot{\mathbf{x}} = \mathbf{F}/m \,, \tag{4.1}$$

where $\mathbf{F}$ is the externally applied force and $\tau = (2/3)e^2/mc^3$ (e is the electron charge, m is the mass). Fermi (1927) proposed modifying the Schrödinger equation as follows:

$$i\frac{\partial \Psi}{\partial t} = \hat{H}\Psi - m\tau\Psi\mathbf{x}\cdot\frac{d^3}{dt^3}\int d^3\mathbf{x}'\ \Psi^*\mathbf{x}'\Psi \,. \tag{4.2}$$

That is, he added an extra 'potential' $-m\tau\mathbf{x}\cdot\frac{d^3}{dt^3}\langle\mathbf{x}\rangle$. Rederiving the Ehrenfest theorem, one finds that $\langle\mathbf{x}\rangle$ then obeys the above Abraham–Lorentz equation.

4.4 The problem of radiation

In Bohr's theory of the atom, the frequency of emitted light corresponded not to the frequency of oscillation of an electron on a Bohr orbit, but to the term difference between two Bohr orbits. No known mechanism could explain the difference between the frequency of oscillation and the frequency of emission. Bohr's theory simply postulated quantum jumps $E_i \rightarrow E_j$ between the stationary states of energy E_i and E_j, accompanied by emission (or absorption) of light of the corresponding frequency $\nu_{ij} = \frac{E_i - E_j}{h}$. In this respect, the Bohr–Kramers–Slater (BKS) theory had the advantage of postulating a collection of virtual oscillators with the observed frequencies. This was also, in a sense, that aspect of the BKS theory that survived into Heisenberg's matrix mechanics.[a]

The idea that wave mechanics could provide a continuous description of the radiation process (as opposed to the picture of quantum jumps) also appears to have been one of the main bones of contention between Schrödinger and the Copenhagen-Göttingen physicists. As Schrödinger wrote to Lorentz on 6 June 1926:

> The frequency discrepancy in the Bohr model, on the other hand, seems to me, (and has indeed seemed to me since 1914), to be something so *monstrous*, that I should like to characterize the excitation of light in this way as really almost *inconceivable*.
>
> *(Przibram 1967, p. 61)*[b]

The reaction by Schrödinger to the BKS theory instead was quite enthusiastic. In part, Schrödinger was well predisposed towards the possibility that energy and momentum conservation be only statistically valid.[c] In part (as argued by de Regt (1997)), this was precisely because the BKS theory provided a mechanism for radiation, in fact a very anschaulich mechanism, albeit 'virtual'. Schrödinger published a paper on the BKS theory, containing an estimate of its energy fluctuations (1924b).[d]

In 1926, with the development of wave mechanics, Schrödinger saw a new possibility of conceiving a mechanism for radiation: the superposition of two waves would involve *two* frequencies, and emitted radiation could be understood as some kind of 'difference tone'. In his first paper on quantisation (1926b), Schrödinger states that this picture would be 'much more pleasing [um vieles sympathischer]'

[a] See Section 1.3 (p. 12) and Chapter 3, especially Section 3.3.1.

[b] The emphasis here is very strong, but cf. the context of this passage in Schrödinger's letter.

[c] Cf. Schrödinger to Pauli, 8 November 1922, in Pauli (1979, pp. 69–71). The idea of abandoning exact conservation laws seems to be derived from Schrödinger's teacher Exner (*ibid.*, and Moore 1989, pp. 152–4). Schrödinger publicly stated this view and allegiance in his inaugural lecture at the University of Zürich (9 December 1922), published as Schrödinger (1929a).

[d] Cf. Darrigol (1992, pp. 247–8) on the problem of the indefinite growth of the energy fluctuations with time (as raised in particular by Einstein). According to Schrödinger, an isolated system would behave in this way, but the problem would disappear for a system coupled to an infinite thermal bath.

than the one of quantum jumps (p. 375), but the idea is rather sketchy: Schrödinger speculates that the energies of the different eigenstates all share a large constant term, and that if the *square* of the frequency is proportional to $mc^2 + E$, then the frequency differences (and therefore the beat frequencies) are approximately given by the hydrogen term differences. The second paper (1926c) refers to radiation only in passing.

Commenting on these papers in his letter to Schrödinger of 27 May, Lorentz pointed out that while beats would arise if the time-dependent wave equation (which Schrödinger did not have yet) were linear, they would still not produce radiation by any known mechanism. Combination tones would arise if the wave equation were non-linear (Przibram 1967, pp. 49–50). As becomes clear in the following letters between Schrödinger and Lorentz, once the charge density of a particle is associated with a quadratic function of ψ, such as $|\psi|^2$, 'difference tones' in the oscillating charge density arise regardless of whether the wave equation is linear or non-linear.

This idea is still the basis of today's semiclassical radiation theory (often used in quantum optics), that is, the determination of classical electromagnetic radiation from the current associated with a charge density proportional to $|\psi|^2$ (for a non-stationary ψ).[a] Schrödinger arrived at this result through his work connecting wave mechanics and matrix mechanics (his 'equivalence paper', 1926d). In fact, Schrödinger showed how to express the elements of Heisenberg's matrices wave mechanically, in particular the elements of the dipole moment matrix, which (by correspondence arguments) were interpreted as proportional to the radiation intensities. Schrödinger now suggested it might be possible 'to give an extraordinarily anschaulich interpretation of the intensity and polarisation of radiation' (1926d, p. 755), by defining an appropriate charge density in terms of ψ. In this paper, he suggested as yet, for a single electron, to use (the real part of) the quadratic function $\psi \frac{\partial \psi^*}{\partial t}$.

The third paper on quantisation (1926f) is concerned with perturbation theory and its application to the Stark effect, but Schrödinger notes in an addendum in proof (footnote on p. 476) that the correct charge density is given by $|\psi|^2$. In the letter to Lorentz of 6 June 1926, he explains in detail how this gives rise to a sensible notion of charge density also for several particles, each contribution being obtained by integrating over the other particles (Przibram 1967, p. 56). This idea is then used and discussed in print (as a '*heuristic hypothesis*') in the fourth paper on quantisation (1926g, p. 118, Schrödinger's italics), where the wave function is also explicitly interpreted in terms of a superposition of all classical configurations

[a] Pauli uses this method (for the case of one particle) in calculating the scattered radiation in the Compton effect during the discussion after Compton's talk (p. 332).

of a system weighted by $|\psi|^2$, and the time constancy of $\int |\psi|^2$ is derived (pp. 135–6).[a]

This is also the picture of the wave function given by Schrödinger in the Solvay report (pp. 411ff.). Schrödinger discreetly skips discussing the view of the wave function as describing only an ensemble. Schrödinger's concern in interpreting the wave function is to understand its manifestation in spacetime. This concern also motivates Schrödinger's discussion of the many-electron atom[b], where he stresses that the spatial charge distributions of the single (non-interacting) electrons already determine the wave function of the interacting electrons.

In the report, Schrödinger starts by rephrasing the question of the meaning of the wave function as that of how a system described by a certain (multi-particle) wave function looks like in three dimensions. He describes this as taking all possible configurations of the classical system simultaneously and weighting them according to $|\psi|^2$. To this picture of the system as a 'snapshot' (as he calls it in the report) or as a 'mollusc' (as he had written to Lorentz) Schrödinger then associates the corresponding electric charge density in 3-space. He is careful to state, however, that this is not an electric charge in the usual sense. For one thing, the electromagnetic field does not exert forces on it. While this charge does describe the *sources* of the field in the semi-classical theory, this coupling of the field to the charges is described as 'provisional': first, because of the problem of radiation reaction (which is not taken into account in the Schrödinger equation), and second, because within a closed system it is inconsistent to model the interaction between different charged particles using the semi-classical field. In particular, the observation of emitted radiation is in principle again a quantum mechanical interaction, and should be described by corresponding potentials in the equation for the total system.

The above picture of the wave function and the question of regarding Schrödinger's formal charge density as a source of classical electromagnetic radiation evidently raised many questions. (Schrödinger introduces the discussion after his report by the remark: 'It would seem that my description in terms of a snapshot was not very fortunate, since it has been misunderstood', p. 425.) It also elicited most of the discussion following Schrödinger's report.[c]

[a] The latter question had also been raised by Lorentz (Przibram 1967, p. 71) and answered in Schrödinger's next letter, not included in Przibram's collection.[8]

[b] Previously unpublished but deriving from methods used in Schrödinger's paper on energy exchange (1927c).

[c] The rest of the discussion includes a few technical questions and comments (contributions by Fowler and De Donder, Born's report on some numerical work on perturbation theory), and some discussion of the 'three-dimensionality' of the many-electron atom. Further discussion of the meaning of Schrödinger's charge densities (and of the meaning of the Schrödinger wave in the context of the transformation theory of matrix mechanics) took place in the general discussion, especially in contributions by Dirac and by Kramers (pp. 445, 448).

A difficulty that is spelled out only in the discussion (contributions by Bohr and by Schrödinger, p. 425), is that using Schrödinger's dipole moment (equation (13) in Schrödinger's talk) to calculate the radiation does not directly yield the correct intensities. As pointed out in the talk, if one evaluates the dipole moment for a superposition $\sum_k c_k \psi_k$, one obtains terms containing the integrals $-e \int q \psi_k \psi_l^* d\tau$. These are the matrix elements of the dipole moment matrix in matrix mechanics, and they can be used within certain limits to calculate the emitted radiation (in particular, as Schrödinger points out, vanishing of the integral implies vanishing of the corresponding spectral line). However, these integrals appear with the coefficients $c_k c_l^*$, whereas both according to Bohr's old quantum theory and to experiment, the intensity of radiation should not depend on the coefficient of the 'final state', say ψ_l. Thus, the use of (13) as a classical dipole moment in the calculation of emitted radiation does not in general yield the correct intensities.

Bohr also drew attention to the fact that by the time of the Solvay conference, Dirac (1927b,c) had already published his treatment of the interaction of the (Schrödinger) electron with the quantised electromagnetic field. Schrödinger replied that he was aware of Dirac's work, but had the same misgivings with q-numbers as he had with matrices: the lack of 'physical meaning', without which he thought the further development of a relativistic theory would be difficult.[a]

If the interpretation of the wave function as a charge density raises problems of principle (as Born and Heisenberg also stress in their report, p. 398), what is the point of suggesting such an interpretation? One possible way of understanding Schrödinger's intentions is to say that he is proposing an anschaulich image of the wave function in terms of the spatial density it defines, *without* in general specifying the dynamical role played by this density. Under certain circumstances, then, this spatial density takes on the dynamical role of a charge density, in particular as a source of radiation. This point of view does not resolve the other problems connected with the wave conception of matter (spreading of the wave packet, Schrödinger's cat, micro-macro question), but it offers a platform from which to work towards their possible resolution. This of course may be an overinterpretation of Schrödinger's position, but it fits approximately with later developments that take Schrödinger's wave conception seriously, for example Bell's (1990) idea of $|\psi|^2$ as 'density of stuff' (in configuration space).

[a] Note that in his treatment of the emission of radiation in *The Physical Principles of the Quantum Theory* (Heisenberg 1930b, pp. 82–4), Heisenberg seems to follow and expand on the discussion of Schrödinger's report. Indeed, Heisenberg first describes two related methods for calculating the radiation, based respectively on calculating the matrix element of the dipole moment of the atom (justified via correspondence arguments), and on calculating the dipole moment of Schrödinger's 'virtual charges' (as he calls them). He then explains precisely the difficulty with the latter discussed here by Bohr and Schrödinger, and goes on to sketch (a variant of) Dirac's treatment of the problem.

4.5 Schrödinger and de Broglie

Thus opens Schrödinger's review paper on wave mechanics for the *Physical Review* (1926h, p. 1049, references omitted):

> The theory which is reported in the following pages is based on the very interesting and fundamental researches of L. de Broglie on what he called 'phase waves' ('ondes de phase') and thought to be associated with the motion of material points, especially with the motion of an electron or [photon].[a] The point of view taken here, which was first published in a series of German papers, is rather that material points consist of, or are nothing but, wave-systems.

This passage both illustrates the well-known fact that Schrödinger arrived at his wave mechanics by developing further the ideas of de Broglie – starting with his paper on gas statistics (Schrödinger 1926a) – and emphasises the main conceptual difference between de Broglie's and Schrödinger's approaches.

While some details of the relation between Schrödinger's and de Broglie's work are discussed in Section 2.3, in this section we wish to raise the question of why Schrödinger should have developed such a different picture of wave mechanics. Indeed, although Schrödinger quotes de Broglie as the rediscoverer of Hamilton's optical-mechanical analogy (1926h, footnote 3 on p. 1052), the two authors apply the analogy in opposite directions: de Broglie treats even the photon as a material particle with a trajectory, while Schrödinger treats even the electron as a pure wave.[b]

As Schrödinger writes in his first paper on quantisation, it was in particular 'reflection on the spatial distribution' of de Broglie's phase waves that gave the impulse for the development of his own theory of wave mechanics (1926b, p. 372). We gain some insight as to what this refers to from a letter from Schrödinger to Landé of 16 November 1925:

> I have tried in vain to make for myself a picture of the phase wave of the electron in the Kepler orbit. Closely neighbouring Kepler ellipses are considered as 'rays'. This, however, gives horrible 'caustics' or the like for the wave fronts.
>
> *(Quoted in Moore 1989, p. 192)*

[a] The text here reads 'proton', which is very likely a misprint, since de Broglie's work indeed focussed on electrons and photons.

[b] As is clear from de Broglie's thesis and as mentioned by de Broglie himself in his report (p. 344), de Broglie had always assumed the picture of trajectories. The above quotation in any case makes it clear that this was Schrödinger's own reading of de Broglie. Already as early as January 1926, Schrödinger writes in his gas theory paper about 'the de Broglie–Einstein undulatory theory of corpuscles in motion, according to which the latter are no more than a kind of "foam crest" ["Schammkaum"] on the wave radiation that constitutes the world background [Weltgrund]' (1926a, p. 95). Other indications that this was well-known are Lorentz's comments to Schrödinger on the construction of electron orbits in his letter of 19 June 1926 coming 'close to de Broglie's arguments' (Przibram 1967, p. 74) and Pauli's reference to Einstein's and de Broglie's 'moving point masses' in his letter to Jordan of 12 April 1926 (Pauli 1979, p. 316).

Thus, one reason for Schrödinger to abandon the idea of trajectories in favour of the pure wave theory might have been that well-behaved trajectories seemed to be incompatible with well-behaved waves. And indeed, in his presentations of the optical-mechanical analogy, Schrödinger states that outside of the geometric limit the notion of 'ray' becomes meaningless (1926b, pp. 495, 507; 1926h, pp. 1052–3). This also seems to be at issue in the exchange between Schrödinger and Lorentz during the discussion of de Broglie's report, where Schrödinger points out that in cases of degeneracy, any arbitrary linear combination of solutions is allowed, and de Broglie's theory would predict very complicated orbits (p. 366).

There are other aspects that could conceivably provide further reasons for Schrödinger's definite abandoning of the trajectories. One possibility is that Schrödinger picked up from de Broglie specifically the idea of particles as singularities of the wave,[a] and was happy to relax it to the idea of wave packets. Indeed, as we have seen above, Schrödinger placed great emphasis on the notion of a wave packet, and if it had been an adequate notion, there would have been no need for a separate notion of a corpuscle in order to explain the particulate aspects of matter.

We have also seen that Schrödinger was acutely aware of the radiation problem, namely of the discrepancy between the orbital frequency of the electron and the frequency of the emitted radiation. While de Broglie in his thesis was able to derive the Bohr orbits from wave considerations, this would in no way seem to alleviate the problem: the frequency of revolution of the electron in de Broglie's theory was the same as in the Bohr model, and would lead to the wrong frequency of radiation if the electron was treated as a classical source. (In the case of degeneracy noted above, the situation would be even more complicated.)[b]

Finally, Schrödinger appeared to be critical of proposals combining waves and particles, for instance as appeared to be done by Born in his collision papers (see Section 4.3). Such misgivings could easily have applied also to de Broglie's theory.

4.6 The conflict with matrix mechanics

In the early discussions on the meaning of quantum theory, the notion of 'Anschaulichkeit' resurfaces time and again. The verb 'anschauen' means 'to look at', and 'anschaulich', which means 'clear', 'vivid' or 'intuitive', has visual connotations that the English word 'intuitive' lacks. Anschaulichkeit of a physical theory is thus a quality of ready comprehensibility that may (but need not[c]) include

[a] Cf. Schrödinger (1926a, p. 99): 'The universal radiation, as "signals" or perhaps singularities of which the particles are meant to occur, is thus something quite essentially more complicated than for instance the wave radiation of Maxwell's theory . . . '.

[b] Note, however, that de Broglie (1924c) had discussed the solution of this problem in the correspondence limit (that is, for the case of high quantum numbers).

[c] Cf. below Heisenberg's use of the term in the uncertainty paper.

a strong component of literal picturability. For Schrödinger, at any rate, it seems that the possibility of grasping a theory through some kind of spatio-temporal intuition was a key component of physical understanding.[a]

It is clear, however, that the conflict between wave mechanics and matrix or quantum mechanics was not, or not only, a philosophical issue, say about the validity of the concepts of spacetime and causality,[b] and that, in the minds of the parties involved, these issues were connected with specifically scientific questions.[c]

The list of these questions is extensive. In his 'equivalence' paper (1926d), Schrödinger states that mathematical equivalence is not the same as physical equivalence, in the sense that two theories can offer quite different possibilities for generalisation and further development.[d] He thinks of two problems in particular: first, the problem of scattering (as mentioned in Section 4.3), second, the problem of radiation (discussed in Section 4.4), in connection with which he then describes the idea of the vibrating charge density. The radiation problem in turn links to further issues at the heart of the debate with Heisenberg and Bohr in particular, on whether there are quantum jumps or whether the process of radiation and the atomic transitions can be described as continuous processes in space and time. In a sense, neither the issue of scattering nor that of radiation resolved the debate in favour of either theory: both theories were modified or reinterpreted in the course of these developments, even though the result (statistical interpretation, Copenhagen interpretation) was not to Schrödinger's taste.

We shall now follow how the conflict evolved, from the beginnings to the time of the Solvay conference, since both sides developed considerably during the crucial period between Schrödinger's first papers and the conference.

4.6.1 Early days

At the time of Schrödinger's first two papers on quantisation, as we have seen in the previous chapter, matrix mechanics was a theory that rejected the notion of electron orbits, indeed the very possibility of a spacetime description, substituting the classical kinematical quantities with matrix quantities; but it kept the postulate of stationary states and of quantum jumps between these states. Matrix mechanics

[a] A good discussion of the role played for Schrödinger by the notion of Anschaulichkeit is given by de Regt (1997, 2001), who argues that Schrödinger's requirement of Anschaulichkeit is derived indirectly from Boltzmann, and is essentially a methodological requirement (as opposed to being a commitment to realism – cf. also the introduction by Bitbol in Schrödinger (1995, fn. 10 on p. 4)).

[b] Note that Kant's conception of space and time is formulated in terms of what he calls the 'Anschauungsformen' (the 'forms of intuition'), so that the discussion of Anschaulichkeit has strong philosophical overtones.

[c] This point has been recently argued also by Perovic (2006).

[d] Similarly, in the discussion after his own report, Schrödinger insists that finding a physical interpretation of the theory is 'indispensable for the further development of the theory' (p. 428).

did not describe the stationary states individually, but only collectively, and allowed one to calculate only transition probabilities for the jumps.

Schrödinger's first paper on quantisation (1926b) contains only a few comments on the possible continuous picture of atomic transitions, as opposed to quantum jumps. (But of course, at this stage, he has very little to say about the problem of radiation, in particular nothing about intensities.) The first time he comments explicitly on the differences between wave mechanics and quantum mechanics is in his second paper. There he writes that wave mechanics offers a way to interpret

> the conviction, more and more coming to the fore today, that *first*: one should deny real significance to the *phase* of the electron motions in the atom; *second*: that one may not even claim that the electron at a certain time is located *on one particular* of the quantum trajectories distinguished by the quantum conditions; *third*: the true laws of quantum mechanics consist not in definite prescriptions for the *individual trajectory*, rather these true laws relate through equations the elements of the whole manifold of trajectories of a system, so that apparently a certain interaction between the different trajectories obtains. (1926c, pp. 508)

As Schrödinger proceeds to say, these claims are in contradiction with the ideas of electron position and electron orbit, but should not be taken as forcing a complete surrender of spatio-temporal ideas. At the time, the mathematical relation between wave mechanics and matrix mechanics was not yet clarified, but Schrödinger hopes there would be a well-defined mathematical relation between the two theories, which could then complement each other. According to Schrödinger, Heisenberg's theory yields the line intensities, and his own offers the possibility of bridging the micro-macro gap.[a] Personally, he finds the conception of emitted frequencies as 'beats' particularly attractive and believes it will provide an anschaulich understanding of the intensity formulas (1926c, pp. 513–14).

Within weeks, the relation between the two theories was clarified independently by Schrödinger (1926d), by Eckart (1926) and by Pauli, in his remarkable letter to Jordan of 12 April 1926 (Pauli 1979, pp. 315–20). This is one of the first documented reactions to Schrödinger's new work from a physicist of the Copenhagen-Göttingen school.[b] In it, Pauli emphasises in fact that in Schrödinger's theory there are no electron orbits, since trajectories are a concept belonging to the geometric limit of the theory. Pauli seems to say that insofar as Schrödinger

[a] Schrödinger's collected works (1984) reproduce this and other papers from Erwin and Anny Schrödinger's own copy of the second edition of *Abhandlungen zur Wellenmechanik* (Schrödinger 1928). In this copy of the book, the passage about the micro-macro bridge is underlined.

[b] Ehrenfest informed Lorentz of the 'equivalence' result (and of Klein's (1926) theory) on 5 May 1926, when Kramers reported them in the colloquium at Leiden. Ehrenfest also mentioned the relation of this result to Lanczos's (1926) work.[9] As noted above (footnote on p. 112), by 27 May Lorentz had received a copy of Schrödinger's paper directly from Schrödinger.

provides a description of individual stationary states, his theory is not in conflict with matrix mechanics, since it also does not contain the concept of electron orbits.[a]

The tone of Schrödinger's remarks and of the comments they elicit changes with the equivalence paper. In a much-quoted footnote, Schrödinger says he had known of Heisenberg's theory but was 'scared away, not to say repelled', by the complicated algebraic methods and the lack of Anschaulichkeit (1926d, p. 735). On 8 June 1926, Heisenberg sends Pauli an equally notorious (but often mis-quoted) comment:

> The more I reflect on the physical part of Schrödinger's theory, the more disgusting [abscheulich] I find it. Imagine the rotating electron whose charge is distributed over the whole space with its axis in a fourth and fifth dimension. What Schrödinger writes of the Anschaulichkeit of his theory "*would scarcely* [*be*] *an appropriate*..." in other words I find it poppycock [Mist].
>
> *(Pauli 1979, p. 328)*

It seems clear that, according to Heisenberg, it is Schrödinger's claim of Anschaulichkeit for his theory that is ludicrous, presumably partly because of the spread-out electron and partly because the waves are in configuration space.[b]

4.6.2 From Munich to Copenhagen

In the summer of 1926, Schrödinger gave a series of talks on wave mechanics in various German universities, in particular, on 21 and 23 July, he talked in Munich at the invitation of Sommerfeld and of Wien.[c] The description of a mechanism for radiation elicited enthusiastic comments by Wien, but criticism from Sommerfeld and from Heisenberg. The discussion was apparently heated, and eventually identified a crucial experiment that would decide between the idea of the continuous mechanism of radiation envisaged in wave mechanics and the idea of quantum jumps. This was incoherent scattering, that is the Raman effect (at that time neither observed nor thus named). According to the quantum prediction (Smekal 1923; Kramers and Heisenberg 1925), incoherent scattering would exist also for atoms in the ground state, because the atoms could be excited by the

[a] Pauli knew very early about Schrödinger's paper, having been informed by Sommerfeld, to whom at Schrödinger's request a copy of the paper had been forwarded by Wien, the editor of *Annalen der Physik* (cf. Pauli 1979, pp. 278, 293). Pauli's analysis in the letter of 12 April appears to be based only on Schrödinger's first paper on quantisation, although at least the official publication date of the second paper was 6 April 1926. See also the letters between Pauli and Schrödinger reproduced in Pauli (1979).

[b] Alternatively, with a 'fourth and fifth dimension', Heisenberg might conceivably be referring to Klein's five-dimensional extension of Schrödinger's equation (Klein 1926). As one example, here is how Moore (1989, p. 221) quotes the passage: 'The more I think of the physical part of the Schrödinger theory, the more abominable I find it. What Schrödinger writes about *Anschaulichkeit* makes scarcely any sense, in other words I think it is bullshit [Mist]'.

[c] Here we follow mainly the account given by Heisenberg (1946).

incident light. According to Schrödinger instead, the effect was due to induced vibrations for the case in which at least two atomic frequencies were already present, and so would not occur in the ground state.[a] Apparently, Sommerfeld and Heisenberg were 'prepared to enter a bet for its existence, while the experimental physicists were against it and Schrödinger took a more wait-and-see attitude' (Heisenberg 1946, p. 5).

In a letter to Pauli of 28 July, Heisenberg gives other specific criticisms of Schrödinger, for throwing overboard 'everything "quantum theoretical": namely photoelectric effect, Franck collisions, Stern–Gerlach effect etc.' (Pauli 1979, p. 338). As a matter of fact, in the letter to Wien of 25 August 1926, quoted above (Section 4.3), Schrödinger admitted that he had great conceptual difficulties with the photoelectric effect (but see below the discussion of Schrödinger, 1929b). Heisenberg further mentioned to Pauli that, together with Schrödinger and Wien, he had discussed Wien's experiments on the decay of luminescence (Wien 1923, ch. XX). This was another point where Schrödinger thought wave mechanics proved superior to matrix mechanics, and Heisenberg encouraged Pauli to calculate and publish the damping coefficients for the hydrogen spectrum.[b]

Heisenberg mentions similar criticisms in his talk on 'Quantum mechanics', given at the 89th meeting of German Scientists and Physicians in Düsseldorf on 23 September 1926 and published in the issue of 5 November of *Naturwissenschaften* (Heisenberg 1926c). This talk could be seen as a public response to the claims that the return to a 'continuum theory' was possible. In it, Heisenberg addresses in particular the problem of Anschaulichkeit: according to Heisenberg, the usual notions of space and time, and in particular their application to physics with the idea that space and matter are in principle continuously divisible, turn out to be mistaken, first of all due to the Unanschaulichkeit of the corpuscular nature of matter, then through the theoretical and experimental considerations leading to the idea of stationary states and quantum jumps (Bohr, Franck–Hertz, Stern–Gerlach), and finally through consideration of radiation phenomena (Planck's radiation formula, Einstein's light quantum, the Compton effect and the Bothe–Geiger experiments). This issue, it is claimed, also relates closely to the question of the degree of 'reality' to be ascribed to material particles or light quanta. Quantum mechanics in its development had thus first of all to free itself from notions of Anschaulichkeit, in order to set up a new kind of kinematics

[a] Note the implied inequivalence of wave and matrix mechanics (despite the recent 'equivalence proofs'). Another such possibility of experimental inequivalence is mentioned in the discussion after Schrödinger's report, with regard to the quadrupole radiation of the atom (p. 426). Cf. also Muller (1997) on the inequivalence of the two theories.

[b] Cf. Born and Heisenberg's report (pp. 395, 397), where spin is considered to be problematic for wave mechanics, and where it is explicitly stated that Dirac (1927c) provides an explanation for the decay experiments.

and mechanics. In discussing the wave theory, Heisenberg considers first de Broglie and Einstein as having developed wave-particle dualism for matter and having suggested the possibility of interference for an ensemble [Schar] of particles. He then explains that Schrödinger found a differential equation for the matter waves that reproduces the eigenvalue problem of quantum (that is, matrix) mechanics. However, according to Heisenberg, the Schrödinger theory fails to provide the link with de Broglie's ideas, that is, it fails to provide an analogy with light waves in ordinary space, because of the need to consider waves in configuration space; the latter kind of waves therefore have only a formal significance. Heisenberg refers to the claim that on the basis of Schrödinger's theory one may be able to return to 'a purely continuous description of the quantum theoretical phenomena', and continues (Heisenberg 1926c, p. 992):

> In developing consequently this point of view one leaves in fact the ground of de Broglie's theory, thus of Q.M. and indeed of all quantum theory and arrives in my opinion at a complete contradiction with experience (law of blackbody radiation; dispersion theory). This route is thus not viable. The actual reality of the de Broglie waves lies rather in the interference phenomena mentioned above, which defy any interpretation on the basis of classical concepts. The extraordinary physical significance of Schrödinger's results lies in the realisation that an anschaulich interpretation of the quantum mechanical formulas contains both typical features of a corpuscular theory and typical features of a wave theory.

After discussing quantum statistics ('which in any case represents a very bizarre further limiting of the reality of the corpuscles', p. 992), Heisenberg concludes (p. 994):

> In our anschaulich interpretation of the physical processes and mathematical formulas there is a dualism between wave theory and corpuscular theory such that many phenomena are described most naturally by a wave theory of light as well as of matter, in particular interference and diffraction phenomena, while other phenomena in turn can be interpreted only on the basis of the corpuscular theory. ... The contradictions of the anschaulich interpretations of different phenomena contained in the current scheme are completely unsatisfactory. For a contradiction-free anschaulich interpretation of the experiments, which in themselves are indeed contradiction-free, some essential trait in our picture of the structure of matter is currently still missing.

After the summer, Schrödinger visited Copenhagen for an intense round of discussions. According to Heisenberg's reconstruction in *Der Teil und das Ganze* (Heisenberg 1969), Schrödinger argued with Bohr precisely about the necessity of finding a mechanism for radiation, while Bohr insisted that quantum jumps were necessary for the derivation of Planck's radiation law, as well as being directly observable in experiments. Heisenberg (1946, p. 6) states that the discussion ended with the recognition 'that an interpretation of wave mechanics without quantum jumps was impossible and that the mentioned crucial experiment [i.e. the Raman effect] in any case would turn out in favour of the quantum jumps'.[a]

Schrödinger in turn admitted his difficulties. In a letter to Wien of 21 October 1926 (quoted in Pauli 1979, p. 339), he wrote: 'It is quite certain that the position of anschaulich images, which de Broglie and I take, has not nearly been developed far enough to account even just for the most important facts. And it is downright probable that here and there a wrong path has been taken that needs to be abandoned'. And, commenting in his preface (dated November 1926) to the first edition of *Abhandlungen zur Wellenmechanik* (1926i) on the fact that the papers were being reprinted unchanged, he invoked 'the impossibility at the current stage of giving an essentially more satisfactory or even definitive new presentation'.

In the meantime, both Bohr and Heisenberg were working to find their own satisfactory interpretation of the theory.[b] In particular as regards Heisenberg, the correspondence with Pauli is very telling.[c] First of all one finds the discussion of Born's collision papers (Born 1926a,b), which led to Heisenberg's paper on fluctuation phenomena (Heisenberg 1926b). As already mentioned in Section 3.4.5, Heisenberg then reports to Pauli about Dirac's transformation theory (Dirac 1927a), which, however, is seen only as an extraordinary formal development. The next letters from Heisenberg to Pauli are from February 1927 and the one of 23 February includes the sketch of the ideas of the uncertainty paper (Pauli 1979, pp. 376–81). This, as was Bohr's simultaneous development of the idea of complementarity, was meant to provide the anschaulich picture that was still missing. Indeed, in the uncertainty paper, Heisenberg (1927, p. 172) formulates Anschaulichkeit as the possibility of arriving at qualitative predictions in simple cases together with formal consistency. With the formulation and application of the uncertainty relations, he was then satisfied to have found such an interpretation.

In a sense, however, the uncertainty paper was also the end of the original notion of 'quantum jumps' as transitions between stationary states of a system. Indeed, as had to be the case given the generalisation of transition probabilities to arbitrary pairs of observables (and as was implied in the Heisenberg–Pauli correspondence), the privileged role of stationary states had to give way. It is somewhat ironic that Heisenberg was to give up quantum jumps only a few months after the discussions with Schrödinger in Copenhagen; nevertheless, quantum mechanics thus wedded to probabilistic transitions between measurements was just as discontinuous as

[a] The well-known quotation 'If this damned quantum jumping is indeed to stay, then I regret having worked on this subject at all' is reported both in Heisenberg (1946) and Heisenberg (1969).

[b] See also Section 3.4.

[c] However, as noted in Pauli (1979, p. 339 and fn. 3 on p. 340), most of Pauli's letters to Heisenberg from this period appear to have been destroyed in the war.

the picture of quantum jumps between stationary states that it replaced, and, for Schrödinger, it was equally unsatisfactory.

Note that, for Bohr at least, wave aspects played a crucial role in the resulting 'Copenhagen' interpretation. Yet, just as in the case of Born's (1926a,b) use of wave mechanics in the discussion of collisions (see Section 3.4.4), Heisenberg appears to have been convinced that the apparent wave aspects could be interpreted entirely in matrix terms. This difference of opinion was reflected in the sometimes tense discussions between Heisenberg and Bohr at the time, in particular on the topic of Heisenberg's treatment of the γ-ray microscope in the uncertainty paper and the corresponding addendum in proof.[a]

4.6.3 Continuity and discontinuity

Between late 1926 and the time of the Solvay conference, Schrödinger continued to work along lines that brought out the attractive features – and sometimes the limitations – of the wave picture. In late 1926 and early 1927, Schrödinger focussed on the 'four-dimensional' theory, with his papers on the Compton effect (1927a), and on the energy-momentum tensor (1927b), while he returned to the 'many-dimensional' theory shortly before the Solvay conference, with his work on energy exchange (1927c) and the treatment of the many-electron atom presented in his report.

The Compton effect is of course a paradigmatic example of a 'discontinuous' phenomenon, but Schrödinger (1927a) gives it a wave mechanical treatment, by analogy with the classical case of reflection of a light wave when it encounters a sound wave (Brillouin 1922).[b] As he remarks in the paper, and as mentioned in the discussion after Compton's report, this treatment relies on the consideration of stationary waves and does not directly describe an individual Compton collision.[c]

The paper on the energy-momentum tensor (1927b), following Gordon (1926), takes the Lagrangian approach to deriving the Klein–Gordon equation, and varies also the electromagnetic potentials, thus deriving the Maxwell equations as well. Schrödinger then considers in particular the energy-momentum tensor of the combined Maxwell and Klein–Gordon fields. As he remarks (Schrödinger 1928, p. x), this is a beautiful formal development of the theory, but it heightens starkly the difficulties with the four-dimensional view, since one cannot insert the

[a] For more on this issue, see Beller (1999, pp. 71–4, 138–41) and Camilleri (2006). For Heisenberg's way out of the difficulty, see Heisenberg (1929, pp. 494–5).

[b] Schrödinger had given a derivation of Brillouin's result by assuming that energy and momentum were exchanged in the form of quanta (1924a). In the paper on the Compton effect he now comments on how, in a sense, he is reversing his own earlier reasoning.

[c] See in particular the remarks by Pauli (p. 332) and the discussion following Schrödinger's contribution (p. 337). Cf. also the closing paragraph of Pauli's letter to Schrödinger of 12 December 1927 (Pauli 1979, p. 366).

electromagnetic potentials thus obtained back into the Klein–Gordon equation. For instance, including the self-field of the electron in the treatment of the hydrogen atom would yield the wrong results, as Schrödinger also mentions in Section II of his report. In this connection, Schrödinger states that: 'The exchange of energy and momentum between the electromagnetic field and "matter" does *not* in reality take place in a continuous way, as the [given] fieldlike expression suggests' (1927b, p. 271).[a] According to Heisenberg (1929), the further development of the four-dimensional theory was indeed purely formal, but provided a basis for the later development of quantum field theory.[b]

The paper on energy exchange (1927c), Schrödinger's last paper before the Solvay conference, marks an attempt to meet the criticism that wave mechanics cannot account for crucial phenomena involving 'quantum jumps'. Following Dirac (1926c), Schrödinger sketches the method of the variation of constants, which he is to use also in the discussion of the many-electron atom in his Solvay report. He then applies it to the system of two atoms in resonance discussed by Heisenberg (1926b) (and by Jordan (1927a)). As discussed in Section 3.4, Heisenberg uses this example to show how in matrix mechanics one can indeed describe change starting from first principles, in particular how one can determine the probabilities for quantum jumps. Now Schrödinger turns the tables around and argues that the treatment of atoms in resonance using wave mechanics shows how one can eliminate quantum jumps from the picture. He argues that the two atoms exchange energies *as if* they were exchanging definite quanta. Indeed, he goes further and suggests that the idea of quantised energy itself should be reinterpreted in terms of wave frequency,[c] and that resonance phenomena are indeed the key to the 'quantum postulates'. Schrödinger then proceeds to formulate statistical considerations about the distributions of the amplitudes of the two systems in resonance, leading to the idea of the squares of the amplitudes as measures of the strength of excitation of an eigenvalue. He then returns to resonance considerations in the case of a system coupled to a heat bath, which he considers would suffice in principle for the derivation of Planck's radiation formula and of all the results of the 'old quantum statistics'.

[a] See again the exchange of letters between Pauli and Schrödinger in December 1926 (Pauli 1979, pp. 364–8).

[b] Heisenberg (1930b) incorporates into his view of quantum theory both the multi-dimensional theory and the four-dimensional theory of Schrödinger's report. The latter is interpreted as a classical wave theory of matter that forms the background for a second quantisation, and includes the backreaction of the self-field via inclusion in the potential (cf. also above, fn. on p. 119).

[c] Schrödinger had expressed the idea of energy as frequency already in his letter to Wien of 25 August 1926: 'What we call the energy of an individual electron is its frequency. Basically it does not move with a certain speed because it has received a certain "shove", but because a dispersion law holds for the waves of which it consists, as a consequence of which a wave packet of this frequency has exactly this speed of propagation' (as quoted in Moore 1989, p. 225).

In the aftermath of the Solvay conference, although in places one finds Schrödinger at least apparently espousing views much closer to those of the Copenhagen-Göttingen school (cf. Moore 1989, pp. 250–1), Schrödinger continued to explore the possibilities of the wave picture.[a] The following is a telling example. In a short paper in *Naturwissenschaften*, Schrödinger (1929b) proposes to illustrate 'how the quantum theory in its newest phase again makes use of continuous spacetime functions, indeed of properties of their *form*, to describe the state and behaviour of a system ...' (p. 487). Schrödinger quotes the example of how photochemical and photoelectric phenomena depend on the form of an impinging wave (that is, on its Fourier decomposition), rather than on local properties of the wave at the point where it impinges on the relevant system. He argues that a continuous picture can be retained, but introduces the idea (which, as he remarks, generalises without difficulty to the case of many-particle wave functions) that the crucial properties of wave functions are in fact properties pertaining to the entire wave. Schrödinger may thus have been the first to introduce the idea of non-localisable properties, as part of the price to pay in order to pursue a wave picture of matter.

[a] Cf. also Bitbol's introduction in Schrödinger (1995), p. 5.

Archival notes

1 Ehrenfest to Lorentz, 30 March 1926, Noord-Hollands Archief, H. A. Lorentz papers, inventory number 20, AHQP-LTZ-11 (in German).
2 Schrödinger to Lorentz, 30 March 1926, AHQP-LTZ-8 (in German).
3 Verschaffelt to Lefébure, 8 April 1926, IIPCS 2573 (in French).
4 Lorentz to Einstein, 6 April 1926, AHQP-86 (in German).
5 Lorentz to Schrödinger, 21 January 1927, AHQP-41, section 9 (in German).
6 Schrödinger to Lorentz, 23 June 1927, AHQP-LTZ-13 (original with Schrödinger's corrections) and AHQP-41, section 9 (carbon copy) (in German).
7 See also Schrödinger to Lorentz, 16 July 1927, AHQP-LTZ-13 (in German).
8 Schrödinger to Lorentz, 23 June 1926, AHQP-LTZ-8 (in German).
9 Ehrenfest to Lorentz, 5 May 1926, AHQP-EHR-23 (in German).

Archival notes

Part II

Quantum foundations and the 1927 Solvay conference

5

Quantum theory and the measurement problem

5.1 What is quantum theory?

For much of the twentieth century, it was widely believed that the interpretation of quantum theory had been essentially settled by Bohr and Heisenberg in 1927. But not only were the 'dissenters' of 1927 – in particular de Broglie, Einstein and Schrödinger – unconvinced at the time: similar dissenting points of view are not uncommon even today. What Popper called 'the schism in physics' (Popper 1982) never really healed. Soon after 1927 it became standard to assert that matters of interpretation had been dealt with, but the sense of puzzlement and paradox surrounding quantum theory never disappeared.

As the century wore on, many of the concerns and alternative viewpoints expressed in 1927 slowly but surely revived. In 1952, Bohm revived and extended de Broglie's theory (Bohm 1952a,b), and in 1993 the de Broglie–Bohm theory finally received textbook treatment as an alternative formulation of quantum theory (Bohm and Hiley 1993; Holland 1993). In 1957, Everett (1957) revived Schrödinger's view that the wave function, and the wave function alone, is real (albeit in a very novel sense), and the resulting 'Everett' or 'many-worlds' interpretation (DeWitt and Graham 1973) gradually won widespread support, especially among physicists interested in quantum gravity and quantum cosmology. Theories even closer to Schrödinger's ideas – collapse theories, with macroscopic objects regarded as wave packets whose spreading is prevented by stochastic collapse – were developed from the 1970s onwards (Pearle 1976, 1979; Ghirardi, Rimini and Weber 1986). As for Einstein's concerns in 1927 about the non-locality of quantum theory (see Chapter 7), re-expressed in the famous EPR paper of 1935, matters came to a head in 1964 with the publication of Bell's theorem (Bell 1964). In the closing decades of the twentieth century, after many stringent experimental tests showed that Bell's inequality was violated by entangled quantum states, non-locality came to be widely regarded as a central fact of the quantum world.

Other concerns, voiced by Schrödinger just a few years after the fifth Solvay conference, also eventually played a central role after decades of obscurity. Schrödinger's 'cat paradox' of 1935 came to dominate discussions about the meaning of quantum theory. And the peculiar 'entanglement' that Schrödinger had highlighted as a key difference between classical and quantum physics (Schrödinger 1935) eventually found its place as a central concept in quantum information theory: as well as being a matter of 'philosophical' concern, entanglement came to be seen as a physical resource to be exploited for technological purposes, and as a central feature of quantum physics that had been strangely under-appreciated for most of the twentieth century.

The interpretation of quantum theory is probably as controversial now as it ever has been. Many workers now recognise that standard quantum theory – centred as it is around the notion of 'measurement' – requires a classical background (containing macroscopic measuring devices), which can never be sharply defined, and which in principle does not even exist. Even so, the operational approach to the interpretation of quantum physics is still being pursued by some, in terms of new axioms that constrain the structure of quantum theory (Hardy 2001, 2002; Clifton, Bub and Halvorson 2003). On the other hand, those who do regard the background problem as crucial tend to assert that everything in the universe – microscopic systems, macroscopic equipment, and even human experimenters – should in principle be described in a unified manner, and that 'measurement' processes must be regarded as physical processes like any other. Approaches of this type include: the Everett interpretation (Everett 1957), which is being subjected to increasing scrutiny at a foundational level (Saunders 1995, 1998; Deutsch 1999; Wallace 2003a,b); the pilot-wave theory of de Broglie and Bohm (de Broglie 1928; Bohm 1952a,b; Bohm and Hiley 1993; Holland 1993), which is being pursued and developed more than ever before (Cushing, Fine and Goldstein 1996; Pearle and Valentini 2006; Valentini 2008a); collapse models (Pearle 1976, 1979, 1989; Ghirardi, Rimini and Weber 1986), which are being subjected to ever more stringent experimental tests (Pearle and Valentini 2006); and theories of 'consistent' or 'decoherent' histories (Griffiths 1984, 2002; Gell-Mann and Hartle 1990; Omnès 1992, 1994; Hartle 1995).

Today, it is simply untenable to regard the views of Bohr and Heisenberg (which in any case differed considerably from each other) as in any sense standard or canonical. The meaning of quantum theory is today an open question, arguably as much as it was in October 1927.

5.2 The measurement problem today

The problem of measurement and the observer is the problem of where the measurement begins and ends, and where the observer begins and ends. ... I think, that – when you analyse this language that the physicists have fallen into, that physics is about the results of observations – you find that on analysis it evaporates, and nothing very clear is being said.
(J. S. Bell (1986, p. 48))

The recurring puzzlement over the meaning of quantum theory often centres around a group of related conceptual questions that usually come under the general heading of the 'measurement problem'.

5.2.1 A fundamental ambiguity

As normally presented in textbooks, quantum theory describes experiments in a way that is certainly practically successful, but seemingly fundamentally ill-defined. For it is usually implicitly or explicitly assumed that there is a clear boundary between microscopic quantum systems and macroscopic classical apparatus, or that there is a clear dividing line between 'microscopic indefiniteness' and the definite states of our classical macroscopic realm. Yet, such distinctions defy sharp and precise formulation.

That quantum theory is therefore fundamentally ambiguous was argued with particular clarity by Bell. For example (Bell 1986, p. 54):

The formulations of quantum mechanics that you find in the books involve dividing the world into an observer and an observed, and you are not told where that division comes – on which side of my spectacles it comes, for example – or at which end of my optic nerve.

The problem being pointed to here is the lack of a precise boundary between the quantum system and the rest of the world (including the apparatus and the experimenter).

A closely related aspect of the 'measurement problem' is the need to explain what happens to the definite states of the everyday macroscopic domain as one goes to smaller scales. Where does macroscopic definiteness give way to microscopic indefiniteness? Does the transition occur somewhere between pollen grains and macromolecules, and if so, where? On which side of the line is a virus?

Nor can quantum 'indefiniteness' or 'fuzziness' be easily confined to the atomic level. For macroscopic objects are made of atoms, and so inevitably one is led to doubt whether rocks, trees or even the Moon have definite macroscopic states, especially when observers are not present. And this in the face of remarkable developments in twentieth-century astrophysics and cosmology, which have traced the origins of stars, galaxies, helium and the other elements, to times long before human observers existed.

The notion of a 'real state of affairs' is familiar from everyday experience: for example, the location and number of macroscopic bodies in a laboratory. Science has shown that there is more to the real state of things than is immediately obvious (for example, the electromagnetic field). Further, it has been shown that the character of the real state of things changes with scale: on large scales we find planets, stars and galaxies, while on small scales we find pollen grains, viruses, molecules and atoms. Nevertheless, at least outside of the quantum domain, the notion of 'real state' remains. The ambiguity emphasised by Bell consists of the lack of a sharp boundary between the 'classical' domain, in which 'real state' is a valid concept, and the 'quantum domain', in which 'real state' is not a valid concept.

Despite decades of effort, this ambiguity remains unresolved within standard textbook quantum theory, and many critics have been led to argue that the notion of real state should be extended, in some appropriate way, into the quantum domain. Thus, for example, Bell (1987, pp. 29–30) writes:

> Theoretical physicists live in a classical world, looking out into a quantum-mechanical world. The latter we describe only subjectively, in terms of procedures and results in our classical domain. This subjective description is effected by means of quantum-mechanical state functions ψ .... The classical world of course is described quite directly – 'as it is'. ... Now nobody knows just where the boundary between the classical and quantum domain is situated. ... A possibility is that we find exactly where the boundary lies. More plausible to me is that we will find that there is no boundary. It is hard for me to envisage intelligible discourse about a world with no classical part – no base of given events ... to be correlated. On the other hand, it is easy to imagine that the classical domain could be extended to cover the whole.

While Bell goes on to argue in favour of adding extra ('hidden') parameters to the quantum formalism, for our purposes the key point being made here is the need to extend the notion of real state into the microscopic domain. This might indeed be achieved by introducing hidden variables, or by other means (for example, the Everett approach). Whatever form the theory may take, the real macroscopic states considered in the rest of science should be part of a unified description of microscopic and macroscopic phenomena – what Bell (1987, p. 30) called 'a homogeneous account of the world'.

There are in fact, as we have mentioned, several well-developed proposals for such a unified or homogeneous (or 'realist') account of the world: the pilot-wave theory of de Broglie (1928) and Bohm (1952a,b); theories of dynamical wave-function collapse (Pearle 1976, 1979, 1989; Ghirardi, Rimini and Weber 1986); and the many-worlds interpretation of Everett (1957). The available proposals that have broad scope assume that the wave function is a real object that is part of the structure of an individual system. At the time of writing, it is not known if realist theories may be constructed without this feature.[a]

5.2.2 Measurement as a physical process: quantum theory 'without observers'

Another closely related aspect of the measurement problem is the question of how quantum theory may be applied to the process of measurement itself. For it seems inescapable that it should be possible (in principle) to treat apparatus and observers as physical systems, and to discuss the process of measurement in purely quantum-theoretical terms. However, attempts to do so are notoriously controversial and apt to result in paradox and confusion.

For example, in the paradox of 'Wigner's friend' (Wigner 1961), an experimenter A (Wigner) possesses a box containing an experimenter B (his friend) and a microscopic system S. Suppose S is initially in, for example, a superposition of energy states

$$|\psi_0\rangle = \frac{1}{\sqrt{2}}\left(|E_1\rangle + |E_2\rangle\right)$$

and that the whole box is initially in a state $|\Psi_0\rangle = |B_0\rangle \otimes |\psi_0\rangle$ (idealising B as initially in a pure state $|B_0\rangle$). Let B perform an ideal energy measurement on S. If A does not carry out any measurement, then from the point of view of A the quantum state of the whole box evolves continuously (according to the Schrödinger equation) into a superposition of states

$$|\Psi(t)\rangle = \frac{1}{\sqrt{2}}\left(|B_1\rangle \otimes |E_1\rangle + |B_2\rangle \otimes |E_2\rangle\right) \tag{5.1}$$

(where $|B_i\rangle$ is a state such that B has found the energy value E_i).

Now, if A wished to, could he (in principle) at later times, by appropriate experiments on the whole box, observe interference effects involving both branches of the superposition in (5.1)? If so, could this be consistent with the point of view of B, according to which the energy measurement had a definite result?

One may well question whether the above scenario is realistic, even in principle (given the resources in our universe). For example, one might question whether a box containing a human observer could ever be sufficiently isolated for environmental decoherence to be negligible. However, if the above 'experimenter B'

[a] For example, according to the stochastic hidden-variables theory of Fényes (1952) and Nelson (1966), the wave function merely provides an emergent description of probabilities. However, despite appearances, it seems that for technical reasons this theory is flawed and does not really reproduce quantum theory: the Schrödinger equation is obtained only for exceptional (nodeless) wave functions (Wallstrom 1994; Pearle and Valentini 2006).

were replaced by an automatic device or machine, the scenario may indeed become realistic, depending on the possibility of isolating the box to sufficient accuracy.[a]

Most scientists agree that macroscopic equipment is subject to the laws of physics, just like any other system, and that it should be possible to describe the operation of such equipment purely in terms of the most fundamental theory available. There is somewhat less consensus over the status of human experimenters as physical systems. Some physicists have suggested, in the context of quantum physics, that a human being cannot be treated as just another physical system, and that human consciousness plays a special role. For example, Wigner (1961) concluded from his paradox that 'the being with a consciousness must have a different role in quantum mechanics than the inanimate measuring device', and that for a system containing a conscious observer 'the quantum mechanical equations of motion cannot be linear'.

Wigner's conclusion, that living beings violate quantum laws, seems increasingly incredible given the impressive progress made in human biology and neuroscience, in which the human organism – including the brain – is treated as (ultimately) a complex electro-chemical system. There is no evidence that human beings are able to violate, for example, the laws of gravity, or of thermodynamics, or basic principles of chemistry, and the conclusion that human beings in particular should be outside the domain of quantum laws seems difficult to accept. An alternative conclusion, of course, is that something is missing from orthodox quantum theory.

Assuming, then, that human experimenters and their equipment may in principle be regarded as physical systems subject to the usual laws, their interaction with microsystems ought to be analysable, and the process of measurement ought to be treatable as a physical process like any other. One can then ask if, over an ensemble of similar experiments, it would be possible in principle for the external experimenter A to observe (at the statistical level) interference effects associated with both terms in (5.1). To deny this possibility would be to claim (with Wigner) that a box containing a human being violates the laws of quantum theory. To accept the possibility would seem to imply that, at time t before experimenter A makes a measurement, there was (at least according to A) no matter of fact about the result of B's observation, notwithstanding the explicit supposition that B had indeed carried out an observation by time t.

It is sometimes said that Wigner's paradox may be evaded by noting that, if the external experimenter A actually performs an experiment on the whole box

[a] It is perhaps worth remarking that, even if decoherence has a role to play here, one has to realise what the problem is in order to understand whether and how decoherence might contribute to a solution (cf. Bacciagaluppi 2005).

that reveals interference between the two branches of (5.1), then this operation will destroy the memory the internal experimenter B had of obtaining a particular result, so that there is no contradiction. But this misses the point. For while it is true that B will then not have any *memory* of having obtained a particular experimental result, the contradiction remains with there being a purported *matter of fact* (at time t) as to the result of B's observation, regardless of whether or not B has subsequently forgotten it. (Note that in this discussion we are talking about matters of fact, not for microsystems, but for macroscopic experimental results.)

Let us examine the reasoning behind Wigner's paradox more closely. We take it that experimenter B agreed beforehand to enter the box and perform an energy measurement on the microscopic system S; and that it was further agreed that after sufficient time had elapsed for B to perform the measurement, A would decide whether or not to carry out an experiment showing interference between the two branches of (5.1). Considering an ensemble of similar experiments, the paradox consists of a contradiction between the following statements concerning the physical state of B just before A decides what to do:

(I) There is no definite state of B, because A can if he wishes perform measurements showing the presence of interference between different states of B.
(II) There is a definite state of B, because B is a human experimenter like any other, and because instead of testing for interference A can simply ask B what he saw.

The argument in (I) is the familiar one from standard quantum theory, applied to the unusual case of a box containing an experimenter. The argument in (II) is unusual: it requires comment and elaboration.

Because B is a human experimenter like any other, we are driven to consider the theoretical possibility that, in the distant future, some 'super-experimenter' could decide to perform an interference experiment on a 'box' containing *us* and our equipment. What would happen to our current (macroscopically recorded) experimental facts – concerning for example the outcome of a spin measurement performed in the laboratory? To be sure, our records of these experiments could one day be erased, but it would be illogical to suppose that the *fact* of these experiments having been carried out (with definite results) could ever be changed. Unless we accept (II), we are in danger of encountering the paradox that facts about what we have done in the laboratory today might later turn out *not* to be facts.

Further support for (II) comes from Wigner's original argument, which centred around the assumed reality of other minds. (Wigner did not regard solipsism as worthy of serious consideration.) From this assumption Wigner inferred that, whatever the circumstances, if one asks a 'friend' what he saw, the answer given by the friend must have been, as Wigner put it, 'already decided in his mind, before I asked him'. But then, if A decides not to perform an interference experiment on the box, and simply asks B what he saw, A is obliged to take B's answer as indicative of the state of B's mind before A asked the question – indeed, before A decided on whether or not to perform an interference experiment. For Wigner, a superposition of the form (5.1) is unacceptable for a system containing a human experimenter or 'friend', because it implies that the friend 'was in a state of suspended animation before he answered my question'.

As Wigner presented it, the argument is based on the assumption that a conscious being will always have a definite state of consciousness. In orthodox quantum theory, of course, one might dismiss as 'meaningless' the question of whether the friend's consciousness contained one impression or the other (B_1 or B_2) before he was asked. However, as Wigner put it, 'to deny the existence of the consciousness of a friend to this extent is surely an unnatural attitude, approaching solipsism'.

Finally, on the topic of Wigner's paradox, it is important to emphasise the distinction between 'matters of fact' on the one hand, and 'memories' (true or false) on the other – a distinction that is comparable to the distinction between facts and opinions, or between truth and belief, or between reality and appearance, distinctions that form part and parcel of the scientific method. Thus, again, while an experimenter's memory of having obtained a certain result might be erased in the future, the *fact* that he once obtained a certain result will necessarily remain a fact: to assert otherwise would be a logical contradiction.

A further, more subtle motivation for treating measurement as a physical process comes from considering the very nature of 'measurement'. As is well known to philosophers, and to experimental physicists, the process of measurement is 'theory-laden'. That is, in order to know how to carry out a measurement correctly, or how to design a specific measuring apparatus correctly, some prior body of theory is required: in particular, one needs some understanding of how the equipment functions, and how it interacts with the system being examined. For this reason, it is difficult to see how the process of quantum measurement can be properly understood, without some prior body of theory that describes the equipment itself and its interaction with the 'system'. And since the equipment usually belongs to the definite macroscopic realm, and the 'system' often does not, a proper understanding seems to require a 'homogeneous account of the world' as discussed above, that is, a theory in which an objective account is provided not only of the macroscopic apparatus, but also of the microsystem and its interaction with the apparatus.

A common conclusion, then, is that a coherent account of quantum measurement requires that quantum theory be somehow extended from a theory of microsystems to a universal physical theory with an unbounded domain of application, with our everyday macroscopic realism being somehow extended to the microscopic level.

Given such a well-defined and universal physical theory, whose subject matter consists of the real states of the world as a whole, it would be possible in principle to use the theory to analyse the process of measurement as a physical process like any other (just as, for example, classical electrodynamics may be used to analyse the process – involving forces exerted by magnetic fields – by which an ammeter measures an electric current).

Thus, for example, pilot-wave theory, or the Everett interpretation, or collapse models, may be applied to situations where a quantum measurement is taking place. If the theory provides an unambiguous account of objective processes in general, it will provide an unambiguous account of the quantum measurement process in particular. The result is a quantum theory 'without observers', in the sense that observers are physical systems obeying the same laws as all other systems, and do not have to be added to the theory as extra-physical elements.

Conclusions as to what is actually happening during quantum measurements will, of course, depend on the details of the theory. For example, consider again Wigner's scenario above. In the Everett interpretation, B's observation within the box has two results, and there is no contradiction if the external experimenter A subsequently observes interference between them. In de Broglie–Bohm theory, B's observation has only one result selected by the actual configuration, but even so the empty wave packet still exists in configuration space, and can in principle re-overlap with (and hence interfere with) the occupied packet if appropriate Hamiltonians are applied.

5.2.3 Quantum cosmology

Further closely related questions, again broadly under the heading of the 'measurement problem', concern the description of the distant past before human beings and other life forms evolved on Earth, and indeed the description of the universe as a whole in epochs before life existed.

While the basic theoretical foundations of big-bang cosmology had already been laid by 1927 (through the work of Friedmann and Lemaître), at that time any suggestion of the need to provide a quantum-theoretical account of the early universe could easily have been dismissed as being of no practical or experimental import. By the 1980s, however, with the development of inflationary cosmology (Guth 1981), the theoretical question became a practical one, with observational implications.

According to our current understanding, the small non-uniformities of temperature observed in the cosmic microwave background originated from classical density perturbations in the early (and approximately homogeneous) universe (Padmanabhan 1993). And according to inflationary theory, those early classical density perturbations originated from quantum fluctuations at even earlier times (Liddle and Lyth 2000). Here we have an example of a cosmological theory in which a 'quantum-to-classical transition' occurred long before life (or even galaxies) developed, and whose details have left an imprint on the sky that can be measured today. This is the measurement problem on a cosmic scale (Kiefer, Polarski and Starobinsky 1998; Perez, Sahlmann and Sudarsky 2006; Valentini 2008b).

But the tension between 'Copenhagen' quantum theory and the requirements of cosmology was felt long before cosmology matured as an experimental science. Thus, for example, in his pioneering work in the 1960s on quantum gravity, when it came to applying the theory to a closed universe DeWitt wrote (DeWitt 1967, p. 1131):

> The Copenhagen view depends on the assumed *a priori* existence of a classical level to which all questions of observation may ultimately be referred. Here, however, the whole universe is the object of inspection; there is no classical vantage point, and hence the interpretation question must be re-argued from the beginning.

DeWitt went on to argue (pp. 1140–2) that, in the absence of a classical level, the Everett interpretation should be adopted. According to DeWitt (p. 1141):

> Everett's view of the world is a very natural one to adopt in the quantum theory of gravity, where one is accustomed to speak without embarrassment of the 'wave function of the universe'.

While DeWitt expresses a preference for the Everett interpretation, for our purposes the central point being made is that, if the whole universe is treated as a quantum object, with no definite (classical) background 'to which all questions of observation may ultimately be referred', then the physics becomes unintelligible unless some form of real state (or ontology) is ascribed to the quantum object. The Everett interpretation provides one way, among others, to do this.[a]

Everett himself, in 1957, had already cited the quantum theory of cosmology as one of his main motivations for going beyond what he called the 'conventional or "external observation" formulation of quantum mechanics' (Everett 1957, p. 454). Everett's general motivation was the need to describe the quantum physics internal to an isolated system, in particular one containing observers. A closed universe was a special case of such a system, and one that arguably would have to be considered as the science of cosmology progressed (as has indeed proved to be the case). Thus Everett wrote (p. 455):

[a] Everett's original formulation was of course open to a number of criticisms, in particular concerning the notion of 'world' and the idea of probability. Such criticisms are arguably being met only through more recent developments (Saunders 1995, 1998; Deutsch 1999; Wallace 2003a,b).

> How is one to apply the conventional formulation of quantum mechanics to the space-time geometry itself? The issue becomes especially acute in the case of a closed universe. There is no place to stand outside the system to observe it. There is nothing outside it to produce transitions from one state to another. ... No way is evident to apply the conventional formulation of quantum mechanics to a system that is not subject to *external* observation. The whole interpretive scheme of that formalism rests upon the notion of external observation. The probabilities of the various possible outcomes of the observation are prescribed exclusively by Process 1 [discontinuous wave function collapse].

In more recent years, similar concerns have motivated the development of a 'generalised quantum mechanics' based on 'consistent' or 'decoherent' histories (Griffiths 1984, 2002; Gell-Mann and Hartle 1990; Omnès 1992, 1994; Hartle 1995), an approach that is also supposed to provide a quantum theory 'without observers', and without a presumed classical background, so as to be applicable to quantum cosmology.

5.2.4 The measurement problem in 'statistical' interpretations of ψ

The measurement problem is often posed simply as the problem of how to interpret a macroscopic superposition of quantum states, such as (pure) states of Schrödinger's cat. This way of posing the measurement problem can be misleading, however, as it usually rests on the implicit assumption (or suggestion) that the quantum wave function ψ is a real physical object identifiable as a complete description of an individual system. It might be that ψ is indeed a real object, but not a complete description (as in pilot-wave theory). Or, ψ might not be a real object at all. Here we focus on the latter possibility.

The quantum wave function ψ might be merely a mathematical tool for calculating and predicting the measured frequencies of outcomes over an ensemble of similar experiments. In which case, it would be immediately wrong to interpret a mathematical superposition of terms in ψ as somehow corresponding to a physical superposition of real states for individual systems, and the 'measurement problem' in the limited sense just mentioned would be a pseudo-problem. This 'statistical interpretation' of quantum theory has been championed in particular by Ballentine (1970).

However, even in the statistical interpretation, the 'measurement problem' in the more general sense remains. For quantum theory is then an incomplete theory that refers only to ensembles, and simply does not fully describe individual quantum systems or their relation to real, individual macroscopic states. The statistical interpretation gives no account of what happens, for example, when an individual electron is being measured: it talks only about the distribution of (macroscopically registered) measurement outcomes over an ensemble of similar experiments. Nor does the statistical interpretation provide any sharp delineation of the boundary between 'macroscopic' objects with an individual (non-ensemble) description and 'microscopic' objects with no such description.[a]

In the statistical interpretation, then, a solution of the measurement problem in the general sense will require the development of a complete description of individual systems. This was Einstein's point of view (Einstein 1949, pp. 671–2):

> The attempt to conceive the quantum-theoretical description as the complete description of the individual systems leads to unnatural theoretical interpretations, which become immediately unnecessary if one accepts the interpretation that the description refers to ensembles of systems and not to individual systems. ... [I]t appears unavoidable to look elsewhere for a complete description of the individual system ....

Einstein was arguably the founder of the statistical interpretation.[b] It should be noted, however, that while Einstein's conclusion about the nature of ψ might turn out to be correct, what seems to have been his main *argument* for this conclusion now appears to be wrong, in that it was based on what now appears to be a false premise – the assumption of locality. For example, in a letter to his friend Michele Besso, dated 8 October 1952, Einstein argued that the 'quantum state' ψ could not be a complete characterisation of the 'real state' of an individual system, on the following grounds:

> A system S_{12}, with known function ψ_{12}, is composed of subsystems S_1 and S_2, which at time t are far away from each other. If one makes a 'complete' measurement on S_1, this can be done in different ways .... From the measurement result *and* the ψ-function ψ_{12}, one can determine ... the ψ-function ψ_2 of the second system. *This will take on different forms*, according to the *kind* of measurement applied to S_1.
>
> But this is in contradiction with assumption (1) [that the quantum state characterises the real state completely], *if one excludes action at a distance*. Then in fact the measurement on S_1 can have no influence on the real state of S_2, and therefore according to (1) can have also no influence on the quantum state of S_2 described by ψ_2.
>
> *(Einstein and Besso 1972, pp. 487–8, emphasis in the original)*

Einstein's argument hinges on the fact that, in a local physics, the measurement made on S_1 can have no effect on the real state of S_2.[c]

With the development of Bell's theorem, however, it seems to be beyond reasonable doubt that quantum physics is *not* local (if one assumes the absence of backwards causation or of many worlds). For if locality is assumed, one may use the EPR argument to infer determinism for the outcomes of quantum measurements

[a] There are different ways of considering a 'statistical' interpretation, depending on one's point of view concerning the nature of probability. In any case, the 'measurement problem' in the general sense still stands.

[b] Cf. Ballentine (1972).

[c] The notion of locality that Einstein uses here is, to be precise, a combination of the principles of 'separability' (that widely separated systems have locally defined real states) and of 'no action at a distance'. Cf. Howard (1990).

at widely separated wings of an entangled state.[a] Following further reasoning by Bell (1964), one may then show that any local and deterministic completion of quantum theory cannot reproduce quantum correlations for all measurements on entangled states. Therefore, locality contradicts quantum theory.[b] Because the premise of Einstein's argument contradicts quantum theory, the argument cannot be used to infer anything about quantum theory or about the nature of ψ. Thus, Einstein's argument does not establish the 'statistical' or 'ensemble' nature of ψ.[c]

[a] 'It is important to note that ... *determinism* ... in the EPR argument ... is not assumed but *inferred* [from locality]. ... It is remarkably difficult to get this point across, that determinism is not a *presupposition* of the analysis' (Bell 1987, p. 143, italics in the original).

[b] See, however, Fine (1999) for a dissenting view.

[c] Einstein's argument above has recently been revived by Fuchs (2002) (who is, however, not explicit about the completeness or incompleteness of quantum theory). Fuchs states (p. 9) that Einstein 'was the first person to say in absolutely unambiguous terms why the quantum state should be viewed as information .... His argument was simply that a quantum-state assignment for a system can be forced to go one way or the other by interacting with a part of the world that should have no causal connection with the system of interest'. Fuchs then quotes at length (p. 10) the above letter by Einstein. Later in the same paper, Fuchs writes (p. 39): 'Recall what I viewed to be the most powerful argument for the quantum state's subjectivity – the Einsteinian argument of [the above letter]. Since [for entangled systems] we can toggle the quantum state from a distance, it must not be something sitting over there, but rather something sitting over here: It can only be our information about the far-away system'. Again, the premise of this Einsteinian argument – locality – is nowadays no longer reasonable (as of course it was in 1952), and so the argument cannot be used to infer the subjective or epistemic nature of the quantum state.

6

Interference, superposition and wave packet collapse

6.1 Probability and interference

According to Feynman, single-particle interference is 'the *only* mystery' of quantum theory (Feynman, Leighton and Sands 1965, ch. 1, p. 1). Feynman considered an experiment in which particles are fired, one at a time, towards a screen with two holes labelled 1 and 2. With both holes open, the distribution P_{12} of particles at the backstop displays an oscillatory pattern of bright and dark fringes. If P_1 is the distribution with only hole 1 open, and P_2 is the distribution with only hole 2 open, then experimentally it is found that $P_{12} \neq P_1 + P_2$. According to the argument given by Feynman (as well as by many other authors), this result is inexplicable by 'classical' reasoning.

By his presentation of the two-slit experiment (as well as by his development of the path-integral formulation of quantum theory), Feynman popularised the idea that the usual probability calculus breaks down in the presence of quantum interference, where it is probability amplitudes (and not probabilities themselves) that are to be added. As pointed out by Koopman (1955), and by Ballentine (1986), this argument is mistaken: the probability distributions at the backstop – P_{12}, P_1 and P_2 – are conditional probabilities with three distinct conditions (both slits open, one or other slit closed), and probability calculus does *not* imply any relationship between these. Feynman's argument notwithstanding, standard probability calculus is perfectly consistent with the two-slit experiment.

In his influential lectures on physics, as well as asserting the breakdown of probability calculus, Feynman claimed that no theory with particle trajectories could explain the two-slit experiment. This claim is still found in many textbooks.[a]

[a] For example, Shankar (1994) discusses the two-slit experiment at length in his chapter 3, and claims (p. 111) that the observed single-photon interference pattern 'completely rules out the possibility that photons move in well-defined trajectories'. Further, according to Shankar (p. 112): 'It is now widely accepted that all particles are described by probability amplitudes $\psi(x)$, and that the assumption that they move in definite trajectories is ruled out by experiment'.

From a historical point of view, it is remarkable indeed that single-particle interference came to be widely regarded as inconsistent with any theory containing particle trajectories – for as we saw in Chapter 2, in the case of electrons this phenomenon was in fact first predicted (by de Broglie) on the basis of precisely such a theory.

As we shall now discuss, in his report at the fifth Solvay conference de Broglie gave a clear and simple explanation for single-particle interference on the basis of his pilot-wave theory; and the extensive discussions at the conference contain no sign of any objection to the consistency of de Broglie's position on this point.

As for Schrödinger's theory of wave mechanics, in which particles were supposed to be constructed out of localised wave packets, in retrospect it is difficult to see how single-particle interference could have been accounted for. It is then perhaps not surprising that, in Brussels in 1927, no specific discussion of interference appears in Schrödinger's contributions.

Born and Heisenberg, on the other hand, do discuss interference in their report, from the point of view of their 'quantum mechanics'. And, they do consider the question of the applicability of probability calculus. The views they present are, interestingly enough, rather different from the views usually associated with quantum mechanics today. In particular, as we shall see below, according to Born and Heisenberg there was (in a very specific sense) *no* conflict between quantum interference and the ordinary probability calculus.

Interference was also considered in the general discussion, in particular by Dirac and Heisenberg – this material will be discussed later, in Section 6.3.

6.1.1 Interference in de Broglie's pilot-wave theory

At the fifth Solvay conference, the subject of interference was addressed from a pilot-wave perspective by de Broglie in his report. In his Section 5, 'The interpretation of interference', de Broglie considered interference experiments with light of a given frequency ν. For a guiding wave Ψ of phase ϕ and amplitude a, de Broglie took the photon velocity to be given by $\mathbf{v} = -\frac{c^2}{h\nu}\nabla\phi$, while the probability distribution was taken to be $\pi = \text{const}\cdot a^2$. As de Broglie had pointed out, the latter distribution is preserved over time by the assumed motion of the photons. Therefore, the usual interference and diffraction patterns follow immediately. To quote de Broglie (p. 351):

> the bright and dark fringes predicted by the new theory will coincide with those predicted by the old [that is, by classical wave optics].

De Broglie also pointed out that his theory gave the correct bright and dark fringes for photon interference experiments, regardless of whether the experiments were performed with an intense or a very feeble source. As he put it (p. 351):

> one can do an experiment of short duration with intense irradiation, or an experiment of long duration with feeble irradiation... if the light quanta do not act on each other the statistical result must evidently be the same.

De Broglie's discussion here addresses precisely the supposed difficulty highlighted much later by Feynman. It is noteworthy that a clear and simple answer to what Feynman thought was 'the only mystery' of quantum mechanics was published as long ago as the 1920s.

Even so, for the rest of the twentieth century, the two-slit experiment was widely cited as proof of the non-existence of particle trajectories in the quantum domain. Such trajectories were thought to imply the relation $P_{12} = P_1 + P_2$, which is violated by experiment. As Feynman put it, on the basis of this argument it should 'undoubtedly' be concluded that '[i]t is *not* true that the electrons go *either* through hole 1 or hole 2' (Feynman, Leighton and Sands 1965, ch. 1, p. 6). Feynman also suggested that, by 1965, there had been a long history of failures to explain interference in terms of trajectories:

> Many ideas have been concocted to try to explain the curve for P_{12} [that is, the interference pattern] in terms of individual electrons going around in complicated ways through the holes. None of them has succeeded.
>
> *(Feynman 1965, ch. 1, p. 6)*

Yet, de Broglie's construction is so simple as to be almost trivial: the quantum probability density $|\psi|^2$ for particle position obeys a continuity equation, with a local probability current; if the trajectories follow the flow lines of the quantum current then, by construction, an incident distribution $|\psi_0|^2$ of particles will necessarily evolve into a distribution $|\psi|^2$ at the backstop – with interference or diffraction, as the case may be, depending on the potential in which the wave ψ evolves.

Not only did Feynman claim, wrongly, that no one had ever succeeded in explaining interference in terms of trajectories; he also gave an argument to the effect that any such explanation was impossible:

> Suppose we were to assume that inside the electron there is some kind of machinery that determines where it is going to end up. That machine must *also* determine which hole it is going to go through on its way. But ... what is inside the electron should not be dependent... upon whether we open or close one of the holes. So if an electron, before it starts, has already made up its mind (a) which hole it is going to use, and (b) where it is going to land, we should find P_1 for those electrons that have chosen hole 1, P_2 for those that have chosen hole 2, *and necessarily* the sum $P_1 + P_2$ for those that arrive through the two holes. There seems to be no way around this.
>
> *(Feynman, Leighton and Sands 1965, ch. 1, p. 10)*

Feynman's argument assumes that the motion of the electron is unaffected by opening or closing one of the holes. This assumption is violated in pilot-wave theory, where the form of the guiding wave behind the two-slit screen does depend on whether or not both slits are open.

A similar assumption is made in the discussion of the two-slit experiment by Heisenberg (1962), in chapter III of his book *Physics and Philosophy*. Heisenberg considers single photons incident on a screen with two small holes and a photographic plate on the far side, and gives the familiar argument that the existence of particle trajectories implies the non-interfering result $P_1 + P_2$. As Heisenberg puts it:

> If [a single photon] goes through the first hole and is scattered there, its probability for being absorbed at a certain point of the photographic plate cannot depend upon whether the second hole is closed or open.
>
> *(Heisenberg 1962)*

This assertion is denied by pilot-wave theory, which provides a simple counter-example to Heisenberg's conclusion that 'the statement that any light quantum must have gone *either* through the first *or* through the second hole is problematic and leads to contradictions'.

Finally, we note that interference was also considered by Brillouin (pp. 367ff.) in the discussion following de Broglie's report, for the case of photons reflected by a mirror. Brillouin drew a figure (p. 368), with a sketch of a photon trajectory passing through an interference region. To our knowledge, plots of trajectories in cases of interference did not appear again in the literature until the pioneering numerical work by Philippidis, Dewdney and Hiley (1979).

6.1.2 Interference in the 'quantum mechanics' of Born and Heisenberg

The subject of interference was considered by Born and Heisenberg, in their report on quantum mechanics (pp. 387ff), for the case of an atom initially in a superposition

$$|\psi(0)\rangle = \sum_n c_n(0)\,|n\rangle \tag{6.1}$$

of energy states $|n\rangle$, with coefficients $c_n(0) = |c_n(0)|\, e^{i\gamma_n}$ and eigenvalues E_n. The Schrödinger equation implies a time evolution

$$c_n(t) = \sum_m S_{nm}(t) c_m(0) \tag{6.2}$$

with (in modern notation) $S_{nm}(t) = \langle n|\, U(t,0)\, |m\rangle$, where $U(t,0)$ is the evolution operator. In the special case where $c_m(0) = \delta_{mk}$ for some k, we have $|c_n(t)|^2 = |S_{nk}(t)|^2$, and Born and Heisenberg interpret $|S_{nk}(t)|^2$ as a transition probability. They also draw the conclusion that 'the $|c_n(t)|^2$ must be the state probabilities' (p. 387).

Born and Heisenberg seem to adopt a statistical interpretation, according to which the system is always in a definite energy state, with jump probabilities $|S_{nk}(t)|^2$ and occupation (or 'state') probabilities $|c_n(t)|^2$ (cf. Section 3.4.6). This is stated quite explicitly (p. 386):

From the point of view of Bohr's theory a system can always be in only *one* quantum state.... According to Bohr's principles it makes no sense to say a system is simultaneously in several states. The only possible interpretation seems to be statistical: the superposition of several eigensolutions expresses that through the perturbation the initial state can go over to any other quantum state....

Note that this is quite different from present-day quantum mechanics, in which a system described by the superposition (6.1) would not normally be regarded as always occupying only one energy state.

At the same time, Born and Heisenberg recognise a difficulty (p. 387):

Here, however, one runs into a difficulty of principle that is of great importance, as soon as one starts from an initial state for which not all the $c_n(0)$ except one vanish.

The difficulty, of course, is that for an initial superposition the final probability distribution is given by

$$|c_n(t)|^2 = \left| \sum_m S_{nm}(t)c_m(0) \right|^2 \tag{6.3}$$

as opposed to

$$|c_n(t)|^2 = \sum_m |S_{nm}(t)|^2 \, |c_m(0)|^2 \tag{6.4}$$

which, as Born and Heisenberg remark, 'one might suppose from the usual probability calculus'. (In standard probability calculus, of course, (6.4) expresses $|c_n(t)|^2$ as a sum over conditional transition probabilities $|S_{nm}(t)|^2$ weighted by the initial population probabilities $|c_m(0)|^2$.)

While Born and Heisenberg refer to (6.3) as the 'theorem of the *interference of probabilities*', they make the remarkable assertion that there is in fact *no* contradiction with the usual rules of probability calculus, and that the $|S_{nm}|^2$ may still be regarded as ordinary probabilities. Further, and equally remarkably, it is claimed that in any case where the state probabilities $|c_n|^2$ are experimentally established, the presence of the unknown phases γ_n makes the interfering expression (6.3) reduce to the non-interfering expression (6.4) (p. 387):

... it should be noted that this 'interference' does not represent a contradiction with the rules of the probability calculus, that is, with the assumption that the $|S_{nk}|^2$ are quite usual probabilities. In fact, ... [(6.4)] follows from the concept of probability ... when and only when the relative number, that is, the probability $|c_n|^2$ of the atoms in the state n, has been *established* beforehand *experimentally*. In this case the phases γ_n are unknown in principle, so that [(6.3)] then naturally goes over to [(6.4)]....

Here, Born and Heisenberg refer to Heisenberg's (recently published) uncertainty paper, which contains a similar claim. There, Heisenberg considers a Stern–Gerlach atomic beam passing through two successive regions of field inhomogeneous in the direction of the beam (so as to induce transitions between energy states without separating the beam into components). If the input beam is in a definite energy state then the beam emerging from the first region will be in a superposition. The probability distribution for energy emerging from the second region will then contain interference – as in (6.3), where the 'initial' superposition (6.1) is now the state emerging from the first region. Heisenberg asserts that, if the energy of an atom is actually measured between the two regions, then because of the resulting perturbation 'the "phase" of the atom changes by amounts that are in principle uncontrollable' (Heisenberg 1927, pp. 183–4), and averaging over the unknown phases in the final superposition yields a non-interfering result.

The same example, with the same phase randomisation argument, is also given by Heisenberg (1930b) in his book *The Physical Principles of the Quantum Theory* (chapter IV, section 2), which was based on lectures delivered at Chicago in 1929. Heisenberg asserts (p. 60) that an energy measurement for an atom in the intermediate region 'will necessarily alter the phase of the de Broglie wave of the atom in state m by an unknown amount of order of magnitude one', so that in applying the (analogue of the) interfering expression (6.3) each term in the sum 'must thus be multiplied by the arbitrary factor $\exp(i\varphi_m)$ and then averaged over all values of φ_m'.

From a modern perspective, this argument seems strange and unfamiliar, and indeed quite wrong. However, the argument makes rather more sense, if one recognises that the 'quantum mechanics' described by Born and Heisenberg is not quantum mechanics as we usually know it today. In particular, the theory as they present it appears to contain no notion of wave packet collapse (or state vector reduction).

The argument given by Born and Heisenberg amounts to saying, in modern language, that if the energies of an atomic population have actually been measured, then one will have a mixture of states of the *superposed* form (6.1), with randomly distributed phases γ_n. Such a mixture is indeed statistically equivalent to a mixture of energy states $|n\rangle$ with weights $|c_n(0)|^2$, because the density operators are the same:

$$\left(\prod_k \frac{1}{2\pi} \int d\gamma_k\right) \sum_{n,m} |c_n(0)|\,|c_m(0)|\, e^{i(\gamma_n - \gamma_m)}\, |n\rangle\, \langle m| = \sum_n |c_n(0)|^2\, |n\rangle\, \langle n|\,. \tag{6.5}$$

However, from a modern point of view, if one did measure the atomic energies and find the value E_n with frequency p_n, the resulting total ensemble would naturally be represented by a density operator $\rho = \sum_n p_n |n\rangle \langle n|$, and there would seem to be no particular reason to rewrite this in terms of the alternative decomposition on the left-hand side of (6.5) (with $|c_n(0)|^2 = p_n$ and random phases $e^{i\gamma_n}$); though of course one could if one wished to. What is more, an actual inconsistency would appear if – having measured the atomic energies – one selected a particular atom that was found to have energy E_m: a subsequent and immediate energy measurement for this particular atom should again yield the result E_m with certainty, as is consistent with the usual representation of the atom by the state $|m\rangle$, and this certainty would be *in*consistent with what appears to be (at least in effect) the proposed representation of the atom by a state $\sum_n |c_n(0)| e^{i\gamma_n} |n\rangle$ with randomised phases γ_n. Indeed, for any subensemble composed of the latter states, all the energy values present in the sum will be possible outcomes of subsequent and immediate energy measurements.[a]

This inconsistency arises, however, *if* one applies the modern notion of state vector collapse – a notion that, upon close examination, appears to be quite absent from the theory presented by Born and Heisenberg. Instead of applying the usual collapse rule, Born and Heisenberg seem to interpret the quantities $|S_{nm}|^2$ as 'quite usual' transition probabilities in all circumstances, even in the presence of interference. On this view, then, an atom that has been found to have energy E_m can be represented by a state $\sum_n |c_n(0)| e^{i\gamma_n} |n\rangle$ with randomised phases γ_n, and the probability of obtaining a value E_n in an immediately successive measurement is given not by $|c_n(0)|^2$ (as would follow from the usual collapse rule) but by the transition probability $|S_{nm}|^2$ – where the latter does indeed approach δ_{nm} as the time interval between the two energy measurements tends to zero, so that the above contradiction does not in fact arise.

Considering now the whole atomic ensemble, if the energies of the atoms have indeed been measured, then using the $|S_{nm}|^2$ as transition probabilities and the $|c_m(0)|^2$ as population probabilities, application of the probability calculus gives the non-interfering result (6.4). As noted by Born and Heisenberg, on their view exactly the *same* result is obtained from the 'interfering' expression (6.3), with random phases γ_m appearing in the coefficients $c_m(0) = |c_m(0)| e^{i\gamma_m}$.

If instead the atomic energies have not been measured, then, according to Born and Heisenberg, the phases γ_m have not been randomised and the expression (6.3) does show interference, in contradiction with the non-interfering expression (6.4).

[a] Of course, if we do not subdivide the atomic ensemble on the basis of the measured energies, the *total* ensemble will be a mixture with density operator $\sum_n |c_n(0)|^2 |n\rangle \langle n|$ and will indeed be indistinguishable from the proposed mixture of states $\sum_n |c_n(0)| e^{i\gamma_n} |n\rangle$ with randomly distributed phases. But there is nothing to prevent an experimenter from selecting atoms according to their measured energies.

How do Born and Heisenberg reconcile the breakdown of (6.4) with their claim that the ordinary probability calculus still holds, with the $|S_{nm}|^2$ being quite ordinary probabilities? The answer seems to be that, if the energies have not been measured, then the population probabilities $|c_m(0)|^2$ are in some sense ill-defined, so that the usual probability formula (6.4) simply cannot be applied: '[(6.4)] follows from the concept of probability . . . when and only when . . . the probability $|c_n|^2$ . . . has been *established* beforehand *experimentally*'.

Born and Heisenberg seem to take an 'operational' view of the population probabilities, in the sense that these are to be regarded as meaningful only when directly measured. And the cited argument in Heisenberg's uncertainty paper suggests that it is operationally impossible to have simultaneously well-defined phase relations and population probabilities in the same experiment. This impossibility was presumably regarded as comparable to the (operational) impossibility of having simultaneously a well-defined position and momentum for a particle. What seems to be at work here, then, is some form of uncertainty relation (or complementarity) between population probabilities and phases: measurement of the former makes the latter ill-defined, and vice versa. Interference requires definite phase relationships, which preclude a well-defined population probability, so that the ordinary probability calculus cannot be applied.[a] If instead the population probability has actually been measured, then the phases are indefinite and averaging over them washes out any interference.

The resulting viewpoint is certainly remarkable. According to Born and Heisenberg, in the presence of interference, the quantities $|S_{nm}|^2$ continue to be quite ordinary (transition) probabilities, while the quantities $|c_m(0)|^2$ cannot be regarded as population probabilities – rendering the formula (6.4) inapplicable. On this view, ordinary probability calculus is not violated; it is simply wrong to assert that the $|c_m(0)|^2$ represent state probabilities in an interfering case.

One may well object to this point of view on the grounds that, even without measuring the energies, for a given preparation of the state (6.1) the coefficients $c_m(0)$ – and hence the values of $|c_m(0)|^2$ – will be known (up to an overall phase). However, presumably, Born and Heisenberg would have had to assert that while the $|c_m(0)|^2$ always exist as mathematical quantities, they cannot be properly interpreted as population probabilities unless the energies have been measured directly.

[a] There is an analogy here with Heisenberg's view of causality, expressed in his uncertainty paper, according to which causality cannot be applied because its premiss is generally false: '. . . in the sharp formulation of the law of causality, "If we know the present exactly, we can calculate the future", it is not the consequent that is wrong, but the antecedent. We *cannot* in principle get to know the present in all [its] determining data' (Heisenberg 1927, p. 197).

From a modern perspective, Born and Heisenberg's treatment of interference is surprising: in modern quantum mechanics, of course, in cases where interference occurs the quantities $|S_{nm}|^2$ would not normally be interpreted as 'quite usual' transition probabilities; while in cases where interference does not occur, the non-interfering result (6.4) (as applied here) would not normally be regarded as arising from the interfering result (6.3) through a process of phase randomisation.

6.2 Macroscopic superposition: Born's discussion of the cloud chamber

Quantum theory is normally understood to allow the 'superposition of distinct physical states'. However, while 'superposition' is well-defined as a mathematical term, it is hard to make sense of when applied to physical states – that is, when the components in a superposition are regarded as simultaneous physical attributes of a single system. The need to understand such 'physical superposition' seems particularly acute when it is considered at the macroscopic level. The difficulty here is closely related to the question of wave packet collapse: how is a mathematical superposition of macroscopically distinct states related to the definite macroscopic states seen in the laboratory?

The Wilson cloud chamber, as used to observe the tracks of α-particles, was discussed at length by Born in the general discussion. The cloud chamber illustrates the measurement problem rather well, and is a good example of how microscopic superpositions can become transferred to the macroscopic domain. It also illustrates how extending the formal quantum description to the environment does not by itself alleviate the measurement problem (despite many claims to the contrary, for example Zurek (1991)).[a] Remarkably, as we shall see, Born asserts that wave packet collapse is not required to discuss the cloud chamber.

The mechanism of the cloud chamber is well known. The α-particles pass through a supersaturated vapour. The passage of the particles causes ionisation, and the vapour condenses around the ions, resulting in the formation of tiny droplets. The droplets scatter light, making the particle tracks visible.

If the emission of an α-particle is undirected, so that the emitted wave function is approximately spherical, how does one account for the approximately straight particle track revealed by the cloud chamber? In the general discussion, Born attributes this question to Einstein, and asserts that to answer it (p. 437)

> ... one must appeal to the notion of 'reduction of the probability packet' developed by Heisenberg.

[a] For a summary of criticisms of environmental decoherence as a solution to the measurement problem, see Bacciagaluppi (2005).

This notion appears in Heisenberg's uncertainty paper, which had been published in May of 1927. In section 3, entitled 'The transition from micro- to macromechanics', Heisenberg had described how a classical electron orbit 'comes into being' through repeated observation of the electron position, using light of wavelength λ. According to Heisenberg, the result of each observation can be characterised by a probability packet of width λ, where the packet spreads freely until the next observation: 'Every determination of position reduces therefore the wave packet back to its original size λ' (Heisenberg 1927, p. 186).

In the case of the cloud chamber, the collapse of the wave packet is applied repeatedly to the α-particle alone. Upon producing visible ionisation, the wave packet of the α-particle collapses, and then starts to spread again, until further visible ionisation is produced, whereupon collapse occurs again, and so on. The probability for the resulting 'trajectory' is concentrated along straight lines, accounting for the observed track in the cloud chamber. As Born puts it (p. 437):

> The description of the emission by a spherical wave is valid only for as long as one does not observe ionisation; as soon as such ionisation is shown by the appearance of cloud droplets, in order to describe what happens afterwards one must 'reduce' the wave packet in the immediate vicinity of the drops. One thus obtains a wave packet in the form of a ray, which corresponds to the corpuscular character of the phenomenon.

Here, the cloud chamber itself – the ionisation, and the formation of droplets from the vapour – is treated as if it were an external 'classical apparatus': only the α-particle appears in the wave function.

6.2.1 Quantum mechanics without wave packet collapse?

Born goes on to consider if wave packet reduction can be *avoided* by treating the atoms of the cloud chamber, along with the α-particle, as a single system described by quantum theory, a suggestion that he attributes to Pauli (p. 437):

> Mr Pauli has asked me if it is not possible to describe the process without the reduction of wave packets, by resorting to a multi-dimensional space, whose number of dimensions is three times the number of all the particles present .... This is in fact possible ... but this does not lead us further as regards the fundamental questions.

Remarkably, Born claims that a treatment without reduction is 'in fact' possible, and goes on to illustrate how, in his opinion, this can be done. As we shall see, Born seems to make use of a 'classical' probability reduction only, without any reduction for the configuration-space wave packet.

As for Born's reference to Pauli, around the time of the Solvay conference Pauli believed that wave packet reduction was needed only for describing subsystems. This is clear from a letter he wrote to Bohr, on 17 October 1927 (one week before the Solvay meeting began), in which Pauli comments on wave packet reduction (Pauli 1979, p. 411):

> This is precisely a point that was not quite satisfactory in Heisenberg [that is, in the uncertainty paper]; there the 'reduction of the packets' seemed a bit mystical. Now however, it is to be stressed that at first such reductions are not necessary if one *includes* in the system all means of measurement. But in order to describe observational results theoretically at all, one has to ask what one can say about just a *part* of the whole system. And then from the complete solution one sees immediately that, in many cases (of course not always), leaving out the means of observation can be formally replaced by such reductions.

Thus, at that time, Pauli thought that the reduction was a formality associated with an effective description of subsystems alone, and that if the apparatus were included in the system then reduction would not be needed at all.

Born, then, presents a multi-dimensional treatment, in which atoms in the cloud chamber are described by quantum theory on the same footing as the α-particle. Born considers the simple case of a model cloud chamber consisting of just two atoms in one spatial dimension. There are two cases, one with both atoms on the same side of the origin (where the α-particle is emitted), and the other with the atoms on opposite sides of the origin. The two 'tubes' in Born's diagram (see his figure, p. 439) represent, for the two cases, the time development of the total (localised) packet in 3-dimensional configuration space. The coordinate x_0 of the α-particle is perpendicular to the page, while x_1, x_2 are the coordinates of the two atoms. In case I, the initial state is localised at $x_0 = 0$, $x_1, x_2 > 0$; in case II it is localised at $x_0 = 0$, $x_1 > 0$, $x_2 < 0$. Each initial packet separates into two packets moving in opposite directions along the x_0-axis. In the first case the two collisions (indicated by kinks in the trajectory of the packet[a]) take place on the same side of the origin; in the second case they take place on opposite sides. Note that, in both cases, both branches of the wave packet – moving in opposite directions – are shown; that is, the complete ('uncollapsed') packets are shown in the figure.

Born remarks (p. 439) that:

> To the 'reduction' of the wave packet corresponds the choice of one of the two directions of propagation $+x_0$, $-x_0$, which one must take as soon as it is established that one of the two points 1 and 2 is hit, that is to say, that the trajectory of the packet has received a kink.

Here Born seems to be saying that, instead of reduction, what takes place is a choice of direction of propagation. But propagation of what? Born presumably does not mean a choice of direction of propagation of the wave packet, for that would amount to wave packet reduction, which Born at the outset has claimed is

[a] Note that the motion occurs in one spatial dimension; a kink in the configuration-space trajectory shows that the corresponding atom has undergone a small spatial displacement.

unnecessary in a multi-dimensional treatment. (And indeed, his figure shows the wave packet propagating in both directions.) Instead, Born seems to be referring to a choice in the direction of propagation of the *system* (which we would represent by a point in configuration space). The wave packet spreads in both directions, and determines the probabilities for the different possible motions of the system. Once the direction of motion of the system is established, by the occurrence of collisions, an ordinary ('classical') reduction of the *probability* distribution occurs – while the wave packet itself is unchanged. In other words, the probabilities are updated but the wave packet does not collapse. This, at least, appears to be Born's point of view.

This may seem a peculiar interpretation – ordinary probabilistic collapse without wave packet collapse – but it is perhaps related to the intuitive thinking behind Born's famous collision papers of the previous year (Born 1926a,b) (papers in which probabilities for results of collisions were identified with squares of scattering amplitudes for the wave function[a]). Born drew an analogy with Einstein's notion of a 'ghost field' that determines probabilities for photons (see Chapter 9). As Born put it:

> In this, I start from a remark by Einstein on the relationship between the wave field and light quanta; he said, for instance, that the waves are there only to show the corpuscular light quanta the way, and in this sense he talked of a 'ghost field'. This determines the probability for a light quantum, the carrier of energy and momentum, to take a particular path; the field itself, however, possesses no energy and no momentum.... Given the perfect analogy between the light quantum and the electron ... one will think of formulating the laws of motion of electrons in a similar way. And here it is natural to consider the de Broglie–Schrödinger waves as the 'ghost field', or better, 'guiding field' ... [which] propagates according to the Schrödinger equation. Momentum and energy, however, are transferred as if corpuscles (electrons) were actually flying around. The trajectories of these corpuscles are determined only insofar as they are constrained by the conservation of energy and momentum; furthermore, only a probability for taking a certain path is determined by the distribution of values of the function ψ.
>
> *(Born 1926b, pp. 803–4)*

Here, Born seems to be suggesting that there are stochastic trajectories for electrons, with probabilities for paths determined by the wave function ψ. It is not clear, though, whether the ψ field is to be regarded as a physical field associated with *individual* systems (as the electromagnetic field usually is), or whether it is to be regarded merely as relating to an ensemble. If the former, then it might make sense to apply collapse to the probabilities without applying collapse to ψ itself. For example, this could happen in a stochastic version of de Broglie–Bohm theory: ψ could be a physical field evolving at all times by the Schrödinger equation (hence

[a] Cf. the discussion in Section 3.4.3.

never collapsing), and instead of generating deterministic particle trajectories (as in standard de Broglie–Bohm theory) ψ could generate probabilistic motions only. Further evidence that Born was indeed thinking along such lines comes from an unpublished manuscript by Born and Jordan, written in 1925, in which they propose a stochastic theory of photon trajectories with probabilities determined by the electromagnetic field (as originally envisaged by Slater) – see Darrigol (1992, p. 253) and Section 3.4.2.[a]

Thus, in his discussion of the cloud chamber, when Born spoke of 'the choice of one of the two directions of propagation', he may indeed have been referring to the possible directions of propagation of the system configuration, with the ψ field remaining in a superposition. On this reading, wave packet reduction for the α-particle would be only an effective description, which properly corresponds to the branching of the total wave function together with a random choice of trajectory (in multi-dimensional configuration space). Unfortunately, however, Born's intentions are not entirely clear: whether this really is what Born had in mind, in 1926 or 1927, is difficult to say. Certainly, in the general discussion at the fifth Solvay conference, Born did maintain that ψ does not really collapse (in this multi-dimensional treatment), so it is difficult to see how he could have thought that ψ gave merely a probability distribution over an ensemble.

The claim that the wave function ψ does not collapse is also found, in effect, in Section II of the report that Born and Heisenberg gave on quantum mechanics. As we saw in Section 6.1.2, in their discussion of interference, an energy measurement is taken to induce a randomisation of the phases appearing in a superposition of energy states, instead of the usual collapse to an energy eigenstate. In their example, after an energy measurement all of the components of the superposition are still present, and the phase relations between them are randomised. In Born's example of the cloud chamber, it seems that, here too, all of the components of the wave function are still present at the end of a measurement. Nothing is said, however, about the relative phases of the components, and we do not know whether or not Born had in mind a similar phase randomisation in this case also.

6.3 Dirac and Heisenberg: interference, state reduction and delayed choice

Another striking feature of quantum theory, as normally understood, is 'interference between alternative histories'. Like superposition, interference is

[a] In this connection, it is interesting to note the following passage from Born's book *Atomic Physics* (Born 1969), which was first published in German in 1933: 'A mechanical process is therefore accompanied by a wave process, the guiding wave, described by Schrödinger's equation, the significance of which is that it gives the probability of a definite course of the mechanical process'. Born's reference to the wave function as 'the guiding wave' shows the lingering influence of the ideas that had inspired him in 1926.

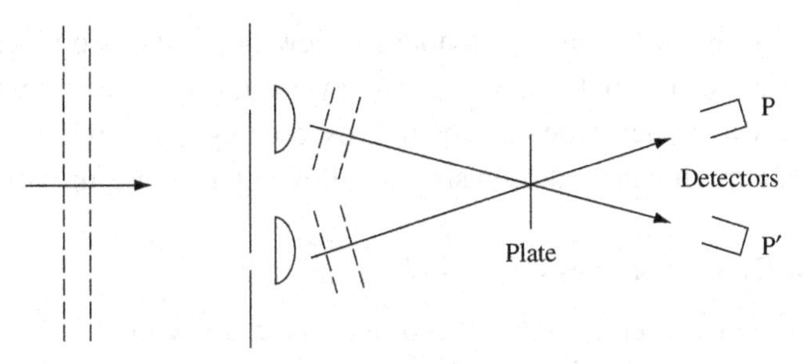

Fig. 6.1. Delayed-choice double-slit experiment.

mathematically well-defined but its physical meaning is ambiguous, and it has long been considered one of the main mysteries of quantum theory (as we saw in Section 6.1). The question of *when* interference can or cannot take place is intimately bound up with the measurement problem, in particular, with the question of when or how definite outcomes emerge from quantum experiments, and with the question of the boundary between the quantum and classical domains.

In modern times, one of the most puzzling aspects of interference was emphasised by Wheeler (1978). In his 'delayed-choice' experiment, it appears that the existence or non-existence of interfering histories in the past is determined by an experimental choice made in the present. One version of Wheeler's experiment – a 'delayed-choice double-slit experiment' with single photons – is shown in Fig. 6.1. A single photon is incident on a screen with two slits. The waves emerging from the slits are focussed (by off-centred lenses) so as to cross each other as shown. The insertion of a photographic plate in the interference region would seem (from the interference pattern) to imply the past existence of interfering trajectories passing through both slits. On the other hand, if no such plate is inserted, then a detection at P or P′seems to imply that the particle passed through the bottom or top slit respectively.[a] Since the plate could have been inserted long after the photon completed most of its journey (or journeys), it appears that an experimental choice now can affect whether or not there was a definite photon path in the past.

A similar point arose in 1927 in the general discussion. Dirac expounded his view that quantum outcomes occur when nature makes a choice. Heisenberg replied that this could not be, because of the possibility of observing interference later on by choosing an appropriate experimental arrangement, leading Heisenberg to conclude that outcomes occur when a choice is made (or brought about) not

[a] This inference is commonly made, usually without explicit justification. Some authors appeal to conservation of momentum for a free particle, but it is not clear how such an argument could be made precise – after all there are no particle trajectories in standard quantum theory. In de Broglie–Bohm theory, the inference is actually wrong: particles detected at P or P′come from the top or bottom slits respectively (Bell 1980, 1987, ch. 14).

by nature but by the *observer*. Heisenberg's view here bears some resemblance to Wheeler's. Dirac, in contrast, seems to say on the one hand that stochastic collapse of the wave packet occurs for microscopic systems, while on the other hand that if the experiment is chosen so as to allow interference then such collapse is postponed.

Here is how Dirac expresses it (pp. 447–8):

According to quantum mechanics the state of the world at any time is describable by a wave function ψ, which normally varies according to a causal law, so that its initial value determines its value at any later time. It may however happen that at a certain time t_1, ψ can be expanded in the form

$$\psi = \sum_n c_n \psi_n ,$$

where the ψ_n's are wave functions of such a nature that they cannot interfere with one another at any time subsequent to t_1. If such is the case, then the world at times later than t_1 will be described not by ψ but by one of the ψ_n's. The particular ψ_n that it shall be must be regarded as chosen by nature.

Note first of all that Dirac regards ψ as describing the state 'of the world' – presumably the whole world. Then, in circumstances where ψ may be expanded in terms of non-interfering states ψ_n, the world is subsequently described by one of the ψ_n (the choice being made by nature, the probability for ψ_n being $|c_n|^2$). Dirac does not elaborate on precisely when or why a decomposition into non-interfering states should exist, nor does he address the question of whether such a decomposition is likely to be unique. Such questions, of course, go to the heart of the measurement problem, and are lively topics of current research.

It is interesting that, in Dirac's view (apparently), there are circumstances in which interference is completely and irreversibly destroyed. For him, the particular ψ_n results from (p. 447):

an irrevocable choice of nature, which must affect the whole of the future course of events.

It seems that according to Dirac, once nature makes a choice of one branch, interference with the other branches is impossible for the whole of the future. A definite collapse has occurred, after which interference between the alternative outcomes is no longer possible, even in principle. This view clearly violates the Schrödinger equation as applied to the whole world: as Dirac states, the wave function ψ of the world 'normally' evolves according to a causal law, but not always.

But Dirac goes further, and recognises that there are circumstances where the choice made by nature cannot have occurred at the point where it might have been expected. Dirac considers the specific example of the scattering of an electron. He first notes that, after the scattering, one must take the wave function to be not the

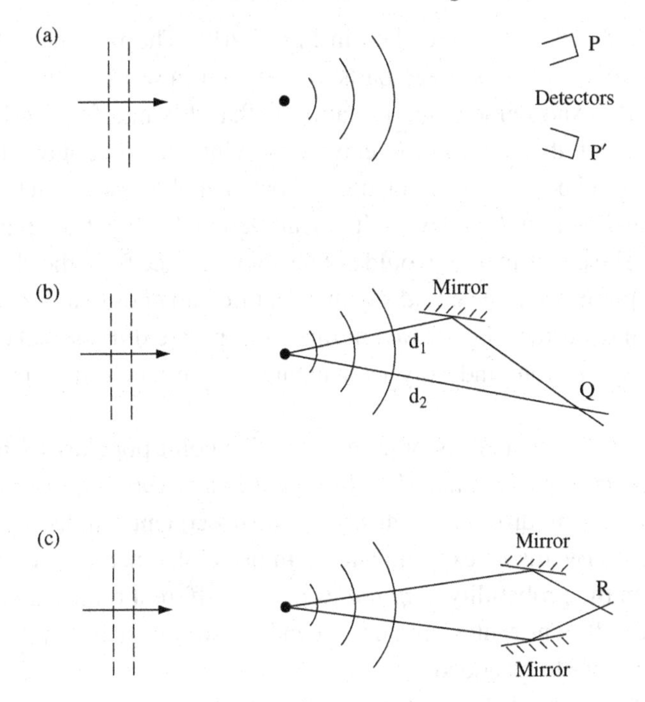

Fig. 6.2. Reconstruction of scattering scenarios discussed by Dirac (Figs. (a), (b)) and Heisenberg (Fig. (c)).

whole scattered wave but a packet moving in a specific direction (that is, one of the ψ_n). He claims (p. 448) that one could infer that nature had chosen this specific direction:

From the results of an experiment, by tracing back a chain of causally connected events one could determine in which direction the electron was scattered and one would thus infer that nature had chosen this direction.

This is illustrated in Fig. 6.2(a). If the electron is detected at P, for example, one may arguably infer a corresponding choice of direction at the time of scattering. (Note that Figs. 6.2(a)–(c) are ours.)

On the other hand, Dirac goes on to make the following observation (p. 448):

If, now, one arranged a mirror to reflect the electron wave scattered in one direction d_1 so as to make it interfere with the electron wave scattered in another direction d_2, one would not be able to distinguish between the case when the electron is scattered in the direction d_2 and when it is scattered in the direction d_1 and reflected back into d_2. One would then not be able to trace back the chain of causal events so far, and one would not be able to say that nature had chosen a direction as soon as the collision occurred, but only [that] at a later time nature chose where the electron should appear.

Dirac's modified scenario is sketched in Fig. 6.2(b). The presence of the mirror leads to interference, at Q, between parts of the electron wave scattered in different directions d_1, d_2. And Dirac's interpretation is that this interference is intimately related to the fact that an experimenter observing the outgoing electron in a direction d_2 'would not be able to distinguish between the case when the electron is scattered in the direction d_2 and when it is scattered in the direction d_1 and reflected back into d_2'. The experimenter would not be able to 'trace back the chain of causal events' to the point where he could say that 'nature had chosen a direction as soon as the collision occurred'. For Dirac, in this case, nature did *not* make a choice at the time of the collision, and only later nature 'chose where the electron should appear'.

What Dirac describes here is precisely the viewpoint popularised by Feynman in his famous lectures (Feynman, Leighton and Sands 1965), according to which if a process occurs by different routes that are subsequently indistinguishable (in the sense that afterwards an experimenter is in principle unable to tell which route was taken) then the probability *amplitudes* for the different routes are to be added; whereas if the different routes are subsequently distinguishable in principle, then the *probabilities* are to be added.[a]

Dirac's presentation of the scattering experiment with the mirror ends with the statement (p. 448):

> The interference between the ψ_n's compels nature to postpone her choice.

In his manuscript, a cancelled version of the sentence begins with 'Thus a *possibility* of interference between . . . ', while another cancelled version begins as 'Thus the *existence* of . . . ' (italics added). Possibly, Dirac hesitated here because he saw that the mirror could be added by the experimenter after the scattering had taken place, leading to difficulties with his view that without the mirror nature makes a choice at the time of scattering. For if, in the absence of the mirror, nature indeed makes a choice at the time of scattering, how could this choice be undone by subsequent addition of the mirror? Whether Dirac really foresaw this difficulty is hard to say. In any case, precisely this point was made by Heisenberg, and Dirac's hesitation here certainly reflects a deep difficulty that lies at the heart of the measurement problem.

Heisenberg makes his point with disarming simplicity (p. 449):

> I do not agree with Mr Dirac when he says that, in the described experiment, nature makes a choice. Even if you place yourself very far away from your scattering material, and if you

[a] Note that Feynman's path-integral formulation of quantum theory – developed in his PhD thesis and elsewhere (Feynman 1942, 1948) – was anticipated by Dirac (1933). Feynman's thesis and Dirac's paper are reprinted in Brown (2005).

measure after a very long time, you are able to obtain interference by taking two mirrors. If nature had made a choice, it would be difficult to imagine how the interference is produced.

What Heisenberg had in mind seems to have been something like the set-up shown in Fig. 6.2(c), where a pair of mirrors is placed far away from the scattering region, causing different parts of the scattered wave to re-overlap and interfere at R. According to Dirac's account, in the absence of any mirrors (Fig. 6.2(a)), upon detection of the particle one might say that nature chose a specific direction at the time of scattering. Heisenberg points out that, by placing mirrors far away (and removing the detectors at P and P′), interference may be observed a long time after the scattering took place.

While Heisenberg does not mention it explicitly, in this example the choice between 'which-way information' on the one hand, or interfering paths on the other, may be made long after the particle has completed most of its journey (or journeys), just as in Wheeler's delayed-choice experiment. Dirac's set-up with no mirrors at all provides which-way information, since detection of the particle at a point far away may be interpreted as providing information on the direction chosen at the time of scattering (Fig. 6.2(a)). Heisenberg's modification, with the two mirrors, demonstrates interference between alternative paths starting from the scattering region (Fig. 6.2(c)). Unlike Wheeler, however, Heisenberg does not explicitly emphasise that the choice of whether or not to add the mirrors could be made at the last moment, long after the scattering takes place. On the other hand, Heisenberg does emphasise that the measurement with the mirrors could be done 'very far away' and 'after a very long time', and notes the contradiction with nature having made a choice at the time of scattering. Thus, Heisenberg's remarks arguably contain the essence of Wheeler's delayed-choice experiment.

Heisenberg goes on to say (pp. 449–50) that, instead of nature making a choice,

I should rather say, as I did in my last paper, that the *observer himself* makes the choice, because it is only at the moment when the observation is made that the 'choice' has become a physical reality and that the phase relationship in the waves, the power of interference, is destroyed.

From the chronology of Heisenberg's publications, here he must be referring to his uncertainty paper (published in May 1927), in which he writes that 'all perceiving is a choice from a plenitude of possibilities' (Heisenberg 1927, p. 197). Heisenberg's statement above that the observer 'makes' the choice seems to be meant in the sense of the observer 'bringing about' the choice. Thus it would seem that, for Heisenberg, a definite outcome occurs – and there is no longer any possibility of interference – only when an experimenter makes an observation. Similar views have been expressed by Wheeler (1986).

One may, however, object to this viewpoint, on the grounds that there is no reason in principle why a more advanced being could not observe interference between the alternative states of the detector registering interference, or, between the alternative states of the human observer watching the detector (cf. the discussion of Wigner's paradox in Section 5.2.2). After all, the detector is certainly just another physical system, built out of atoms. And as far as we can tell, human observers can likewise be treated as physical systems built out of atoms. To say that 'the power of interference' is 'destroyed' when and only when a human observer intervenes is to make a remarkable assertion to the effect that human beings, unlike any other physical systems, have special properties by virtue of which they cannot be treated by ordinary physical laws but generate deviations from those laws. As we have already mentioned, there is no evidence that human beings are able to violate, for example, the laws of gravity or of thermodynamics, and it would be remarkable if they were indeed able to violate the laws of quantum physics.

It is interesting to note that, while for Heisenberg the human observer seems to play a crucial role at the end of a quantum experiment, for Dirac the human observer – and his 'free will' – seems to play a crucial role at the beginning, in the preparation stage. For as Dirac puts it (p. 446, Dirac's italics): 'The disturbances that an experimenter applies to a system to observe it are directly under his control, and are acts of free will by him. *It is only the numbers that describe these acts of free will that can be taken as initial numbers for a calculation in the quantum theory*' (cf. the discussion about determinism in Section 8.2).

Returning to Dirac's view of quantum outcomes, Heisenberg's objection certainly causes a difficulty. If a choice – or collapse to a particular ψ_n – really does occur around the time of scattering, then a 'delayed interference experiment' of the form described by Heisenberg should show *no* interference, and Dirac's view would amount to a violation of the quantum formalism along the lines of dynamical models of wave function collapse (Pearle 1976, 1979, 1989; Ghirardi, Rimini and Weber 1986). And Dirac's caveats concerning the possibility of tracing back a chain of causal events do not lead to a really satisfactory position either. As in Feynman's view that interference occurs only for paths that are subsequently indistinguishable, the question is begged as to the precise definition of subsequently distinguishable or subsequently indistinguishable paths: for in a delayed-choice set-up, it appears to be at the later whim of the experimenter to decide whether certain paths taken in the past are subsequently distinguishable or not. This procedure correctly predicts the experimental results (or statistics thereof), but it has the peculiar consequence that whether or not there is a matter of fact about the past depends on what the experimenter does in the present.

Finally, as discussed in Section 6.1.1, interference was considered from a pilot-wave perspective by de Broglie in his report and by Brillouin in the discussion that followed. De Broglie did not comment, however, on the exchange between Dirac and Heisenberg. From a modern point of view it is clear that, in his theory, the particle trajectory does take one particular route after a scattering process, while at the same time there are portions of the scattered wave travelling along the alternative routes. An 'empty' part of the wave can be subsequently reflected by a mirror, and if the reflected wave later re-overlaps with the part of the wave carrying the particle, then in the interference zone the particle is indeed affected by both components. Similarly, de Broglie's theory provides a straightforward account of Wheeler's delayed-choice experiment, without present actions influencing the past in any way (Bell 1980, 1987, ch. 14; Bohm, Dewdney and Hiley 1985).

6.4 Further remarks on Born and Heisenberg's quantum mechanics

As we saw in Section 6.1.2, Born and Heisenberg's report contains some remarkable comments about the nature of interference (in Section II, 'Physical interpretation'). These comments are perhaps related to a conceptual transition that seems to occur at around this point in their presentation. In the earlier part of their Section II, Born and Heisenberg describe a theory in which probabilistic transitions occur between possessed values of energy; while later in the same section, in their discussion of arbitrary observables, they emphasise probabilistic transitions from one measurement to the next (still noting the presence of interference). Earlier in that section they explicitly assert that a system always occupies a definite energy state at any one time, while in the later treatment of arbitrary observables nothing is said about whether a system always possesses definite values or not. This is perhaps not surprising, given that the discussion of interference (for the case of energy measurements) made it clear that taking the quantities $|c_n(t)|^2 = |\langle n|\psi(t)\rangle|^2$ to be population probabilities for energies E_n led to a difficulty in the presence of interference. As we saw in Section 6.1.2, Born and Heisenberg resolved the difficulty by asserting that unmeasured population probabilities are somehow not applicable or meaningful. This does not seem consistent with the view they expressed earlier, that atoms always have definite energy states even when these are not measured. How could an ensemble of atoms have definite energy states, without the energy distribution being meaningful?

Consideration of interference, then, was likely to force a shift away from the view that atoms are always in definite stationary states. Later, in his book of 1930, Heisenberg did in fact explicitly deny that atoms are always in such states. Considering again the example from his uncertainty paper, of atoms passing through two successive regions of inhomogeneous field (see Section 6.1.2),

Heisenberg notes that if the energies are not actually measured in the intermediate region, then, because of the resulting 'interference of probabilities',

> it is not reasonable to speak of the atom as having been in a stationary state between F_1 and F_2 [that is, in the intermediate region].
>
> *(Heisenberg 1930b, p. 61)*

As we have already noted, again in Section 6.1.2, Born and Heisenberg's discussion of interference seems to dispense with the standard collapse postulate for quantum states. Upon performing an energy measurement, instead of the usual collapse to a single energy eigenstate $|n\rangle$, Born and Heisenberg in effect replace the superposition $\sum_n |c_n(0)|\, e^{i\gamma_n}\, |n\rangle$ by a similar expression with randomised phases γ_n. And the justification given for this appears to be some form of uncertainty relation or complementarity between population probabilities and phases: if the former have been measured, then the latter are ill-defined, and vice versa. On this view, the definite phase relationships associated with interference preclude the possibility of speaking of a well-defined population probability, so that the usual formulas of probability calculus cannot be properly applied; on the other hand, if the population probability has been measured experimentally, then the phases are ill-defined, and averaging over the random phases destroys interference.

One crucial point is not entirely clear, however. Was the phase randomisation thought to occur only upon measurement of energy, or upon measurement of any arbitrary observable?

The phase randomisation explicitly appealed to by Born and Heisenberg takes the following form: for a quantum state

$$|\psi(t)\rangle = \sum_E |E\rangle\, \langle E\,|\psi(t)\rangle = \sum_E |E\rangle\, \langle E\,|\psi(0)\rangle\, e^{-iEt} \tag{6.6}$$

an energy measurement induces a random change in each phase factor $e^{-iEt} \rightarrow e^{-iEt}e^{i\gamma(E)}$, where each $\gamma(E)$ is random on the unit circle. This procedure might be generalised to, for example, measurements of position as follows: for a state

$$|\psi\rangle \propto \sum_{\mathbf{x}} |\mathbf{x}\rangle\, \langle \mathbf{x}\,|\psi\rangle \propto \sum_{\mathbf{x}} \left(\sum_{\mathbf{p}} |\mathbf{p}\rangle\, e^{-i\mathbf{p}\cdot\mathbf{x}} \right) \langle \mathbf{x}\,|\psi\rangle \tag{6.7}$$

(writing as if $\mathbf{x}$ and $\mathbf{p}$ were discrete, for simplicity) one might suppose that a position measurement induces a random change $e^{-i\mathbf{p}\cdot\mathbf{x}} \rightarrow e^{-i\mathbf{p}\cdot\mathbf{x}}e^{i\gamma(\mathbf{x})}$, resulting in a state

$$\sum_{\mathbf{x}} |\mathbf{x}\rangle\, \langle \mathbf{x}\,|\psi\rangle\, e^{i\gamma(\mathbf{x})} \tag{6.8}$$

with random relative phases. Averaging over the random phases $\gamma(\mathbf{x})$ would then destroy interference between different positions, just as in the case of energy measurements.

However, we have found no clear evidence that Born or Heisenberg considered any such generalisation.[a] It then seems possible that the phase randomisation argument for the suppression of interference was to be applied to the case of energy measurements only. On the other hand, there is a suggestive remark by Heisenberg in the general discussion (p. 449), quoted in the last section. When expressing his view that definite outcomes occur only when an experimenter makes an observation, Heisenberg refers to an example where position measurements are made at the end of a scattering process (see Fig. 6.2(c)), and he states that it is only when the observation is made that 'the phase relationship in the waves, the power of interference, is destroyed'. This might be read as suggesting that the waves continue to exist, but may or may not have the capacity to interfere – depending on whether or not the phase relations have been randomised by the position measurement. If Heisenberg did take such a view, his use of wave packet reduction for position measurements in the uncertainty paper would have to be interpreted as some sort of effective description.

Even in the case of energy measurements, the status of the phase randomisation argument is not clear. After all, Born and Heisenberg assert that the time-dependent Schrödinger equation itself (which they use to discuss interference) is only phenomenological, and applicable to subsystems only. Fundamentally, they have a time-independent theory for a closed system. Presumably, the phase randomisation for energy measurements was also seen as phenomenological only, with the measured system being treated as a subsystem.

As we saw in Section 6.3, in the general discussion Dirac describes what is recognisably the process of wave packet reduction. Born and Heisenberg, in contrast, seem to speak only of the ordinary reduction (or conditionalisation) of

[a] Remarkably, in the well-known book on quantum theory by Bohm (1951), instead of the standard collapse rule one finds precisely the above phase-randomisation postulate, generalised in fact to any observable A (p. 600): '... whenever a measurement of any variable is carried out, the interaction between the system under observation and the observing apparatus always multiplies each part of the wave function corresponding to a definite value of A by a random phase factor, $e^{i\alpha_a}$. Thus, if the wave function is $\sum_a c_a \psi_a(x)$ before the measurement [where the ψ_a are eigenfunctions of A with eigenvalues a], it is changed into $\sum c_a e^{i\alpha_a} \psi_a(x)$. The random phase factors cause interference between difference [*sic*] $\psi_a(x)$ to be destroyed'. Similarly, for the specific case of position measurements, Bohm writes (p. 122): '... each part of the wave function corresponding to a definite position of the electron at the time of the measurement is changed during the course of the interaction between electron and observing apparatus in such a way that it is multiplied by an unpredictable and uncontrollable phase factor, $e^{i\alpha}$'. Bohm attempts (p. 602) to justify the appearance of such random phase factors, but his argument seems obscure. In his preface (p. v) Bohm acknowledges, as an important influence on his book, lectures on quantum theory delivered by Oppenheimer at Berkeley. It is intriguing to note that Oppenheimer obtained his PhD in 1927 at Göttingen, under Born himself. However, we have not attempted to trace this possible line of influence.

'probability functions' as it appears in standard probability calculus. As they put it, near the end of Section III of their report (p. 394):

> the result of each measurement can be expressed by the choice of appropriate initial values for probability functions . . .. Each new experiment replaces the probability functions valid until now with new ones, which correspond to the result of the observation . . ..

On the other hand, Born and Heisenberg's contributions at the Solvay conference do not seem sufficiently clear or complete to warrant definite conclusions as to what they believed concerning the precise relationship between probabilities and the wave function; sometimes it is unclear whether or not they mean to draw a distinction between probability distributions on the one hand and wave functions on the other.

It may be that, in October 1927, Born and Heisenberg had in some respects not yet reached a definitive point of view, perhaps partly because of the different perspectives that Born and Heisenberg each brought to the subject. Born's recent thinking (in 1926) had been influenced by Einstein's idea of a guiding field, while Heisenberg's recent thinking (in his uncertainty paper) had been influenced by the operational approach to physics.

Concerning the question of wave packet collapse, it should also be remembered that Pauli seems to have played an important role in Born and Heisenberg's thinking at the time. In particular, as we saw in Born's discussion of the cloud chamber (Section 6.2.1), Pauli had been critical of Heisenberg's use of the reduction of the wave packet in the uncertainty paper (in a discussion of classical electron orbits), and Born – who by his own account was following Pauli's suggestion – tried to show that such reduction was unnecessary.

7

Locality and incompleteness

7.1 Einstein's 1927 argument for incompleteness

A huge literature arose out of the famous 'EPR' paper by Einstein, Podolsky and Rosen (1935), entitled 'Can quantum-mechanical description of physical reality be considered complete?'. The EPR paper argued, on the basis of (among other things) the absence of action at a distance, that quantum theory must be incomplete.[a] It is less well-known that a much simpler argument, leading to the same conclusion, was presented by Einstein eight years earlier in the general discussion at the fifth Solvay conference (pp. 440ff.).

Einstein compares and contrasts two views about the nature of the wave function ψ, for the specific case of a single electron. According to view I, ψ represents an ensemble (or 'cloud') of electrons, while according to view II, ψ is a complete description of an individual electron. Einstein argues that view II is incompatible with locality, and that to avoid this, in addition to ψ there should exist a localised particle (along the lines of de Broglie's theory). Thus, according to this reasoning, if one assumes locality, then quantum theory (as normally understood today) is incomplete.

The conclusion of Einstein's argument in 1927 is the same as that of EPR in 1935, even if the form of the argument is rather different. Einstein considers electrons striking a screen with a small hole that diffracts the electron wave, which on the far side of the screen spreads out uniformly in all directions and strikes a photographic film in the shape of a hemisphere with large radius (see Einstein's figure, p. 440). Einstein's argument against view II is then as follows:

If $|\psi|^2$ were simply regarded as the probability that at a certain point a given particle is found at a given time, it could happen that *the same* elementary process produces an

[a] Note that, as pointed out by Fine (1986) and discussed further by Howard (1990), the logical structure of the EPR paper (which was actually written by Podolsky) is more complicated and less direct than Einstein had intended.

action *in two or several* places on the screen. But the interpretation, according to which $|\psi|^2$ expresses the probability that *this* particle is found at a given point, assumes an entirely peculiar mechanism of action at a distance, which prevents the wave continuously distributed in space from producing an action in *two* places on the screen.

The key point here is that, if there is no action at a distance, and if the extended field ψ is indeed a complete description of the physical situation, then if the electron is detected at a point P on the film, it could happen that the electron is also detected at another point Q, or indeed at any point where $|\psi|^2$ is non-zero. Upon detection at P, it appears that a 'mechanism of action at a distance' prevents detection elsewhere.

Einstein's argument is so concise that its point is easily missed, and one might well dismiss it as arising from an elementary confusion about the nature of probability. (Indeed, Bohr comments that he does not 'understand what precisely is the point' Einstein is making.) For example, it might be thought that, since we are talking about a probability distribution for just one particle, it is a matter of pure logic that only one detection can occur.[a] But this would be to beg the question concerning the nature of ψ. Einstein's wording above attempts to convey a distinction between probability for a 'given' particle (leading to the possibility of multiple detections) and probability for '*this*' particle (leading to single detection only). The wording is not such as to convey the distinction very clearly, perhaps indicating an inadequate translation of Einstein's German into French.[b] But from the context, the words 'probability that *this* particle is found' are clearly being used to express the assumption that in this case ψ indeed expresses the probability for just *one* particle detection.

As shown by Hardy (1995), Einstein's argument may be readily put into the same rigorous form as the later EPR argument. (See Norsen (2005) for a careful and extensive discussion.) Hardy simplifies Einstein's example, and considers a single particle incident on a beam splitter (Fig. 7.1), so that there are only two points P_1, P_2 at which the particle might be detected. One may then adopt the following sufficient condition, given by EPR, for the existence of an element of reality:

If, without in any way disturbing a system, we can predict with certainty (that is, with probability equal to unity) the value of a physical quantity, then there exists an element of physical reality corresponding to this physical quantity.

(Einstein, Podolsky and Rosen 1935, p. 777)

[a] Cf. Shimony (2005).

[b] Unfortunately, the full German text of Einstein's contribution to the general discussion seems to have been lost; the Einstein archives contain only a fragment, consisting of just the first four paragraphs (AEA 16-617.00).

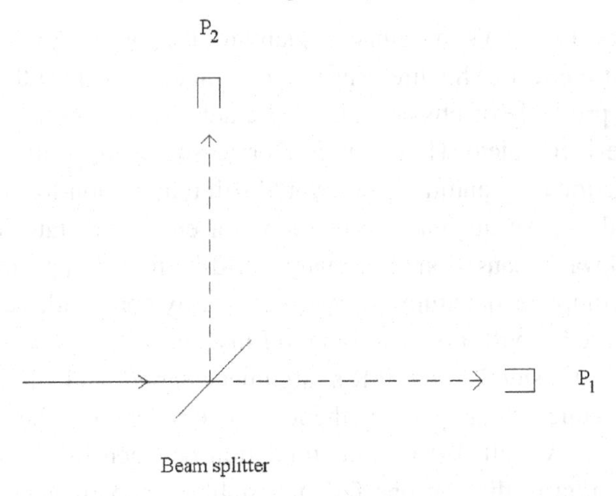

Fig. 7.1. Hardy's simplified version of Einstein's argument.

Now, if a detector is placed at P_1, either it will fire or it will not. In either case, from the state of the detector at P_1 one could deduce with certainty whether or not a detector placed at P_2 would fire. Such deductions could be made for any individual run of the experiment. Even though the outcome at P_1 cannot be predicted in advance, in each case the outcome allows us to infer the existence of a definite element of reality at P_2. If locality holds, an element of reality at P_2 cannot be affected by the presence or absence of a detector at P_1. Therefore, even if no detector is placed at P_1, there must still be an element of reality at P_2 corresponding to detection *or* no detection at P_2. Since ψ is a superposition of detection *and* no detection at P_2, ψ contains nothing corresponding to the deduced element of reality at P_2. Therefore, ψ is not a complete description of a single particle.[a]

Thus, 'the essential points in the EPR argument had already been made by Einstein some eight years earlier at the fifth Solvay conference' (Hardy 1995, p. 600).

Einstein concludes that:

In my opinion, one can remove this objection only in the following way, that one does not describe the process solely by the Schrödinger wave, but that at the same time one localises the particle during the propagation. I think that Mr de Broglie is right to search in this direction. If one works solely with the Schrödinger waves, interpretation II of $|\psi|^2$ implies to my mind a contradiction with the postulate of relativity.

In other words, for Einstein, action at a distance can be avoided only by admitting that the wave function is incomplete.

[a] Note that in this argument the incompleteness of quantum theory is *inferred* from the assumption of locality. Cf. Bell (1987, p. 143).

According to Einstein's argument, quantum theory is either non-local or incomplete. For the rest of his life, Einstein continued to believe that locality was a fundamental principle of physics, and so he adhered to the view that quantum theory must be incomplete. However, further reasoning by Bell (1964) showed that any completion of quantum theory would still require non-locality, in order to reproduce the details of quantum correlations for entangled states (assuming the absence of backwards causation or of many worlds[a]). It then appears that, whether complete or incomplete, quantum theory is necessarily non-local, a conclusion that would surely have been deeply disturbing to Einstein.

It is ironic that Einstein's (and EPR's) argument started out by holding steadfast to locality and deducing that quantum theory is incomplete. But then the argument, as carried further by Bell, led to a contradiction between locality and quantum correlations, so that in the end one fails to establish incompleteness and instead establishes non-locality (with completeness or incompleteness remaining an open question).

7.2 A precursor: Einstein at Salzburg in 1909

In September 1909, at a meeting in Salzburg, Einstein gave a lecture entitled 'On the development of our views concerning the nature and constitution of radiation' (Einstein 1909). Einstein summarised what he saw as evidence for the dual nature of radiation: he held that light had both particle and wave aspects, and argued that classical electromagnetic theory would have to be abandoned. It seems to have gone unnoticed that one of Einstein's arguments at Salzburg was essentially the same as the argument he presented at the 1927 Solvay conference (though applied to light quanta instead of to electrons).

Einstein began his lecture by noting that the phenomena of interference and diffraction make it plain that, at least in some respects, light behaves like a wave. He then went on to describe how, in other respects, light behaves as if it consisted of particles. In experiments involving the photoelectric effect, it had been found that the velocity of the photoelectrons was independent of the radiation intensity. According to Einstein, this was more consistent with 'Newton's emission theory of light' than with the wave theory. Einstein also discussed pressure fluctuations in blackbody radiation, and showed that these contained two terms, which could be naturally identified as contributions from particle-like and wave-like aspects of the radiation.

[a] Bell's argument assumes that there is no common cause between the hidden variables (defined at the time of preparation) and the settings of the measuring apparatus. It also assumes that there is no backwards causation, so that the hidden variables are not affected by the future outcomes or apparatus settings. Further, the derivation of the Bell inequalities assumes that a quantum measurement has only one outcome, and therefore does not apply in the many-worlds interpretation.

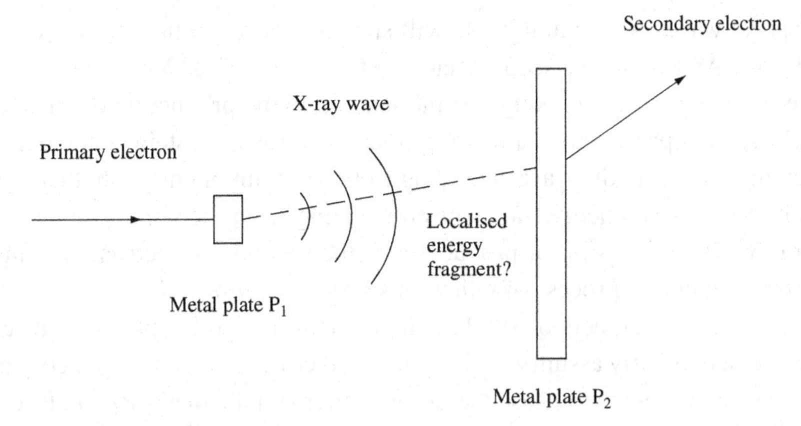

Fig. 7.2. Figure based on Einstein's 1909 argument for the existence of localised light quanta. Assuming the principle of local action, the delocalised X-ray wave can produce an electron (of energy comparable to that of the primary electron) in a small region of the second plate only if, in addition to the wave, there is a localised energy fragment propagating in space from P_1 to P_2.

Of special interest here is another argument Einstein gave for the existence of localised light quanta. Einstein considered a beam of electrons ('primary cathode rays') incident upon a metal plate P_1 and producing X-rays (see Fig. 7.2). These X-rays, in turn, strike a second metal plate P_2 leading to the production of electrons ('secondary cathode rays') from P_2. Experimentally, it had been found that the velocity of the secondary electrons had the same order of magnitude as the velocity of the primary electrons. Further, the available evidence suggested that the velocity of the secondary electrons did not depend at all on the distance between the plates P_1 and P_2, or on the intensity of the primary electron beam, but only on the velocity of the primary electrons. Assuming this to be strictly true, Einstein then asked what would happen if the primary intensity were so small, or the area of the plate P_1 so small, that one could consider just one electron striking the plate, as in Fig. 7.2. According to Einstein,

> we will have to assume that on P_2 (as a result of the impinging of the above electron on P_1) either nothing is being produced or that a secondary emission of an electron occurs on it with a velocity of the same order of magnitude as of the electron impinging on P_1. In other words, the elementary radiation process seems to proceed such that it does not, as the wave theory would require, distribute and scatter the energy of the primary electron in a spherical wave propagating in all directions. Rather, it seems that at least a large part of this energy is available at some location of P_2 or somewhere else.
>
> *(Einstein 1909, English translation, p. 388)*

Einstein's argument, then, is that according to the wave theory the point of emission of the X-ray from the first plate must be the source of waves spreading out in space, waves whose amplitude will spread over the region occupied by the second plate. And yet, in the second plate, *all* the energy of the X-ray becomes concentrated in the vicinity of a single point, leading to the production of an electron with velocity comparable to that of the primary electron.[a] Einstein concluded from this that, in addition to the wave spreading from the point of emission, there seems also to be a localised energy fragment propagating from the point of emission of the X-ray wave to the point of production of the secondary electron. As Einstein put it: 'the elementary process of radiation seems to be *directed*'.

Now, Einstein's argument of 1909 implicitly assumes a principle of local action, similar to that explicitly assumed in his published critique of quantum theory at the 1927 Solvay conference. Because the distance between the plates P_1 and P_2 can be arbitrarily large, the wave impinging on P_2 can be spread over an arbitrarily large area. The production of an electron in a highly localised region of P_2 can then be accounted for only if, in addition to the delocalised wave, there is a localised energy fragment propagating through space – for otherwise, there would have to be some mechanism by means of which energy spread out over arbitrarily large regions of space suddenly becomes concentrated in the neighbourhood of a single point.

It should be quite clear, then, that Einstein's 1927 argument for the existence of localised electrons (accompanying de Broglie–Schrödinger waves) was identical in form to one of his 1909 arguments for the existence of localised light quanta (accompanying electromagnetic waves). In his 1927 argument, the small hole in the screen (see his figure) acts as a source for an electron wave, which spreads over the area of the photographic film – just as, in the 1909 argument, the point where the primary electron strikes the first plate acts as a source for an X-ray wave, which spreads over the area of the second plate. Both arguments depend crucially on the assumption (implicit in 1909, explicit in 1927) that there is no action at a distance.

As we shall discuss further in Chapter 9, by 1927 Einstein had already spent over 20 years trying to reconcile localised energy quanta – which he had postulated in 1905 – with the wave aspect of radiation. And for much of that time, he had been more or less alone in his belief in the existence of such quanta. It is then perhaps not so surprising that at the 1927 Solvay meeting Einstein was able to raise such a penetrating critique of the view that the wave function is a complete description of a single electron: from his long and largely solitary experience pondering the wave–particle duality of light, Einstein could immediately see that, in the analogous case of electron waves, the principle of local action entailed the existence of localised particles moving through space, in addition to the wave function.

[a] Cf. Compton's report, p. 309: 'It is clearly impossible that all the energy of an X-ray pulse which has spread out in a spherical wave should spend itself on this [small region]'.

The meeting in Salzburg took place four years before Bohr published his model of the atom. After 1913, one might have simply rephrased Einstein's argument in terms of atomic transitions. Consider an atom A that makes a transition from an initial stationary state with energy E_i to a final stationary state with lower energy E_f. At a later time, the energy $E_i - E_f$ lost by atom A may be *wholly* absorbed by an arbitrarily distant atom B, if there exists an appropriate transition from the initial state of B to a final state corresponding to an energy increase $E_i - E_f$. This process may seem unmysterious, if one imagines atom A emitting a photon, or 'localised energy quantum', which somehow propagates through space from A to B. However, if one tries to make do without the photon concept, and represents the electromagnetic field in terms of (classical) waves only – which spread out in all directions from A – then it is hard to understand how the energy lost by A may be wholly transferred to B: instead, one would expect the energy to spread out in space like the waves themselves, so that the energy density becomes diluted.

We have laboured this point because the power of Einstein's simple argument seems to have been generally missed, not only in 1909, but also in 1927, and for decades afterwards. Indeed, it appears that Einstein's point did not start to become widely appreciated until the late twentieth century (see, again, Norsen (2005)).

In retrospect, it seems quite puzzling that Einstein's simple argument should have taken so long to be understood. A perhaps related puzzle, emphasised by Pais (1982, pp. 382–6), is why Einstein's light-quantum hypothesis itself should have been largely ignored by so many physicists until the advent of the Compton effect in 1923. Even after Millikan's experimental confirmation of Einstein's photoelectric equation in 1916, 'almost no one but Einstein himself would have anything to do with light-quanta' (Pais 1982, p. 386).[a]

We do not wish to suggest, of course, that Einstein's locality argument should today be regarded as establishing the existence of localised photons: for the implicit premise of Einstein's argument – the principle of locality – today seems to be ruled out by Bell's theorem. Our point, rather, is that prior to Bell's work (and certainly in 1909) Einstein's arguments were indeed compelling and should have been taken more seriously.

7.3 More on non-locality and relativity

At the end of his long contribution to the general discussion (in which he argued for the incompleteness of quantum theory), Einstein objected to the multi-dimensional representation in configuration space, on the grounds that (p. 442)

[a] Though according to Brillouin's recollections of 1962, the situation was rather different in France, where Einstein's light quantum was accepted (by Langevin, Perrin and Marie Curie) much earlier than it was elsewhere (Mehra and Rechenberg 1982a, p. 580).

> ... the feature of forces of acting only at small *spatial* distances finds a less natural expression in configuration space than in the space of three or four dimensions.

As Einstein himself stated, he was here adding another argument against what he called view II (the view that ψ is a complete description of an individual system), a view that he claimed is 'essentially tied to a multi-dimensional representation (configuration space)'.

Einstein's point seems to be that, if physics is fundamentally grounded in configuration space, there will be no reason to expect physics to be characterised by *local* action. This objection should be seen in the context of Einstein's concerns, in the period 1926–7, over the non-separability of Schrödinger's wave mechanics for many-body systems (Howard 1990, pp. 83–91; cf. Section 12.2).

A certain form of classical locality survives, of course, in modern quantum theory and quantum field theory, in the structure of the Hamiltonian or Lagrangian, a structure that ensures the absence of controllable non-local signals at the statistical level. But even so, we understand today that, in fact, quantum physics is characterised by non-locality. And the non-locality may indeed be traced to the fact that, unlike classical theory, quantum theory is not grounded in ordinary three-dimensional space.

The setting for standard quantum theory is Hilbert space, whose tensor-product structure allows for entanglement and associated non-local effects. In the pilot-wave formulation of quantum theory, the setting is configuration space (in which the pilot wave propagates), and in general the motions of spatially separated particles are non-locally connected. Further, Bell's theorem shows that, if we leave aside backwards causation or many worlds, then quantum theory is in some sense non-local under any interpretation or formulation. As Ballentine once pointed out, while discussing the significance of Bell's theorem:

> Perhaps what is needed is not an explanation of nonlocality, but an explanation of locality. Why, if locality is not true, does it work so well in so many different contexts?
>
> *(Ballentine 1987, pp. 786–7)*

Einstein's fear, that there would be difficulties with locality in quantum physics, has certainly been borne out by subsequent developments. In standard quantum theory, there appears to be a peaceful but uneasy 'coexistence' with relativity. While from a pilot-wave (or more generally, from a deterministic hidden-variables) point of view, statistical locality appears as an accidental feature of the 'quantum equilibrium' state (Valentini 1991b, 2002a).

In his main contribution to the general discussion, Dirac (p. 445) also notes that 'the general theory of the wave function in many-dimensional space necessarily involves the abandonment of relativity', but he suggests that this problem might be solved by 'quantising 3-dimensional waves' (that is, by what we would now call quantum field theory). And de Broglie in his report, when considering the pilot-wave dynamics of many-body systems, notes that unlike in the case of a single particle 'it does not appear easy to find a wave Ψ that would define the motion of the system taking Relativity into account' (p. 353), a difficulty that has persisted in pilot-wave theory right up to the present day (see, for example, Berndl *et al.* (1996)).

In 1927, then, there was a fairly broad recognition that the fundamental use of configuration space did not bode well for consistency with relativity.

8

Time, determinism and the spacetime framework

8.1 Time in quantum theory

By 1920, the spectacular confirmation of general relativity, during the solar eclipse of 1919, had made Einstein a household name. Not only did relativity theory (both special and general) upset the long-received Newtonian ideas of space and time, it also stimulated a widespread 'operationalist' attitude to physical theories. Physical quantities came to be seen as inextricably interwoven with our means of measuring them, in the sense that any limits on our means of measurement were taken to imply limits on the definability, or 'meaningfulness', of the physical quantities themselves. In particular, Einstein's relativity paper of 1905 – with its operational analysis of simultaneity – came to be widely regarded as a model for the new operationalist approach to physics.

Not surprisingly, then, as the puzzles continued to emerge from atomic experiments, in the 1920s a number of workers suggested that the concepts of space and time would require still further revision in the atomic domain. Thus, Campbell (1921, 1926) suggested that the puzzles in atomic physics could be removed if the concept of time was given a purely statistical significance: 'time, like temperature, is a purely statistical conception, having no meaning except as applied to statistical aggregates' (quoted in Beller 1999, p. 97). In an operationalist vein, Campbell considered 'clocks' based on (random) radioactive decays. He suggested that it might be possible to construct a theory that did not involve time at all, and in which 'all the experiments on which the prevailing temporal conceptions are based can be described in terms of statistics' (quoted in Beller 1999, p. 98). On the other hand, Senftleben (1923) asserted that Planck's constant h set limits to the definability of the concepts of space and time, and concluded that spacetime must be discontinuous. According to Beller (1999, pp. 96–101), both Campbell and Senftleben had a significant influence on Heisenberg in his formulation of the uncertainty principle.[a] Certainly, in a letter to Pauli of 28 October 1926, Heisenberg expresses views very similar to Campbell's (Pauli 1979, p. 350):

I have for all that a hope in a later solution of more or less the following kind (but one should not say something like this aloud): that space and time are really only statistical concepts, such as, say, temperature, pressure etc. in a gas. I mean that spatial and temporal concepts are meaningless for *one* corpuscle and that they make more and more sense the more particles are present. I often try to get further in this direction, but until now it will not work.

Be that as it may, at the 1927 Solvay conference, in their report on quantum mechanics, Born and Heisenberg seem to express the remarkable view that temporal changes do not occur at all for closed systems, and that the time-dependent Schrödinger equation emerges only as an effective and approximate description for subsystems. How these views related to Campbell's, or indeed if they did at all, is not clear.

In their Section II, 'Physical interpretation', Born and Heisenberg begin with the following statement (p. 383):

The most noticeable defect of the original matrix mechanics consists in the fact that at first it appears to give information not about actual phenomena, but rather only about possible states and processes. ... it says nothing about when a given state is present ... matrix mechanics deals only with closed periodic systems, and in these there are indeed no changes. In order to have true processes ... one must direct one's attention to a *part* of the system ....

From a modern point of view, the original matrix mechanics did not contain the notion of a general state $|\Psi\rangle$ for a system (not even a static, Heisenberg-picture state). The only states that appeared in the theory were the stationary states $|E_i\rangle$ (cf. Chapter 3). Even as regards stationary states, there seems to have been no notion of *initial* state ('it says nothing about when a given state is present'). Instead, the matrices provided a collective representation of all the energy eigenstates of a closed system with Hamiltonian H. In modern notation, the matrices consisted of matrix elements of (Heisenberg-picture) observables $\Omega(t) = e^{(i/\hbar)Ht}\Omega(0)e^{-(i/\hbar)Ht}$ in the energy basis:

$$\langle E_i |\Omega(t)| E_j\rangle = \langle E_i |\Omega(0)| E_j\rangle e^{(i/\hbar)(E_i - E_j)t} . \tag{8.1}$$

As noted in Chapter 3, the formal mathematics of matrix mechanics then seems to represent an atomic system somewhat in the manner of the Bohr–Kramers–Slater (BKS) theory (cf. Chapter 9), with each matrix element corresponding to a virtual oscillator of frequency $\nu_{ij} = (E_i - E_j)/h$.

[a] For Campbell's influence on Bohr's formulation of complementarity, see Beller (1999, pp. 135–7) and also Mehra and Rechenberg (2000, pp. 189–90).

The matrix formalism, without a notion of initial state, amounts to a static description.[a] However, Born and Heisenberg add an intuitive physical picture to the formalism, to the effect that a subsystem of a larger (closed) system is in fact in one stationary state at any one time and performs random, indeterministic 'quantum jumps' between such states (cf. Chapter 3).

Born and Heisenberg then go on to say that '[t]he clumsiness of the matrix theory in the description of processes developing in time can be avoided' (p. 385) by introducing what we would now call the time-dependent Schrödinger equation. Here, it might appear that their view is that the mentioned 'defect of the original matrix mechanics' is removed by generalisation to a time-dependent theory. However, they add (p. 385) that:

> Essentially, the introduction of time as a numerical variable reduces to thinking of the system under consideration as coupled to another one and neglecting the reaction on the latter. But this formalism is very convenient . . . .

These words give, instead, the impression that the time-dependent theory is regarded as only emergent in some approximation; the time-dependent Schrödinger equation seems to have no fundamental status.

Even so, this 'convenient' formalism 'leads to a further development of the statistical view'. They include a time-dependent external perturbation in the (time-dependent) Schrödinger equation and show how to calculate the time development of any initial wave function. Born and Heisenberg then argue that, following Bohr's original (1913) theory of stationary states, a system can be in only one energy eigenstate at any one time, leading to the interpretation of a superposition $|\Psi\rangle = \sum_n c_n(t)\,|E_i\rangle$ as a statistical mixture, with state probabilities $|c_n|^2$. The time evolution of the wave function then describes transition probabilities from initial to final stationary states. This might seem clear enough, but a difficulty is then raised concerning the interpretation of a case where the initial wave function is already a superposition, resulting in 'interference of probabilities' at later times (see Section 6.1.2).

Let us now consider what Schrödinger had to say, in his report on wave mechanics, concerning time in quantum theory. (De Broglie's report does not contain any special remarks on this subject.) In his report, Schrödinger first presents (or derives from a variational principle) what we would now call the time-independent Schrödinger equation for a non-relativistic many-body system with coordinates $q_1, q_2, \ldots, q_n$. After noting that the eigenfunctions ψ_k, with eigenvalues E_k, may be identified with Bohr's stationary states, Schrödinger addresses the question of time. He first points out (pp. 408–9) that the time-independent

[a] Even if one added a notion of initial state, because the only allowed states are energy eigenstates the description of a closed system would still be static.

theory might be regarded as sufficient, providing as it does a description of stationary states, together with expressions for jump probabilities between them:

One can take the view that one should be content in principle with what has been said so far .... The single stationary states of Bohr's theory would then in a way be described by the eigenfunctions ψ_k, which *do not contain time at all.* One ... can form from them ... quantities that can be aptly taken to be *jump probabilities* between the single stationary states.

Here, the jump probabilities are to be obtained from matrix elements such as (in modern notation)

$$\langle k \,|Q_i|\, k' \rangle = \int dq \; q_i \psi_k^* \psi_{k'} \tag{8.2}$$

which can all be calculated from the eigenfunctions ψ_k.

Schrödinger suggests further that interacting systems could be treated in the same way, by regarding them as one single system.

Schrödinger then goes on to discuss this point of view and its relation with the ideas of Campbell (p. 409):

On this view the *time variable* would play absolutely no role in an isolated system – a possibility to which N. Campbell ... has recently pointed. Limiting our attention to an isolated system, we would not perceive the passage of time in it any more than we can notice its possible progress in space .... What we would notice would be merely a sequence of discontinuous transitions, so to speak a cinematic image, but without the possibility of comparing the time intervals between the transitions.

According to these ideas, then, time does not exist at the level of isolated atomic systems, and our usual (macroscopically defined) time emerges only from the statistics of large numbers of transitions between stationary states. As Schrödinger puts it:

Only secondarily, and in fact with increasing precision the more extended the system, would a *statistical* definition of time result from *counting* the transitions taking place (Campbell's 'radioactive clock'). Of course then one cannot understand the jump probability in the usual way as the probability of a transition calculated relative to unit time. Rather, a *single* jump probability is then utterly meaningless; only with *two* possibilities for jumps, the probability that the one may happen *before* the other is equal to *its* jump probability divided by the sum of the two.

Schrödinger claims that this is the only consistent view in a theory with quantum jumps, asserting that '[e]ither all changes in nature are discontinuous or not a single one'.

Having sketched a timeless view of isolated systems with discrete quantum jumps, Schrödinger states that such a discrete viewpoint 'still poses great difficulties', and he goes on to develop his own theory of time-dependent quantum states, in which (continuous) time evolution does play a fundamental role even at the level of a single atomic system. Here, a general time-dependent wave function $\psi(q,t)$ – a solution of the time-dependent Schrödinger equation, with arbitrary initial conditions – is regarded as the description of the continuous time development of a single isolated system.

From a contemporary perspective, it is clear that quantum theory as we know it today is rather less radical than some expected it to be in the 1920s, especially concerning the concepts of space and time. Both non-relativistic quantum mechanics and relativistic quantum field theory take place on a classical spacetime background; time and space are continuous and well-defined, even for closed systems. The evolution operator $U(t,t_0)$ provides a continuous time evolution $|\Psi(t)\rangle = U(t,t_0)\,|\Psi(t_0)\rangle$ for any initial quantum state, with respect to an 'external' time parameter t (even in quantum field theory, in a given inertial frame). While Schrödinger's and de Broglie's interpretations of the wave function did not gain widespread acceptance, their view of time in quantum theory coincides with the one generally accepted today.[a]

In contrast, the views on time expressed by Born and Heisenberg are somewhat reminiscent of views put forward by some later workers in the context of canonical quantum gravity. There, the wave functional $\Psi[^{(3)}\mathcal{G}]$ on the space of 3-geometries $^{(3)}\mathcal{G}$ contains no explicit time parameter, and obeys a 'timeless' Schrödinger equation $\mathcal{H}\Psi = 0$ (the Wheeler–DeWitt equation, with Hamiltonian density operator $\mathcal{H}$). It is claimed that 'time' emerges only phenomenologically, through the analysis of interaction with quantum clocks, or by the extraction of an effective time variable from the 3-metric (the radius of an expanding universe being a popular choice) (DeWitt 1967). However, closer analysis reveals a series of difficulties with such proposals: for example, it is difficult to ensure the emergence of a well-behaved time parameter t such that only one physical state is associated with each value of t. (See, for example, Unruh and Wald (1989).) Despite some 50 years of effort, including the technical progress made in recent years using 'loop' variables to solve the equations (Rovelli 2004), the 'problem of time' in canonical quantum gravity remains unresolved.[b]

[a] We are of course referring here to the standard theories as presented in textbooks. The literature contains a number of proposals, along operational lines, calling for a 'quantum spacetime' that incorporates quantum-theoretical limits on the construction of rods and clocks. A statistical approach to causal structure, somewhat reminiscent of Campbell's statistical view of time, has recently been proposed by Hardy (2005).

[b] Barbour (1994a,b) has proposed a timeless formulation of classical and quantum physics. As applied to quantum gravity, the viability of Barbour's scheme seems to depend on unproven properties of the Wheeler–DeWitt equation.

8.2 Determinism and probability

In the published text, the first section of the general discussion bears the title 'Causality, determinism, probability'. Lorentz's opening remarks are mainly concerned with the importance of having a clear and definite picture of physical processes (see Section 8.3). He ends by addressing the question of determinism (p. 433):

> ... I think that this notion of probability should be placed at the end, and as a conclusion, of theoretical considerations, and not as an *a priori* axiom, though I may well admit that this indeterminacy corresponds to experimental possibilities. I would always be able to keep my deterministic faith for the fundamental phenomena....

Lorentz seems to demand that the fundamental phenomena be deterministic, and that indeterminism should be merely emergent or effective. Probabilities should not be axiomatic, and some theoretical explanation is needed for the experimental limitations encountered in practice. This view would nowadays be usually associated with deterministic hidden-variables theories, such as de Broglie's pilot-wave dynamics (though it might also be associated with the many-worlds interpretation of Everett).

De Broglie's basic equations – the guidance equation and Schrödinger equation – are certainly deterministic. De Broglie in his report, and Brillouin in the subsequent discussion, give examples of how these equations determine the trajectories (during interference and diffraction, atomic transitions and elastic scattering). As regards probabilities, de Broglie pointed out that if an ensemble of systems with initial wave function Ψ_0 begins with a Born-rule distribution $P_0 = |\Psi_0|^2$, then as Ψ evolves, the system dynamics will maintain the distribution $P = |\Psi|^2$ at later times. However, nothing was said about how the initial distribution might arise in the first place. Subsequent work has shown that, in de Broglie's theory, the Born-rule distribution can arise from the complex evolution generated by the dynamics itself, much as thermal distributions arise in classical dynamics, thereby providing an example of the kind of theoretical explanation that Lorentz wished for (Bohm 1953; Valentini 1991a, 1992, 2001; Valentini and Westman 2005).[a] On this view, the initial ensemble considered by de Broglie corresponds to a special 'quantum equilibrium' state analogous to thermal equilibrium in classical physics.

[a] Bohm (1953) considered the particular case of an ensemble of two-level molecules and argued that external perturbations would drive it to equilibrium. A general argument for relaxation was not given, however, and soon afterwards Bohm and Vigier (1954) modified the dynamics by adding random fluctuations that drive any system to equilibrium. This move to a stochastic theory seems unnecessary: a general H-theorem argument, analogous to the classical coarse-graining H-theorem, has been given (Valentini 1991a, 1992, 2001), and numerical simulations show a very efficient relaxation – with an exponential decay of the coarse-grained H-function – on the basis of the purely deterministic de Broglie–Bohm theory (Valentini and Westman 2005).

Lorentz goes on to say (p. 433):

Could a deeper mind not be aware of the motions of these electrons? Could one not keep determinism by making it an article of faith? Must one necessarily elevate indeterminism to a principle?

Here, again, we now know that de Broglie's theory provides an example of what Lorentz seems to have had in mind. For in principle, the theory allows the existence of 'quantum non-equilibrium' distributions $P \neq |\Psi|^2$ (Valentini 1991b, 1992), just as classical physics allows the existence of non-thermal distributions (not uniformly distributed on the energy surface in phase space). Such distributions violate many of the standard quantum constraints. In particular, an experimenter possessing particles with a distribution P much narrower than $|\Psi|^2$ would be able to use those particles to perform 'subquantum' measurements on ordinary systems, measurements that are more accurate than those normally allowed by the uncertainty principle. An experimenter possessing such 'non-equilibrium particles' would in fact be able to use them to observe the (normally invisible) details of the trajectories of ordinary particles (Valentini 2002b; Pearle and Valentini 2006). From this point of view, there is indeed no need to 'elevate indeterminism to a principle' – for the current experimental limitations (embodied in the uncertainty principle) are not built into the laws of physics; rather, they are merely contingent features of the quantum equilibrium state $P = |\Psi|^2$.

In the general discussion, as we saw in Section 6.3, Dirac expressed the view that quantum outcomes occur when nature makes a choice, a view countered by Heisenberg who claimed that the 'choice' is in some sense really made by the observer. As Lorentz noted at the end of this exchange, the view that nature makes a choice amounts to a fundamental indeterminism, while at the same time, Dirac and Heisenberg had radically different views about the meaning of this indeterminism.

Dirac also gave an argument for why quantum theory had to be indeterministic. In his view, the indeterminism was necessary because of the inevitable disturbance involved in setting up an initial quantum state (pp. 446–7):

I should now like to express my views on determinism and the nature of the numbers appearing in the calculations of the quantum theory.... In the classical theory one starts from certain numbers describing completely the initial state of the system, and deduces other numbers that describe completely the final state. This deterministic theory applies only to an isolated system.

But, as Professor Bohr has pointed out, an isolated system is by definition unobservable. One can observe the system only by disturbing it and observing its reaction to the disturbance. Now since physics is concerned only with observable quantities the deterministic classical theory is untenable.

Dirac's argument seems unsatisfactory. First of all, as a general philosophical point, the claim that 'physics is concerned only with observable quantities' is not realistic. As is well known to philosophers of science as well as to experimentalists, observation is 'theory-laden': in order to carry out observations (or measurements) some body of theory is required in order to know how to carry out a *correct* observation (for example, some knowledge is required of how the system being measured interacts with the apparatus, in order to design a correctly functioning apparatus). Thus, some body of theory is necessarily conceptually prior to observation, and it is logically impossible to base physical theory on 'observables' only. More specifically, Dirac claims that classical determinism is untenable because of the disturbance involved in observing a system; but there are many cases in classical physics where experimenters can use their knowledge of the interactions involved to compensate for the disturbance caused by the measurement. Disturbance per se cannot be a reason for indeterminism.

It may well be that Dirac had in mind the kind of 'irreducible' or 'uncontrollable' disturbance that textbooks commonly associate with the uncertainty principle. However, it is interesting that, in fact, Dirac goes on to say that the disturbances applied by an experimenter *are* under his control (p. 446, Dirac's italics):

> In the quantum theory one also begins with certain numbers and deduces others from them. ... The disturbances that an experimenter applies to a system to observe it are directly under his control, and are acts of free will by him. *It is only the numbers that describe these acts of free will that can be taken as initial numbers for a calculation in the quantum theory.* Other numbers describing the initial state of the system are inherently unobservable, and do not appear in the quantum theoretical treatment.

The 'disturbances' refer to the experimental operations that the experimenter chooses to apply to the system, and indeed these are normally regarded as freely controlled by the experimenter (at least in some effective sense). The sense in which the word 'disturbance' is being used here is quite different from the textbook sense of uncontrollable disturbance associated with quantum uncertainty. Dirac seems to regard quantum-theoretical eigenvalues as representing the extent to which an experimenter can controllably manipulate a system. Macroscopic operations are under our control, and through these we can prepare an initial state specified by particular eigenvalues. For Dirac, these initial numbers represent 'acts of free will' in the form of laboratory operations. (The final remark about 'other numbers' that are unobservable, and that do not appear in quantum theory, is intriguing, and might be taken as suggesting that there are other degrees of freedom that cannot be controlled by us.)

A view quite different from Dirac's is expressed by Born at the very end of the general discussion. According to Born, the constraints on the preparation of an initial quantum state are *not* what distinguishes quantum from classical mechanics, for in classical physics too (p. 469)

. . . the precision with which the future location of a particle can be predicted depends on the accuracy of the measurement of the initial location.

For Born, the difference rather lies in the law of propagation of probability packets:

It is then not in this that the manner of description of quantum mechanics, by wave packets, is different from classical mechanics. It is different because the laws of propagation of packets are slightly different in the two cases.

8.3 Visualisability and the spacetime framework

With hindsight, from a contemporary perspective, perhaps the most characteristic feature of quantum physics is the apparent absence of visualisable processes taking place within a spacetime framework. From Bell's theorem, it appears that any attempt to provide a complete description of quantum systems (within a single world) will require some form of non-locality, leading to difficulties with relativistic spacetime. At the time of writing, we possess only one hidden-variables theory of broad scope – the pilot-wave theory of de Broglie and Bohm – and in this theory there is a field on *configuration* space (not 3-space) that affects the motion of quantum systems.[a] Arguably, then, pilot-wave theory does not really fit into a spacetime framework: the physics is grounded in configuration space, and the interactions encoded in the pilot wave take place outside of 3-space. Instead of a non-local hidden-variables theory, one might prefer to have a complete account of quantum behaviour in terms of many worlds: there too, one leaves behind ordinary spacetime as a basic framework for physics, since the totality of what is real cannot be mapped onto a single spacetime geometry. Generally speaking, whatever one's view of quantum theory today, the usual spacetime framework seems too restrictive, and unable to accomodate (at least in a natural way) the phenomena associated with quantum superposition and entanglement.[b]

We saw in Section 8.1 that the quantum or matrix mechanics of Born and Heisenberg certainly did not provide an account of physical systems in a spacetime framework. In contrast, the initial practical success of Schrödinger's wave mechanics in 1926 had led some workers to think that an understanding in terms of (wave processes in) space and time might be possible after all. The resulting tension between Schrödinger on the one hand, and Bohr, Heisenberg and Pauli on the other (in the year preceding the Solvay meeting) has been described at length in Section 4.6, where we saw that for Schrödinger the notion

[a] As already mentioned in Section 5.2.1, attempts to construct hidden-variables theories without an ontological wave function, along the lines pioneered by Fényes (1952) and Nelson (1966), seem to fail (Wallstrom 1994; Pearle and Valentini 2006).

[b] A possible exception here is some version of quantum theory with dynamical wave function collapse.

of 'Anschaulichkeit' – in the sense of visualisability in a spacetime framework – played a key role.

The clash between quantum physics and the spacetime framework was a central theme of the fifth Solvay conference. There was a notable tension between those participants who still hoped for a spacetime-based theory and those who insisted that no such theory was possible. These differences are especially apparent in the general discussion where, as we have already discussed in Section 7.3, difficulties were raised concerning locality and relativity.

Lorentz, in his opening remarks at the first session of the general discussion, seems to set the tone for one side of the debate, by speaking in favour of space and time as a basic framework for physics (p. 432):

> We wish to make a representation of the phenomena, to form an image of them in our minds. Until now, we have always wanted to form these images by means of the ordinary notions of time and space. These notions are perhaps innate; in any case, they have developed from our personal experience, by our daily observations. For me, these notions are clear and I confess that I should be unable to imagine physics without these notions. The image that I wish to form of phenomena must be absolutely sharp and definite, and it seems to me that we can form such an image only in the framework of space and time.

Lorentz's commitment to processes taking place in space and time was shared by de Broglie and Schrödinger, even though both men had found themselves unable to avoid working in terms of configuration space. As we saw in Section 2.4, in his report de Broglie presented his pilot-wave dynamics, with a guiding field in configuration space, only as a makeshift; he hoped that his pilot-wave dynamics would turn out to be an effective theory only, and that underlying it would be a theory of wave fields in 3-space with singularities representing particle motion (the double-solution theory). Schrödinger, too, despite working with a many-body wave equation in configuration space, hoped that the physical content of his theory could be ultimately interpreted in terms of processes taking place in 3-space (see Chapter 4).

In contrast, some of the other participants, in particular Bohr and Pauli, welcomed – indeed insisted upon – the break with the spacetime framework. As Bohr put it in the general discussion, after Einstein's remarks on locality and completeness (p. 442):

> The whole foundation for [a] causal spacetime description is taken away by quantum theory, for it is based on [the] assumption of observations without interference.

In support of Bohr's contention, Pauli (pp. 443–5) provided an intriguing argument to the effect that interactions between particles cannot be understood in a spacetime framework. Specifically, Pauli based his argument on the quantum-theoretical account of the long-range interactions known as van der Waals forces.

Pauli begins his argument by saying (in agreement with Bohr) that the use of multi-dimensional configuration space is

> only a technical means of formulating mathematically the laws of mutual action between several particles, actions which certainly do not allow themselves to be described simply, in the ordinary way, in space and time.

Here, on the one hand, configuration space has only mathematical significance. But on the other hand, as Einstein feared (see Section 7.3), there is according to Pauli no explicit account of local action in space and time. Pauli adds that the multi-dimensional method might one day be replaced with what we would now call quantum field theory. He then goes on to say that, in any case, in accordance with Bohr's point of view, no matter what 'technical means' are used to describe 'the mutual actions of several particles', such actions 'cannot be described in the ordinary manner in space and time' (p. 444).

Pauli then illustrates his point with an example. He considers two widely separated hydrogen atoms, each in their ground state, and asks what their 'energy of mutual action' might be. According to the usual description in space and time, says Pauli, for large separations there should be no mutual action at all. And yet (p. 444),

> when one treats the same question by the multi-dimensional method, the result is quite different, and in accordance with experiment.

What Pauli is referring to here is the problem of accounting for van der Waals forces between atoms and molecules. Classically, molecules with dipole moments tend to align, resulting in a mean interaction energy $\propto 1/R^6$ (where R is the distance between two molecules). However, many molecules exhibiting van der Waals forces have zero dipole moment; and while the classical orientation effect becomes negligible at high temperatures, van der Waals forces do not. The problem of explaining van der Waals forces was finally solved by quantum theory, beginning with the work of Wang in 1927.[a] (Note that the effect here is quite distinct from that of exchange forces resulting from the Pauli exclusion principle. The latter forces are important only when the relevant electronic wave functions have significant overlap, whereas van der Waals forces occur between neutral atoms even when they are so far apart that their charge clouds have negligible overlap.)

A standard textbook calculation of the van der Waals force, between two hydrogen atoms (1 and 2) separated by a displacement **R**, proceeds as follows.[b] The unperturbed energy eigenstate is the product $\Psi = \psi_0^1\psi_0^2$ of the two ground

[a] For a detailed review, see Margenau (1939).
[b] See, for example, Schiff (1955, pp. 176–80).

states. If $R = |\mathbf{R}|$ is much larger than the Bohr radius, the classical electrostatic potential between the atoms is

$$V = \frac{e^2}{R^3}\left(\mathbf{r}_1 \cdot \mathbf{r}_2 - 3\frac{(\mathbf{r}_1 \cdot \mathbf{R})(\mathbf{r}_2 \cdot \mathbf{R})}{R^2}\right), \tag{8.3}$$

where $\mathbf{r}_1, \mathbf{r}_2$ are respectively the positions of electrons 1, 2 relative to their nuclei. In the state $\Psi = \psi_0^1 \psi_0^2$ the mean values of $\mathbf{r}_1, \mathbf{r}_2$ both vanish, so that the expectation value of V vanishes. However, the presence of V perturbs the ground state of the system, to a new and entangled state Ψ', satisfying $(H_0 + V)\Psi' = (2E_0 + \Delta E)\Psi'$, where H_0 is the unperturbed Hamiltonian, E_0 is the ground-state energy of hydrogen, and ΔE is the energy perturbation from which one deduces the van der Waals force. Standard perturbation methods show that ΔE has a leading term proportional to $1/R^6$, accounting for the (attractive) van der Waals force.

Pauli regarded this result – which had been obtained by Wang from the Schrödinger equation in configuration space – as evidence that in quantum theory there are interactions that cannot be described in terms of space and time. The result may be roughly understood classically, as Pauli points out, by imagining that in each atom there is an oscillating dipole moment that can induce (and so interact with) a dipole moment in the other atom. But such understanding is only heuristic. In a proper treatment, using 'multi-dimensional wave mechanics', the correct result is obtained by methods that, according to Pauli, cannot be understood in a spacetime framework.[a]

Pauli's point seems to have been that, because the perturbed wave function Ψ' is entangled, multi-dimensional configuration space plays a crucial role in bringing about the correct result, which therefore cannot be properly understood in terms of 3-space alone. If this argument seems unwarranted, it ought to be remembered that, before Wang's derivation in 1927, there had been a long history of failed attempts to explain van der Waals forces classically (Margenau 1939).

Disagreements over both the usefulness and the tenability of the spacetime framework are also apparent elsewhere in the general discussion. For example, Kramers (p. 450) asks: 'What advantage do you see in giving a precise value to the velocity v of the photons?' To this de Broglie replies (in a spirit similar to that of Lorentz above):

> This allows one to imagine the trajectory followed by the photons and to specify the meaning of these entities; one can thus consider the photon as a material point having a position and a velocity.

[a] According to Ehrenfest, Bohr also gave an argument (in a conversation with Einstein) against a spacetime description when treating many-particle problems. See Section 1.5.

Kramers was unconvinced:

> I do not very well see, for my part, what advantage there is, for the description of experiments, in making a picture where the photons travel along well-defined trajectories.

In this exchange, Kramers had suggested that de Broglie's theory could not explain radiation pressure from a single photon, thereby questioning the tenability (and not just the usefulness) of the spacetime framework for the description of elementary interactions (cf. Section 10.4).

Later in the general discussion, another argument against de Broglie's theory is provided by Pauli. As we shall discuss at length in Section 10.2, Pauli claims on the basis of an example that pilot-wave theory cannot account for the discrete energy exchange taking place in inelastic collisions. It is interesting to note that, according to Pauli, the root of the difficulty lies in the attempt to construct a deterministic particle dynamics in a spacetime framework (p. 463):

> . . . this difficulty . . . is due directly to the condition assumed by Mr de Broglie, that in the individual collision process the behaviour of the particles should be completely determined and may at the same time be described completely by ordinary kinematics in spacetime.

9

Guiding fields in 3-space

In this chapter, we address proposals (by Einstein and by Bohr, Kramers and Slater) according to which quantum events are influenced by 'guiding fields' in 3-space. These ideas led to a predicted violation of energy-momentum conservation for single events, in contradiction with experiment. The contradiction was resolved only by the introduction of guiding fields in configuration space. All this took place before the fifth Solvay conference, but nevertheless forms an important background to some of the discussions that took place there.

9.1 Einstein's early attempts to formulate a dynamical theory of light quanta

Since the publication of his light-quantum hypothesis in 1905, Einstein had been engaged in a solitary struggle to construct a detailed theory of light quanta, and to understand the relationship between the quanta on the one hand and the electromagnetic field on the other.[a] Einstein's efforts in this direction were never published. We know of them indirectly: they are mentioned in letters, and they are alluded to in Einstein's 1909 lecture in Salzburg. Einstein's published papers on light quanta continued for the most part in the same vein as his 1905 paper: using the theory of fluctuations to make deductions about the nature of radiation, without giving details of a substantial theory. Einstein was essentially alone in his dualistic view of light, in which localised energy fragments coexisted with extended waves, until the work of de Broglie in 1923 – which extended the dualism to all particles, and made considerable progress towards a real theory (see Chapter 2).

[a] In 1900 Planck had, of course, effectively introduced a quantisation in the interaction between radiation and matter; but it was Einstein in 1905 who first proposed that radiation itself (even in free space) consisted, at least in part, of spatially localised energy quanta. In 1918, Einstein wrote to his friend Besso: 'I do not doubt anymore the *reality* of radiation quanta, although I still stand quite alone in this conviction' (original italics, as quoted in Pais (1982, p. 411)).

A glimpse of Einstein's attempts to formulate a dynamical theory of light quanta may be obtained from a close reading of his 1909 Salzburg lecture. There, as we saw in Section 7.2, Einstein marshalled evidence that light waves contain localised energy fragments. He suggested that the electromagnetic field is associated with singular points at which the energy is localised, and he offered the following remarkable (if heuristic) picture:

> I more or less imagine each such singular point as being surrounded by a field of force which has essentially the character of a plane wave and whose amplitude decreases with the distance from the singular point. If many such singularities are present at separations that are small compared with the dimensions of the field of force of a singular point, then such fields of force will superpose, and their totality will yield an undulatory field of force that may differ only slightly from an undulatory field as defined by the current electromagnetic theory of light.
>
> *(Einstein 1909, English translation, p. 394)*

Here, each light quantum is supposed to have an extended field associated with it, and large numbers of quanta with their associated fields are supposed to yield (to a good approximation) the electromagnetic field of Maxwell's theory. In other words, the electromagnetic field as we know it emerges from the collective behaviour of large numbers of underlying fields associated with individual quanta.

A similar view is expressed in a letter from Einstein to Lorentz written a few months earlier, in May 1909:

> I conceive of the light quantum as a point that is surrounded by a greatly extended vector field, that somehow diminishes with distance. Whether or not when several light quanta are present with mutually overlapping fields one must imagine a simple superposition of the vector fields, that I cannot say. In any case, for the determination of events, one must have equations of motion for the singular points in addition to the differential equations for the vector field.
>
> *(Quoted in Howard 1990, p. 75)*

From the last sentence, it is clear that Einstein's conception was supposed to be deterministic.

Einstein's view in 1909, then, is remarkably reminiscent of de Broglie's pilot-wave theory as well as of his theory of the 'double solution' (from which de Broglie hoped pilot-wave theory would emerge, see Chapter 2). Einstein seems to have thought of each individual light quantum as being accompanied by some kind of field in 3-space that affects the motion of the quantum.

Fascinating reactions to Einstein's ideas appear in the recorded discussion that took place after Einstein's lecture. Stark pointed out a phenomenon that seemed to speak in favour of localised energy quanta in free space: 'even at great distances, up to 10 m, electromagnetic radiation that has left an X-ray tube for the surrounding space can still achieve concentrated action on a single electron'

(Einstein 1909, English translation, p. 397). Stark's point here is, again, Einstein's locality argument; as we noted at the end of Section 7.2, in retrospect it seems puzzling that this simple and compelling argument was not widely understood much earlier, but here Stark clearly appreciates it. However – and this is probably why the argument did not gain currency – doubts were raised as to how such a theory could explain interference. Planck spoke as follows:

> Stark brought up something in favor of the quantum theory, and I wish to bring up something against it; I have in mind the interferences at the enormous phase differences of hundreds of thousands of wavelengths. When a quantum interferes with itself, it would have to have an extension of hundreds of thousands of wavelengths. This is also a certain difficulty.
>
> *(Einstein 1909, English translation, p. 397)*

To this objection, Einstein gives a most interesting reply:

> I picture a quantum as a singularity surrounded by a large vector field. By using a large number of quanta one can construct a vector field that does not differ much from the kind of vector field we assume to be involved in radiations. I can well imagine that when rays impinge upon a boundary surface, a separation of the quanta takes place, due to interaction at the boundary surface, possibly according to the phase of the resulting field at which the quanta reach the interface. . . . I do not see any fundamental difficulty in the interference phenomena.
>
> *(Einstein 1909, English translation, p. 398)*

Again, the similarity to de Broglie's later ideas is striking. Einstein seems to think that the associated waves can affect the motions of the quanta, in such a way as to account for interference.

It should be noted, though, that in this exchange it is somewhat unclear whether the subject is interference for single photons or for many photons. Einstein talks about interference in terms of the collective behaviour of many quanta, rather than in terms of one quantum at a time. While single-photon interference with very feeble light was observed by Taylor (1909) in the same year, Stark at least seems not to know of Taylor's results, for at this point (just before Einstein's reply) he interjects that 'the experiments to which Mr Planck alluded involve very dense radiation . . .. With radiation of very low density, the interference phenomena would most likely be different' (Einstein 1909, English translation, p. 397). On the other hand, Planck may well have thought of the light-quantum hypothesis as implying that light quanta would move independently, like the molecules of an ideal gas (in which case even for intense radiation one could consider the motion of each quantum independently). Einstein countered precisely such a view at the beginning of his reply to Planck, where he states that 'it must not be assumed that radiations consist of non-interacting quanta; this would make it impossible to explain the phenomena of interference'. This might be read as implying that interactions among different quanta are essential, and that interference would not occur with one photon at a time. However, Einstein may simply have meant that the light quanta cannot be thought of as free particles: they must be accompanied by a wave as well, in order to explain interference. (This last reading fits with Einstein's discussion, in the lecture, of thermal fluctuations in radiation: these cannot be obtained from a gas of free and independent particles alone; a wave-like component is also needed.)

While there is some uncertainty over the details of Einstein's proposal, in retrospect, given our present understanding of how de Broglie's pilot-wave theory provides a straightforward explanation of particle interference (see Section 6.1.1), Einstein's reply to Planck seems very reasonable. However, it appears that yet another of Einstein's arguments was not appreciated by his contemporaries. Seven years later, after having verified Einstein's photoelectric equation experimentally, Millikan nevertheless completely rejected Einstein's light-quantum hypothesis, which he called 'reckless ... because it flies in the face of the thoroughly established facts of interference' (Millikan 1916, p. 355).

9.2 The failure of energy-momentum conservation

It appears that Einstein was still thinking along similar lines in the 1920s, though again without publishing any detailed theory. As we saw at the end of Section 6.2.1, in his collision papers of 1926 Born notes the analogy between his own work and Einstein's ideas: 'I start from a remark by Einstein ... ; he said ... that the waves are there only to show the corpuscular light quanta the way, and in this sense he talked of a "ghost field" ' (Born 1926b, pp. 803–4). Born gives no reference to any published paper of Einstein's, however.

How Einstein's thinking at this time compared with that in 1909 is hard to say. Certainly, in 1909, he thought of the electromagnetic field as being built up from the collective behaviour of large numbers of vector fields associated with individual quanta. In such a scenario, it seems plausible that in the right circumstances the intensity of the emergent electromagnetic field could act as, in effect, a probability field. (Unlike Born, Einstein would have regarded a purely probabilistic description as a makeshift only.)

Just one year before Born's collision papers, in 1925, Einstein gave a colloquium in Berlin where he indeed discussed the idea that every particle (including electrons, following de Broglie) was accompanied by a 'Führungsfeld' or guiding field (Pais 1982, p. 441; Howard 1990, p. 72). According to Wigner, who was present at the colloquium:

> Yet Einstein, though in a way he was fond of it, never published it. He realized that it is in conflict with the conservation principles: at a collision of a light quantum and an electron for instance, both would follow a guiding field. But these guiding fields give only

the probabilities of the directions in which the two components, the light quantum and the electron, will proceed. Since they follow their directions independently, . . . the momentum and the energy conservation laws would be obeyed only statistically .... This Einstein could not accept and hence never took his idea of the guiding field quite seriously.

(Wigner 1980, p. 463)

In the early 1920s, then, Einstein was still thinking along lines that are reminiscent of de Broglie's work, but he never published these ideas because they conflicted with the conservation laws for individual events. The difficulty Einstein faced was overcome only by the introduction (through the work of de Broglie, Schrödinger, and also Born) of a guiding field in configuration space – a *single* (and generally entangled) field that determined probabilities for all particles collectively. In Einstein's approach, where each particle had its own guiding field, the possibility of entanglement was precluded, and the correlations were not strong enough to guarantee energy-momentum conservation for single events.

As a simplified model of what Einstein seems to have had in mind, consider two (distinguishable) particles 1 and 2 moving towards each other in one dimension, with equal and opposite momenta p and $-p$ respectively. Schematically, let us represent this with an initial wave function $\psi(x_1, x_2, 0) = \psi_1^+(x_1, 0)\psi_2^-(x_2, 0)$, where ψ_1^+ and ψ_2^- are broad packets (approximating plane waves) moving along $+x$ and $-x$ respectively. Let the packets meet in a region centred around the origin, where we imagine that an elastic collision takes place with probability $1/2$. At large times, we assume that $\psi(x_1, x_2, t)$ takes the schematic and entangled form

$$\psi(x_1, x_2, t) = \frac{1}{\sqrt{2}} \left(\psi_1^+(x_1, t)\psi_2^-(x_2, t) + \psi_1^-(x_1, t)\psi_2^+(x_2, t) \right) , \tag{9.1}$$

with the first branch corresponding to the particles having moved freely past each other and the second corresponding to an elastic collision reversing their motions. The initial state has zero total momentum. In the final state, both possible outcomes correspond to zero total momentum. Thus, in quantum theory, whatever the outcome of an individual run of the experiment, momentum is always conserved. Now imagine if, instead of using a single wave function $\psi(x_1, x_2, t)$ in configuration space, we made use of two 3-space waves $\psi_1(x, t)$, $\psi_2(x, t)$ (one for each particle). It would then seem natural that, during the collision, the initial 3-space wave $\psi_1(x, 0) = \psi_1^+(x, 0)$ for particle 1 would evolve into

$$\psi_1(x, t) = \psi_1^+(x, t) + \psi_1^-(x, t) , \tag{9.2}$$

while the initial 3-space wave $\psi_2(x, 0) = \psi_2^-(x, 0)$ for particle 2 would evolve into

$$\psi_2(x, t) = \psi_2^-(x, t) + \psi_2^+(x, t) . \tag{9.3}$$

If the amplitude of each 3-space wave determined the probabilities for the respective particles, there would then be *four* (equiprobable) possible outcomes for the scattering experiment, the two stated above, together with one in which both particles move to the right and one in which both particles move to the left. These possibilities would correspond, in effect, to a final (configuration-space) wave function of the separable form

$$(\psi_1^+ + \psi_1^-)(\psi_2^- + \psi_2^+) = \psi_1^+\psi_2^- + \psi_1^+\psi_2^+ + \psi_1^-\psi_2^- + \psi_1^-\psi_2^+ . \quad (9.4)$$

The final total momenta for the four (equiprobable) possible outcomes are 0, $+2p$, $-2p$, 0. The total momentum would not be generally conserved for individual outcomes, but conservation would hold on average. Note the fundamental difference between (9.1) and (9.4): the former is entangled, the latter is not.

Einstein considered such a failure of the conservation laws reason enough to reject the idea. A theory bearing some resemblance to this scheme was, however, proposed and published by Bohr, Kramers and Slater (1924a,b).

In the Bohr–Kramers–Slater (BKS) theory, there are no photons. Associated with an atom in a stationary state is a 'virtual radiation field' containing all those frequencies corresponding to transitions to and from other stationary states. The virtual field determines the transition probabilities for the atom itself, and also contributes to the transition probabilities for other, distant atoms. However, the resulting correlations between widely separated atoms are not strong enough to yield energy-momentum conservation for individual events (cf. Compton's account of the BKS theory in his report, p. 303).[a]

To illustrate this, consider again (in modern language) an atom A emitting a photon of energy $E_i - E_f$ that is subsequently absorbed by a distant atom B. In the BKS theory, the transition at B is not directly caused by the transition at A (as it would be in a simple 'semiclassical' picture of the emission, propagation and subsequent absorption of a localised light quantum). Rather, the virtual radiation field of A contains terms corresponding to transitions of A, and this field contributes to the transition probabilities at B. Yet, if B actually undergoes a transition corresponding to an energy increase $E_i - E_f$, atom A is *not* constrained to make a transition corresponding to an energy decrease $E_i - E_f$. The connection between the atoms is merely statistical, and energy conservation holds only on average, not for individual processes.[b]

Einstein objected to the BKS theory – in a colloquium, and in private letters and conversations (Mehra and Rechenberg 1982a, pp. 553–4; Pais 1982, p. 420;

[a] For a detailed account of the BKS theory, see Darrigol (1992, ch. 9).

[b] Note that Slater's original theory did contain photons, whose motions were guided (statistically) by the electromagnetic field. The photons were removed at the instigation of Bohr and Kramers (Mehra and Rechenberg 1982a, pp. 543–7).

Howard 1990, pp. 71–4) – partly for the same reason he had not published his proposals: that, as he believed, energy-momentum conservation would hold even in the case of elementary interactions between widely separated systems.[a] Einstein's expectation was subsequently confirmed by the Bothe–Geiger and Compton–Simon experiments (Bothe and Geiger 1925b; Compton and Simon 1925).

In the Bothe–Geiger experiment, by means of counter coincidences, Compton scattering (that is, relativistic electron-photon scattering) was studied for individual events, to see if the outgoing scattered photon and the outgoing recoil electron were produced simultaneously. Strict temporal coincidences were expected on the basis of the light-quantum hypothesis, but not on the basis of the BKS theory. Such coincidences were in fact observed by Bothe and Geiger.[b] In the Compton–Simon experiment, Compton scattering was again studied, with the aim of verifying the conservation laws for individual events. Such conservation was in fact observed by Compton and Simon.[c] As a result of these experiments, it was widely concluded that the BKS theory was wrong.[d]

It appears that, even at the time of the fifth Solvay conference, Einstein was still thinking to some extent in terms of guiding fields in 3-space. Evidence for this comes from an otherwise incomprehensible remark Einstein made during his long contribution to the general discussion. There, Einstein compared two interpretations of the wave function ψ for a single electron. On Einstein's view I, ψ represents an ensemble – or 'cloud' – of electrons, while on his view II ψ is a complete description of an individual electron. As we discussed at length in Section 7.1, Einstein argued that interpretation II is inconsistent with locality. Now, Einstein also made the following remark concerning the conservation of energy and momentum according to interpretations I and II (p. 441):

> The second conception goes further than the first.... It is only by virtue of II that the theory contains the consequence that the conservation laws are valid for the elementary process; it is only from II that the theory can derive the result of the experiment of Geiger and Bothe....

As currently understood, of course, a purely 'statistical' interpretation of the wave function (cf. Section 5.2.4) would yield correct predictions, in agreement

[a] In a letter to Ehrenfest dated 31 May 1924, Einstein wrote: 'This idea [the BKS theory] is an old acquaintance of mine, but one whom I do not regard as a respectable fellow' (Howard 1990, p. 72). Einstein listed five criticisms, including the violation of the conservation laws, and a difficulty with thermodynamics (Mehra and Rechenberg 1982a, p. 553).

[b] See Compton's account of 'Bothe and Geiger's coincidence experiments', p. 319, and also Mehra and Rechenberg (1982a, pp. 609–12).

[c] See Compton's account of 'Directional emission of scattered X-rays', pp. 320–1, and also Mehra and Rechenberg (1982a, p. 612).

[d] Cf. Bohr's remarks in the discussion of Compton's report, p. 328.

with the conservation laws for elementary processes (such as scattering). Why, then, did Einstein assert that interpretation I would conflict with such elementary conservation? It must surely be that, for whatever reason, he was thinking of interpretation I as tied specifically to wave functions in 3-space, resulting in a failure of the conservation laws for single events as discussed above. In contrast, specifically regarding interpretation II, Einstein explicitly asserted (p. 442) that it is 'essentially tied to a multi-dimensional representation (configuration space)'.

The conflict between Einstein's ideas about guiding fields in 3-space and energy-momentum conservation was, as we have mentioned, resolved only by the introduction of a guiding field in configuration space – with the associated entanglement and non-separability that Einstein was to find so objectionable. Einstein's early worries about non-separability (long before the EPR paper) have been extensively documented by Howard (1990). Defining 'separability' as the idea that 'spatio-temporally separated systems possess well-defined real states, such that the joint state of the composite system is wholly determined by these two separate states' (Howard 1990, p. 64), Howard highlights the fundamental difficulty Einstein faced in having to choose, in effect, between separability and energy-momentum conservation:

> But as long as the 'guiding' or 'virtual fields' [determining the probabilities for particle motions, or for atomic transitions] are assigned separately, one to each particle or atom, one cannot arrange *both* for the merely probabilistic behavior of individual systems *and* for correlations between interacting systems sufficient to secure strict energy-momentum conservation in all individual events. As it turned out, it was only Schrödinger's relocation of the wave fields from physical space to configuration space that made possible the assignment of joint wave fields that could give the strong correlations needed to secure strict conservation.
>
> *(Howard 1990, p. 73)*

Here, then, is a remarkable historical and physical connection between non-separability or entanglement on the one hand, and energy-momentum conservation on the other. The introduction of probability waves in configuration space, with generic entangled states, finally made it possible to secure energy-momentum conservation for individual emission and scattering events, at the price of introducing a fundamental non-separability into physics.

10

Scattering and measurement in de Broglie's pilot-wave theory

At the fifth Solvay conference, some questions that are closely related to the quantum measurement problem (as we would now call it) were addressed in the context of pilot-wave theory, in both the discussion following de Broglie's report and the general discussion. Most of these questions concerned the treatment of scattering (elastic and inelastic); they were raised by Born and Pauli, and replies were given by Brillouin and de Broglie. Of special interest is the famous – and widely misunderstood – objection by Pauli concerning inelastic scattering. Another question closely related to the measurement problem was raised by Kramers, concerning the recoil of a single photon on a mirror.

In this chapter, we shall first outline the pilot-wave theory of scattering, as currently understood, and examine the extensive discussions of scattering – in the context of de Broglie's theory – that took place at the conference.

We shall see that de Broglie and Brillouin correctly answered the query raised by Born concerning elastic scattering. Further, we shall see that Pauli's objection concerning the inelastic case was both more subtle and more confused than is generally thought; in particular, Pauli presented his example in terms of a misleading optical analogy (that was originally given by Fermi in a more restricted context). Contrary to a widespread view, de Broglie's reply to Pauli *did* contain the essential points required for a proper treatment of inelastic scattering; at the same time, Fermi's misleading analogy confused matters, and neither de Broglie in 1927 nor Bohm in 1952 saw what the true fault with Pauli's example was. (As we shall also see, a proper pilot-wave treatment of Pauli's example was not given until 1956, by de Broglie.)

We shall also outline the pilot-wave theory of quantum measurement, again as currently understood, and we shall use this as a context in which to examine the question raised by Kramers, of the recoil of a single photon on a mirror, which de Broglie was unable to answer.

10.1 Scattering in pilot-wave theory

Let us first consider elastic scattering by a fixed potential associated with some scattering centre or region. An incident particle may (for a pure quantum state) be represented by a freely evolving wave packet $\psi_{\rm inc}(\mathbf{x}, t)$ that is spatially finite (that is, limited both longitudinally and laterally), and that has mean momentum $\hbar\mathbf{k}$. During the scattering, the wave function evolves into $\psi = \psi_{\rm inc} + \psi_{\rm sc}$, where $\psi_{\rm sc}$ is the scattered wave. At large distances from the scattering region, and off the axis (through the scattering centre) parallel to the incident wave vector $\mathbf{k}$, only the scattered wave $\psi_{\rm sc}$ contributes to the particle current density $\mathbf{j}$ – where $\mathbf{j}$ is often used in textbook derivations of the scattering cross section.

As is well known, the mathematics of a fully time-dependent calculation of the scattering of a finite packet may be simplified by resorting to a time-independent treatment in which $\psi_{\rm inc}$ is taken to be an infinitely extended plane wave $e^{i\mathbf{k}\cdot\mathbf{x}}$. At large distances from the scattering region, the wave function $\psi = e^{i\mathbf{k}\cdot\mathbf{x}} + \psi_{\rm sc}$ (a time-independent eigenfunction of the total Hamiltonian) has the asymptotic form

$$\psi = e^{ikz} + f(\theta, \phi)\frac{e^{ikr}}{r} \tag{10.1}$$

(taking the z-axis parallel to $\mathbf{k}$, using spherical polar coordinates centred on the scattering region, and ignoring overall normalisation). The scattering amplitude f gives the differential cross section $d\sigma/d\Omega = |f(\theta, \phi)|^2$.

In the standard textbook derivation of $d\sigma/d\Omega$, the current density $\mathbf{j}$ is used to calculate the rate of probability flow into an element of solid angle $d\Omega$, where $\mathbf{j}$ is taken to be the current associated with $\psi_{\rm sc}$ only, even though $\psi_{\rm sc}$ overlaps with $\psi_{\rm inc} = e^{i\mathbf{k}\cdot\mathbf{x}}$. This is justified because the plane wave $e^{i\mathbf{k}\cdot\mathbf{x}}$ is, of course, merely an abstraction used for mathematical convenience: a real incident wave will be spatially limited, and will not overlap with the scattered wave at the location of the particle detector (which is assumed to be located off the axis of incidence, so as not to be bathed in the incident beam).

The above standard discussion of scattering may readily be recast in pilot-wave terms, where

$$\mathbf{v} = \frac{\mathbf{j}}{|\psi|^2} = \frac{\hbar}{m}\operatorname{Im}\frac{\nabla\psi}{\psi} = \frac{\nabla S}{m} \tag{10.2}$$

(with $\psi = |\psi|\, e^{(i/\hbar)S}$) is interpreted as the actual velocity field of an ensemble of particles with positions distributed according to $|\psi|^2$. The differential cross section $d\sigma/d\Omega$ measures the fraction of incident particles whose actual trajectories end (asymptotically) in the element of solid angle $d\Omega$.

That a real incident packet is always spatially finite is, of course, an elementary point known to every student of wave optics. This (often implicit) assumption is essential to introductory textbook treatments of the scattering of light, whether by a Hertzian dipole or by a diffraction grating. If the incident wave were a literally infinite plane wave, then the scattered wave would of course overlap with the incident wave everywhere, and no matter where a detector was placed it would be affected by the incident wave as well as by the scattered wave.

Certainly, the participants at the fifth Solvay conference – many of whom had extensive laboratory experience – were aware of this simple point. We emphasise this because, as we shall see, the finiteness of incident wave packets played a central role in the discussions that took place regarding scattering in de Broglie's theory.

Let us now consider inelastic scattering: specifically, the scattering of an electron by a hydrogen atom initially in the ground state. The atom can become excited by the collision, in which case the outgoing electron will have lost a corresponding amount of energy. Let the scattering electron have position $\mathbf{x}_s$ and the atomic electron have position $\mathbf{x}_a$. In a time-dependent description, the total wave function evolves into

$$\Psi(\mathbf{x}_s, \mathbf{x}_a, t) = \phi_0(\mathbf{x}_a)e^{-iE_0t/\hbar}\psi_{inc}(\mathbf{x}_s, t) + \sum_n \phi_n(\mathbf{x}_a)e^{-iE_nt/\hbar}\psi_n(\mathbf{x}_s, t) . \quad (10.3)$$

Here, the first term is an initial product state, where ϕ_0 is the ground-state wave function of hydrogen with ground-state energy E_0 and ψ_{inc} is a (finite) incident packet. The scattering terms have components as shown, where the ϕ_n are the nth excited states of hydrogen and the ψ_n are outgoing wave packets. It may be shown by standard techniques that, asymptotically, the nth outgoing packet ψ_n is centred on a radius $r_n = (\hbar k_n/m)t$ from the scattering region, where k_n is the outgoing wave number fixed by energy conservation.

Because the outgoing (asymptotic) packets ψ_n expand with different speeds, they eventually become widely separated in space. The actual scattered electron with position $\mathbf{x}_s(t)$ can occupy only one of these non-overlapping packets, say ψ_i, and its velocity will then be determined by ψ_i alone. Further, the motion of the atomic electron will be determined by the corresponding ϕ_i alone, and after the scattering the atom will be (in effect) in an energy eigenstate ϕ_i.

We are using two well-known properties of pilot-wave dynamics: (a) If a wave function $\Psi = |\Psi| e^{(i/\hbar)S}$ is a superposition of terms $\Psi_1 + \Psi_2 + \ldots$ having no overlap in configuration space, then the phase gradient ∇S at the occupied point of configuration space (which gives the velocity of the actual configuration) reduces to ∇S_i, where $\Psi_i = |\Psi_i| e^{(i/\hbar)S_i}$ is the occupied packet. (b) If the occupied packet Ψ_i is a product over certain configuration components, then the velocities of those components are determined by the associated factors in the product.

Applying (a) and (b) to the case discussed here, once the ψ_n have separated the total wave function Ψ becomes a sum of non-overlapping packets, where only one packet $\Psi_i = \phi_i(\mathbf{x}_a)e^{-iE_it/\hbar}\psi_i(\mathbf{x}_s, t)$ can contain the actual configuration $(\mathbf{x}_a, \mathbf{x}_s)$. The velocity of the scattered electron is then given by $\dot{\mathbf{x}}_s = (\hbar/m)\,\mathrm{Im}(\nabla\psi_i/\psi_i)$, while the velocity of the atomic electron is given by $\dot{\mathbf{x}}_a = (\hbar/m)\,\mathrm{Im}(\nabla\phi_i/\phi_i)$. Thus, there takes place an effective 'collapse of the wave packet' to the state $\phi_i\psi_i$.

It is straightforward to show that, if the initial ensemble of $\mathbf{x}_s$, $\mathbf{x}_a$ has distribution $|\Psi|^2$, the probability for ending in the ith packet – that is, the probability for the atom to end in the state ϕ_i – will be given by $\int d^3\mathbf{x}_s\ |\psi_i(\mathbf{x}_s, t)|^2$, in accordance with the usual quantum result.

Further, the effective 'collapse' to the state $\phi_i\psi_i$ is for all practical purposes irreversible. As argued by Bohm (1952a, p. 178), the scattered particle will subsequently interact with many other degrees of freedom – making it very unlikely that distinct states $\phi_i\psi_i$, $\phi_j\psi_j$ $(i \neq j)$ will interfere at later times, as this would require the associated branches of the total wave function to overlap with respect to *every* degree of freedom involved.

Again, for mathematical convenience, one often considers the limit in which the incident wave ψ_{inc} is unlimited (a plane wave). This makes the calculation of Ψ easier. In this limit, the outgoing wave packets ψ_n become unlimited too, and overlap with each other everywhere: *all* the terms in (10.3) then overlap in every region of space. If one naively assumed that this limit corresponded to a real situation, the outgoing electron would never reach a constant velocity because it would be guided by a superposition of overlapping terms, rather than by a single term in (10.3). Similarly, after the scattering, the atomic electron would not be guided by a single eigenfunction ϕ_i, and the atom would not finish in a definite energy state. In any real situation, of course, ψ_{inc} will be limited in space and time, and at large times the outgoing packets ψ_n will separate: the trajectory of the scattered electron will be guided by only one of the ψ_n and (in regions outside the path of the incident beam) will *not* be affected by ψ_{inc}. (Note that longitudinal finiteness of the incident wave ψ_{inc} leads to a separation of the outgoing waves ψ_n from each other, while lateral finiteness of ψ_{inc} ensures that ψ_{inc} does not overlap with the ψ_n in regions off the axis of incidence.)

As we shall see, apart from the practical irreversibility of the effective collapse process, the above 'pilot-wave theory of scattering' seems to have been more or less understood by de Broglie (and perhaps also by Brillouin) in October 1927. A detailed treatment of scattering was given by Bohm in his first paper on de Broglie–Bohm theory (Bohm 1952a).

In Bohm's second paper, however, appendix B gives a misleading account of the de Broglie–Pauli encounter at the fifth Solvay conference (Bohm 1952b, pp. 191–2), and this seems to be the source of the widespread misunderstandings concerning this encounter. Citing the proceedings of the fifth Solvay conference, Bohm wrote the following:

De Broglie's suggestions met strong objections on the part of Pauli, in connection with the problem of inelastic scattering of a particle by a rigid rotator. Since this problem is conceptually equivalent to that of inelastic scattering of a particle by a hydrogen atom, which we have already treated . . ., we shall discuss the objections raised by Pauli in terms of the latter example.

Bohm then describes 'Pauli's argument': taking the incoming particle to have a plane wave function, all the terms in (10.3) overlap, so that 'neither atom nor the outgoing particle ever seem to approach a stationary energy', contrary to what is observed experimentally. According to Bohm:

Pauli therefore concluded that the interpretation proposed by de Broglie was untenable. De Broglie seems to have agreed with the conclusion, since he subsequently gave up his suggested interpretation.

Bohm then gives what he regards as his own, original answer to Pauli's objection:

. . . as is well known, the use of an incident plane wave of infinite extent is an excessive abstraction, not realizable in practice. Actually, both the incident and outgoing parts of the ψ-field will always take the form of bounded packets. Moreover, . . . all packets corresponding to different values of n will ultimately obtain classically describable separations. The outgoing particle must enter one of these packets, . . . leaving the hydrogen atom in a definite but correlated stationary state.

By way of conclusion, Bohm writes:

Thus, Pauli's objection is seen to be based on the use of the excessively abstract model of an infinite plane wave.

At this point, in the light of what we said above about the use of plane waves in elementary wave optics and in scattering theory, it is natural to ask how a physicist of Pauli's abilities could have made the glaring mistake that Bohm claims he made. In fact, as we shall see in the next section, Pauli's objection was *not* based on a failure to appreciate the importance of the finiteness of initial wave packets. On the contrary, Pauli's objection shows some understanding of the crucial role played by limited packets in pilot-wave theory. What really happened is that Pauli's objection involved a peculiar and misleading analogy with optics, according to which the incident packet appeared to be *necessarily* unlimited, in conditions such as to prevent the required separation into non-overlapping components.

10.2 Elastic and inelastic scattering: Born and Brillouin, Pauli and de Broglie

In the discussion following de Broglie's report, Born suggests (p. 364) that de Broglie's guidance equation will fail for an *elastic* collision between an electron and an atom. Specifically, Born asks if the electron speed will be the same before and after the collision, to which de Broglie simply replies that it will.

Later in the same discussion, Brillouin (pp. 367ff.) gives an extensive and detailed presentation, explaining how de Broglie's theory accounts for the elastic scattering of a photon from a mirror. Brillouin is quite explicit about the role played by the finite extension of the incident packet. In his Fig. 2 (see p. 368), Brillouin shows an incident photon trajectory (at angle θ to the normal) guided by an incoming and *laterally limited* packet. The packet is reflected by the mirror, producing an outgoing packet that is again laterally limited. Near the mirror there is an interference zone, where the incoming and outgoing packets overlap. As Brillouin puts it:

> Let us draw a diagram for the case of a limited beam of light falling on a plane mirror; the interference is produced in the region of overlap of two beams.

Brillouin sketches the photon trajectory, which curves away from the mirror as it enters the interference zone, moves approximately parallel to the mirror while in the interference zone, and then moves away again, eventually settling into a rectilinear motion guided by the outgoing packet (Brillouin's Fig. 2). Brillouin describes the trajectory thus:

> . . . at first a rectilinear path in the incident beam, then a bending at the edge of the interference zone, then a rectilinear path parallel to the mirror, with the photon travelling in a bright fringe and avoiding the dark interference fringes; then, when it comes out, the photon retreats following the direction of the reflected light beam.

Here we have a clear description of an incident photon, guided by a finite packet and moving uniformly towards the mirror, with the packet then undergoing interference and scattering, while the photon is eventually carried away – again with a uniform motion – by a finite outgoing packet. (The ingoing and outgoing motions of the photon are strictly uniform, of course, only in the limit where the guiding packets become infinitely broad.)

Despite his description in terms of finite packets, however, in order to calculate the precise motion of the photon in the interference zone Brillouin uses the standard device of treating the incoming packet as an infinite plane wave. In this (abstract) approximation, he shows that the photon moves parallel to the mirror with a speed $v = c \sin \theta$. Because the incoming packet is in reality limited, a photon motion parallel to the mirror is (approximately) realised only in the interference zone, as Brillouin sketches in his accompanying Fig. 2.

While Brillouin's figure shows a laterally limited packet, it is clear from subsequent discussion that the incident packet was implicitly regarded as limited *longitudinally* as well (as of course it must be in any realistic situation). For in the general discussion the question of photon reflection by a mirror was raised again, and Einstein asked (p. 451) what happens in de Broglie's theory in the case of normal incidence ($\theta = 0$), for which the formula $v = c \sin\theta$ predicts that the photons will have zero speed. Piccard responded to Einstein's query, and pointed out that, indeed, the photons are stationary in the limiting case of normal incidence. The meaning of this exchange between Einstein and Piccard is clear: if a longitudinally limited packet were incident normally on the mirror, the incident packet would carry the photon towards the mirror; in the region where the incident and reflected packets overlap, the photon would be at rest; once the packet has been reflected, the photon will be carried away by the outgoing packet. The discussion of the case of normal incidence implicitly assumes that the incident packet is longitudinally limited.

Thus, Brillouin's example of photon reflection by a mirror illustrates the point that the use of plane waves was for calculational convenience only, and that, when it came to the discussion of real physical examples, it was clear to all (and hardly worth mentioning explicitly) that incident waves were in reality limited in extent (in all directions). Brillouin's mathematical use of plane waves parallels their use in the general theory of scattering sketched in Section 10.1.

Elastic scattering is also discussed in de Broglie's report, for the particular case of electrons incident on a fixed, periodic potential – the potential generated by a crystal lattice. This case is especially interesting in the present context, because it involves the separation of the scattered wave into non-overlapping packets – the interference maxima of different orders, well-known from the theory of X-ray diffraction – with the particle entering just one of these packets (a point that is relevant to a proper understanding of the de Broglie–Pauli encounter). Here is how de Broglie describes the diffraction of an electron wave by a crystal lattice (p. 357):

> ... the wave Ψ will propagate following the general equation, in which one has to insert the potentials created by the atoms of the crystal considered as centres of force. One does not know the exact expression for these potentials but, because of the regular distribution of atoms in the crystal, one easily realises that the scattered amplitude will show maxima in the directions predicted by Mr von Laue's theory. Because of the role of pilot wave played by the wave Ψ, one must then observe a selective scattering of the electrons in these directions.

Again, as in Brillouin's discussion, there is no need to mention explicitly the obvious point that the incident wave Ψ will be spatially limited.

Let us now turn to the inelastic case. This was discussed by de Broglie in his report, in particular in the final part, which includes a review of recent experiments involving the inelastic scattering of electrons by atoms of helium. De Broglie noted that, according to Born's calculations (using his statistical interpretation of the wave function), the differential cross section should show maxima as a function of the scattering angle (p. 357):

> ... Mr Born has studied ... the collision of a narrow beam of electrons with an atom. According to him, the curve giving the number of electrons that have suffered an inelastic collision as a function of the scattering angle must show maxima and minima ....

As de Broglie then discussed in detail, such maxima had been observed experimentally by Dymond (though the results were only in qualitative agreement with the predictions). Having summarised Dymond's results, de Broglie commented (p. 359):

> The above results must very probably be interpreted with the aid of the new Mechanics and are to be related to Mr Born's predictions.

We are now ready for a close examination of Pauli's objection in the general discussion, according to which there is a difficulty with de Broglie's theory in the case of inelastic collisions. That there might be such a difficulty was in fact already suggested by Pauli a few months earlier, in a letter to Bohr dated 6 August 1927, already quoted in Chapter 2. As well as noting the exceptional quality of de Broglie's 'Structure' paper (de Broglie 1927b), in his letter Pauli states that he is suspicious of de Broglie's trajectories and cannot see how the theory could account for the discrete energy exchange seen in individual inelastic collisions between electrons and atoms (Pauli 1979, pp. 404–5). (Such discrete exchange had been observed, of course, in the Franck–Hertz experiment.) This is essentially the objection that Pauli raises less than three months later in Brussels.

In the general discussion (p. 462), Pauli begins by stating his belief that de Broglie's theory works for elastic collisions:

> It seems to me that, concerning the statistical results of scattering experiments, the conception of Mr de Broglie is in full agreement with Born's theory in the case of elastic collisions ....

This preliminary comment by Pauli is significant. For if, as Bohm asserted in 1952, Pauli's objection was based on 'the use of the excessively abstract model of an infinite plane wave' (Bohm 1952b, p. 192), then Pauli would have regarded de Broglie's theory as problematic *even in the elastic case*. For in an elastic collision, if the incident wave ψ_{inc} is infinitely extended (a plane wave), then any part of the outgoing region will be bathed in the incident wave: the scattered particle will inevitably be affected by both parts of the superposition $\psi_{\text{inc}} + \psi_{\text{sc}}$, and will never settle down to a constant speed. Since Pauli agreed that the outgoing speed *would* be constant in the elastic case – as de Broglie had asserted (in reply to Born) in the discussion following his report – Pauli presumably understood that the finite incident packet would not affect the scattered particle.

Pauli goes on to claim that de Broglie's theory will not work for inelastic collisions, in particular for the example of scattering by a rotator. To understand Pauli's point, it is important to distinguish between the real physical situation being discussed, and the optical analogy used by Pauli – an analogy that had been introduced by Fermi as a convenient (and as we shall see limited) means to solve the scattering problem. Pauli's objection is framed in terms of Fermi's analogy, and therein lies the confusion.

The real physical set-up consists of an electron moving in the (x, y)-plane and colliding with a rotator. The latter is a model scattering centre with one rotational degree of freedom represented by an angle φ.[a] Pauli took the initial wave function to be

$$\psi_0(x, y, \varphi) \propto e^{(i/\hbar)(p_x x + p_y y + p_\varphi \varphi)} , \tag{10.4}$$

with p_φ restricted to $p_\varphi = m\hbar$ $(m = 0, 1, 2, \ldots)$. The inelastic scattering of an electron by a rotator had been treated by Fermi (1926) using an analogy with optics, according to which the (time-independent) scattering of an electron in two spatial dimensions by a rotator is mathematically equivalent to the (time-independent) scattering of a (scalar) light wave $\psi(x, y, \varphi)$ in three spatial dimensions by an infinite diffraction grating, with φ interpreted as a third spatial coordinate ranging over the whole real line $(-\infty, +\infty)$. The infinite 'grating' constitutes a periodic potential, arising mathematically from the periodicity associated with the original variable φ. Similarly, the function $\psi(x, y, \varphi)$ is necessarily unlimited along the φ-axis. By construction, then, both the incident wave and the grating are unlimited along the φ-axis. Fermi's analogy is useful, because the different spectral orders for diffracted beams emerging from the grating correspond to the possible final (post-scattering) energy states of the rotator. However, as we shall see, Fermi's analogy has only a very limited validity.

Pauli, then, presents his objection in terms of Fermi's optical analogy. He says (p. 463):

> It is, however, an essential point that, in the case where the rotator is in a stationary state before the collision, the incident wave is unlimited in the direction of the [φ-]axis. For this reason, the different spectral orders of the grating will always be superposed at each point of configuration space. If we then calculate, according to the precepts of Mr de Broglie, the angular velocity of the rotator after the collision, we must find that this velocity is not constant.

In Fermi's three-dimensional analogy, for an incident beam unlimited along φ (as well as along x and y) the scattered waves will indeed overlap everywhere. The

[a] Classically, a rotator might consist of a rigid body free to rotate about a fixed axis. The quantum rotator was known to have a discrete spectrum of quantised energy levels (corresponding to quantised states of angular momentum), and was sometimes considered as a useful and simple model of a quantum system.

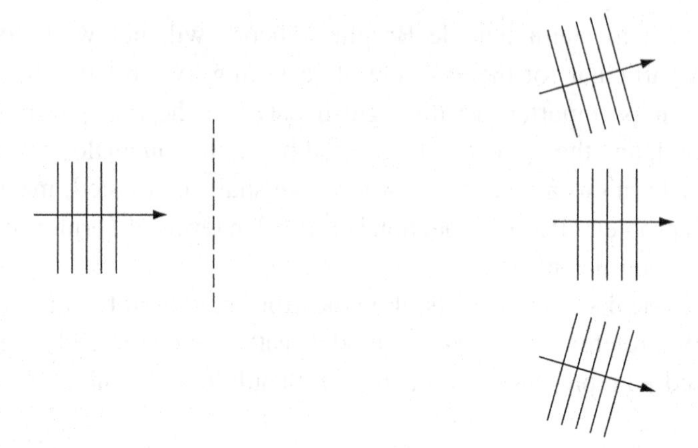

Fig. 10.1. Scattering of a laterally limited wave by a finite diffraction grating, showing the separation of the first-order beams from the zeroth-order beam.

final configuration will then be guided by a superposition of all the final energy states, and the final velocity of the configuration will not be constant. It then appears (according to Pauli) that the final angular velocity of the rotator will not be constant, contrary to what is expected for a stationary state, and that there will be no definite outcome for the scattering experiment (that is, no definite final energy state for the rotator).

In the usual discussion of diffraction gratings in optics, it is of course assumed that both the grating and the incident beam are laterally limited, so that the emerging beams separate as shown in Fig. 10.1 (where only the zeroth-order and first-order beams are drawn). But in Fermi's analogy, there can be no such lateral limitation and no such separation. It might be thought that, in the case of no lateral limitation, separation of an optical beam would nevertheless take place if the incident wave were longitudinally limited. However, as shown in Fig. 10.2, symmetry dictates that there will be no beams beyond the zeroth order.[a] Thus, if one accepts Fermi's analogy with the scattering of light by an infinite grating, even if one takes a longitudinally limited incident light wave, there will still be no separation and the difficulty remains. The only way to obtain a separation of the scattered beams is through a lateral localisation along φ – and according to Pauli's argument this is impossible.

According to Fermi's analogy, then, it is inescapable that (for an initial stationary state) the incident wave $\psi(x, y, \varphi)$ is unlimited on the φ-axis, and one cannot avoid

[a] Geometrically, the presence of diverging higher-order beams (as in Fig. 10.1) would define a preferred central point on the grating. Roughly speaking, if one considers Fig. 10.1 in the limit of an infinite grating (and of a laterally unlimited incident beam), the higher-order emerging beams are 'pushed off to infinity', resulting in Fig. 10.2.

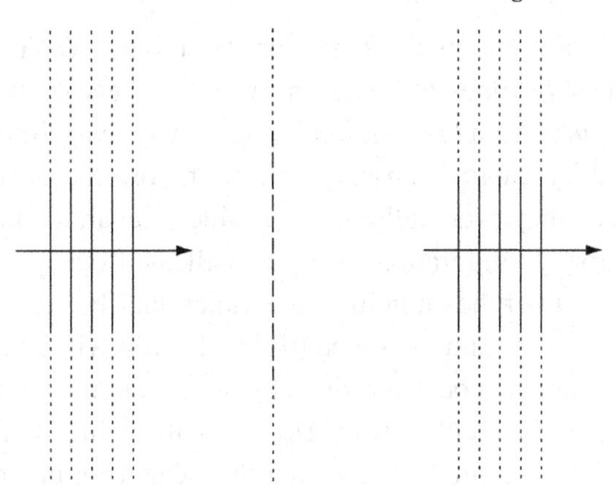

Fig. 10.2. Scattering of a laterally unlimited wave by an infinite diffraction grating.

the conclusion that after the scattering the rotator need not be in a definite energy eigenstate. There seems to be no way out of Pauli's difficulty, and de Broglie's pilot-wave theory appears to be untenable.

Today, one might answer Pauli's objection by considering the measuring apparatus used to detect the outgoing scattered particle (or, used to measure the energy of the atom). By including the degrees of freedom of the apparatus in the quantum description, one could obtain a separation of the total wave function into non-overlapping branches, resulting in a definite quantum outcome (cf. Section 10.3). However, in 1927 the measuring apparatus was not normally considered to be part of the quantum system; and even today, in the pilot-wave theory of scattering (sketched in Section 10.1), a definite outcome is generally guaranteed by the separation of packets for the scattered particle. As we shall now show, the usual pilot-wave theory of scattering in fact suffices in Pauli's example too.

To see how Pauli's objection would normally be met today, note first that in Pauli's example the real physical situation consists of an electron moving in two spatial dimensions x, y and colliding with a rotator whose angular coordinate φ ranges over the unit circle (from 0 to 2π). In de Broglie's dynamics, the wave function $\psi(x, y, \varphi, t)$ yields velocities $\dot{x}, \dot{y}$ for the electron and $\dot{\varphi}$ for the rotator, where these velocities are given by the quantum current divided by $|\psi|^2$. As we saw in the case of inelastic scattering by an atom, the final wave function will take the form

$$\psi(x, y, \varphi, t) = \phi_0(\varphi)e^{-iE_0t/\hbar}\psi_{\text{inc}}(x, y, t) + \sum_n \phi_n(\varphi)e^{-iE_nt/\hbar}\psi_n(x, y, t), \tag{10.5}$$

where now the ϕ_n are stationary states for the rotator. Once again, for finite (localised) ψ_{inc}, at large times the outgoing wave packets ψ_n will be centred on a radius $r_n = (\hbar k_n/m)t$ from the scattering region, where again k_n is the outgoing wave number fixed by energy conservation. As before, because the ψ_n expand with different speeds they eventually become widely separated in space. The total (electron-plus-rotator) configuration (x, y, φ) will then occupy only *one* branch $\phi_i(\varphi)e^{-iE_i t/\hbar}\psi_i(x, y, t)$ of the outgoing wave function. The scattered electron will then be guided by ψ_i only, and the speed of the electron will be constant. Further, the motion of the rotator will be determined by the stationary state ϕ_i, and will also be uniform. As long as the incident wave ψ_{inc} is localised in x and y, the final wave function separates in configuration space and the scattering process has a definite outcome. Pauli's objection therefore has a straightforward answer.

On Fermi's analogy, however, Pauli's example *seems* to be equivalent to the scattering of a laterally unlimited light wave in three-dimensional space by an infinite grating, for which no separation of beams can take place. Since, for the original system, we have seen that the final state does separate, it is clear that Fermi's analogy must be mistaken in some way.

To see what is wrong with Fermi's analogy, one must examine Fermi's original paper. There, Fermi introduces the coordinates

$$\xi = \sqrt{m}x, \quad \eta = \sqrt{m}y, \quad \zeta = \sqrt{J}\varphi, \tag{10.6}$$

and writes the time-independent Schrödinger equation – for the combined rotator-plus-electron system of total energy E – in the form

$$\frac{\partial^2\psi}{\partial\xi^2} + \frac{\partial^2\psi}{\partial\eta^2} + \frac{\partial^2\psi}{\partial\zeta^2} + \frac{2}{\hbar^2}(E - V)\psi = 0\,, \tag{10.7}$$

where the potential energy V is a periodic function of ζ with period $2\pi\sqrt{J}$. Fermi then considers an optical analogue of the wave equation (10.7). As Fermi puts it (Fermi 1926, p. 400):

In order to see the solution of [(10.7)], we consider the optical analogy for the wave equation [(10.7)]. In the regions far from the ζ-axis, where V vanishes, [(10.7)] is the wave equation in an optically homogeneous medium; in the neighbourhood of the ζ-axis, the medium has an anomaly in the refractive index, which depends periodically on ζ. Optically this is nothing more than a linear grating of period $2\pi\sqrt{J}$.

Fermi then goes on to consider a plane wave striking the grating, and relates the outgoing beams of different spectral orders to different types of collisions between the electron and the rotator.

Now, if we include the time dependence $\psi \propto e^{-(i/\hbar)Et}$, we may write

$$\frac{2}{\hbar^2}E\psi = -\frac{1}{c^2}\frac{\partial^2\psi}{\partial t^2} \qquad (10.8)$$

with $c \equiv \sqrt{E/2}$, and (10.7) indeed coincides with the wave equation of scalar optics for the case of a given frequency. However – and here is where Fermi's analogy breaks down – because the 'speed of light' c depends on the energy E (or frequency E/h), the time evolution of the system in regions far from the ζ-axis is in general *not* equivalent to wave propagation in 'an optically homogeneous medium'. The analogy holds only for one energy E at a time (which is all Fermi needed to consider). In a realistic case where the (finite) incident electron wave is a sum over different momenta – and as we have said, it was understood that in any realistic case the incident wave would indeed be finite and therefore equal to such a sum – the problem of scattering by the rotator *cannot* be made equivalent to the scattering of light by a grating in an otherwise optically homogeneous medium: on the contrary, the 'speed of light' c would have to vary as the square root of the frequency.

Pauli's presentation does not mention that Fermi's optical analogy is valid only for a single frequency. Since, as we have seen, the finiteness of realistic incident packets was implicit in all these discussions, the impression was probably given that Fermi's analogy holds generally. A finite (in x and y) electron wave incident on the rotator would then be expected to translate into a longitudinally finite but laterally infinite (along ζ) light wave incident on an infinite grating, with a resulting lack of separation in the final state. Since, however, the quantity c is frequency-dependent, in general the analogy with light is invalid and the conclusion unwarranted.

What really happens, then, in the two-dimensional scattering of an electron by a rotator, if one interprets the angular coordinate φ as a third spatial axis à la Fermi? The answer is found in a detailed discussion of Pauli's objection given by de Broglie, in a book published in 1956, whose chapter 14 bears the title 'Mr Pauli's objection to the pilot-wave theory' (de Broglie 1956). There, de Broglie gives what is in fact the first proper analysis of Pauli's example in terms of pilot-wave theory (with one spatial dimension suppressed for simplicity). The result of de Broglie's analysis can be easily seen by reconsidering the expression (10.5), and allowing φ to range over the real line, with $\psi(x, y, \varphi, t)$ regarded as a periodic function of φ with period 2π. Because ψ separates (as we have seen) into packets that are non-overlapping with respect to x and y, the time evolution of ψ will be as sketched in Fig. 10.3 (where we suppress y) – a figure that we have adapted from de Broglie's book (p. 176). The figure shows a wave moving along the x-axis towards the rotator at $x = 0$. The wave is longitudinally limited (finite along x) and laterally

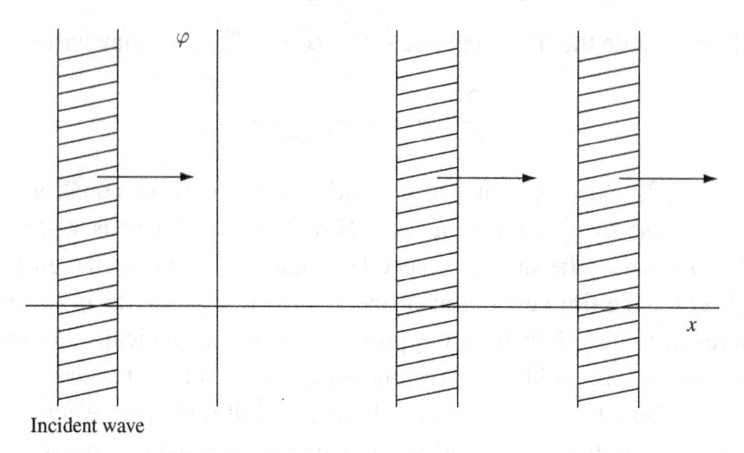

Fig. 10.3. True evolution of the electron-rotator wave function in Pauli's example. Adapted from de Broglie (1956, p. 176).

unlimited (infinite along φ), and separates as shown into similar packets moving at different speeds after the scattering. (Only two of the final packets are drawn.)

The different speeds of the packets after the collision correspond, of course, to the different possible kinetic energies of the electron after the collision. Note that the crucial separation into packets moving at different speeds would be ruled out if one were mistakenly to accept Fermi's analogy with light scattering 'in an optically homogeneous medium' – for the outgoing packets would then all have to move with the same speed (that of light).

A complete reply to Pauli's original objection should then make two points: (1) Because of the frequency-dependent 'speed of light' $c \equiv \sqrt{E/2}$, Fermi's analogy with optics is of very limited validity and cannot be applied to a real case with a finite incident wave. (2) For a finite incident electron wave, a separation of packets does in fact take place with respect to the spatial coordinates of the electron.

Let us now examine how de Broglie replied to Pauli in October 1927, in the general discussion. As will become clear, de Broglie understood the general separation mechanism required to yield a definite outcome, but he was misled by the (generally false) optical analogy and phrased his answer in terms of it.

De Broglie replies by first pointing out the importance of having a laterally limited incident wave, to avoid overlap among the diffracted beams, and to avoid overlap between these and the incident wave (p. 463):

> The difficulty pointed out by Mr Pauli has an analogue in classical optics. One can speak of the beam diffracted by a grating in a given direction only if the grating and the incident wave are laterally limited, because otherwise all the diffracted beams will overlap and be bathed in the incident wave. In Fermi's problem, one must also assume the wave ψ to be limited laterally in configuration space.

De Broglie then notes that, if one can assume ψ to be limited laterally, then the system velocity will become constant once the diffracted waves have separated (from each other and from the incident beam):

> ... the velocity of the representative point of the system will have a constant value, and will correspond to a stationary state of the rotator, as soon as the waves diffracted by the φ-axis will have separated from the incident beam.

What de Broglie is describing here is precisely the separation into non-overlapping packets in configuration space, which the pilot wave must undergo in order for the scattering experiment to have a definite outcome – just as we have discussed in Section 10.1 above.

However, in Pauli's example, it appeared that the incident ψ could not be limited laterally. On this point, de Broglie claims that Fermi's configuration space is artificial, having been formed 'by rolling out along a line the cyclic variable φ'. But as de Broglie's treatment of 1956 shows, extending the range of φ from the unit circle to the real line does not really make any difference: it simply distributes copies of the configuration space along the φ-axis (see Fig. 10.3).

Pauli's presentation did not mention the frequency-dependent speed $c \equiv \sqrt{E/2}$, and in his reply de Broglie did not mention it either. Indeed, it seems that neither de Broglie in 1927 nor Bohm in 1952 noticed this misleading aspect of Pauli's objection.

Note that finiteness of the incident wave with respect to x and y had not been questioned by anyone. As we have seen, the finite spatial extension of realistic wave packets was implicitly assumed by all. Pauli had raised the impossibility of finiteness *specifically* with respect to the φ-axis, and de Broglie responded on this specific point alone. While de Broglie was misled by Fermi's analogy, even so his remarks contain the key point, later developed in detail by Bohm – that finiteness of the initial packet will ensure a separation into non-overlapping final packets in configuration space. As de Broglie himself put it in his book of 1956, in reference to his 1927 reply to Pauli: 'Thus I had indeed realised... that the answer to Mr Pauli's objection had to rest on the fact that the wave trains are always limited, an idea that has been taken up again by Mr Bohm in his recent papers' (de Broglie 1956, p. 176).

Finally, we point out that – leaving aside the misleading nature of Fermi's optical analogy – de Broglie's audience may well have understood his description of wave packets separating in configuration space, for Born had already described something very similar for the Wilson cloud chamber, earlier in the general discussion (pp. 437ff.). As we saw in Section 6.2, Born discussed the formation of a track in a cloud chamber in terms of a branching of the total wave function in a multi-dimensional configuration space (formed by all the particles involved).

In particular, for the simple case of an α-particle interacting with two atoms in one dimension, Born described an initially localised packet that separates into two non-overlapping branches in (three-dimensional) configuration space – as sketched in Born's figure (p. 439). By the time de Broglie came to reply to Pauli, then, the audience was already familiar with the idea of a wave function evolving in configuration space and developing non-overlapping branches. Thus, it seems more likely than not that this aspect of de Broglie's reply to Pauli would have been understood.[a]

10.3 Quantum measurement in pilot-wave theory

The general theory of quantum measurement in pilot-wave theory was first developed by Bohm in his second paper on the theory (Bohm 1952b). Bohm understood that the degrees of freedom associated with the measurement apparatus were simply extra coordinates that should be included in the total system (as in, for example, Born's 1927 discussion of the cloud chamber). Thus, denoting by x the coordinates of the 'system' and y the coordinates of the 'apparatus', the dynamics generates a trajectory $(x(t), y(t))$ for the total configuration, guided by a wave function $\Psi(x, y, t)$ (where again the velocity field $(\dot{x}, \dot{y})$ is given by the quantum current of Ψ divided by $|\Psi|^2$). Bohm showed how, in the circumstances corresponding to a quantum measurement, Ψ separates into non-overlapping branches and the system coordinate $x(t)$ is eventually guided by an effectively 'reduced' packet for the system.

As a simple example, let the system have an initial wave function

$$\psi_0(x) = \sum_n c_n \phi_n(x) , \tag{10.9}$$

where the ϕ_n are eigenfunctions of some Hermitian operator Q with eigenvalues q_n. Suppose an experiment is performed that in quantum theory would be called 'a measurement of the observable Q'. This might be done by coupling the system to a 'pointer' with coordinate y and initial wave function $g_0(y)$, where g_0 is narrowly peaked around $y = 0$. An appropriate coupling might be described by an interaction Hamiltonian $H = aQP_y$, where a is a coupling constant and P_y is the momentum operator conjugate to y. If we neglect the rest of the Hamiltonian,

[a] It may also be worth noting that, in his report, when presenting the pilot-wave theory of many-body systems, de Broglie explicitly asserts (p. 352) that his probability formula in configuration space 'fully accords ... with the results obtained by Mr Born for the collision of an electron and an atom, and by Mr Fermi for the collision of an electron and a rotator'. Because the printed version of de Broglie's lecture was already complete by the time the conference took place, this remark could not have been added after de Broglie's clash with Pauli. (De Broglie wrote to Lorentz on 11 October 1927 – AHQP-LTZ-11, in French – that he had received the proofs of his report from Gauthier-Villars and corrected them.)

for an initial product wave function $\Psi_0(x, y) = \psi_0(x)g_0(y)$ the Schrödinger equation

$$i\hbar\frac{\partial\Psi}{\partial t} = aQ\left(-i\hbar\frac{\partial}{\partial y}\right)\Psi \tag{10.10}$$

has the solution

$$\Psi(x, y, t) = \sum_n c_n\phi_n(x)g_0(y - aq_n t) . \tag{10.11}$$

Because g_0 is localised, Ψ evolves into a superposition of terms that *separate* with respect to y (in the sense of having negligible overlap with respect to y). As we saw in the case of scattering, the final configuration (x, y) can be in only one 'branch' of the superposition, say $\phi_i(x)g_0(y - aq_i t)$, which will guide $(x(t), y(t))$ thereafter. And because the active branch is a product in x and y, the velocity of x will be determined by $\phi_i(x)$ alone. In effect, at the end of the quantum measurement, the system is guided by a 'reduced' wave function $\phi_i(x)$. Thus, the pointer plays the same role in the pilot-wave theory of measurement as the scattered particle does in the pilot-wave theory of scattering.

It is also readily shown that if the initial ensemble of x, y has a distribution $|\Psi_0(x, y)|^2$, then the probability of x, y ending in the branch $\phi_i(x)g_0(y - aq_i t)$ is given by $|c_i|^2$, in agreement with the standard Born rule.

Finally, again as we saw in the case of scattering, the effective 'collapse' to the state $\phi_i(x)g_0(y - aq_i t)$ is for all practical purposes irreversible. As noted by Bohm (1952b, p. 182), the apparatus coordinate y will subsequently interact with many other degrees of freedom, making it very unlikely that distinct states $\phi_i(x)g_0(y - aq_i t)$, $\phi_j(x)g_0(y - aq_j t)$ $(i \neq j)$ will interfere later on, because the associated branches of the total wave function would have to overlap with respect to *every* degree of freedom involved.

10.4 Recoil of a single photon: Kramers and de Broglie

We have sketched the pilot-wave theory of quantum measurement to emphasise that when one leaves the limited domain of particle scattering (by atoms, or by fixed obstacles such as mirrors or diffracting screens), it can become essential to describe the apparatus itself in terms of pilot-wave dynamics. In some situations, the coordinates of the apparatus *must* be included in the total wave function (for the 'supersystem' consisting of system plus apparatus), to ensure that the total wave function separates into non-overlapping branches in configuration space.

As we shall see in Chapter 11, it seems that this point was not properly appreciated by de Broglie in 1927. Indeed, most theoreticians at the time simply applied quantum theory (in whatever form they preferred) to microscopic systems only. Macroscopic apparatus was usually treated as a given classical background. However, Born's treatment of the cloud chamber in the general discussion shows that the key insight was already known: the apparatus is made of atoms, and should ultimately be included in the wave function, which will develop a branching structure as the measurement proceeds. All that was needed, within pilot-wave theory, was to carry through the details properly for a general quantum measurement – something that de Broglie did not see in 1927 and that Bohm did see in 1952.

Now, the need to include macroscopic equipment in the wave function is relevant to a problem raised by Kramers in the general discussion, concerning the recoil of a mirror due to the reflection of a *single* photon. The discussion had returned to the question of how de Broglie's theory accounts for radiation pressure on a mirror (a subject that had been discussed at the end of de Broglie's lecture). As Kramers put it (p. 450):

> But how is radiation pressure exerted in the case where it is so weak that there is only one photon in the interference zone? . . . And if there is only one photon, how can one account for the sudden change of momentum suffered by the reflecting object?

Neither de Broglie nor Brillouin were able to give an answer. De Broglie claimed that pilot-wave theory in its current form was able to give only the mean pressure exerted by an ensemble (or 'cloud') of photons.

What was missing from de Broglie's understanding was that, in such a case, the position of the mirror would have to be treated by pilot-wave dynamics and included in the wave function. Schematically, let x_{p} be the position of the photon (on an axis normal to the surface of the mirror) and let x_{m} be the position of the reflecting surface. Let us treat the mirror as a very massive but free body, with initial wave function $\phi(x_{\mathrm{m}}, 0)$ localised around $x_{\mathrm{m}} = 0$ (at $t = 0$). The incident photon initially has a localised wave function $\psi_{\mathrm{inc}}(x_{\mathrm{p}}, 0)$ directed towards the mirror, with mean momentum $\hbar k$. Roughly, if the photon packet strikes the mirror at time t_0, then the initial total wave function $\Psi(x_{\mathrm{p}}, x_{\mathrm{m}}, 0) = \psi_{\mathrm{inc}}(x_{\mathrm{p}}, 0)\phi(x_{\mathrm{m}}, 0)$ will evolve into a wave function of the schematic form

$$\Psi(x_{\mathrm{p}}, x_{\mathrm{m}}, t) \sim \psi_{\mathrm{ref}}(x_{\mathrm{p}}, t)\phi(x_{\mathrm{m}} - (\Delta p/M)(t - t_0), 0)e^{(i/\hbar)(\Delta p)x_{\mathrm{m}}} , \qquad (10.12)$$

where ψ_{ref} is a reflected packet directed away from the mirror, $\Delta p = 2\hbar k$ is the momentum transferred to the mirror, M is the mass of the mirror, and $\phi(x_{\mathrm{m}} - (\Delta p/M)(t - t_0), 0)$ is a packet (whose spreading we ignore) moving to the right with speed $\Delta p/M$. The actual coordinates x_{p}, x_{m} will be guided by Ψ in accordance with de Broglie's equation, and the position $x_{\mathrm{m}}(t)$ of the mirror will follow the moving packet. The recoil of the mirror can therefore be accounted for.

Thus, while de Broglie had the complete pilot-wave dynamics of a many-body system, he seems not to have understood that it is sometimes necessary to include the coordinates of macroscopic equipment in the pilot-wave description. Otherwise, he might have been able to answer Kramers.

On the other hand, in ordinary quantum theory too, a proper explanation for the recoil of the mirror would also have to treat the mirror as part of the quantum system. If the mirror is regarded as a classical object, then quantum theory would strictly speaking be as powerless as pilot-wave theory. Perhaps this is why Brillouin made the following remark (p. 450):

> No theory currently gives the answer to Mr Kramers' question.

11

Pilot-wave theory in retrospect

As we discussed in Section 2.4, in his Solvay lecture of 1927 de Broglie presented the pilot-wave dynamics of a non-relativistic many-body system, and outlined some simple applications of his 'new dynamics of quanta' (to interference, diffraction and atomic transitions). Further, as we saw in Section 10.2, contrary to a widespread misunderstanding, in the general discussion de Broglie's reply to Pauli's objection contained the essential points needed to treat inelastic scattering (even if Fermi's misleading optical analogy confused matters): in particular, de Broglie correctly indicated how definite quantum outcomes in scattering processes arise from a separation of wave packets in configuration space. We also saw in Section 10.4 that de Broglie was unable to reply to a query from Kramers concerning the recoil of a single photon on a mirror: to do so, he would have had to introduce a joint wave function for the photon and the mirror.

De Broglie's theory was revived by Bohm 25 years later (Bohm 1952a,b) (though with the dynamics written in terms of a law for acceleration instead of a law for velocity). Bohm's truly new and very important contribution was a pilot-wave account of the general quantum theory of measurement, with macroscopic equipment (pointers, etc.) treated as part of the quantum system. In effect, in 1952, Bohm provided a detailed derivation of quantum phenomenology from de Broglie's dynamics of 1927 (albeit with the dynamical equations written differently).

Despite this success, until about the late 1990s most physicists still believed that hidden-variables theories such as de Broglie's could not possibly reproduce the predictions of quantum theory (even for simple cases such as the two-slit experiment, cf. Section 6.1). Or, they believed that such theories had been disproved by experiments testing EPR-type correlations and demonstrating violations of Bell's inequality. As just one striking example of the latter belief, in the early 1990s James T. Cushing, then a professor of physics and of philosophy at the University of Notre Dame, submitted a research proposal to the US National Science Foundation 'for theoretical work to be done, within the framework of Bohm's version of quantum theory, on some foundational questions in quantum mechanics' (Cushing 1996, p. 6), and received the following evaluation:

> The subject under consideration, the rival Copenhagen and causal (Bohm) interpretations of the quantum theory, has been discussed for many years and in the opinion of several members of the Physics Division of the NSF, the situation has been settled. The causal interpretation is inconsistent with experiments which test Bell's inequalities. Consequently ... funding ... a research program in this area would be unwise.
>
> *(Cushing 1996, p. 6)*

This is ironic because, as even a superficial reading of Bell's original papers shows, the non-local theory of de Broglie and Bohm was a primary motivation for Bell's work on his famous inequalities (Bell 1964, 1966). Bell knew that pilot-wave theory was empirically equivalent to quantum theory, and wanted to find out if the non-locality was a peculiarity of this particular model, or if it was a general feature of all hidden-variables theories. Bell's conclusion was that local theories have to satisfy his inequality, which is inconsistent with EPR-type correlations, and that therefore any viable theory must be non-local – like pilot-wave theory. There was never any question, in experimental tests of Bell's inequality, of testing the non-local theory of de Broglie and Bohm; rather, it was the class of local theories that was being tested.[a]

Despite this and other misunderstandings, in recent years the pilot-wave theory of de Broglie and Bohm, with particle trajectories guided by a physically real wave function, has gained wide acceptance as an alternative (though little used) formulation of quantum theory. While it is still occasionally asserted that any theory with trajectories (or other hidden variables) must disagree with experiment, such erroneous claims have become much less frequent.[b] This change in attitude seems to have been largely a result of the publication in 1987 of Bell's influential book on the foundations of quantum theory (Bell 1987), several chapters of which consisted of pedagogical explanations of pilot-wave theory as an objective and deterministic account of quantum phenomena. Leaving aside the question of

[a] See chapter 9 of Cushing (1994) for an extensive discussion of the generally hostile reactions to and misrepresentations of Bohm's 1952 papers.

[b] For example, in his book *The Elegant Universe*, Greene (2000) asserts that not only are quantum particle trajectories unmeasurable (because of the uncertainty principle), their very existence is ruled out by experiments testing the Bell inequalities: '... theoretical progress spearheaded by the late Irish physicist John Bell and the experimental results of Alain Aspect and his collaborators have shown convincingly that ... [e]lectrons – and everything else for that matter – cannot be described as simultaneously being at such-and-such location *and* having such-and-such speed' (p. 114). This is corrected a few years later in *The Fabric of the Cosmos*, where, commenting on 'Bohm's approach' Greene (2005) writes that it 'does not fall afoul of Bell's results because ... possessing definite properties forbidden by quantum uncertainty is *not* ruled out; only locality is ruled out ... ' (p. 206, original italics).

whether or not pilot-wave theory (or de Broglie–Bohm theory) is closer to the truth about the quantum world than other formulations, it is a remarkable fact that it took about three-quarters of a century for the theory to become widely accepted as an internally consistent alternative.

Even today, however, there are widespread misconceptions not only about the physics of pilot-wave theory, but also about its history. It is generally recognised that de Broglie worked along pilot-wave lines in the 1920s, that Bohm developed and extended the theory in 1952, and that Bell publicised the theory in his book in 1987. But the full extent of de Broglie's contributions in the 1920s is usually not recognised.

A careful examination of the proceedings of the fifth Solvay conference changes our perception of pilot-wave theory, both as a physical theory and as a part of the history of quantum physics.

11.1 Historical misconceptions

Many of the widespread misconceptions about the history of pilot-wave theory are conveniently summarised in the following extract from the book by Bohm and Hiley (1993, pp. 38–9):

> The idea of a 'pilot wave' that guides the movement of the electron was first suggested by de Broglie in 1927, but only in connection with the one-body system. De Broglie presented this idea at the 1927 Solvay Congress where it was strongly criticised by Pauli. His most important criticism was that, in a two-body scattering process, the model could not be applied coherently. In consequence de Broglie abandoned his suggestion. The idea of a pilot wave was proposed again in 1952 by Bohm in which an interpretation for the many-body system was given. This latter made it possible to answer Pauli's criticism ....[a]

As we have by now repeatedly emphasised, in his 1927 Solvay report (pp. 351–3) de Broglie did in fact present pilot-wave theory in configuration space for a many-body system, not just the one-body theory in 3-space; and further, as we saw in Section 10.2, de Broglie's reply to Pauli's criticism contained the essential ideas needed for a proper rebuttal.[b] As we shall see below, the claim that de Broglie abandoned his theory because of Pauli's criticism is also not true.

Contrary to widespread belief, then, the many-body theory with a guiding wave in configuration space is originally due to de Broglie and not Bohm; and in 1927, de Broglie did understand the essentials of the pilot-wave theory of scattering. Thus the main content of Bohm's first paper of 1952 (Bohm 1952a) – which presents the

[a] Similar historical misconceptions appear in Cushing (1994, pp. 118–21, 149).

[b] To our knowledge, Bonk (1994) is the only other author to have noticed that de Broglie's reply to Pauli was indeed along the right lines.

dynamics (though in terms of acceleration), with applications to scattering – was already known to de Broglie in 1927.

Note that the theory was regarded as only provisional, by both de Broglie in 1927 and Bohm in 1952. In particular, as we saw in Sections 2.3.2 and 2.4, de Broglie regarded the introduction of a pilot wave in configuration space as a provisional measure. Bohm, on the other hand, suggested that the basic principles of the theory would break down at nuclear distances of order 10^{-13} cm (Bohm 1952a, pp. 178–9).

As we have said, in contrast with de Broglie's presentation of 1927, Bohm's dynamics of 1952 was based on acceleration, not velocity. For de Broglie, the basic law of motion for particles with masses m_i and wave function $\Psi = |\Psi|\, e^{(i/\hbar)S}$ was the guidance equation

$$m_i \frac{d\mathbf{x}_i}{dt} = \nabla_i S \,, \tag{11.1}$$

whereas for Bohm, the basic law of motion was the Newtonian equation

$$\frac{d\mathbf{p}_i}{dt} = m_i \frac{d^2\mathbf{x}_i}{dt^2} = -\nabla_i (V + Q) \,, \tag{11.2}$$

where

$$Q \equiv -\sum_i \frac{\hbar^2}{2m_i} \frac{\nabla_i^2 |\Psi|}{|\Psi|} \tag{11.3}$$

is the 'quantum potential'.

Taking the time derivative of (11.1) and using the Schrödinger equation yields precisely (11.2). For Bohm, however, (11.1) was not a law of motion but rather a constraint $\mathbf{p}_i = \nabla_i S$ to be imposed on the *initial* momenta (Bohm 1952a, p. 170). This initial constraint happens to be preserved in time by (11.2), which Bohm regarded as the true law of motion. Indeed, Bohm suggested that the initial constraint $\mathbf{p}_i = \nabla_i S$ could be relaxed, leading to corrections to quantum theory:

> ... this restriction is not inherent in the conceptual structure ... it is quite consistent in our interpretation to contemplate modifications in the theory, which permit an arbitrary relation between $\mathbf{p}$ and $\nabla S(\mathbf{x})$.
>
> *(Bohm 1952a, pp. 170–1)*

For de Broglie, in contrast, there was never any question of relaxing (11.1): he regarded (11.1) as the basic law of motion for a new form of particle dynamics, and indeed for him (11.1) embodied – as we saw in Chapter 2 – the unification of the principles of Maupertuis and Fermat, a unification that he regarded as the guiding principle of his new dynamics. (De Broglie did mention in passing, however, the alternative formulation in terms of acceleration, both in his 'Structure' paper (cf. Section 2.3.1) and in his Solvay report (p. 350).)

Some authors seem to believe that the 'recasting' of Bohm's second-order dynamics into first-order form was due to Bell (1987). But in fact, Bell's (pedagogical) presentation of the theory – based on the guidance equation for velocity, and ignoring the notion of quantum potential – was identical to de Broglie's original presentation. De Broglie's first-order dynamics of 1927 is sometimes referred to as 'Bohmian mechanics'. As already noted in Chapter 2, this is a confusing misnomer: firstly because of de Broglie's priority, and secondly because Bohm's dynamics of 1952 was actually second-order in time.

Another common historical misconception concerns the reception of de Broglie's theory at the Solvay conference. It is usually said that de Broglie's ideas attracted hardly any attention. It is difficult to understand how such an impression originated, for even a cursory perusal of the proceedings reveals that de Broglie's theory was extensively discussed, both after de Broglie's lecture and during the general discussion. Nevertheless, in his classic account of the historical development of quantum theory, Jammer asserts that when de Broglie presented his theory:

> It was immediately clear that nobody accepted his ideas .... In fact, with the exception of some remarks by Pauli ... de Broglie's causal interpretation was not even further discussed at the meeting. Only Einstein once referred to it *en passant*.
>
> *(Jammer 1966, p. 357)*[a]

It appears to have escaped Jammer's attention that the general discussion contains extensive and varied comments on many aspects of de Broglie's theory (including a query by Einstein about the speed of photons), as does the discussion after de Broglie's lecture, and that support for de Broglie's ideas was expressed by Brillouin and by Einstein.

In a later historical study, again, Jammer (1974, pp. 110–11) writes – in reference to the fifth Solvay conference – that de Broglie's theory 'was hardly discussed at all', and that 'the only serious reaction came from Pauli'. In the same study, Jammer quotes extensively from de Broglie's report and from the general discussion, apparently without noticing the extensive discussions of de Broglie's theory that appear both after de Broglie's report and in the general discussion.

But Jammer is by no means the only historian to have given short shrift to de Broglie's major presence at the 1927 Solvay conference.

In his book *The Solvay Conferences on Physics*, Mehra (1975, p. xvi) quotes de Broglie himself as saying, with reference to his presentation of pilot-wave theory in 1927, that 'it received hardly any attention'. But these words were written by de Broglie some 46 years later (de Broglie 1974), and de Broglie's recollection (or

[a] Jammer adds footnoted references to pp. 280 and 256 respectively of the original proceedings, where Pauli's objection (involving Fermi's treatment of the rotator) appears, and where, in his main contribution to the general discussion, Einstein comments that in his opinion de Broglie 'is right to search in this direction' (p. 441).

misrecollection) after nearly half a century is belied by the content of the published proceedings.[a]

In volume 6 of their monumental *The Historical Development of Quantum Theory*, Mehra and Rechenberg (2000, pp. 246–50) devote several pages to the published proceedings of the 1927 Solvay conference, focussing on the general discussion – mainly on the comments by Einstein, Dirac and Heisenberg – as well as on the unpublished comments by Bohr. The rest of the general discussion is summarised in a single sentence (p. 250):

> After the Einstein–Pauli–Dirac–Heisenberg exchange, the general discussion turned to more technical problems connected with the description of photons and electrons in quantum mechanics, as well as with the details of de Broglie's recent ideas.

It is added that 'though these points possess some intrinsic interest, they do not throw much light on the interpretation debate'.

As we shall see in Section 12.1, it would appear that, for Mehra and Rechenberg, as indeed for most commentators, the 'interpretation debate' centred mainly around private or semi-private discussions between Bohr and Einstein, and that the many pages of published discussions were of comparatively little interest. Thus the remarkable downplaying of the discussion of de Broglie's theory, as well as of other ideas, is coupled with a strong tendency – on the part of many authors – to portray (incorrectly) the 1927 conference as focussed primarily on the confrontation between Bohr and Einstein.

We have already noted the widespread historical misconceptions concerning the de Broglie–Pauli encounter in the general discussion. A related misconception concerns de Broglie's thinking in the immediate aftermath of the Solvay conference. It is often asserted that, soon after the conference, de Broglie abandoned his theory primarily because of Pauli's criticism. This is not correct. In his book *An Introduction to the Study of Wave Mechanics* (de Broglie 1930), which was published just three years after the Solvay meeting, de Broglie gives three main reasons for why he considers his pilot-wave theory to be unsatisfactory. First, de Broglie considers (p. 120) a particle incident on an imperfectly reflecting mirror, and notes that if the particle is found in the transmitted beam then the reflected part of the wave must disappear (this being 'a necessary consequence of the interference principle'). De Broglie concludes that 'the wave is not a physical phenomenon in the old sense of the word. It is of the nature of a symbolic representation of a probability ... '. Here, de Broglie did not understand how pilot-wave theory accounts for the effective (and practically irreversible) collapse of the wave packet, by means of a separation into non-overlapping branches involving many degrees

[a] In fact, several commentators have drawn erroneous conclusions about de Broglie, by relying on mistaken 'recollections' written by de Broglie himself decades later.

of freedom (cf. Sections 10.1 and 10.3). Second, de Broglie notes (pp. 121, 133) that a particle in free space guided by a superposition of plane waves would have a rapidly varying velocity and energy, and he cannot see how this could be consistent with the outcomes of quantum energy measurements, which would coincide strictly with the energy eigenvalues present in the superposition. To solve this second problem, de Broglie would have had to apply pilot-wave dynamics to the process of quantum measurement itself – including the apparatus in the wave function if necessary – as done much later by Bohm (see Section 10.3). This question of energy measurement bears some similarity to that raised by Pauli, and perhaps Pauli's query set de Broglie thinking about this problem. But even so, de Broglie in effect gave the essence of a correct reply to Pauli's query, and the problem of energy measurement was posed by de Broglie himself. Third, in applying pilot-wave theory to photons, de Broglie finds (p. 132) that in some circumstances (specifically, in the interference zone close to an imperfectly reflecting mirror) the photon trajectories have superluminal speeds, which he considers unacceptable. De Broglie's book does not mention Pauli's criticism.

It is also often claimed that, when de Broglie abandoned his pilot-wave theory (soon after the Solvay conference), he quickly adopted the views of Bohr and Heisenberg. Thus, for example, Cushing (1994, p. 121) writes: 'By early 1928 he [de Broglie] had decided to adopt the views of Bohr and Heisenberg'.[a] But de Broglie's book of 1930, in which the above difficulties with pilot-wave theory are described, contains a 'General introduction' that is 'the reproduction of a communication made by the author at the meeting of the British Association for the Advancement of Science held in Glasgow in September, 1928' (de Broglie 1930, p. 1). This introduction therefore gives an overview of de Broglie's thinking in late 1928, almost a year after the fifth Solvay conference. While de Broglie makes it clear (p. 7) that in his view it is 'not possible to regard the theory of the pilot-wave as satisfactory', and states that the 'point of view developed by Heisenberg and Bohr ... appears to contain a large body of truth', his concluding paragraph shows that he was still not satisfied:

> To sum up, the physical interpretation of the new mechanics remains an extremely difficult question ... the dualism of waves and particles must be admitted .... Unfortunately the profound nature of the two members in this duality and the precise relation existing between them still remain a mystery.
>
> *(de Broglie 1930, p. 10)*

At that time, de Broglie seems to have accepted the formalism of quantum theory, and the statistical interpretation of the wave function, but he still thought that an

[a] As evidence for this, Cushing cites later recollections by de Broglie in his book *Physics and Microphysics* (de Broglie 1955), which was originally published in French in 1947. Again, de Broglie's recollections decades later do not seem a reliable guide to what actually happened circa 1927.

adequate physical understanding of wave-particle duality had yet to be reached. Yet another year later, in December 1929, doubts about the correct interpretation of quantum theory could still be discerned in de Broglie's Nobel Lecture (de Broglie 1999):

> Is it even still possible to assume that at each moment the corpuscle occupies a well-defined position in the wave and that the wave in its propagation carries the corpuscle along in the same way as a wave would carry along a cork? These are difficult questions and to discuss them would take us too far and even to the confines of philosophy. All that I shall say about them here is that nowadays the tendency in general is to assume that it is not constantly possible to assign to the corpuscle a well-defined position in the wave.

This brings us to another historical misconception concerning de Broglie's work. Nowadays, de Broglie–Bohm theory is often presented as a 'completion' of quantum theory.[a] Critics sometimes view this as an arbitrary addition to or amendment of the quantum formalism, the trajectories being viewed as an additional 'baggage' being appended to an already given formalism. Regardless of the truth or otherwise of pilot-wave theory as a physical theory, such a view certainly does not do justice to the historical facts. For the elements of pilot-wave theory – waves guiding particles via de Broglie's velocity formula – were already in place in de Broglie's thesis of 1924, before either matrix or wave mechanics existed. And it was by following the lead of de Broglie's thesis that Schrödinger developed the wave equation for de Broglie's matter waves. While Schrödinger dropped the trajectories and considered only the waves, nevertheless, historically speaking the wave function ψ and the Schrödinger equation both grew out of de Broglie's phase waves.[b]

The pilot-wave theory of 1927 was the culmination of de Broglie's independent work from 1923, with a major input from Schrödinger in 1926. There is no sense in which de Broglie's trajectories were ever 'added to' some pre-existing theory. And when Bohm revived the theory in 1952, while it may have seemed to Bohm's contemporaries (and indeed to Bohm himself) that he was adding something to quantum theory, from a historical point of view Bohm was simply reinstating what had been there from the beginning.

The failure to acknowledge the priority of de Broglie's thinking from 1923 is visible even in the discussions of 1927. In the discussion following de Broglie's lecture, Pauli (p. 365) presents what he claims is the central idea of de Broglie's theory:

> I should like to make a small remark on what seems to me to be the mathematical basis of Mr de Broglie's viewpoint concerning particles in motion on definite trajectories. His

[a] An often-cited motivation for introducing the trajectories is, of course, to solve the measurement problem.
[b] See also the discussion in Section 4.5.

> conception is based on the principle of conservation of charge . . . if in a field theory there exists a conservation principle . . . it is always formally possible to introduce a velocity vector . . . and to imagine furthermore corpuscles that move following the current lines of this vector.

Pauli's assertion that de Broglie's theory is based on the conservation of charge makes sense only for one particle: for a many-body system, de Broglie's velocity field is associated with conservation of probability in configuration space, whereas conservation of charge is always tied to 3-space. Still, the point remains that in any theory with a locally conserved probability current, it is indeed possible to introduce particle trajectories following the flow lines of that current. This way of presenting de Broglie's theory then makes the trajectories look like an addendum to a pre-existing structure: given the Schrödinger equation with its locally conserved current, one can add trajectories if one wishes. But to present the theory in this way is a major distortion of the historical facts and priorities. The essence of de Broglie's dynamics came before Schrödinger's work, not after. Further, de Broglie obtained his velocity law not from the Schrödinger current (which was unknown in 1923 or 1924) but from his postulated relation between the principles of Maupertuis and Fermat. And finally, while de Broglie's trajectory equation did not in fact owe anything to the Schrödinger equation, again, the latter equation arose out of considerations (of the optical-mechanical analogy) that had been initiated by de Broglie. It seems rather clear that the historical priority of de Broglie's work was being downplayed by Pauli's remarks, as it has been more or less ever since.

A related historical misconception concerns the status of the Schrödinger equation in pilot-wave theory. It is sometimes argued that this equation has no natural place in the theory, and that therefore the theory is artificial. For example, commenting on what he calls 'Bohm's theory', Polkinghorne (2002, pp. 55, 89) writes:

> There is an air of contrivance about it that makes it unappealing. For example, the hidden wave has to satisfy a wave equation. Where does this equation come from? The frank answer is out of the air or, more accurately, out of the mind of Schrödinger. To get the right results, Bohm's wave equation must be the Schrödinger equation, but this does not follow from any internal logic of the theory and it is simply an *ad hoc* strategy designed to produce empirically acceptable answers. . . . It is on these grounds that most physicists find the greatest difficulty with Bohmian ideas . . . the *ad hoc* but necessary appropriation of the Schrödinger equation as the equation for the Bohmian wave has an unattractively opportunist air to it.

Polkinghorne's comments are a fair criticism of Bohm's 1952 reformulation of de Broglie's theory. For as we have seen, Bohm based his presentation on the Newtonian equation of motion (11.2) for acceleration, with a 'quantum potential' Q determined by the wave function Ψ through (11.3): according to Bohm, Ψ

generates a 'quantum force' $-\nabla_i Q$, which accounts for quantum effects. From this Newtonian standpoint, the wave equation for Ψ does indeed have nothing to do with the internal logic of the theory: it is then fair to say that, in Bohm's formulation, the Schrödinger equation is 'appropriated' for a purpose quite foreign to the origins of that equation. However, Polkinghorne's critique does not apply to pilot-wave theory in its original de Broglian formulation, as a new form of dynamics in which particle velocities are determined by guiding waves (rather than particle accelerations being determined by Newtonian forces). For as a matter of historical fact, the Schrödinger equation *did* follow from the internal logic of de Broglie's theory. After all, Schrödinger set out in the first place to find the general wave equation for de Broglie's waves; and his derivation of that equation owed much to the optical-mechanical analogy, which was a key component of de Broglie's approach to dynamics.[a] It cannot be said that de Broglie 'appropriated' the Schrödinger equation for a purpose foreign to its origins, when the original purpose of the Schrödinger equation was in fact to describe de Broglie's waves.

11.2 Why was de Broglie's theory rejected?

One might ask why de Broglie's theory did not gain widespread support soon after 1927. This question has been considered by Bonk (1994), who applies Bayesian reasoning to some of the discussions at the fifth Solvay conference, in an attempt to understand the rapid acceptance of the 'Copenhagen' interpretation. It has also been suggested that were it not for certain historical accidents, de Broglie's theory might have triumphed in 1927 and emerged as the dominant interpretation of quantum theory (Cushing 1994, ch. 10). It is difficult to evaluate how realistic Cushing's 'alternative historical scenario' might have been. Here, we shall simply highlight two points that are usually overlooked, and which are relevant to any evaluation of why de Broglie's theory did not carry the day.

Our first point is that, because de Broglie did in fact give a reply to Pauli's criticism that contained the essence of a correct rebuttal – and because in contrast de Broglie completely failed to reply to the difficulty raised by Kramers – the question of whether or not de Broglie made a convincing case at the fifth Solvay conference (and if not, why not) should be reconsidered.

Our second point is that there was a good technical reason for why 'standard' quantum theory had an advantage over pilot-wave theory in 1927. By the use of a simple 'collapse postulate' for microsystems, it was generally possible to account

[a] Nowadays, it is common in textbooks to motivate the free-particle Schrödinger equation as the simplest equation satisfied by a plane de Broglie wave $e^{i(\mathbf{k}\cdot\mathbf{x}-\omega t)}$ with the non-relativistic dispersion relation $\hbar\omega = (\hbar k)^2/2m$. This 'derivation' is just as natural in pilot-wave theory as it is in standard quantum theory.

for quantum measurement outcomes without having to treat the measurement process (including the apparatus) quantum-mechanically. In contrast, it was easy to find examples where in pilot-wave theory it was essential to use the theory itself to analyse the measurement process: as we saw in the last section, in his book of 1930 de Broglie could not see how, for a particle guided by a superposition of energy eigenfunctions, an energy measurement would give one of the results expected from quantum theory. Agreement with quantum theory requires an analysis of the measurement process in terms of pilot-wave dynamics, with the apparatus included as part of the system, as shown by Bohm (1952b). In contrast, in ordinary quantum theory it usually suffices in practice simply to apply a collapse rule to the microscopic system alone. Such a collapse rule is of course merely pragmatic, and defies precise formulation (there being no sharp boundary between 'microscopic' and 'macroscopic', cf. Section 5.2); yet, in ordinary laboratory situations, it yields predictions that may be compared with experiment.

These two points should be taken into account in any full evaluation of why de Broglie's theory was rejected in 1927 and shortly thereafter.

Further relevant material, that seems to have never been considered before, consists of comments by Heisenberg on the possibility of a deterministic pilot-wave interpretation. These comments do not appear in the proceedings of the fifth Solvay conference, nor are they directed at de Broglie's theory. Rather, they appear in a letter Heisenberg wrote to Einstein a few months earlier, on 10 June 1927, and they concern *Einstein's* version of pilot-wave theory. Heisenberg's remarks could just as well have been directed at de Broglie's theory, however. Both Einstein's theory and Heisenberg's comments thereon are discussed in the next section.

11.3 Einstein's alternative pilot-wave theory (May 1927)

As we saw in Section 2.3.1, de Broglie first arrived at pilot-wave theory in a paper published in *Journal de Physique* in May 1927 (de Broglie 1927b). In the same month, Einstein proposed what in retrospect appears to be an alternative version of pilot-wave theory, with particle trajectories determined by the many-body wave function but in a manner different from that of de Broglie's theory. This new theory was described in a paper entitled 'Does Schrödinger's wave mechanics determine the motion of a system completely or only in the sense of statistics?', which was presented on 5 May 1927 at a meeting of the Prussian Academy of Sciences. On the same day Einstein wrote to Ehrenfest that '. . . in a completely unambiguous way, one can associate definite movements with the solutions [of the Schrödinger equation]' (Howard 1990, p. 89). However, on 21 May, before the paper appeared in print, Einstein withdrew it from publication (Kirsten and Treder 1979, p. 135; Pais 1982, p. 444). The paper remained unpublished, but its contents are nevertheless known from a manuscript version in the Einstein archive – see Howard (1990, pp. 89–90) and Belousek (1996).[a]

Einstein's unpublished version of pilot-wave theory has some relevance to his argument for the incompleteness of quantum theory, given in the general discussion. As we saw in Section 7.1, according to Einstein's argument, locality requires 'that one does not describe the [diffraction] process solely by the Schrödinger wave, but that at the same time one localises the particle during the propagation' (p. 441). Einstein added: 'I think that Mr de Broglie is right to search in this direction' – without mentioning that he himself had recently made an attempt in the same direction.

It is quite possible that, before abandoning his version of pilot-wave theory, Einstein had considered presenting it at the fifth Solvay conference. Indeed, had Einstein been happy with his new theory, there is every reason to think he would have presented it a few months later in Brussels. As discussed in Section 1.3, Lorentz had asked Einstein to give a report on particle statistics, and while Einstein had agreed to do so, he was reluctant. Less than a month after withdrawing his pilot-wave paper, Einstein withdrew his commitment to speak at the Solvay conference, writing to Lorentz on 17 June: '... I kept hoping to be able to contribute something of value in Brussels; I have now given up that hope. ... I did not take this lightly but tried with all my strength ... ' (quoted from Pais 1982, p. 432). It then seems indeed probable that, instead of (or in addition to) speaking about particle statistics, Einstein had hoped to present something like his version of pilot-wave theory.

Let us now describe what Einstein's proposal was. (For more detailed presentations, see Belousek (1996) and Holland (2005).)

Einstein's starting point is the time-independent Schrödinger equation[b]

$$-\frac{\hbar^2}{2}\nabla^2\Psi + V\Psi = E\Psi \tag{11.4}$$

for a many-body system with potential energy V, total energy E, and wave function Ψ on an n-dimensional configuration space. Einstein considers (11.4) to define a

[a] Archive reference: AEA 2-100.00 (in German); currently available online at http:// www.alberteinstein.info/db/ViewDetails.do?DocumentID=34338.

[b] The particle masses make no explicit appearance, because they have been absorbed into the configuration-space metric $g_{\mu\nu}$. Einstein is following Schrödinger's usage. In his second paper on wave mechanics, Schrödinger (1926c) introduced a non-Euclidean metric determined by the kinetic energy (see also Schrödinger's report, p. 407). The Laplacian ∇^2 is then understood in the Riemannian sense, and the Schrödinger equation indeed takes the form (11.4) with no masses appearing explicitly. Note that in analytical mechanics also it is sometimes convenient to write the kinetic energy $\frac{1}{2}M_{\mu\nu}\dot{q}^\mu\dot{q}^\nu$ as $\frac{1}{2}(ds/dt)^2$ where $ds^2 = M_{\mu\nu}dq^\mu dq^\nu$ is a line element with metric $M_{\mu\nu}$ (Goldstein 1980, pp. 369–70).

kinetic energy

$$E - V = -\frac{\hbar^2}{2}\frac{\nabla^2\Psi}{\Psi}, \tag{11.5}$$

which may also be written as

$$E - V = \frac{1}{2}(ds/dt)^2 = \frac{1}{2}g_{\mu\nu}\dot{q}^\mu\dot{q}^\nu, \tag{11.6}$$

where $ds^2 = g_{\mu\nu}dq^\mu dq^\nu$ is the line element in configuration space and $\dot{q}^\mu$ is the system velocity. The theory is expressed in terms of arbitrary coordinates q^μ with metric $g_{\mu\nu}$. The Laplacian is then given by $\nabla^2\Psi = g^{\mu\nu}\nabla_\mu\nabla_\nu\Psi$, where ∇_μ is the covariant derivative. Einstein writes $\nabla_\mu\nabla_\nu\Psi$ as $\Psi_{\mu\nu}$ (which he calls the 'tensor of Ψ-curvature'), and seeks unit vectors A^μ that extremise $\Psi_{\mu\nu}A^\mu A^\nu$, leading to the eigenvalue problem

$$(\Psi_{\mu\nu} - \lambda g_{\mu\nu})A^\nu = 0, \tag{11.7}$$

with n real and distinct solutions $\lambda_{(\alpha)}$. At each point, the $A^\mu_{(\alpha)}$ define a local orthogonal coordinate system (with Euclidean metric at that point). In this coordinate system, both $\Psi_{\mu\nu}$ and $g_{\mu\nu}$ are diagonal, with components $\bar{\Psi}_{\alpha\beta} = \lambda_{(\alpha)}\delta_{\alpha\beta}$ and $\bar{g}_{\alpha\beta} = \delta_{\alpha\beta}$. The two expressions (11.5) and (11.6) for the kinetic energy then become

$$E - V = -\frac{\hbar^2}{2}\frac{1}{\Psi}\sum_\alpha \bar{\Psi}_{\alpha\alpha} \tag{11.8}$$

and

$$E - V = \frac{1}{2}\sum_\alpha \bar{\dot{q}}^2_\alpha, \tag{11.9}$$

respectively. Einstein then introduces the hypothesis that these two expressions match term by term, so that

$$\bar{\dot{q}}^2_\alpha = -\frac{\hbar^2}{\Psi}\bar{\Psi}_{\alpha\alpha}. \tag{11.10}$$

Using $\bar{\Psi}_{\alpha\alpha} = \lambda_{(\alpha)}$, and transforming back to the original coordinate system, where $\dot{q}^\mu = \sum_\alpha \bar{\dot{q}}_\alpha A^\mu_{(\alpha)}$, Einstein obtains the final result

$$\dot{q}^\mu = \hbar\sum_\alpha \pm\sqrt{\frac{-\lambda_{(\alpha)}}{\Psi}}A^\mu_{(\alpha)}, \tag{11.11}$$

which expresses $\dot{q}^\mu$ in terms of $\lambda_{(\alpha)}$ and $A^\mu_{(\alpha)}$, where at each point in configuration space $\lambda_{(\alpha)}$ and $A^\mu_{(\alpha)}$ are determined by the local values of the wave function Ψ and its derivatives. (The ambiguity in sign is, according to Einstein, to be expected for quasiperiodic motions.)

Thus, according to (11.11), the system velocity $\dot{q}^\mu$ is locally determined (up to signs) by Ψ and its derivatives. This is Einstein's proposed velocity law, to be compared and contrasted with de Broglie's.

The manuscript contains an additional note 'added in proof', in which Einstein asserts that the theory he has just outlined is physically unacceptable, because it predicts that for a system composed of two independent subsystems, with an additive Hamiltonian and a product wave function, the velocities for one subsystem generally depend on the instantaneous coordinates of the other subsystem. However, Einstein adds that Grommer has pointed out that this problem could be avoided by replacing Ψ with $\ln \Psi$ in the construction of the velocity field. According to Einstein: 'The elaboration of this idea should occasion no difficulty...' (Howard 1990, p. 90). Despite Einstein's apparent optimism that Grommer's modification would work, it has been assumed (Howard 1990, p. 90; Cushing 1994, p. 128) that Einstein withdrew the paper because he soon realised that Grommer's suggestion did not work. But it has been shown by Holland (2005) that the replacement $\Psi \rightarrow \ln \Psi$ does in fact remove the difficulty raised by Einstein. Holland shows further, however, that there are *other* difficulties with Einstein's theory – with or without Grommer's modification. It is not known if Einstein recognised these other difficulties, but if he did, they would certainly have been convincing grounds on which to abandon the scheme altogether.

The difficulties raised by Holland are as follows. First, the theory applies only to a limited range of quantum states – real stationary states with $E \geq V$. Second, the system velocity $\dot{q}^\mu$ is defined only in a limited domain of configuration space (for example, without Grommer's modification, $\lambda_{(\alpha)}/\Psi$ must be negative for $\dot{q}^\mu$ to be real). Third, even where $\dot{q}^\mu$ is defined, the continuity equation $\nabla_\mu(|\Psi|^2 \dot{q}^\mu) = 0$ is generally not satisfied (where $|\Psi|^2$ is time-independent for a stationary state), so that the velocity field $\dot{q}^\mu$ does not map an initial Born-rule distribution into a final Born-rule distribution. This last feature removes any realistic hope that the theory could reproduce the predictions of standard quantum theory (Holland 2005).[a]

At the end of his manuscript (before the note added in proof), Einstein writes: '... the assignment of completely determined motions to solutions of Schrödinger's differential equation is ... just as possible as is the assignment of determined motions to solutions of the Hamilton–Jacobi differential equation in classical mechanics' (Belousek 1996, pp. 441–2). Given the close relationship

[a] As Holland (2005) also points out, from a modern point of view the mutual dependence of particle motions for product states, which Einstein found so unacceptable, need not be a real difficulty. We now know that a hidden-variables theory must be non-local, and there is no reason why in some theories the underlying non-locality could not exist for factorisable quantum states as well as for entangled ones. Indeed, Holland gives an example of just such a theory.

between the Hamilton–Jacobi function and the phase S of Ψ (the latter reducing to the former in the short-wavelength limit), it is natural to ask why Einstein did not consider de Broglie's velocity field – proportional to the phase gradient ∇S – which is a straightforward generalisation of the Hamilton–Jacobi velocity formula. As well as being much simpler than Einstein's, de Broglie's velocity field immediately satisfies Einstein's desired separability of particle motions for product states.[a] Why did Einstein instead propose what seems a much more complicated and unwieldy scheme to generate particle velocities from the wave function?

Perhaps Einstein did not adopt de Broglie's velocity field simply because, for the real wave functions Einstein considered, the phase gradient $\nabla S = \hbar\,\mathrm{Im}(\nabla\Psi/\Psi)$ vanishes; though it is not clear why Einstein thought one could restrict attention to such wave functions. As for the seemingly peculiar construction that Einstein did adopt, it should be noted that Einstein was (as he himself states) following Schrödinger in his use of a non-Euclidean metric determined by the kinetic energy (Schrödinger 1926c). Further details of Einstein's construction may have been related to another idea Einstein was pursuing: that quantisation conditions could arise from a generally covariant and 'overdetermined' field theory that constrains initial states, with particles represented by singularities (Pais 1982, pp. 464–8), an idea that, in retrospect, seems somewhat reminiscent of de Broglie's double-solution theory (Section 2.3.1).

It would of course have been most interesting to see how physicists would have reacted had Einstein in fact published his paper or presented it at the fifth Solvay conference. It so happens that Heisenberg had heard about Einstein's theory through Born and Jordan, and – on 19 May, just two days before Einstein withdrew the paper – wrote to Einstein asking about it. On 10 June 1927, Heisenberg wrote to Einstein again, this time with detailed comments and arguments against what Einstein was (or had been) proposing.[b]

At the beginning of this letter, after thanking Einstein for his 'friendly letter', Heisenberg says he would like to explain why he believes indeterminism is necessary and not just possible. He characterises Einstein as thinking that, while all experiments will agree with the statistical quantum theory, nevertheless it will be possible to talk about definite particle trajectories. Heisenberg then outlines an objection. He considers free electrons with a constant and very low velocity – hence large de Broglie wavelength λ – striking a grating with spacing comparable to λ. He remarks that, in Einstein's theory, the electrons will be scattered in discrete spatial directions, and that if the initial position of a particle were known one could

[a] It is not known whether or not Einstein noticed the non-locality of de Broglie's theory for entangled wave functions.

[b] Heisenberg to Einstein, 19 May and 10 June 1927, AEA 12-173.00 and 12-174.00 (both in German).

calculate where the particle will hit the grating. Heisenberg then asserts that one could set up an obstacle at that point, so as to deflect the particle in an arbitrary direction, independently of the rest of the grating. Heisenberg says that this could be done, if the forces between the particle and the obstacle act only at short range (over distances much smaller than the spacing of the grating). Heisenberg then adds that, in actual fact, the electron will be scattered in the usual discrete directions regardless of the obstacle. Heisenberg goes on to say that one could escape this conclusion if one 'sets the motion of the particle again in direct relation to the behaviour of the waves'. But this means, says Heisenberg, that the size of the particle – or the range of its interaction – depends on its velocity. Heisenberg asserts that making such assumptions actually amounts to giving up the word 'particle' and leads to a loss in understanding of why the simple potential energy e^2/r appears in the Schrödinger equation or in the matrix Hamiltonian function. On the other hand, Heisenberg agrees that: 'If you use the word "particle" so liberally, I take it to be very well possible that one can define also particle trajectories'. But then, adds Heisenberg, one loses the simplicity of quantum theory, according to which the particle motion takes place classically (to the extent that one can speak about motion in quantum theory). Heisenberg notes that Einstein seems willing to sacrifice this simplicity for the sake of maintaining causality. He remarks further that, in Einstein's conception, many experiments are still determined only statistically, and 'we could only console ourselves with the fact that, while for us, because of the uncertainty relation $p_1 q_1 \sim h$, the principle of causality would be meaningless, yet the good Lord would know in addition the position of the particle and thereby could retain the validity of the causal law'. Heisenberg adds that he finds it unattractive to try to describe more than just the 'connection between experiments' [Zusammenhang der Experimente].

It is likely that Heisenberg had similar views of de Broglie's theory, and that if he had commented on de Broglie's theory at the fifth Solvay conference he would have said things similar to the above.

Heisenberg's objection concerning the electron and the grating seems to be based on inappropriate reasoning taken from classical physics (a common feature of objections to pilot-wave theory even today), and Heisenberg agrees that if the motion of the particle is strictly tied to that of the waves, then it should be possible to obtain consistency with observation (as we now know is indeed the case for de Broglie's theory). Even so, Heisenberg seems to think that such a highly non-classical particle dynamics would lack the simplicity and intelligibility of certain classical ideas that quantum theory preserves. Finally, Heisenberg is unhappy with a theory containing unobservable causal connections. Similar objections to pilot-wave theory are considered in the next section.

11.4 Objections: in 1927 and today

It is interesting to observe that many of the objections to pilot-wave theory that are commonly heard today were already voiced in 1927.

Regarding the existence or non-existence of de Broglie's trajectories, Brillouin – in the discussion following de Broglie's lecture (p. 365) – could just as well have been replying to a present-day critic of de Broglie–Bohm theory when he said that:

> Mr Born can doubt the real existence of the trajectories calculated by L. de Broglie, and assert that one will never be able to observe them, but he cannot prove to us that these trajectories do not exist. There is no contradiction between the point of view of L. de Broglie and that of other authors ....

Here we have a conflict between those who believe in hidden entities because of the explanatory role they play, and those who think that what cannot be observed in detail should play no theoretical role. Such conflicts are not uncommon in the history of science: for example, a similar polarisation of views occurred in the late nineteenth century regarding the reality of atoms (which the 'energeticists' regarded as metaphysical fictions).

The debate over the reality of the trajectories postulated by de Broglie has been sharpened in recent years by the recognition that, from the perspective of pilot-wave theory itself, our inability to observe those trajectories is not a fundamental constraint built into the theory, but rather an accident of initial conditions with a Born-rule probability distribution $P = |\Psi|^2$ (for an ensemble of systems with wave function Ψ). The statistical noise associated with such 'quantum equilibrium' distributions sets limits to what can be measured, but for more general 'non-equilibrium' distributions $P \neq |\Psi|^2$, the uncertainty principle is violated and observation of the trajectories becomes possible (Valentini 1991b, 1992, 2002b; Pearle and Valentini 2006). Such non-equilibrium entails, of course, a departure from the statistical predictions of quantum theory, which are obtained in pilot-wave theory only as a special 'equilibrium' case. Arguably, the above disagreement between Born and Brillouin might nowadays turn on the question of whether one is willing to believe that quantum physics is merely the physics of a special statistical state.

Another objection sometimes heard today is that a velocity law different from that assumed by de Broglie is equally possible. In the discussion following de Broglie's lecture, Schrödinger (p. 366) raised the possibility of an alternative particle velocity defined by the momentum density of a field. Pauli pointed out that, for a relativistic field, if the particle velocity were obtained by dividing the momentum density by the energy density, then the resulting trajectories would differ from those obtained by de Broglie, who assumed a velocity defined (in the case of a single particle) by the ratio of current density to charge density. Possibly, then as now, the existence of alternative velocities may have been interpreted as casting doubt on the reality of the velocities actually assumed by de Broglie (though the true velocities would become measurable in the presence of quantum non-equilibrium).

In addition to the criticism that the trajectories cannot be observed, today it is also often objected that the trajectories are rather strange from a classical perspective. The peculiar nature of de Broglie's trajectories was addressed in the discussion following de Broglie's lecture (p. 366ff.), and again in the general discussion (pp. 451, 460ff.). It was, for example, pointed out that the speed of an electron could be zero in a stationary state, and that for general atomic states the orbits would be very complicated. At the end of his discussion of photon reflection by a mirror, Brillouin (pp. 369–70) argued that de Broglie's non-rectilinear photon paths (in free space) were necessary in order to avoid a paradox posed by Lewis, in which in the presence of interference it appeared that photons would collide with only one end of a mirror, causing it to rotate, even though from classical electrodynamics the mean radiation pressure on the mirror is expected to be uniform (see Brillouin's Fig. 3, p. 369).

This last example of Brillouin's recalls present-day debates involving certain kinds of quantum measurements, in which the trajectories predicted by pilot-wave theory have counter-intuitive features that some authors have labelled 'surreal' (Englert *et al.* 1992; Aharonov and Vaidman 1996), while other authors regard these features as perfectly understandable from within pilot-wave theory itself (Valentini 1992, p. 24; Dewdney, Hardy and Squires 1993; Dürr *et al.* 1993). A key question here is whether it is reasonable to expect a theory of subquantum dynamics to conform to classical intuitions about measurement (given that it is the underlying dynamics that should be used to analyse the measurement process).

Pilot-wave theory is sometimes seen as a return to classical physics (welcomed by some, criticised by others). But in fact, de Broglie's velocity-based dynamics is a new form of dynamics that is quite distinct from classical theory; therefore, it is to be expected that the behaviour of the trajectories will *not* conform to classical expectations. As we saw in detail in Chapter 2, de Broglie did indeed originally regard his theory as a radical departure from the principles of classical dynamics. It was Bohm's later revival of de Broglie's theory, in an unnatural pseudo-Newtonian form, that led to the widespread and mistaken perception that de Broglie–Bohm theory constituted a return to classical physics. In more recent years, de Broglie's original first-order dynamics has again become recognised as a new form of dynamics in its own right (Bell 1987; Dürr, Goldstein and Zanghì 1992; Valentini 1992).

12

Beyond the Bohr–Einstein debate

The fifth Solvay conference is usually remembered for the clash that took place between Bohr and Einstein, supposedly concerning in particular the possibility of breaking the uncertainty relations. It might be assumed that this clash took the form of an official debate that was the centrepiece of the conference. However, no record of any such debate appears in the published proceedings, where both Bohr and Einstein are in fact relatively silent.

The available evidence shows that in 1927 the famous exchanges between Bohr and Einstein actually consisted of informal discussions, which took place semi-privately (mainly over breakfast and dinner), and which were overheard by just a few of the participants, in particular Heisenberg and Ehrenfest. The historical sources for this consist, in fact, entirely of accounts given by Bohr, Heisenberg and Ehrenfest. These accounts essentially ignore the extensive formal discussions appearing in the published proceedings.

As a result of relying on these sources, the perception of the conference by posterity has been skewed on two counts. First, at the fifth Solvay conference there occurred much more that was memorable and important besides the Bohr–Einstein clash. Second, as shown in detail by Howard (1990), the real nature of Einstein's objections was in fact misunderstood by Bohr, Heisenberg and Ehrenfest – for Einstein's main target was not the uncertainty relations, but what he saw as the non-separability of quantum theory.

Below we shall indicate how these misunderstandings arose, summarise what now appear to have been Einstein's true concerns, and end by urging physicists, philosophers and historians to reconsider what actually took place in Brussels in October 1927, bearing in mind the deep questions that we still face concerning the nature of quantum physics.

12.1 The standard historical account

According to Heisenberg, the discussions that took place at the fifth Solvay conference 'contributed extraordinarily to the clarification of the physical foundations of the quantum theory' and indeed led to 'the outward completion of the quantum theory, which now can be applied without worries as a theory closed in itself' (Heisenberg 1929, p. 495). For Heisenberg, and perhaps for others in the Copenhagen-Göttingen camp, the 1927 conference seems to have played a key role in finalising the interpretation of the theory.

However, the perception that the interpretation had been finalised proved to be mistaken. As we have shown at length in Chapter 5, the interpretation of quantum theory is today still an open question, and deep concerns as to its meaning have stubbornly persisted. Further, as we have seen throughout Part II of this book, many of today's fundamental concerns were voiced (often at considerable length) at the fifth Solvay conference. Given that these concerns are still very much alive, it has evidently been a mistake to allow recollections of private discussions between Bohr and Einstein to overshadow our historical memory of the rest of the conference.

Note that, as we saw in Chapter 1 (p. 14), while Mehra (1975, p. 152) and also Mehra and Rechenberg (2000, p. 246) state that the general discussion was a discussion following 'Bohr's report', in fact Bohr did not present a report at the conference, nor was he invited to give one. This misunderstanding seems to have arisen because, at Bohr's request, a translation of his Como lecture appears in the published proceedings, to replace his remarks in the general discussion; and this has no doubt contributed to the common view that Bohr played a central role at the conference, when in fact it is clear from the proceedings that at the official meetings both he and Einstein played a rather marginal role.[a] Further, it seems rather clear that the standard (and unbalanced) version of events was propagated in particular by Bohr and Heisenberg, especially through their writings decades later.

There is very little independent evidence from the time as to what was said between Bohr and Einstein. Thus, for example, Mehra and Rechenberg (2000, p. 251) note the 'little evidence of the Bohr–Einstein debate in the official conference documents', and rely on eyewitness reports by Ehrenfest, Heisenberg and Bohr to yield 'a fairly consistent historical picture of the great epistemological debate between Bohr and Einstein' (p. 256).

One frequently cited piece of contemporary evidence is a description of the conference written by Ehrenfest a few days later, in a letter to his students and

[a] Since Bohr had not been invited to give a report, one might also question the propriety of his request that his remarks in the general discussion be replaced by a translation of a rather lengthy paper he was already publishing elsewhere.

associates in Leiden. This letter is cited at length by Mehra and Rechenberg (2000, pp. 251–3). An extract reads:

Bohr towering completely over everybody. . . . step by step defeating everybody. . . . It was delightful for me to be present during the conversations between Bohr and Einstein. . . . Einstein all the time with new examples. In a certain sense a sort of Perpetuum Mobile of the second kind to break the UNCERTAINTY RELATION. Bohr . . . constantly searching for the tools to crush one example after the other. Einstein . . . jumping out fresh every morning. . . . I am almost without reservation pro Bohr and contra Einstein.

This letter has often been taken as representative of the conference. However, there is a marked contrast with the published proceedings (in which Bohr and Einstein are mostly silent), a contrast which has not been taken into account.[a]

After examining the published proceedings, Mehra and Rechenberg (2000, pp. 250–6) go on to consider at greater length recollections by Heisenberg and Bohr – written decades after the conference – concerning the discussions between Bohr and Einstein. With hindsight, again given that the interpretation of quantum theory is today an open question, it would be desirable to have a more balanced view of the conference, focussing more on the content of the published proceedings, and rather less on these later recollections by just two of the participants.

Here is an extract from Heisenberg's recollection, written some 40 years later (Heisenberg 1967, p. 107):

The discussions were soon focussed upon a duel between Einstein and Bohr . . . . We generally met already at breakfast in the hotel, and Einstein began to describe an ideal experiment in which he thought the inner contradictions of the Copenhagen interpretation were especially clearly visible. Einstein, Bohr and I walked together from the hotel to the conference building, and I listened to the lively discussion between those two people whose philosophical attitudes were so different, . . . at lunch time the discussions continued between Bohr and the others from Copenhagen. Bohr had usually finished the complete analysis of the ideal experiment by late afternoon and would show it to Einstein at the supper table. Einstein had no good objection to this analysis, but in his heart he was not convinced.

From this account, the Bohr–Einstein clash appears indeed to have been a private discussion, with a few of Bohr's close associates in attendance. And yet, there has been a marked tendency to portray this discussion as the centrepiece of the whole conference. Thus, for example, in the preface to Mehra's book *The Solvay Conferences on Physics*, Heisenberg wrote the following about the 1927 conference (Mehra 1975, pp. v–vi):

[a] We have attempted to compare Ehrenfest's contemporary account with that in letters written by other participants soon after the conference, but have found nothing significant.

> Therefore the discussions at the 1927 Solvay Conference, from the very beginning, centred around the paradoxa of quantum theory.... Einstein therefore suggested special experimental arrangements for which, in his opinion, the uncertainty relations could be evaded. But the analysis carried out by Bohr and others during the Conference revealed errors in Einstein's arguments. In this situation, by means of intensive discussions, the Conference contributed directly to the clarification of the quantum-theoretical paradoxa.

Heisenberg says nothing at all about the alternative theories of de Broglie and Schrödinger, or about the views of Lorentz or Dirac (for example), or about the other extensive discussions recorded in the proceedings.

As for the text of Mehra's book on the Solvay conferences, the chapter devoted to the fifth Solvay conference contains a summary of the general discussion, which says nothing about the published discussions beyond providing a list of the participants. Mehra's summary states that a debate took place between Bohr and Einstein, and that the famous Bohr–Einstein dialogue began here in 1927. Mehra then adds an appendix, reproducing Bohr's famous essay 'Discussion with Einstein on epistemological problems in atomic physics' (Bohr 1949), written more than 20 years after the conference took place. Once again, the published discussions are made to appear rather insignificant compared to (Bohr's recollection of) the informal discussions between Bohr and Einstein.

The above essay by Bohr is in fact the principal and most detailed historical source for the Bohr–Einstein debate. This essay was Bohr's contribution to the 1949 festschrift for Einstein's seventieth birthday. It is reprinted as the very first paper in Wheeler and Zurek's (1983) influential collection *Quantum Theory and Measurement*, as well as elsewhere. It gives a detailed account of Bohr's discussions with Einstein at the fifth Solvay conference (as well as at the sixth Solvay conference of 1930). According to Bohr, the discussions in 1927 centred around, among other things, a version of the double-slit experiment, in which according to Einstein it was possible to observe interference while at the same time deducing which path the particle had taken, a claim conclusively refuted by Bohr.

Were Bohr's recollections accurate? Jammer certainly thought so:

> Bohr's masterly report of his discussions with Einstein on this issue, though written more than 20 years after they had taken place, is undoubtedly a reliable source for the history of this episode.
>
> *(Jammer 1974, p. 120)*

Though as Jammer himself adds (p. 120): 'It is, however, most deplorable that additional documentary material on the Bohr–Einstein debate is extremely scanty'.

We now know that, as we shall now discuss, Bohr's recollection of his discussions with Einstein did not properly capture Einstein's true intentions, essentially because, at the time, no one understood what Einstein's principal concern was: the non-separability of quantum theory.

12.2 Towards a historical revision

Separability – the requirement that the joint state of a composite of spatially separated systems should be determined by the states of the component parts – was a condition basic to Einstein's field-theoretic view of physics (as indeed was the absence of action at a distance). As already mentioned in Section 9.2, Howard (1990) has shown in great detail how Einstein's concerns about the failure of separability in quantum theory date back to long before the famous EPR paper of 1935.[a] There is no doubt that, by 1909, Einstein understood that if light quanta were treated like the spatially independent molecules of an ideal gas, then the resulting fluctuations were inconsistent with Planck's formula for blackbody radiation; and certainly, in 1925, Einstein was concerned that Bose–Einstein statistics entailed a mysterious interdependence of photons. Also in 1925, as we discussed in Section 9.2, Einstein's theory of guiding fields in 3-space – in which spatially separated systems each had their own guiding wave, in accordance with Einstein's separability criterion – conflicted with energy-momentum conservation for single events. Howard (pp. 83–91) argues further that, in spring 1927, Einstein must have realised that Schrödinger's wave mechanics in configuration space violated separability, because a general solution to the Schrödinger equation for a composite system could *not* be written as a product over the components.[b] In other words, Einstein objected to what we would now call entanglement, and concluded that the wave function in configuration space could not represent anything physical.

Einstein's concerns about separability continued up to and beyond the fifth Solvay conference. While it seems that Einstein did have some early doubts about the validity of the uncertainty relations, Howard's reconstruction shows that Einstein's main concern lay elsewhere. The primary aim of the famous thought experiments that Einstein discussed with Bohr, in 1927 and subsequently, was not to defeat the uncertainty relations but to highlight the (for him disturbing) feature of quantum theory – that spatially separated systems cannot be treated independently. As Howard (p. 94) puts it, regarding the 1927 Solvay conference:

> But if the uncertainty relations *really* were the main sticking point *for Einstein*, why did Einstein not say so in the published version of his remarks, or anywhere else for that matter in correspondence or in print in the weeks and months following the Solvay meeting?

[a] See also Fine (1986, ch. 3).

[b] Howard's argument here is somewhat circumstantial, appealing in part to the difficulty with separability that Einstein had with his own hidden-variables amendment of wave mechanics (cf. Section 11.3).

We have indeed seen in Chapter 7 that Einstein's criticism in the general discussion concerned locality and incompleteness (just like the later EPR argument), not the uncertainty relations. Bohr, in his reply, states that he does not 'understand what precisely is the point' Einstein is making. It seems rather clear that, indeed, in 1927 Bohr did not understand Einstein's point, and it is remarkable that what is most often recalled about the fifth Solvay conference was in fact largely a misunderstanding.

According to Bohr's later recollections (Bohr 1949), at the fifth Solvay conference Einstein proposed a version of the two-slit experiment in which measurement of the transverse recoil of a screen with a single slit would enable one to deduce the path of a particle through a second screen with two slits, while at the same time observing interference on the far side of the second screen. (Consideration of this experiment must have taken place informally, not in the official discussions.) This experiment has been analysed in detail by Wootters and Zurek (1979), who show the crucial role played by quantum non-separability between the particle and the first screen. While the evidence is somewhat sketchy in this particular instance, according to Howard the main point that concerned Einstein in this experiment was precisely such non-separability.

That separability was indeed Einstein's central concern is clearer in the later 'photon-box' thought experiment he discussed with Bohr at the sixth Solvay conference of 1930, involving weighing a box from which a photon escapes. Again, Bohr discusses this experiment at length in his recollections, where according to him it was yet another of Einstein's attempts to circumvent the uncertainty relations. Specifically, according to Bohr, Einstein's intention was to beat the energy-time uncertainty relation, by measuring both the energy of the emitted photon (by weighing the box before and after) and its time of emission (given by a clock controlling the shutter releasing the photon). On Bohr's account, this attempt failed, ironically, because of the time dilation in a gravitational field implied by Einstein's own general theory of relativity.

However, it seems that in fact the photon-box experiment was (like Einstein's published objection of 1927) really a form of the later EPR argument for incompleteness. This is shown by a letter Ehrenfest wrote to Bohr on 9 July 1931, just after Ehrenfest had visited Einstein in Berlin (Howard 1990, pp. 98–9). Ehrenfest reports that Einstein said he did not invent the photon-box experiment to defeat the uncertainty relations (which he had for a long time no longer doubted), but 'for a totally different purpose'. Ehrenfest then explains that Einstein's real intention was to construct an example in which the choice of measurement at one location would enable an experimenter to predict *either* one *or* the other of two incompatible quantities for a system that was far away at the time of the measurement. In the example at hand, if the escaped photon is reflected back towards the box after having travelled a great distance, then the time of its return may be predicted with certainty if the experimenter checks the clock reading (while the photon is still far away); alternatively, the energy (or frequency) of the returning photon may be predicted with certainty if, instead, the experimenter chooses to weigh the box (again while the photon is still far away). Because the two possible operations take place while the photon is at a great distance, the assumptions of separability and locality imply that both the time and energy of the returning photon are in reality determined in advance (even if in practice an experimenter cannot carry out both predictions simultaneously), leading to the conclusion that quantum theory is incomplete.[a]

The true thrust of Einstein's argument was not appreciated at the time, perhaps because Bohr and his associates tended to identify the existence of physical quantities with their experimental measurability: if two quantities could not be measured simultaneously in the same experiment, they did not exist simultaneously in the same experiment. With this attitude in mind, it would be natural to mistake Einstein's claim of simultaneous existence for a claim of simultaneous measurability.

We feel that Howard's reappraisal of the Bohr–Einstein debate, as well as being of great intrinsic interest, also provides an instructive example of how the history of quantum physics should be reconsidered in the light of our modern understanding of quantum theory and its open problems. There was certainly much more to the fifth Solvay conference than the Bohr–Einstein clash, and a similar reappraisal of other crucial encounters at that time seems overdue.

If the history of quantum theory is written on the assumption that Bohr, Heisenberg and Born were right, and that de Broglie, Schrödinger and Einstein were wrong, the resulting account is likely to be unsatisfactory: opposing views will tend to be misunderstood or underestimated, supporting views over-emphasised, and valid alternative approaches ignored.

A reconsideration of the fifth Solvay conference certainly entails a re-evaluation of de Broglie's pilot-wave theory as a coherent but (until very recently) 'forgotten' formulation of quantum theory. Schrödinger's ideas, too, seem more plausible today, in the light of modern collapse models. One should also reconsider what Born and Heisenberg's 'quantum mechanics' really was, in particular as concerns the role of time and the collapse of the wave function.

[a] The reasoning here is similar to that in Einstein's own (and simpler) version of the EPR argument, which first appears in a letter from Einstein to Schrödinger of 19 June 1935, one month after the EPR paper was published (Fine 1986, ch. 3). The argument – which Einstein repeated and refined between 1936 and 1949 – runs essentially as follows. A complete theory should associate one and only one theoretical state with each real state of a system; in an EPR-type experiment on correlated systems, depending on what measurement is carried out at one wing of the experiment, quantum theory associates different wave functions with what must (assuming locality) be the same real state at the other distant wing. Therefore, quantum theory is incomplete. For a detailed discussion, see Howard (1990).

There is no longer a definitive, widely accepted interpretation of quantum mechanics; it is no longer clear who was right and who was wrong in October 1927. Therefore, it seems particularly important at this time to return to the historical sources and re-evaluate them. We hope that physicists, philosophers and historians will reconsider the significance of the fifth Solvay conference, both for the history of physics and for current research in the foundations of quantum theory.

There is no longer a definitive widely accepted [illegible] of quantum mechanics [illegible] [illegible] [illegible] [illegible] We hope that our [illegible] [illegible] [illegible] [illegible] [illegible]

Part III

The proceedings of the 1927 Solvay conference

INSTITUT INTERNATIONAL DE PHYSIQUE SOLVAY

ÉLECTRONS ET PHOTONS

RAPPORTS ET DISCUSSIONS

DU

CINQUIÈME CONSEIL DE PHYSIQUE

TENU A BRUXELLES DU 24 AU 29 OCTOBRE 1927

SOUS LES AUSPICES

DE L'INSTITUT INTERNATIONAL DE PHYSIQUE SOLVAY

Publiés par la Commission administrative de l'Institut.

PARIS

GAUTHIER-VILLARS ET Cie, ÉDITEURS

LIBRAIRES DU BUREAU DES LONGITUDES, DE L'ÉCOLE POLYTECHNIQUE

Quai des Grands-Augustins, 55.

1928

Photograph taken in Brussels
on 27 October 1927

H. A. LORENTZ

H. A. Lorentz †

Hardly a few months have gone by since the meeting of the fifth physics conference in Brussels, and now I must, in the name of the scientific committee, recall here all that meant to the Solvay International Institute of Physics he who was our chairman and the moving spirit of our meetings. The illustrious teacher and physicist, H. A. Lorentz, was taken away in February 1928 by a sudden illness, when we had just admired, once again, his magnificent intellectual gifts which age was unable to diminish in the least.

Professor Lorentz, of a simple and modest demeanour, nevertheless enjoyed an exceptional authority, thanks to the combination of rare qualities in a harmonious whole. Theoretician with profound views – eminent teacher in the highest forms of instruction and tirelessly devoted to this task – fervent advocate of all international scientific collaboration – he found, wherever he went, a grateful circle of pupils, disciples and those who carried on his work. Ernest Solvay had an unfailing appreciation of this moral and intellectual force, and it was on this that he relied to carry through a plan that was dear to him, that of serving Science by organising conferences composed of a limited number of physicists, gathered together to discuss subjects where the need for new insights is felt with particular intensity. Thus was born the Solvay International Institute of Physics, of which Ernest Solvay followed the beginnings with a touching concern and to which Lorentz devoted a loyal and fruitful activity.

All those who had the honour to be his collaborators know what he was as chairman of these conferences and of the preparatory meetings. His thorough knowledge of physics gave him an overall view of the problems to be examined. His clear judgement, his fair and benevolent spirit guided the scientific committee in the choice of the assistance it was appropriate to call upon. When we then were gathered together at a conference, one could only admire without reservations the mastery with which he conducted the chairmanship. His shining intellect dominated the discussion and followed it also in the details, stimulating it or preventing it from drifting, making sure that all opinions could be usefully expressed, bringing out the final conclusion as far as possible. His perfect knowledge of languages allowed him to interpret, with equal facility, the words uttered by each one. Our chairman appeared to us, in fact, gifted with an invincible youth, in his passion for scientific truth and in the joy he had in comparing opinions, sometimes with a shrewd smile on his face, and even a little mischievousness when confronted with an unforeseen aspect of the question. Respect and affection went to him spontaneously, creating a cordial and friendly atmosphere, which facilitated the common work and increased its efficiency.

True creator of the theoretical edifice that explains optical and electromagnetic phenomena by the exchange of energy between electrons contained in matter and radiation viewed in accordance with Maxwell's theory, Lorentz retained a devotion to this classical theory. All the more remarkable is the flexibility of mind with which he followed the disconcerting evolution of the quantum theory and of the new mechanics.

The impetus that he gave to the Solvay institute will be a memory and an example for the scientific committee. May this volume, faithful report of the work of the recent physics conference, be a tribute to the memory of he who, for the fifth and last time, honoured the conference by his presence and by his guidance.

M. CURIE

Fifth physics conference

The fifth of the physics conferences, provided for by article 10 of the statutes of the international institute of physics founded by Ernest Solvay, held its sessions in Brussels on the premises of the institute from 24 to 29 October 1927.

The following took part in the conference:

Mr H. A. LORENTZ †, of Haarlem, *Chairman.*

Mrs P. CURIE, of Paris; Messrs N. BOHR, of Copenhagen; M. BORN, of Göttingen; W. L. BRAGG, of Manchester; L. BRILLOUIN, of Paris; A. H. COMPTON, of Chicago; L.-V. DE BROGLIE, of Paris; P. DEBYE, of Leipzig; P. A. M. DIRAC, of Cambridge; P. EHRENFEST, of Leiden; A. EINSTEIN, of Berlin; R. H. FOWLER, of Cambridge; Ch.-E. GUYE, of Geneva; W. HEISENBERG, of Copenhagen; M. KNUDSEN, of Copenhagen; H. A. KRAMERS, of Utrecht; P. LANGEVIN, of Paris; W. PAULI, of Hamburg; M. PLANCK, of Berlin; O. W. RICHARDSON, of London; [E. SCHRÖDINGER, of Zurich;] C. T. R. WILSON, of Cambridge, *Members.*

Mr J.-É. VERSCHAFFELT, of Gent, fulfilled the duties of *Secretary.*

Messrs Th. DE DONDER, E. HENRIOT and Aug. PICCARD, professors at the University of Brussels, attended the meetings of the conference as guests of the scientific committee, Mr Ed. HERZEN, professor at the École des Hautes Études de Bruxelles, as representative of the Solvay family.

Professor I. LANGMUIR, of Schenectady (U.S. of America), visiting Europe, attended the meetings as a guest.

Mr Edm. van AUBEL, member of the scientific committee, and Mr H. DESLANDRES, director of the Meudon observatory, invited to participate in the conference meetings, had been excused.

Sir W. H. BRAGG, member of the scientific committee, who had handed in his resignation before the meetings and requested to be excused, also did not attend the sessions.

The administrative commission of the institute was composed of:

Messrs Jules BORDET, professor at the University of Brussels, appointed by H. M. the King of the Belgians; Armand SOLVAY, engineer, manager of Solvay and Co.; Maurice BOURQUIN, professor at the University of Brussels; Émile HENRIOT, professor at the University of Brussels; Ch. LEFÉBURE, engineer, appointed by the family of Mr Ernest Solvay, *Administrative Secretary*.

The scientific committee was composed of:

Messrs H. A. LORENTZ†, professor at the University of Leiden, *Chairman*; M. KNUDSEN, professor at the University of Copenhagen, *Secretary*; W. H. BRAGG, professor at the University of London, president of the Royal Institution; Mrs Pierre CURIE, professor at the Faculty of Sciences of Paris; Messrs A. EINSTEIN,[a] professor, in Berlin; Charles-Eug. GUYE, professor at the University of Geneva; P. LANGEVIN, professor at the Collège de France, in Paris; O. W. RICHARDSON, professor at the University of London; Edm. van AUBEL, professor at the University of Gent.

Sir W. H. BRAGG, resigning member, was replaced by Mr B. CABRERA, professor at the University of Madrid.

To replace its late chairman, the scientific committee chose Professor P. LANGEVIN.

[a] Chosen in replacement of Mr H. Kamerlingh Onnes, deceased.

The intensity of X-ray reflection[a]

BY MR W. L. BRAGG

1. – THE CLASSICAL TREATMENT OF X-RAY DIFFRACTION PHENOMENA

The earliest experiments on the diffraction of X-rays by crystals showed that the directions in which the rays were diffracted were governed by the classical laws of optics. Laue's original paper on the diffraction of white[1] radiation by a crystal, and the work which my father and I initiated on the reflection of lines[2] in the X-ray spectrum, were alike based on the laws of optics which hold for the diffraction grating. The high accuracy which has been developed by Siegbahn and others in the realm of X-ray spectroscopy is the best evidence of the truth of these laws. Advance in accuracy has shown the necessity of taking into account the very small refraction of X-rays by the crystal, but this refraction is also determined by the classical laws and provides no exception[3] to the above statement.

The first attempts at crystal analysis showed further that the strength of the diffracted beam was related to the structure of the crystal in a way to be expected by the optical analogy. This has been the basis of most work on the analysis of crystal structure. When monochromatic X-rays are reflected from a set of crystal planes, the orders of reflection are strong, weak, or absent in a way which can be accounted for qualitatively by the arrangement of atoms[4] parallel to these planes. In the analysis of many structures, it is not necessary to make a strict examination of the strength of the diffracted beams. Slight displacements of the atoms cause the intensities of the higher orders to fluctuate so rapidly, that it is possible to fix

[a] We follow Bragg's original English typescript, from the copy in the Richardson collection, AHQP-RDN, document M-0059 (indexed as 'unidentified author' in the microfilmed catalogue). Obvious typos are corrected mostly tacitly and some of the spelling has been harmonised with that used in the rest of the volume. Discrepancies between the original English and the published French are endnoted (*eds.*).

the atomic positions with high accuracy by using a rough estimate of the relative intensity of the different orders.

When we attack the problem of developing an accurate quantitative theory of intensity of diffraction, many difficulties present themselves. These difficulties are so great, and the interpretation of the experimental results has often been so uncertain, that it has led[5] to a natural distrust of deductions drawn from intensity measurements. Investigators of crystal structures have relied on qualitative methods,[6] since these were in many cases quite adequate. The development of the quantitative analysis has always interested me personally, particularly as a means of attacking the more complicated crystalline structures, and it would seem that at the present time the technique has reached a stage when we can rely on the results. It is my purpose in this paper to attempt a critical survey of the present development of the subject. It is of considerable interest because it is our most direct way of analysing atomic and molecular structure.

In any X-ray examination of a crystalline body, what we actually measure is a series of samples[7] of the coherent radiation scattered in certain definite directions by the unit of the structure. This unit is, in general, the element of pattern of the crystal, while in certain simple cases it may be a single atom.

In the examination of a small body by the microscope, the objective receives the radiation scattered in different directions by the body, and the information about its structure, which we get by viewing the final image, is contained at an earlier stage[8] in these scattered beams. Though the two cases of microscopic and X-ray examination are so similar, there are certain important differences. The scattered beams in the microscope can be combined again to form an image, and in the formation of the image the phase relationship between beams scattered in different directions plays an essential part. In the X-ray problem, since we can only measure the intensity of scattering in each direction, this phase relationship cannot be determined experimentally, though in many cases it can be inferred.[9] Further, the microscope receives the scattered beams over a continuous range of directions, whereas the geometry of the crystalline structure limits our examination to certain directions of scattering. Thus we cannot form directly an image of the crystalline unit which is being illuminated by X-rays. We can only measure experimentally the strength of the scattered beams, and then build up an image piece by piece from the information we have obtained.

It is important to note that in the case of X-ray examination all work is being carried out at what is very nearly the theoretical limit of the resolving power of our instruments. The range of wavelength which it is convenient to use lies between 0.6 Å and 1.5 Å. This range is of sufficiently small wavelength for work with the details of crystal structure, which is always on a scale of several Ångström units, but the wavelengths are inconveniently great for an examination into atomic structure. It is unfortunate from a practical point of view that there is no convenient steady source of radiation between the K lines of the metal palladium, and the very much shorter K lines of tungsten. This difficulty will no doubt be overcome, and a technique of 'ultraviolet' X-ray microscopy will be developed, but at present all the accurate work on intensity of reflection has been done with wavelengths in the neighbourhood of 0.7 Å.

We may conveniently[10] divide the process of analysis into three stages.

a) The experimental measurement of the intensities of the diffracted beams.
b) The reduction of these observations, with the aid of theoretical formulae, to measurements of the amplitudes of the waves scattered by a single unit of the structure, when a wave train of given amplitude falls on it.
c) The building up of the image, or deduction of the form of the unit, from these measurements of scattering in different directions.

2. – History of the use of quantitative methods

The fundamental principles of a mathematical analysis of X-ray reflection were given in Laue's original paper [1], but the precise treatment of intensity of reflection may be said to have been initiated by Darwin [2] with two papers in the *Philosophical Magazine* early in 1914, in which he laid down the basis for a complete theory of X-ray reflection based on the classical laws of electrodynamics.[11] The very fundamental and independent treatment of the whole problem by Ewald [3], along quite different lines, has confirmed Darwin's conclusions in all essentials. These papers established the following important points.

1. Two formulae for the intensity of X-ray reflection can be deduced, depending on the assumptions which are made. The first of these has since come to be known as the formula for the 'ideally imperfect crystal' or 'mosaic crystal'.[a] It holds for a crystal in which the homogeneous blocks are so small that the reduction in intensity of a ray passing through each block, and being partly reflected by it, is wholly accounted for by the ordinary absorption coefficient. This case is simple to treat from a mathematical point of view, and in actual fact many crystals approach this physical condition of a perfect mosaic.

   The second formula applies to reflection by an ideally perfect crystal. Here ordinary[12] absorption plays no part in intensity of reflection. This is perfect over a finite range of glancing angles, all radiation being reflected within this range. The range depends on the efficiency of the atom planes in scattering. The second formula is entirely different from the first, and leads to numerical results of a different order of magnitude.

[a] I believe we owe to Ewald the happy suggestion of the word 'mosaic'.

2. The actual intensity of reflection in the case of rocksalt is of the order to be expected from the imperfect crystal formula.
3. The observed rapid decline in intensity of the high orders is only partly accounted for by the formula for reflection, and must be due in addition to the spatial distribution of scattering matter in the atoms (electron distribution).
4. When a crystal is so perfect that it is necessary to allow for the interaction of the separate planes, the transmitted beam is extinguished more rapidly than corresponds to the true absorption of the crystal (extinction).
5. There exists a refractive index for both crystalline and amorphous substances, slightly less than unity, which causes small deviations from the law $n\lambda = 2d \sin\theta$.

Another important factor in intensity of reflection had been already examined theoretically by Debye [4], this being the diminution in intensity with rising temperature due to atomic movement. Though subsequent work has put Debye's and Darwin's formulae in modified and more convenient forms, the essential features were all contained in these early papers.

On the experimental side, the first accurate quantitative measurements were made by W. H. Bragg [5].[13] The crystal was moved with constant angular velocity through the reflecting position, and the total amount of reflected radiation measured. He showed that the reflection[14] from rocksalt for a series of faces lay on a smooth curve when plotted against the sine of the glancing angle, emphasising that a definite physical constant was being measured. This method of measurement has since been widely used. The quantity $\frac{E\omega}{I}$, where E is the total energy of radiation[15] reflected, ω the angular velocity of rotation, and I the total radiation falling on the crystal face per second, is independent of the experimental arrangements, and is a constant for a given reflection from a mosaic crystal; it is generally termed the '*integrated reflection*'.[16] It is related in a simple way to the energy measurements from a powdered crystal, which have also been employed for accurate quantitative work. W. H. Bragg's original measurements were comparisons[17] of this quantity for different faces, not absolute measurements in which the strength of an incident beam was considered.

W. H. Bragg further demonstrated the existence of the extinction effect predicted by Darwin, by passing X-rays through a diamond crystal set for reflection and obtaining an increased absorption. He made measurements of the diminution in intensity of reflection[18] with rising temperature predicted by Debye, and observed[19] by Laue, and showed that the effect was of the expected order. In the Bakerian Lecture in 1915 [6] he described measurements in the intensity of a very perfect crystal, calcite, which seemed to show that the intensity was proportional to the scattering power of the atomic planes and not to the square of the power (this is to be expected from the formula for reflection by a perfect crystal). In the same address he proposed the use of the Fourier method of interpreting the measurements[20] which has been recently used with such success by Duane, Havighurst, and Compton, and which is dealt with in the fourth section of this summary.[21] At about the same time, Debye and Compton independently discussed the influence of electronic distribution in the atom on the intensity of reflection.

The next step was made by Compton [7] in 1917. Darwin's formula for the mosaic crystal was deduced by a different method, and was applied to the interpretation of W. H. Bragg's results with rocksalt. Compton concluded that the electronic distribution in the atoms was of the type to be expected from Bohr's atomic model. Compton then published the first measurements of the *absolute* intensity of reflexion. A monochromatic beam of X-rays was obtained by reflection from a crystal, and this was reflected by a second rotating crystal (rocksalt and[22] calcite). The absolute value of the integrated reflection $\frac{E\omega}{I}$ was found to be of the right order for rocksalt when calculated by the imperfect crystal formula, but to be very low for calcite indicating strong extinction or a wrong formula, in the second case.

In 1921 and 1922 I published with James and Bosanquet a series of measurements on rocksalt in which we tried to obtain a high accuracy. We made absolute measurements of intensity for the strongest reflections,[a] and compared the weaker reflections with them. Our main contributions in these papers were a more accurate set of measurements of integrated reflection for a large number of planes, and a method for estimating and correcting for the effect of extinction. As Darwin showed in a paper in 1922 [9] on the theoretical interpretation of our results, we only succeeded in correcting for extinction of the kind he termed 'secondary' and not for 'primary' extinction.[b] Since then measurements by Havighurst [10], by Harris, Bates and McInnes[23] [11] and by Bearden [12] have been made on the reflecting power of powdered sodium chloride when extinction is absent. Their measurements have agreed with ours very closely indeed, confirming one's faith in intensity measurements, and showing that we were fortunate in choosing a crystal for our examination where primary extinction was very small. In the same papers we tried to make a careful analysis of the results in order to find how much information about atomic structure could be legitimately deduced from them, and we published curves showing the electron distribution in sodium and chlorine[24] atoms.

In this discussion, I have refrained from any reference to the question of reflection by 'perfect' crystals. The formula for reflection by such crystals was

[a] In our paper we failed to give due acknowledgement to Compton's absolute measurements in 1917 of which we were not aware at the time.

[b] Primary extinction is an excessive absorption of the beam which is being reflected in each homogeneous block of crystal, secondary extinction a statistical excessive absorption of the beam in the many small blocks of a mosaic crystal.

first obtained by Darwin, and has been arrived at independently by Ewald. The reflection by such crystals has been examined amongst others by Bergen Davis[25] and Stempel [13], and by Mark [14] and predictions of the theory have been verified. It is not considered here, because I wish to confine the discussion to those cases where a comparison of the intensity of incident and reflected radiation leads to accurate quantitative estimates of the distribution of scattering matter. This ideal can be attained with actual crystals,[26] when they are of the imperfect or mosaic type, though allowance for extinction is sometimes difficult in the case of the stronger reflections. On the other hand, it is far more difficult to know what one is measuring in the case of crystals which approximate to the perfect type. It is a fortunate circumstance that mathematical formulae can be applied most easily to the type of imperfect crystal more common in nature.

3. – Results of quantitative analysis

For the sake of conciseness, only one of the many intensity formulae will be given here, for it illustrates the essential features of them all. Let us suppose that the integrated reflection is being measured when X-rays fall on the face of a rotating crystal of the mosaic type. We then have

$$\rho = \frac{Q}{2\mu}\frac{1+\cos^2 2\theta}{2}.$$

(a) μ is the effective absorption coefficient, which may be greater than the normal coefficient, owing to the existence of extinction at the reflecting angle.

(b) The factor $\frac{1+\cos^2 2\theta}{2}$ is the 'polarisation factor', which arises because the incident rays are assumed to be unpolarised.

(c)

$$Q = \left(\frac{Ne^2}{mc^2}F\right)^2 \frac{\lambda^3}{\sin 2\theta},$$

where e and m are the electronic constants,[27] c the velocity of light, λ the wavelength used, N the number of scattering units per unit volume, and θ the glancing angle.

(d) F is the quantity we are seeking to deduce. It represents the scattering power of the crystal unit in the direction under consideration, measured in terms of the scattering power of a single[28] electron according to the classical formula of J. J. Thomson. It is defined by Ewald as the 'Atomfaktor'[29] when it applies to a single atom.

Formulae applicable to other experimental arrangements (the powder method for instance) are very similar, and contain the same quantity Q. Our measurements of reflection thus lead to values of Q, and so of F, since all other quantities in the formulae are known. Measurements on a given crystal yield a series of values for F, and all the information that can be found out about this crystalline or atomic structure is represented by these values. They are the same for the same crystal whatever wavelength is employed (since F is a function of $\frac{\sin\theta}{\lambda}$), though of course with shorter wavelength we have the advantage of measuring a much greater number of these coefficients (increased resolving power).

At this stage the effect of the thermal agitation of the atom will be considered as influencing the value of F. If we wish to make deductions about atomic structure, the thermal agitation must be taken into account. Allowance for it is a complicated matter, because not only do some atoms move more than others, but also they change their relative mean positions as the temperature alters in the more complex crystals.

This will be dealt with more fully below.

A series of examples will now be given to show that these quantitative formulae, when tested, lead to results which indicate that the theory is on the right lines. It is perhaps more convincing to study the results obtained with very simple crystals, though I think that the success of the theory in analysing highly complex structures is also very strong evidence, because we have covered such a wide range of substances.

In the simple crystals, where the positions of the atoms are definite, we can get the scattering power of individual atoms. The results should both indicate the correct number of electrons in the atom, and should outline an atom of about the right size. When F is plotted against $\frac{\sin\theta}{\lambda}$ its value should tend to the number, N, of electrons in the atom for small values of $\frac{\sin\theta}{\lambda}$, and should fall away as $\frac{\sin\theta}{\lambda}$ increases, at a rate which is reasonably explained by the spatial extension of the atom. In Fig. 1, the full lines give F curves obtained experimentally by various observers. The dotted lines are F curves calculated for the generalised atomic model of Thomas [15], of appropriate atomic number. The Thomas atomic model, which has been shown for comparison, is most useful as it gives us the approximate electronic distribution in an atom of any atomic number. Thomas calculates an ideal distribution of electrons in an atom of high atomic number. He assumes spherical symmetry for the atom, and supposes that 'electrons are distributed uniformly in the six-dimensional phase-space for the motion of an electron, at the rate of two for each h^3 of (six) volume'.[30] He thus obtains an ideal electron atmosphere around the nucleus, the constants of which can be simply adjusted[31] so as to be suitable for any given nuclear charge. It is of course to be expected that the lower the atomic weight, the more the actual distribution of scattering matter will depart from this arrangement, and will reflect the idiosyncrasies of the particular atom in question. The figure will show, however, that the actual curves are very similar to those calculated for Thomas' models. In particular, it will be clear that they tend to maximum values not far removed from the number of electrons in the atom in each case. The general agreement between the observed and calculated F curves must mean that our

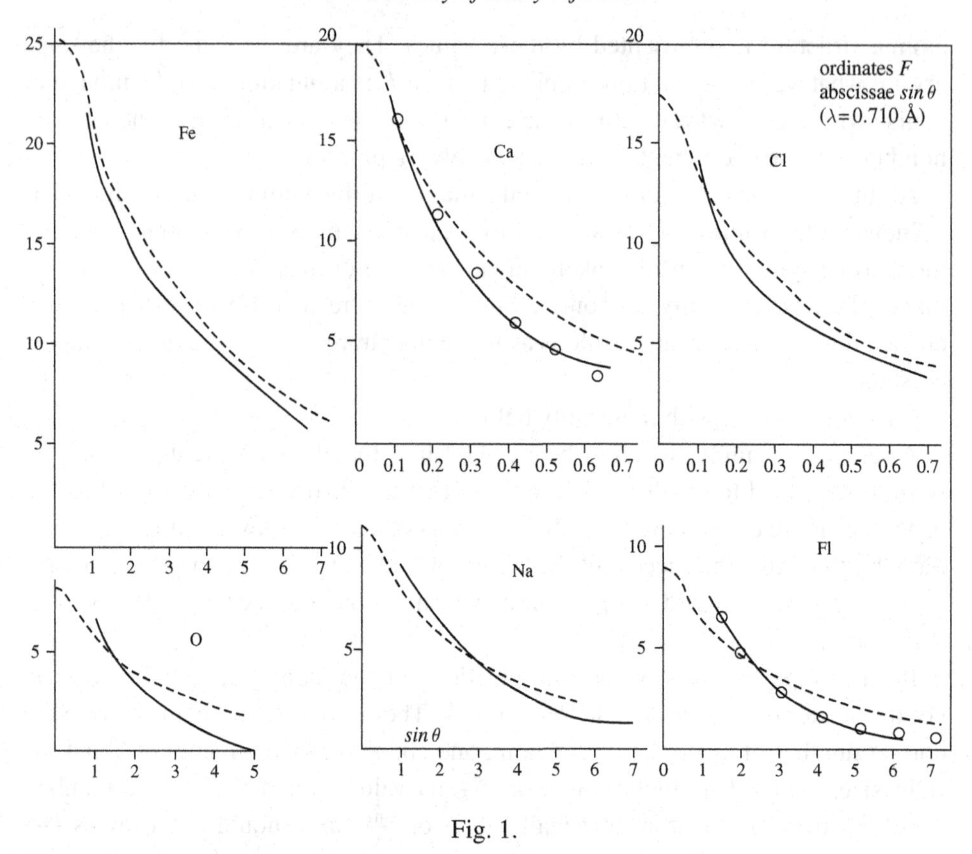

Fig. 1.

measurements of F are outlining a picture of the atom. The agreement holds also for other atomic models than those of Thomas, which all lead to atoms with approximately the same spatial extension and electronic distribution, as is well known.

All these measurements of F necessitate absolute values for the integrated reflection. It is not necessary to measure these directly in each case. When any one reflection has been measured in absolute value (by comparison of incident and reflected radiation), other crystals may be compared with it. The standard which has been used in every case, as far as I am aware, is the rocksalt crystal. Absolute measurements on this have been made by Compton [7], by Bragg, James and Bosanquet [8], and by Wasastjerna [18] which agree satisfactorily with each other.

4. – Interpretation of measurements of F

In interpreting these measurements of scattering power, we may either calculate the scattering of a proposed atomic model and compare it with the observed F

curve, or we may use the observations to calculate the distribution of scattering matter directly. The latter method is the more attractive, and in the hands of Duane, Havighurst, and Compton it has yielded highly interesting 'images' of the atomic structure seen by X-rays. There is a close analogy between the examination of a series of parallel planes by means of X-rays, and the examination of a diffraction grating, by a microscope, which is considered in Abbe's theory of microscopic vision.[a] The objective of the microscope may be considered as receiving a limited number of orders of spectra from the grating. These spectra in their turn build up the image viewed by the eyepiece, and the perfection of this image depends on the number of spectra received. The strength of each spectral order depends on the magnitude of the corresponding coefficient in that Fourier series which represents the amplitude of the light transmitted at each point of the grating. The extension of this well-known optical principle to the X-ray field was suggested by W. H. Bragg [6] in 1915. He had formed the conclusion[32] that the amplitudes of the scattered wave from rocksalt were inversely proportional to the square of the order of reflection, and he showed that[33] 'the periodic function which represents the density of the medium must therefore be of the form[34]

$$\text{const} + \frac{\cos 2\pi \frac{x}{d}}{1^2} + \frac{\cos 4\pi \frac{x}{d}}{2^2} + \cdots + \frac{\cos 2n\pi \frac{x}{d}}{n^2} + \cdots\text{'}$$

and in this way built up a curve showing the periodic density of the rocksalt grating. The method was not applied, however, to the much more accurate measurements which are now available until recently, when Duane and Havighurst showed how much could be done with it. Duane independently arrived at a more general formula of the same type, giving the density of scattering matter at any point in the whole crystal as a triple Fourier series, whose coefficients depend on the intensity of reflection from planes of all possible indices. Havighurst applied this principle to our measurements of rocksalt, and to measurements which he has made on other crystals, and obtained a picture of the relative density of scattering matter along certain lines in these crystals. Compton made the further step of putting the formulae in a form which gives the absolute density of electronic distribution (assuming the scattering to be by electrons obeying the classical laws). Compton gives a very full discussion of the whole matter in his book *X-rays and Electrons*.[35] It is not only an extremely attractive way of making clear just what has been achieved by the X-ray analysis, but also the most direct method of determining the structure.

[a] See for instance the discussion of this theory and of A. B. Porter's experiments to illustrate it in Wood's *Optics*, chapter VIII.

The formula for the distribution of scattering matter in parallel sheets, for a crystal with a centre of symmetry, is given by Compton as follows

$$P_z = \frac{Z}{a} + \frac{2}{a} \sum_1^{\infty} F_n \cos \frac{2\pi n z}{a}.$$

Here z is measured perpendicularly to the planes which are spaced a distance a apart. $P_z dz$ is the amount of scattering matter between planes at distances z and $z + dz$, and Z $(= \int_0^a P_z dz)$ is the total scattering matter of the crystal unit. This is a simplified form of Duane's formula for a Fourier series of which the general term is

$$A_{n_1 n_2 n_3} \sin\left(\frac{2\pi n_1 x}{a_1} - \delta_{n_1}\right) \sin\left(\frac{2\pi n_2 y}{a_2} - \delta_{n_2}\right) \sin\left(\frac{2\pi n_3 z}{a_3} - \delta_{n_3}\right),$$

$A_{n_1 n_2 n_3}$ being proportional to the amplitude of the scattered wave from the plane $(n_1 n_2 n_3)$.

Another Fourier series, due to Compton, gives the radial distribution of scattering matter, i.e. the values of U_n where $U_n dr$ is the amount of scattering matter between radii r and $r + dr$

$$U_n = \frac{8\pi r}{a^2} \sum_1^{\infty} n F_n \sin \frac{2\pi n r}{a},$$

where a is chosen so that values of F occur at convenient intervals on the graph for F.

If we know the values of F for a given atom over a sufficiently wide range, we can build up an image of the atom either as a 'sheet distribution' parallel to a plane, or as a radial distribution of scattering matter around the nucleus. In using these methods of analysis, however, it is very necessary to remember that we are working right at the limit of resolving power of our instruments, and in fact are attempting a more ambitious problem than in the corresponding optical case. In A. B. Porter's experiments to test Abbe's theory, he viewed the image of a diffraction grating and removed any desired group of diffracted rays by cutting them off with a screen. The first order gives blurred lines, four or five orders give sharper lines with a fine dark line down the centre, eight orders give two dark lines down the centre of each bright line and so forth. These imperfect images are due to the absences of the higher members in building up the Fourier series. In exactly the same way we get false detail in our X-ray image, owing to ignorance of the values of the higher members in the F curve. Similarly, the fine structure which actually exists may be glossed over, since by using a wavelength of 0.7 Å, we cannot hope to 'resolve' details of atomic structure on a scale of less than half this value.

The ignorance of the values of higher members of the Fourier series matters much less in the curve of sheet distribution than in that for radial distribution, since the latter converges far more slowly. Examples of the Fourier method of analysis are given in the next paragraph.

As opposed to this method of building up an image from the X-ray results,[36] we may make an atomic model and test it by calculating an F curve for it which can be compared with that obtained experimentally. This is the most satisfactory method of testing models arrived at by other lines of research, for nothing has to be assumed about the values of the higher coefficients F. It is of course again true that our test only applies to details of the proposed model on a scale comparable with the wavelength we are using. Since we can reflect X-rays right back from an atomic plane, we may get a resolving power for a given wavelength with the X-ray method twice as great as the best the microscope can yield.

It is perhaps worth mentioning the methods I used with James and Bosanquet in our determination of the electronic distribution in sodium and chlorine in 1922. We tried to avoid extrapolations of the F curve beyond the limit of experimental investigation. We divided the atom arbitrarily into a set of shells, with an unknown number of electrons in each shell. These unknowns were evaluated by making the scattering due to them fit the F curve over the observed range, this being simply done by solving simultaneous linear equations. We found we got much the same type of distribution however the shells were chosen, and that a limit to the electronic distribution at a radius of about 1.1 Å in sodium and 1.8 Å in chlorine was clearly indicated. Our distribution corresponds in its general outline to that found by the much more direct Fourier analysis, as the examples in paragraph 7 will show.

5. – EXAMPLES OF ANALYSIS

We owe to Duane [20] the appreciation of the very attractive way in which the Fourier analysis represents the results of X-ray examinations. It has the great merit of representing, in the form of a single curve, the information yielded by all orders of reflection from a given plane, or from the whole crystal. It is of course only an alternative way of interpreting the results, and the deductions we can make about atomic or molecular structures depend in the end on the extent to which we can trust our experimental observations, and not on the method of analysis we use. The Fourier method is so direct however, and its significance so easy to grasp, that Duane's introduction of it marks a great advance in technique of analysis.

I have reserved to paragraph 7 the more difficult problem of the arrangement of scattering matter in the atoms themselves, and the examples given here are of a

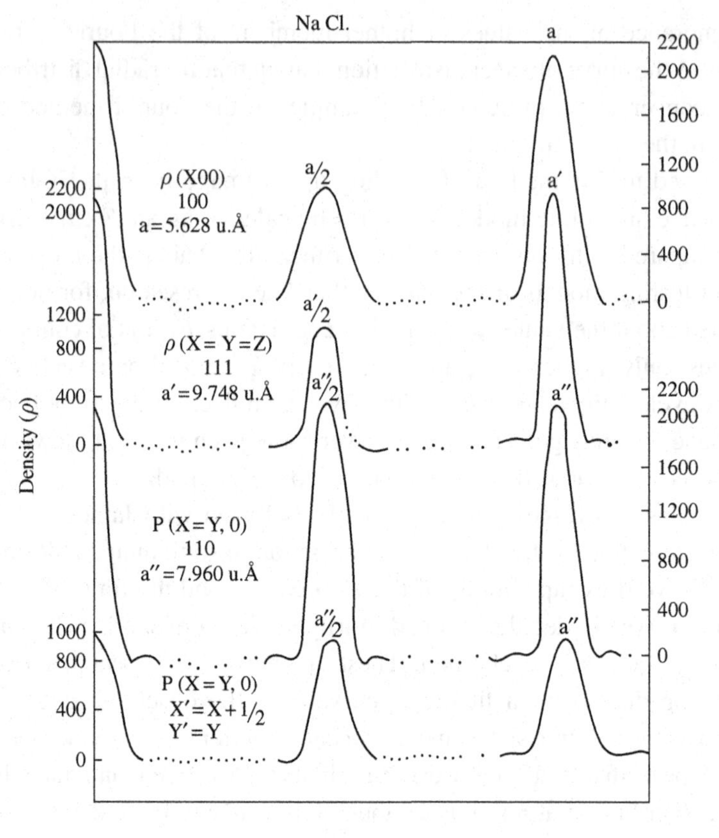

Fig. 2.

simpler character. They illustrate the application of analysis to the general problem of the distribution of scattering matter in the whole crystal, when we are not so near the limit of resolving power. The curves in Fig. 2 represent the first application of the new method of Fourier analysis to accurate data, carried out by Havighurst [21] in 1925. He used our determinations of F for sodium chloride, and Duane's three-dimensional Fourier series, and calculated the density of scattering matter along a cube edge through sodium and chlorine centres, along a cube diagonal through the same atoms, and along two face diagonals chosen so as to pass through chlorine atoms alone or sodium atoms alone in the crystal. The atoms show as peaks in the density distribution. In the other examples, the formula for distribution in sheets has been applied to some results we have obtained in our work on crystal structure at Manchester. I have given them because I feel they are convincing evidence of the power of quantitative measurements, and show that all methods of interpretations lead to the same results.

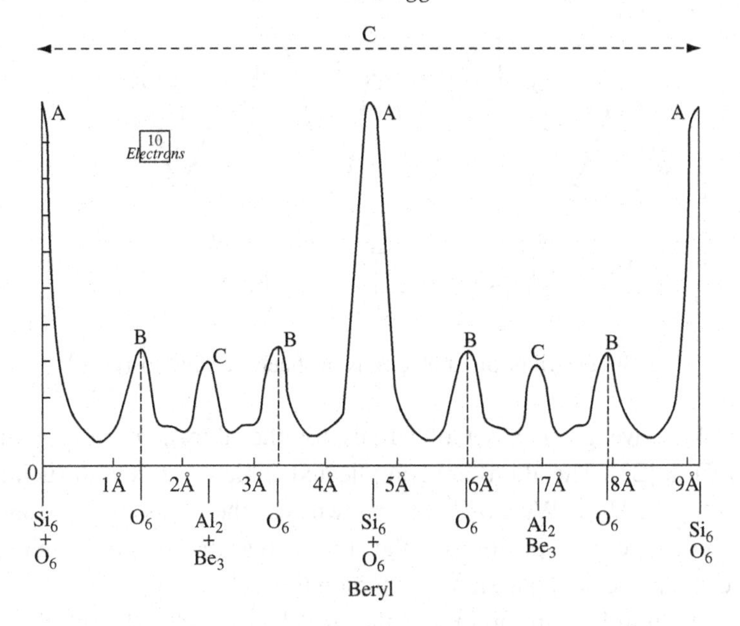

Fig. 3a. Distribution of electrons in sheets parallel to 0001.[37]

Mr West and I [22] recently analysed the hexagonal crystal beryl, $Be_3Al_2Si_6O_{18}$,[38] which has a structure of some complexity, depending on seven parameters. We obtained the atomic positions by the usual method of analysis, using more or less known *F* curves for the atoms in the crystal, and moving them about till we explained the observed *F*s due to the crystal unit. Fig. 3 shows the reinterpretations of this result by the Fourier method. Fig. 3a gives the electron density in sheets perpendicular to the principal axis of the crystal, which is of a very simple type. The particular point to be noted is the correspondence between the position of the line B in the figure and the hump of the Fourier analysis. The line B marks the position of a group of oxygen atoms which lies between two other groups A and C fixed by symmetry, the position of B being fixed by a parameter found by familiar methods of crystal analysis. The hump represents the same group fixed by the Fourier analysis, and it will be seen how closely they correspond. In Figs. 3b and 3c more complex sets of planes are shown. The dotted curve represents the interpretation of our results by Fourier analysis. The full curve is got by adding together the humps due to the separate atoms shown below, the position of these having been obtained by our X-ray analysis and their sizes by the aid of the curve in Fig. 3a in which the contribution of the atoms can be separated out. The correspondence between the two shows that the older methods and the Fourier analysis agree. It is to be noted that the crystal had first to be analysed by the older methods, in order that the sizes of the Fourier coefficients might be known.

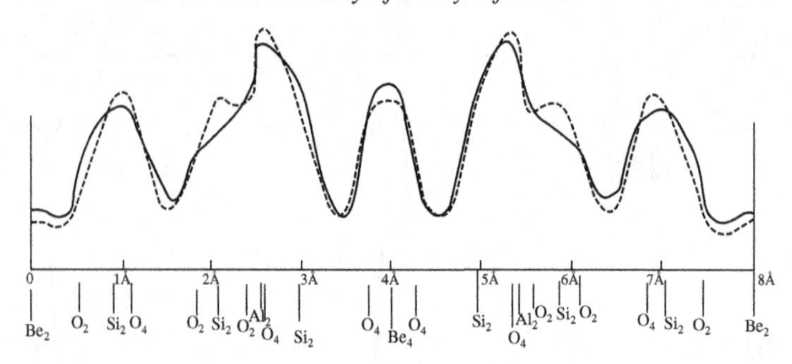

Fig. 3b. Distribution of electrons in sheets parallel to $10\bar{1}0$.[39]

In Fig. 4 I have given a set of curves for the alums, recently analysed by Professor Cork [23]. The alums are complicated cubic crystals with such formulae as $KAl(SO_4)_2.12\ H_2O$. Wyckoff[40] has shown that the potassium and aluminium atoms[41] occupy the same positions in the cubic cell as the sodium atom in rocksalt. Now we can replace the potassium by ammonium, rubidium, caesium, or thallium, and the aluminium by chromium, or other trivalent metals. Though the positions of the other atoms in the crystals are not yet known, they will presumably be much the same in all these crystals. If we represent by a Fourier series the quantitative measurements of the alums, we would expect the density of scattering matter to vary from crystal to crystal at the points occupied by the metal atoms, but to remain constant elsewhere. The curves show this in the most interesting[42] way.

The effect of heat motion on the movements of the atoms has already been mentioned. It was first treated theoretically by Debye [4]. Recently Waller [24] has recalculated Debye's formula, and has arrived at a modified form of it. Debye found that the intensities of the interference maxima in a simple crystal should be multiplied by a factor e^{-M}, where

$$M = \frac{6h^2}{\mu k \Theta} \frac{\varphi(x)}{x} \frac{\sin^2 \theta}{\lambda^2},$$

$$x = \frac{\Theta}{T} = \frac{\text{characteristic temperature of crystal}}{\text{absolute temperature}}.$$

Without going into further detail, it is sufficient to note that Waller's formula differs from Debye's by making the factor e^{-2M}, not e^{-M}. James and Miss Firth [25] have recently carried out a series of measurements for rocksalt between the temperatures 86° abs. and 900° abs. They find that Waller's formula is very closely followed up to 500° abs., though at higher temperatures the decline in intensity is even more rapid, as is perhaps to be expected owing to the crystal becoming more loosely bound. I have given the results of the measurements in Figs. 5 and 6, both as an example of the type of information which can be got from X-ray measurements,

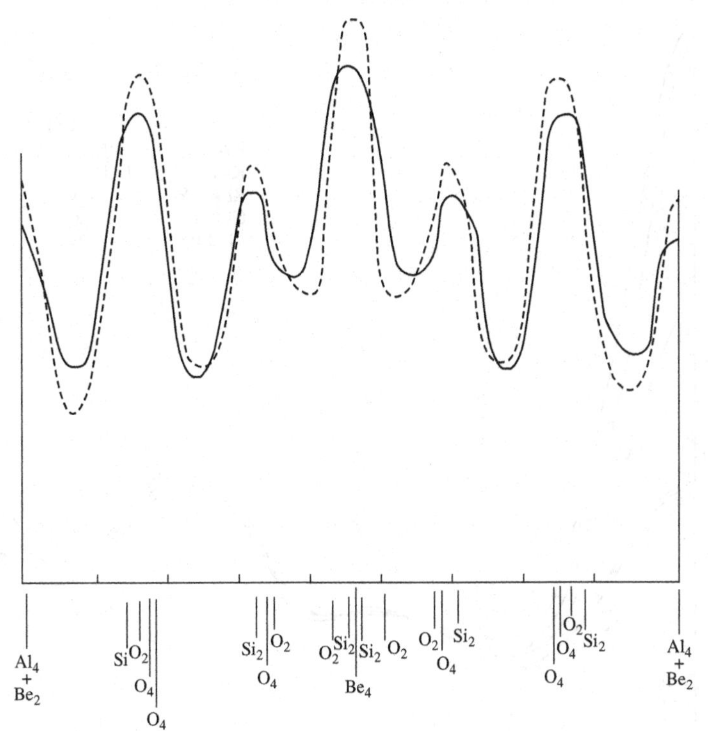

Fig. 3c. Distribution of electrons in sheets parallel to $11\bar{2}0$.[43]

and because these actual figures are of interest as a set of careful and accurate measurements of scattering power.

Fig. 5 shows the F curves for sodium and chlorine at different temperatures. The rapid decline in intensity for the higher orders will be realised when it is remembered that they are proportional to F^2. The curve for absolute zero is an extrapolation from the others, following the Debye formula as modified by Waller.

In Fig. 6 the same results are interpreted by the Fourier analysis. The curve at room temperatures for NaCl is practically identical with the interpretation of our earlier figures by Compton, in his book *X-rays and Electrons*,[44] though the figures on which it is based should be more accurate.[45] The curves show the manner in which the sharply defined peaks due to Cl and Na at low temperatures become diffuse owing to heat motion at the higher temperatures.

Several interesting points arise in connection with this analysis. In the first place, James and Firth[46] find that the heat factor is different for sodium and chlorine, the sodium atoms moving with greater average amplitudes than the chlorine atoms. This has a very interesting bearing on the crystal dynamics which is being further investigated by Waller. To a first approximation both atoms are affected equally by the elastic waves travelling through the crystal, but in a further approximation

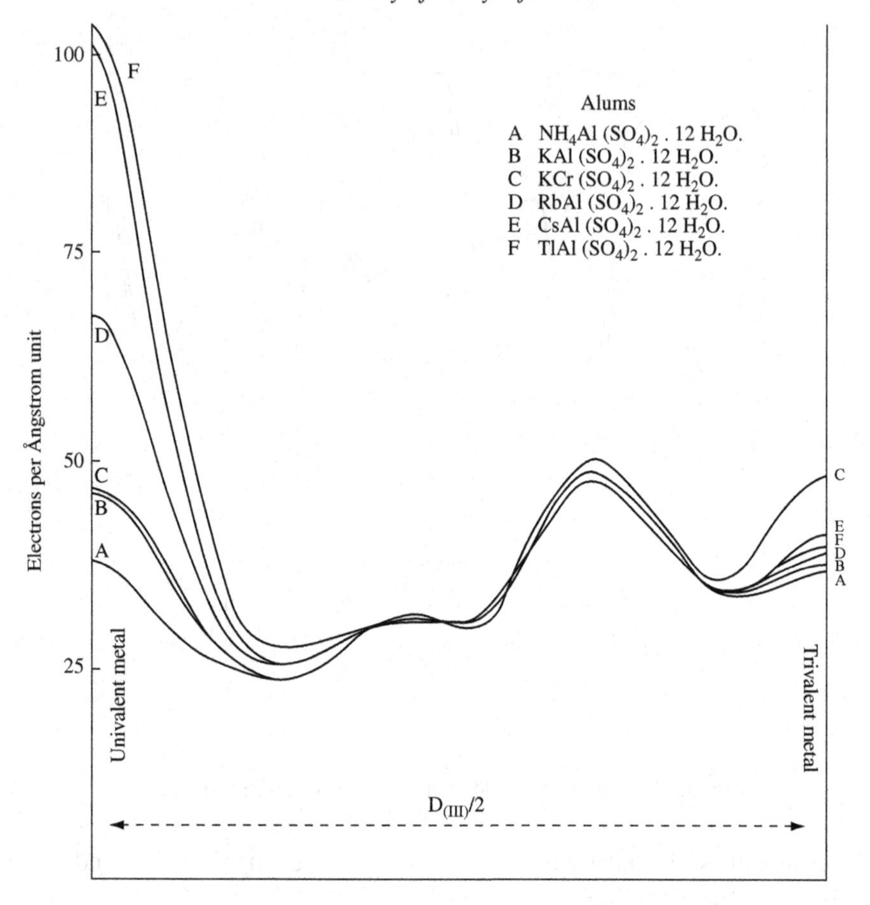

Fig. 4.

it can be seen that the sodium atoms are more loosely bound than the chlorine atoms. If an atom of either kind were only fixed in position by the six atoms immediately surrounding it, Waller has shown that there would be no difference between the motions of a sodium atom between six chlorine atoms, or a chlorine atom between six sodium atoms. However, the chlorine is more firmly pinned in position because it has in addition twelve large chlorine neighbours, whereas the sodium atom is much less influenced by the twelve nearest sodium atoms. Hence arises the difference in their heat motions. It is important to find the correct method for reducing observations to absolute zero, and this difference in heat motion must be satisfactorily analysed before this is possible.

In the second place, the accuracy which can be attained by the experimental measurements holds out some hope that we may be able to test directly whether there is zero-point energy[47] or not. This is being investigated by James and Waller. If a reliable atomic model is available, it would seem that the measurements

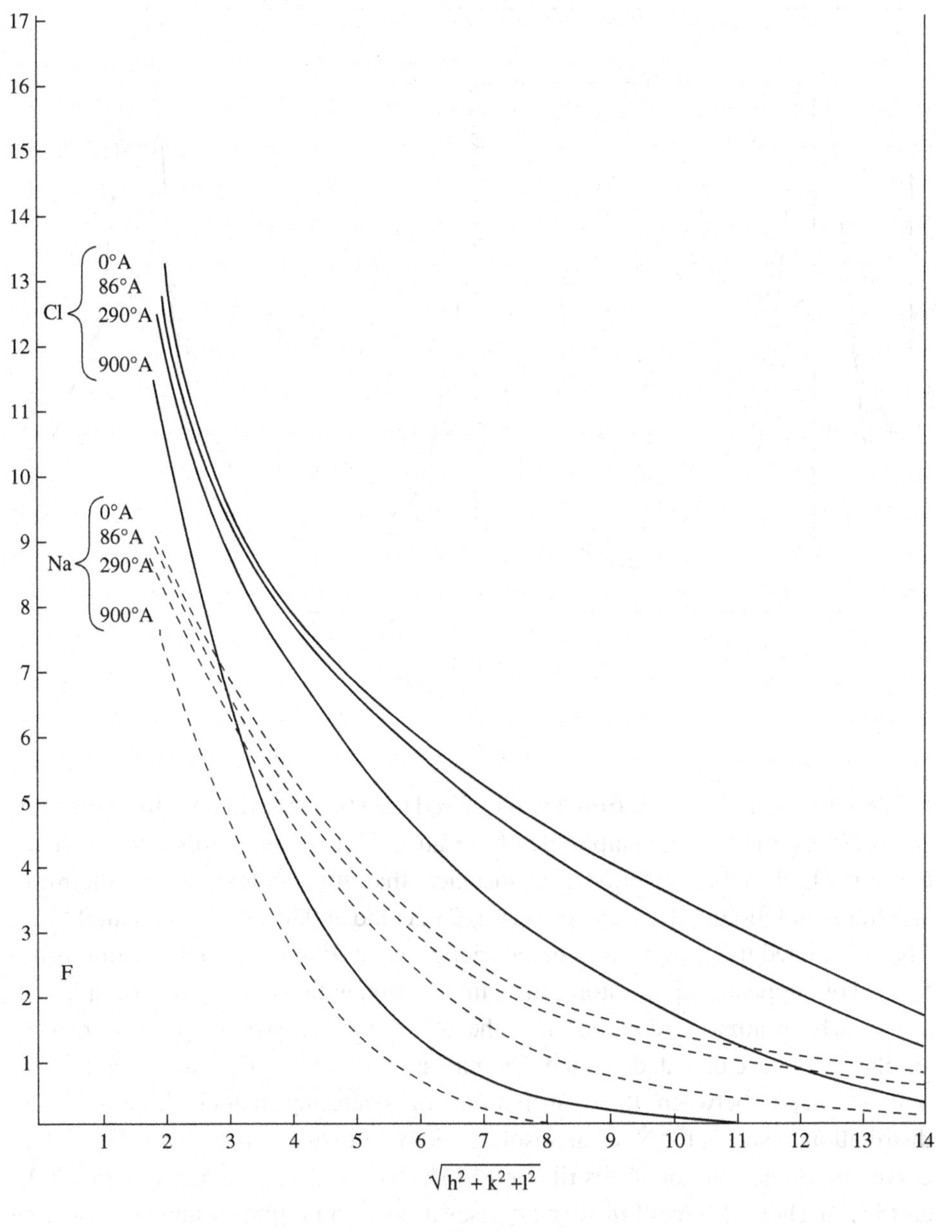

Fig. 5.

can tell whether there is vibration at absolute zero or not, for the theoretical diminution in intensity due to the vibration is much larger than the experimental error in measuring F. I feel considerable diffidence in speaking of the question of zero-point energy, and would like to have the advice of the mathematical physicists present.

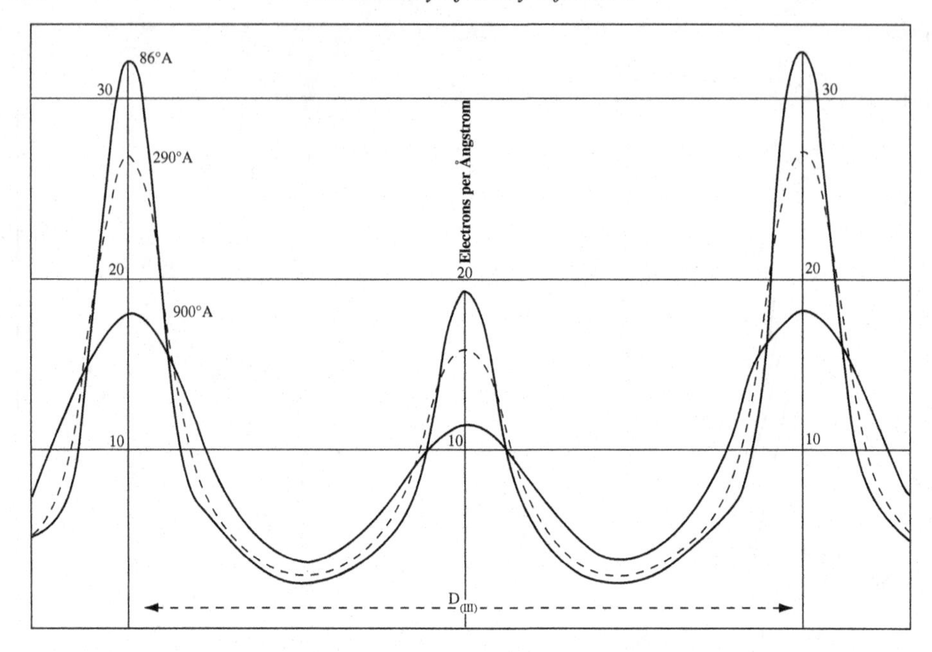

Fig. 6.

We may calculate, either from the measured heat factor or directly from the Fourier analyses, the average amplitude of vibration for different temperatures. James and Firth find by both methods, for instance, that at room temperature the mean amplitude of vibration for both atoms is 0.21 Å, and at 900° abs. it is about 0.58 Å. They examined the form which the Fourier curve at 0° abs. assumes when it is deformed by supposing all the atoms to be in vibration with the same mean amplitude.

It has been already remarked that the observed F curves for atoms are very similar to those calculated for the Thomas atomic model. The same comparison may be made between the distributions of scattering matter. In Fig. 7 the distribution in sheets for NaCl at absolute zero is shown as a full curve. The dotted curve shows the horizontal distribution in sheets for atoms of atomic number 17 and 11. In Thomas' model the density rises towards an infinite value very close to the nucleus, and this is represented by the very sharp peaks at the atomic centres in the dotted curve. We would not expect the observed distribution to correspond to the actual Thomas distribution at these points. Throughout the rest of the crystal the distribution is very similar. The comparison is interesting, because it shows how delicate a matter it is to get the fine detail of atomic structure from the observations. Thomas' distribution is quite continuous and takes no account of K, L and M sets of electrons. The slight departures of the observed curve from the smooth Thomas

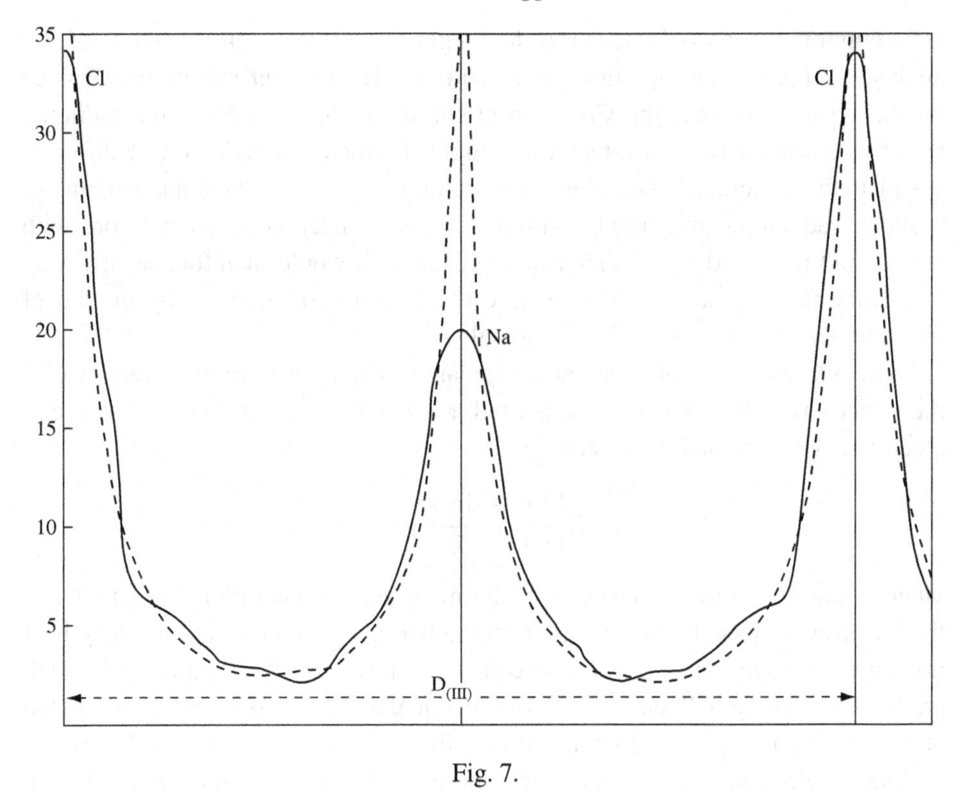

Fig. 7.

curve represent the experimental evidence for the existence of all the individual features of the atom.

6. – The mechanism of X-ray scattering

Before going on to discuss the application of the analysis to atomic structure, it is necessary to consider what is being measured when a distribution of scattering matter is deduced from the X-ray results. The classical treatment regards the atom as containing a number of electrons, each of which scatters radiation according to the formula of J. J. Thomson. Since a vast number of atoms contribute to the reflection by a single crystal plane, we should obtain a picture of the *average electronic distribution*. The quantity F should thus tend to a maximum value, at small angles of scattering, equal to the number of electrons in the atom, and should fall away owing to their spatial distribution as $\frac{\sin\theta}{\lambda}$ increases. The observed[48] F curves are of this character, as has been seen. When interpreted as an atomic distribution, they give atoms containing the correct number of electrons, and this seems satisfactory from the classical viewpoint. On the other hand, the evidence of the Compton effect would appear at first sight to cast doubt on the whole of our analysis. What we are measuring is essentially the *coherent* radiation diffracted by the crystal, whereas the Compton effect shows that a part of the radiation which is scattered is of different wavelength. Further, this radiation of different wavelength is included with the coherent radiation, when the total amount of scattered radiation is measured, and found to agree under suitable conditions with the amount predicted by J. J. Thomson's formula. It would therefore seem wrong to assume that we obtain a true picture of electronic distribution by the aid of measurements on the coherent radiation alone.

Even before the advent of the new mechanics, Compton's original treatment of the effect which he discovered suggested a way out of this difficulty. The recoil electron is given an amount of energy

$$2\frac{h^2}{m}\frac{\nu'}{\nu}\left(\frac{\sin\theta}{\lambda}\right)^2,$$

where ν and ν' are the frequencies of the modified and unmodified radiations. If the electron is ejected from the atom the radiation is modified in wavelength, if not coherent waves are scattered. Since there is little modified scattering at small angles, the F curve will tend to a maximum equal to the number of electrons in the atom, and any interpretation of the curve will give an atom containing the correct number of electrons. As $\frac{\sin\theta}{\lambda}$ increases, more and more of the scattered radiation will be modified, and in calculating the F curve this must be taken into account. However, if $\frac{\nu'}{\nu}$ is not far from unity, the F curve will remain a function of $\frac{\sin\theta}{\lambda}$, since whatever criterion is applied for the scattering of modified or unmodified radiation, it will depend on the energy imparted to the scattering electron, which is itself a function of $\frac{\sin\theta}{\lambda}$. Our X-ray analysis would thus give us an untrue picture of the atom, but one which is consistently the same whatever wavelength is employed. Williams [27] and Jauncey [28] have recalculated F curves from atomic models using this criterion, and found a better fit to the experimental curves when the Compton effect was taken into account. (Examples of this closer approximation will be found in the paper by Williams [27] in 1926. See also a discussion by Kallmann and Mark [26].)[49]

The point at issue is illustrated by the curves in Fig. 8. Three F curves for chlorine are plotted in the figure. The dotted line represents the observed F curve (James and Firth). The continuous line is the F curve calculated from Hartree's [29] atomic model for chlorine. It shows a hump at a value of $\sin\theta$ of 0.4, which is not present in the observed curve. This hump arises from the fact that the outer electrons in the chlorine model give negative values for F just short of this point,[50] and positive values again at the point itself. All atomic models calculated with electronic orbits show similar irregularities which are not actually observed. When,

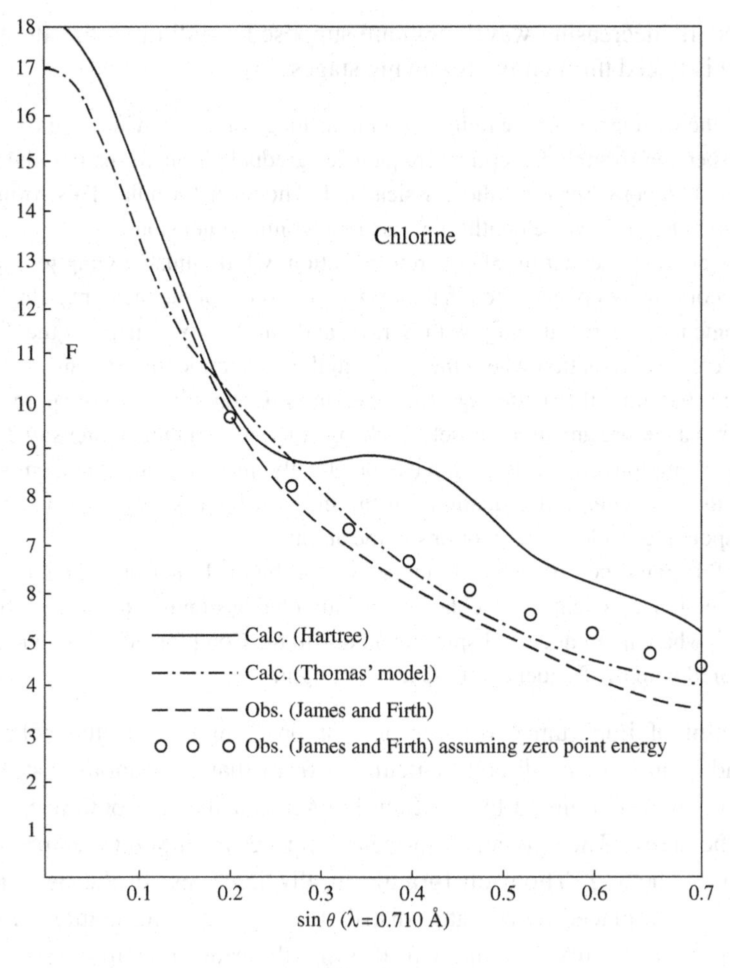

Fig. 8.

however, the Compton effect is taken into consideration, these outer electrons are found to give a very small contribution to the F curve at the large angles where the humps[51] occur, because they scatter so much modified radiation. The allowance for the Compton effect smooths out the hump, and leads to F curves much more like those observed. The third curve shows the F curve due to the continuous Thomas distribution and is a close fit to the observed curve.

I have quoted from a note by Dr Ivar Waller, in the following tentative summary of the interpretation which the new mechanics gives us of this phenomenon.[a] In a recent letter to *Nature* [30], Waller discusses the transition for the whole range from ordinary dispersion into Compton effect. His note only refers to scattering by a single electron, but it can probably be extended to many-electron atoms. Waves

[a] Space forbids a reference to the many theoretical papers which have contributed towards this interpretation.

of continually decreasing wavelength are supposed to fall upon the atom, and the transition is traced through the following stages.

a) While the wavelength of the radiation remains long compared with atomic dimensions, the dispersion formula for optical frequencies gradually transforms into the scattering for free electrons given by the classical J. J. Thomson formula. This formula holds approximately to[52] wavelengths approaching atomic dimensions.
b) At this point the scattering of coherent radiation will diminish, owing to interference, and become more concentrated in the forward direction of the incident light. This is the phenomenon we are studying, with X-rays, and our F curves map out the distribution of the coherent radiation where the wavelength is of atomic dimensions.
c) At the same time, the scattering of incoherent radiation will become appreciable, and approximate more and more closely in change of wavelength and intensity distribution to the Compton effect. It will have practically merged into the Compton effect when the momentum of a quantum of the incident light is large compared with that corresponding to electronic motions in the atom.
d) Up to this point the Thomson formula holds for the total intensity of light scattered in any direction, coherent and incoherent radiation being summed together. It first ceases to hold, when the frequency displacement due to the Compton effect is no longer small compared with the frequency of the incident light.

The point of importance for our present problem is that 'the coherent part of the radiation is to be directly calculated from that continuous distribution of electricity which is defined by the Schrödinger density-distribution in the initial state of the atom'. The classical treatment supposes each point electron to scatter according to the J. J. Thomson formula in all directions. In the new treatment, the electron is replaced by a spatial distribution of scattering matter, and so each electron has an 'F curve' of its own. It will still scatter coherent radiation in all directions, but its amount will fall away from that given by the classical formula owing to interference as $\frac{\sin\theta}{\lambda}$ increases, and this decline will be much more rapid for the more diffuse outer electrons than for the concentrated inner electrons. The total amount of radiation T scattered in any direction by the electron is given by the Thomson formula. A fraction f^2T will be coherent, and will be calculated by the laws of interference from the Schrödinger distribution, and the remainder, $(1-f^2)T$, will be incoherent. Thus the total coherent radiation will be F^2T where F is calculated from the Schrödinger distribution for the whole atom. An amount $(N-\sum f^2)T$ will be scattered with change of wavelength. Our measurements of X-ray diffraction, if this be true, can be trusted to measure the Schrödinger continuous distribution of electricity in the crystal lattice.

A very interesting point arises in the case where characteristic absorption frequencies of the scattering atom are of shorter wavelength than the radiation which is being scattered. In general, this has not been so when careful intensity measurements have been made since atoms of low atomic weight have alone been investigated. On the classical analogy, we would expect a reversal in phase of the scattered radiation, when an electron has a characteristic frequency greater than that of the incident light. A fascinating[53] experiment by Mark and Szilard [31] has shown that something very like this takes place. They investigated the (111) and (333) reflections of RbBr, which are extremely weak because Rb and Br oppose each other and are nearly equal in atomic number. They found that these 'forbidden' reflections were indeed absent when the soft Cu_K or hard Ba_K radiation was used, but that Sr_K radiation was appreciably reflected ($Sr_{K\alpha}$ λ 0.871 Å;[54] absorption edges of Rb_K and Br_K, 0.814 Å and 0.918 Å). The atoms are differentiated because a reversal of phase in scattering by the K electrons takes place in the one case and not in the other.

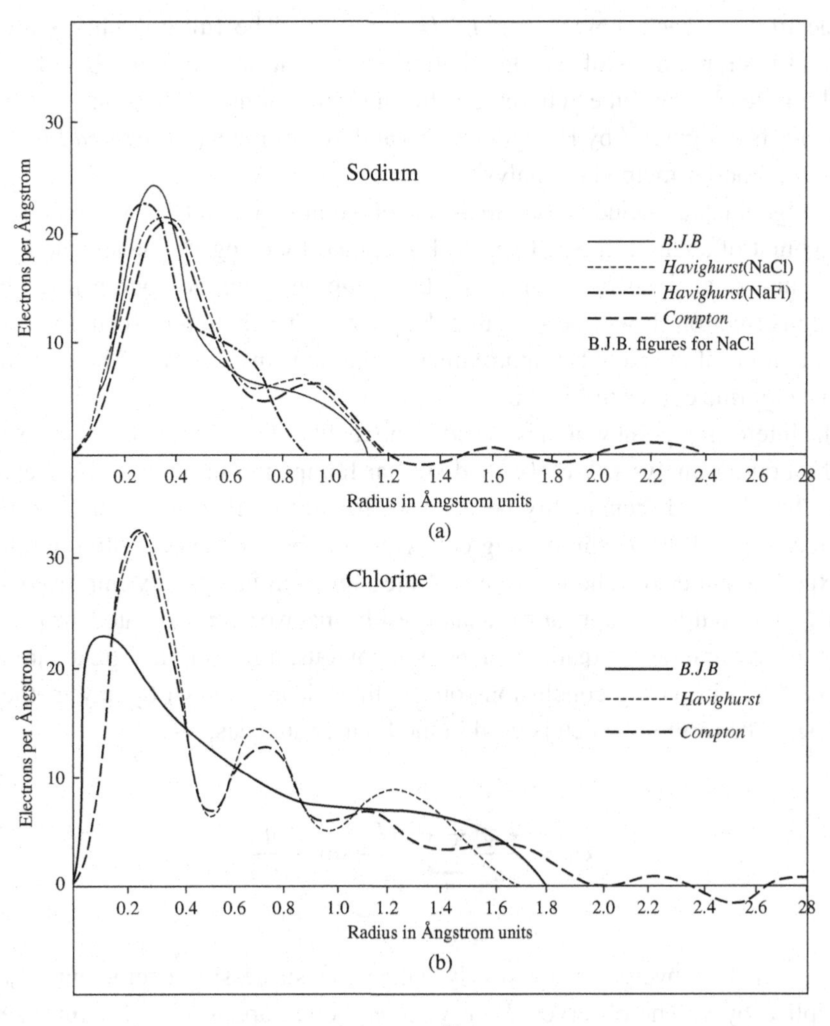

Fig. 9.[55]

7. – THE ANALYSIS OF ATOMIC STRUCTURE BY X-RAY INTENSITY MEASUREMENTS

It has been seen that the intensity measurements assign the correct number of electrons to each atom in a crystal, and indicate a spatial extension of the atoms of the right order. In attempting to make the further step of deducing the arrangement of the electrons in the atom, the limitations of the method begin to be very apparent.

In all cases where analysis has been attempted, the atom has been treated as spherically symmetrical. The analysis is used to determine the amount of scattering matter $U_n dr$ between radii r and $r+dr$. All methods of analysis give a distribution of the same general type. I have given, for instance, a series of analyses of sodium and chlorine in Fig. 9. In these figures, U_n is plotted as ordinate against r as abscissa. The total area of the curve in each case is equal to the number of electrons in each atom, since $\int_0^\infty U_n dr = N$. The full-line curves are our original interpretations of the distribution in sodium and chlorine, based on our 1921 figures.[56] The other curves are the interpretations of the same or closely similar sets of figures[57] by Havighurst [32] and by Compton (*X-rays and Electrons*) using the Fourier method of analysis.

In Fig. 9a are included our analysis of sodium in NaCl, two analyses by Havighurst of sodium in NaCl and NaF obtained by using Duane's triple Fourier series, and an analysis of our figures[58] by Compton using the Fourier formula for radial distribution. It will be seen that the general distribution of scattering matter and the limits of the atom are approximately the same in each case. The same holds for the chlorine curves in Fig. 9b.

The interesting point which is raised is the reality of the humps which are shown by the Fourier analysis. We obtained similar humps in our analysis by means of shells but doubted their reality because we found that if we smoothed them out and recalculated the F curve, it agreed with the observed curve within the limits of experimental error. The technique of measurement has greatly improved since then, and it would even appear from later results that we over-estimated the possible errors of our first determinations of F. It is obvious, however, that great care must still be taken in basing conclusions on the finer details shown by any method of analysis. The formula which is used in the Fourier analysis,

$$U_n = \frac{4\pi r}{a} \sum_1^\infty \frac{2nF_n}{a} \sin \frac{2\pi nr}{a},$$

is one which converges very slowly, since the successive coefficients F_n are multiplied by n. The observed F curve must be extrapolated to a point when F

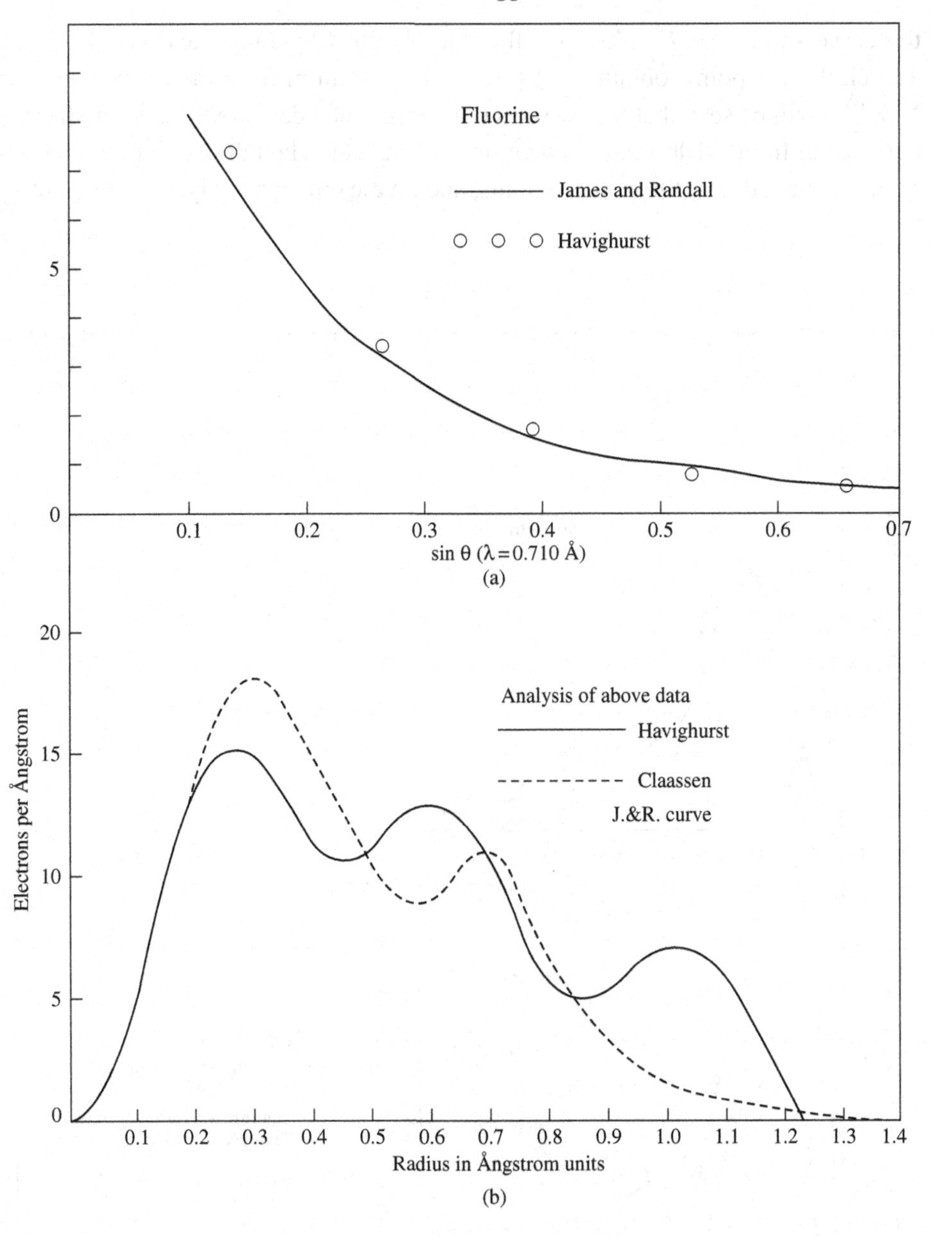

Fig. 10.

is supposed to fall to zero, and the precise form of the curve reacts very sensitively to the way in which this extrapolation is carried out.

The curves in Fig. 10 will illustrate the extent to which the analysis can be considered to give us information about the actual atomic distribution. In Fig. 10a the curve shows the F values for fluorine obtained by James and Randall [17]. The circles are points obtained[59] by Havighurst from measurements on CaF, LiF, NaF;[60] it will be seen that the two sets of experimental data are in very satisfactory agreement. In Fig. 10b I have shown on the one hand Havighurst's interpretations of the F curve drawn through his points, and on the other an analysis carried out by

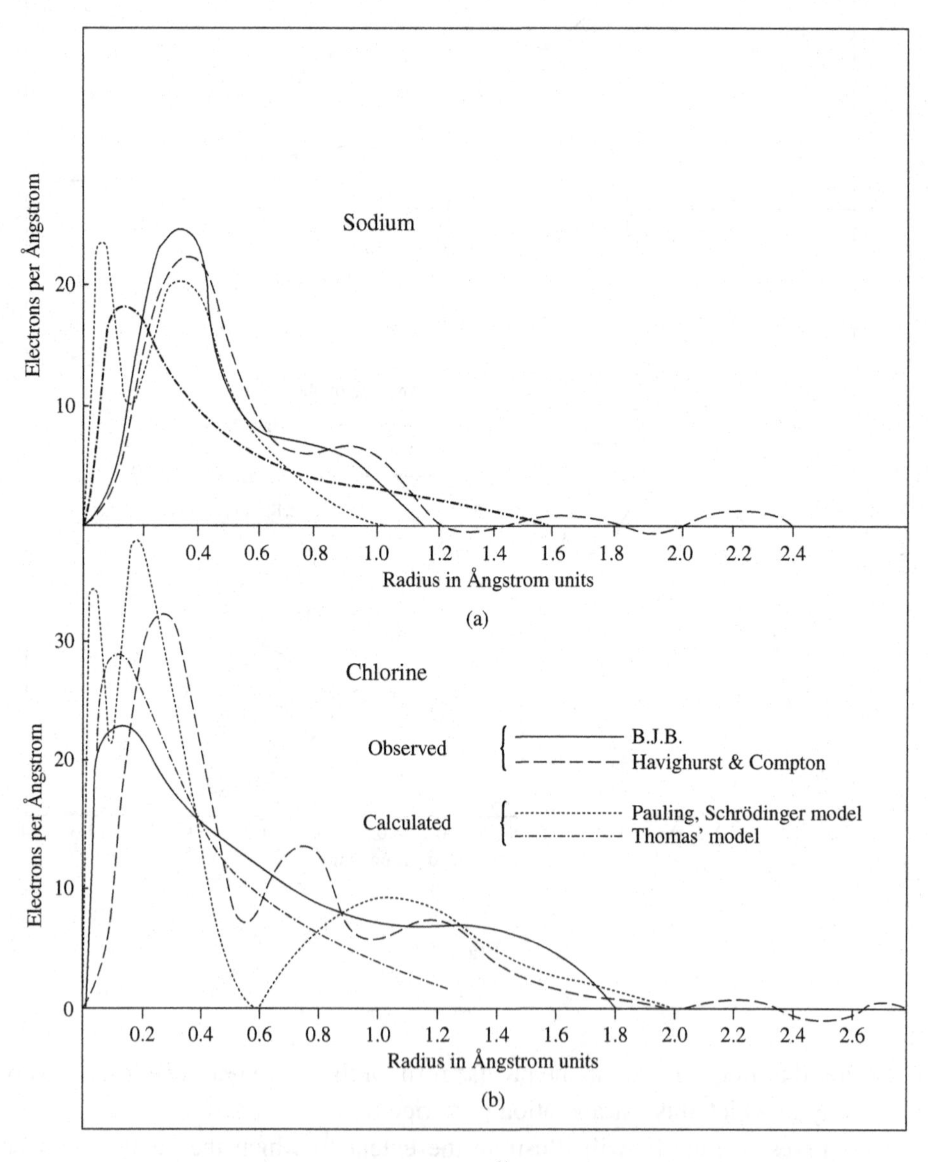

Fig. 11.[61]

Claassen [16] of James and Randall's using the Fourier method. The distributions are the same in their main outlines, but the peaks occur in quite different places.

Compton (*X-rays and Electrons*, p. 167) in discussing his diagrams of radial distribution has remarked that slight differences in the F curves lead to wide differences in details of the curves, and that too much confidence should not be placed on these details. Havighurst [32] discusses the significance of the analysis very fully in his paper on electron distribution in the atoms. Our data are not yet sufficiently accurate or extensive. Nevertheless, we are so near to attaining an accuracy of a satisfactory order, and the results of the analysis seem to indicate so clearly its fundamental correctness, that it appears to be well worth while to pursue enquiry further. Work with shorter wavelengths, and at low temperatures, when heat motion is small and a large range of F values can be measured, should yield us accurate pictures of the atomic structure itself. Given accurate data,[62] the Fourier method of analysis provides a direct way of utilising them.

The radial distribution of scattering power outlined in this way is in general agreement with any reasonable atomic model. We have seen, in particular, that the F curves, and therefore the radial distributions, of Thomas' model[63] are in approximate accord with those actually observed. If it is true that the scattering of coherent radiation is to be calculated in all cases by the Schrödinger density distribution, we should test our model against this distribution.

An interesting attempt along these lines has been recently made by Pauling [33]. He has used certain simplifying assumptions to obtain an approximate Schrödinger density-distribution for many-electron atoms. I have shown in Fig. 11 four sets of curves. The radial electron distributions deduced by Havighurst and by Compton are shown as one curve since they are very similar. The figure shows also our first analysis of electron distribution. Matched against these are plotted the generalised distribution of the Thomas model, and the Schrödinger density distribution calculated by Pauling.

We have obviously not yet reached a point when we can be satisfied with the agreement between theory and experiment, yet the success attained so far is a distinct encouragement to further investigation.

8. – The refraction of X-rays

At Professor Lorentz's[64] suggestion I have added a very brief note on the refraction of X-rays, since the phenomenon is so intimately connected with the question of intensity of reflection and scattering, and is another example of the successful application of classical laws. The diffraction phenomenon dealt with above (intensity of reflection) arises from the scattering of coherent radiation in all directions by the atoms of a crystal. The refractive index may be considered as being due to the scattering in the forward direction of coherent radiation, which interferes with the primary beam. The arrangement of the scattering matter plays no part, so that the body may be crystalline or amorphous. The measurement of the refractive index is thus a direct measure of the amount of coherent radiation scattered in the forward direction of the incident beam.

1. Darwin [2] appears to have been first in pointing out that theory assigns a refractive index for X-rays differing from unity by about one part in a million. He predicted that a very slight departure from the law of reflection

$$n\lambda = 2d \sin \theta_0$$

would be found, the actual angle θ being given by Darwin's formula

$$\theta - \theta_0 = \frac{1 - \mu}{\sin \theta \cos \theta}\,.$$

Ewald's [34] independent treatment of X-ray reflection leads to an equivalent result, though the problem is approached along quite different lines.

As is well known, the first experimental evidence of an index of refraction was found in a departure from the reflection laws. Stenström [35] observed differences in the apparent wavelength of soft X-rays (3 Å) as measured in the different orders, which were explained by Ewald's laws of X-ray reflection. The increased accuracy of X-ray spectroscopy has shown that similar deviations from the simple law of reflection exist for harder rays, though the deviations are much smaller than in the ordinary X-ray region.[65] Thus the deviations have been detected for hard rays by Duane and Patterson [36] and by Siegbahn and Hjalmar [37]. It is difficult to measure the refractive index by means of these deviations in the ordinary way, since they are so small, but Davis [38, 39] developed a very ingenious way of greatly increasing the effect. A crystal is ground so that the rays reflected by the atomic planes enter or leave a face at a very fine glancing angle, and thus suffer a comparatively great deflection.

Compton [40] discovered the total reflection of X-rays, and measured the index of refraction in this way. The refractive index is slightly less than unity, hence X-rays falling at a very fine glancing angle on a plane surface of a body are totally reflected, none of the radiation passing into the body. Compton showed that, although the refractive index is so nearly unity, yet the critical glancing angle is quite appreciable.

Finally, the direct effect of refraction by a prism has been observed by Larsson, Siegbahn and Waller [41]. X-rays entered one face of a glass prism at a very fine glancing angle, and suffered a measurable deflection. They obtained in this way a dispersion spectrum of X-rays.

2. In all cases where the frequency of the X-radiation is great compared with any frequency characteristic of the atom, the refractive index measured by any of these methods is in close accord[66] with the formula

$$1 - \mu = \frac{ne^2}{2\pi m\nu^2},$$

where n is the number of electrons per unit volume in the body, e and m are the electronic constants, and ν the frequency of the incident radiation. The formula follows directly from the classical Drude-Lorentz theory of dispersion, in the limiting case where the frequency of the radiation is large compared with the 'free periods' of the electrons in the atom. It can be put in the form [42][67]

$$1 - \mu = 2.71 \times 10^{-6} \frac{\rho Z}{A} \lambda^2,$$

where λ is the wavelength in Ångström units of the incident radiation, ρ the density of the substance, Z and A the average atomic number and atomic weight of its constituents (for all light atoms Z/A is very nearly 0.5).[68] Expressed in this form, the order of magnitude of $1 - \mu$ is easily grasped. The critical glancing angle θ for total reflection is given by

$$\cos\theta = \mu,$$

whence

$$\theta = \sqrt{\frac{ne^2}{\pi m\nu^2}}.$$

Expressing θ in minutes of arc, and λ in Ångström units as before,

$$\theta = 8.0\,\lambda\sqrt{\frac{\rho Z}{A}}.$$

Measurements of refractive index have been made by Compton and by Doan using the method of total reflection, by Davis, Hatley and Nardroff using reflection in a crystal, and by Larsson, Siegbahn and Waller with a prism. A variety of substances has been examined, and wavelengths between 0.5 and 2 Å have been used. The accuracy of the experimental determination of $1 - \mu$ is of the order of one to five per cent. As long as the critical frequencies of the atom have not been approached, the results have agreed with the above formula within experimental error. Just as in the measurements of intensity of reflection the F curves approach a limit at small angles equal to the number of electrons in the atom, so these measurements of refractive index when interpreted by classical theory lead to a very accurate numbering of the electrons in the scattering units.

3. A highly interesting field is opened up by the measurements of refractive index for wavelengths in the neighbourhood of a critical frequency of the atom. It is a striking fact that the simple dispersion formula

$$\mu - 1 = \frac{e^2}{2\pi m} \sum_1^n \frac{n_s}{\nu_s^2 - \nu^2}$$

still gives values for the refractive index agreeing with experiment in this region, except when the critical frequency is very closely approached indeed. Davis and von Nardroff reflected $\mathrm{Cu}_{\mathrm{K}\alpha}$ and $\mathrm{Cu}_{\mathrm{K}\beta}$ X-rays[69] from iron pyrites, and found that the refractive indices could be reproduced by substituting constants in the formulae corresponding to two K electrons in iron with the frequency of the K absorption edge.[70] R. L. Doan [44] has recently made a series of measurements by the total reflection method. His accurate data support the conclusion that the Drude-Lorentz theory of dispersion represents the facts, 'not only in regions remote from the absorption edge,[71] but also in some instances in which the radiation approaches the natural frequencies of certain groups of electrons'. The existence of two K electrons[72] is very definitely indicated. Kallmann and Mark [43] have gone more deeply into the form of the dispersion curve in the neighbourhood of the critical frequencies. The change in scattering power of an atom as the frequency of the scattered radiation passes through a critical value is of course another aspect of this anomalous dispersion; the experiment of Mark and Szilard which showed this effect has been described above. There is ample evidence that measurements of refractive index will in future prove to be a most fruitful means of investigating the response of the atom to incident radiation of frequency very near each of its own characteristic frequencies.

References[a]

[1] [W. Friedrich, P. Knipping and] M. v. Laue, *Bayr. Akad. d. Wiss. Math. phys. Kl.* (1912), 303.
[2] C. G. Darwin, *Phil. Mag.*, **27** (1914), 315, 675.
[3] P. P. Ewald, *Ann. d. Phys.*, **54** ([1917]), 519.
[4] P. Debye, *Ann. d. Phys.*, **43** (1914), 49.
[5] W. H. Bragg, *Phil. Mag.*, **27** (1914), 881.
[6] W. H. Bragg, *Phil. Trans. Roy. Soc.* [*A*], **215** (1915), 253.
[7] A. H. Compton, *Phys. Rev.*, **9** (1917), 29; **10** (1917), 95.
[8] W. L. Bragg, R. W. James and C. H. Bosanquet, *Phil. Mag.*, **41** (1921), 309; **42** (1921), 1; **44** (1922), 433.
[9] C. G. Darwin, *Phil. Mag.*, **43** (1922), 800.
[10] R. J. Havighurst, *Phys. Rev.*, **28** (1926), n. 5, 869, 882.
[11] L. Harris, S. J. Bates and D. A. MacInnes,[73] *Phys. Rev.*, **28** (1926), 235.
[12] J. A. Bearden, *Phys. Rev.*, **27** (1926), 796; **29** (1927), 20.
[13] Bergen Davis and W. M. Stempel, *Phys. Rev.*, [**17**] (1921), 608.
[14] H. Mark, *Naturwiss.*, **13** (1925), n. 49/50, 1042.
[15] L. H. Thomas, *Proc. Camb. Phil. Soc.*, **23** (1927), 542.
[16] A. Claassen, *Proc. Phys. Soc. London*, **38** [pt] 5 (1926), 482.
[17] R. W. James and J. T. Randall, *Phil. Mag.* [(7)], **1** (1926), 1202.
[18] J. A. Wasastjerna, *Comm. Fenn.*, **2** (1925), 15.
[19] W. L. Bragg, C. G. Darwin and R. W. James, *Phil. Mag.* (7), **1** (1926), 897.
[20] W. Duane, *Proc. Nat. Acad. Sci.*, **11** (1925), 489.
[21] R. J. Havighurst, *Proc. Nat. Acad. Sci.*, **11** (1925), 502.
[22] W. L. Bragg and J. West, *Roy. Soc. Proc. A*, **111** (1926), 691.
[23] [J. M.] Cork, *Phil. Mag.* [(7), **4** (1927), 688].
[24] I. Waller, *Upsala Univ. Årsskr. 1925*, [11]; *Ann. d. Phys.*, **83** (1927), 153.
[25] R. W. James and E. Firth, *Roy. Soc. Proc.* [*A*, **117** (1927), 62].
[26] [H.] Kallmann and H. Mark, *Zeit. f. Phys.*, **26** (1926), [n.] 2, [120].
[27] E. J. Williams, *Phil. Mag.* [(7)], **2** (1926), 657.
[28] [G. E. M.] Jauncey, *Phys. Rev.*, **29** (1927), 605.
[29] D. R. Hartree, *Phil. Mag.*, **50** (1925), 289.
[30] I. Waller, *Nature*, [**120**] (July 1927), [155].
[31] H. Mark and L. Szilard, *Zeit. f. Phys.*, **33** (1925), 688.
[32] R. J. Havighurst, *Phys. Rev.*, **29** (1927), 1.
[33] L. Pauling, *Roy. Soc. Proc. A*, **114** (1927), 181.
[34] P. P. Ewald, *Phys. Zeitsch.*, **21** (1920), 617; *Zeitschr. f. Physik*, **2** (1920), 332.
[35] W. Stenström, Exper[imentelle] Unters[uchungen] d[er] Röntgenspektra. Dissertation, Lund (1919).
[36] [W.] Duane and [R. A.] Patterson, *Phys. Rev.*, **16** (1920), [526].[74]
[37] M. Siegbahn, *Spektroskopie der Röntgenstrahlen* [(Berlin: Springer, 1924)].
[38] [C. C. Hatley and Bergen Davis], *Phys. Rev.*, **23** (1924), 290.[75]
[39] B[ergen] Davis and R. von Nardroff, *Phys. Rev.*, **23** (1924), 291.
[40] A. H. Compton, *Phil. Mag.*, **45** (1923), 1121.
[41] [A.] Larsson, [M.] Siegbahn and [I.] Waller, *Naturwiss.*, **12** ([1924]), 1212.
[42] I. Waller, [Theoretische Studien zur] Interferenz- und Dispersionstheorie der Röntgenstrahlen. [Dissertation, Upsala (1925)].

[a] The style of the references has been modernised and uniformised (*eds.*).

[43] H. Kallmann and H. Mark, *Ann. d. Physik*, **82** (1927), 585.
[44] R. L. Doan, *Phil. Mag.* [(7), **4** n.] 20 (1927), [100].

A very complete account of work on intensity of reflection is given by Compton in *X-rays and Electrons* and by Ewald in volume 24 of the *Handbuch der Physik* by H. Geiger and K. Scheel,[76] *Aufbau der festen Materie und seine Erforschung durch Röntgenstrahlen*, section 18.

Discussion of Mr Bragg's report

MR DEBYE. – To what extent can you conclude that there exists an energy at absolute zero?

MR BRAGG. – Waller and James have recently submitted a paper to the Royal Society in which they discuss the relation between the influence of temperature on the intensity of reflection (Debye effect) and the elastic constants of a crystal. Using the experimentally determined value of the Debye coefficient, they deduce the scattering by an atom at rest from the scattering by the atom at the temperature of liquid air (86° abs.). The curve deduced for the scattering by a perfectly motionless atom can of course take two forms, according to whether or not, in interpreting the results of the experiment, one assumes the existence of an energy at absolute zero.

If one assumes the existence of such an energy, the curve deduced from the experimental results agrees with that calculated by Hartree by applying Schrödinger's mechanics. The agreement is really very good for sodium as well as for chlorine. On the other hand, the curve that one obtains if one does not assume any energy at absolute zero deviates considerably from the calculated curve by an amount that exceeds the possible experimental error.

If these experimental results[a] are confirmed by new experiments, they provide a direct and convincing proof of the existence of an energy at absolute zero.

MR DEBYE. – Would the effect not be larger if one did the experiments with diamond?

MR BRAGG. – In the case of diamond, it is difficult to interpret the results obtained using a single crystal, because the structure is very perfect and the 'extinction' is strong. One would have to work with diamond powder. But I cannot say if it would be easy to find that there exists an energy at absolute zero in diamond; I should consider it further.

MR FOWLER. – Here is how Hartree calculates the atomic fields. Starting from Thomas' atomic field, taken as a first approximation, he calculates the Schrödinger functions for an electron placed in this field, then the density of charge in the atom corresponding to the Schrödinger functions, and then the corresponding atomic field, which will differ from that of Thomas. By successive approximations one modifies the field until the calculations yield the field which served as a starting

[a] *Note added 5 April 1928.* The results to which allusion is made here have just been published in detail by Messrs James, Waller and Hartree in a paper entitled: 'An investigation into the existence of zero-point energy in the rock-salt lattice by an X-ray diffraction method' (*Proc. Roy. Soc. A*, **118** (1928), 334).

point. This method gives very good values for the levels corresponding to X-rays and to visible light, and leads to the atom that Mr Bragg considered for comparison with experiments.

MR HEISENBERG. – How can you say that Hartree's method gives exact results, if it has not given any for the hydrogen atom? In the case of hydrogen the Schrödinger functions must be calculated with the aid of his differential equation, in which one introduces only the electric potential due to the nucleus. One would not obtain correct results if one added to this potential the one coming from a charge distribution by which one had replaced the electron. One may then obtain exact results only by taking the charge density of all the electrons, except the one whose motion one wishes to calculate. Hartree's method is certainly very useful and I have no objection to it, but it is essentially an approximation.

MR FOWLER. – I may add to what I have just said that Hartree is always careful to leave out the field of the electron itself in each state, so that, when he considers an L electron, for example, the central part of the field of the whole atom is diminished by the field of an L electron, as far as this may be considered as central. Hartree's method would then be entirely exact for hydrogen and in fact he has shown that it is extremely close to being exact for helium. (One finds a recent theoretical discussion of Hartree's method, by Gaunt, in *Proc. Cambr. Phil. Soc.*, **24** (1928), 328.)

MR PAULI. – In my opinion one must not perform the calculations, as in wave mechanics, by considering a density $|\psi(x, y, z)|^2$ in three-dimensional space,[77] but must consider a density in several dimensions

$$|\psi(x_1, y_1, z_1, \ldots, x_N, y_N, z_N)|^2,$$

which depends on the N particles in the atom. For sufficiently short waves the intensity of coherent scattered radiation is then proportional to[78]

$$\int \ldots \int \sum_1^N e^{\frac{2\pi i}{\lambda}(\overrightarrow{n_d}-\overrightarrow{n_u},\, \overrightarrow{r_k})} |\psi(x_1, \ldots, z_N)|^2 \, dx_1 \ldots dz_N,$$

where λ is the wavelength of the incident radiation, $\overrightarrow{n_u}$ a unit vector in the direction of propagation, and $\overrightarrow{n_d}$ the corresponding unit vector for the scattered radiation; the sum must be taken over all the particles. The result that one obtains by assuming a three-dimensional density cannot be rigorously exact; it can only be so to a certain degree of approximation.

MR LORENTZ. – How have you calculated the scattering of radiation by a charge distributed over a region comparable to the volume occupied by the atom?

MR BRAGG. – To interpret the results of observation as produced by an average distribution of the scattering material, we applied J. J. Thomson's classical formula for the amplitude of the wave scattered by a single electron.

MR COMPTON. – If we assume that there is always a constant ratio between the charge and mass of the electron, the result of the classical calculation of reflection by a crystal is exactly the same, whether the charge and mass are assumed concentrated in particles (electrons) or distributed irregularly in the atom. The intensity of reflection is determined by the average density of the electric charge in different parts of the atom. That may be represented either by the probability that a point charge occupies this region or by the volume density of an electric charge distributed in a continuous manner through this region.

MR KRAMERS. – The use that one may make of the simple Thomas model of the atom in the search for the laws of reflection is extremely interesting. It would perhaps not be superfluous to investigate what result would be obtained for the electron distribution if, instead of restricting oneself to considering a single centre of attraction, one applied Thomas' differential equation to an infinity of centres distributed as in a crystal grating. Has anyone already tried to solve the problem of which Mr Bragg has just spoken, of the calculation of the general distribution of the electronic density around the nucleus of a heavy atom, in the case where there are many nuclei, as in a crystal?

MR BRAGG. – No, no one has yet attacked this problem, which I only mentioned because it is interesting.

MR DIRAC. – Do the scattering curves depend on the phase relations between the oscillations of different atoms?

MR BRAGG. – No, because the results of our experiments give only the average scattering produced in each direction by a very large number of atoms.

MR DIRAC. – What would happen if you had two simple oscillators performing harmonic vibrations? Would they produce a different scattering when in phase than when out of phase?

MR BORN. – The correct answer to the question of scattering by an atom is contained in the remark by Mr Pauli. Strictly speaking there is no three-dimensional charge distribution that may describe exactly how an atom behaves; one always has to consider the total configuration of all the electrons in the space of $3n$ dimensions. A model in three dimensions only ever gives a more or less crude approximation.

MR KRAMERS asks a question concerning the influence of the Compton effect on the scattering.

MR BRAGG. – I have already said something on that subject in my report.[a] Assuming a model of the atom of the old type, Jauncey and Williams have used the criterion that the wavelength is modified when the recoil of the scattering electron is sufficient to take it entirely outside the atom. Williams was the first to apply this criterion to scattering curves obtained with crystals. He pointed out that while the speed of the recoil electron depends on both the scattering angle and the wavelength, any criterion one uses is a function of $\frac{\sin\theta}{\lambda}$, just as the interference effects depend on $\frac{\sin\theta}{\lambda}$. This implies that the existence of the Compton effect modifies the scattering curve such that we can always assign the same scattering curve to no matter what type of atom, whatever the wavelength may be.

MR FOWLER. – If I have understood properly, Mr Bragg uses theoretical calculations by Waller that have not yet been published. When light is scattered by an atom in accordance with the interpretation given by Mr Waller by means of the new mechanics, the *total* amount of scattered light is given exactly by J. J. Thomson's classical formula (except for very hard γ-rays). This light is composed of the coherent scattered radiation and of the modified light (Compton scattering). In the theorem of the reflection of X-rays only the coherent scattered light must be used, and indeed it is; and this light is given exactly by the F curves like those proposed by Hartree. These F curves for atomic scattering are obviously given simply by the classical scattering for each electron, diminished by interference.

MR BRAGG. – I should like to develop Mr Fowler's remark by recalling Waller and Wentzel's conclusions briefly sketched in my report. The scattering by one of the electrons in an atom partly remains the same and partly is modified. Within certain limits the total amount of scattered radiation is given by J. J. Thomson's formula. A fraction f^2 of this amount is not modified, f being a coefficient smaller

[a] Cf. Bragg's report, Section 6 (*eds.*).

than 1, depending on the interference of the spatial distribution of the charge according to Schrödinger and calculated according to the classical laws of optics. The remaining fraction $1 - f^2$ is modified.[79]

MR LORENTZ. – It is, without doubt, extremely noteworthy that the total scattering, composed of two parts of quite different origin, agrees with Thomson's formula.

MR KRAMERS makes two remarks:

1. As Mr Bragg has pointed out the importance of there being interest in having more experimental data concerning the refrangibility of X-rays in the neighbourhood of the absorption limit, I should like to draw attention to experiments performed recently by Mr Prins in the laboratory of Professor Coster at Groningen. By means of his apparatus (the details of the experiments and the results obtained are described in a paper published recently in *Zeitschrift für Physik*, **47** (1928), [479]), Mr Prins finds in a single test the angle of total reflection corresponding to an extended region of frequencies. In the region of the absorption limit of the metal, he finds an abnormal effect, which consists mainly of a strong decrease in the angle of total reflection on the side of the absorption limit located towards the short wavelengths. This effect is easily explained taking into account the influence of absorption on the total reflection, without it being necessary to enter into the question of the change in refrangibility of the X-rays. In fact, the absorption may be described by considering the refractive index n as a complex number, whose imaginary part is related in a simple manner to the absorption coefficient. Introducing this complex value for n in the well-known formulas of Fresnel for the intensity of reflected rays, one finds that the sharp limit of total reflection disappears, and that the manner in which the intensity of reflected rays depends on the angle of incidence is such that the experiment must give an 'effective angle of total reflection' that is smaller than in the case where there is no absorption and that decreases as the absorption increases.

According to the atomic theory one would also expect to find, in the region of the absorption limit, anomalies in the real part of the refractive index, producing a similar though less noticeable decrease of the effective angle of total reflection on the side of the absorption edge directed towards the large wavelengths. Mr Prins has not yet succeeded in showing that the experiments really demonstrate this effect.[a]

[a] Continuing his research Mr Prins has established (February 1928) the existence of this effect, in agreement with the theory.

The theory of these anomalies in the real part of the refractive index constitutes the subject of my second remark.

2. Let us consider plane and polarised electromagnetic waves, in which the electric force can be represented by the real part of $Ee^{2\pi i\nu t}$, striking an atom which for further simplicity we shall assume to be isotropic. The waves make the atom behave like an oscillating dipole, giving, by expansion in a Fourier series, a term with frequency ν. Let us represent this term by the real part of $Pe^{2\pi i\nu t}$, where P is a complex vector having the same direction as the vector E to which it is, moreover, proportional. If we set

$$\frac{P}{E} = f + ig \,, \tag{1}$$

where f and g are real functions of ν, the real and imaginary parts of the refractive index of a sample of matter are related in a simple way to the functions f and g of the atoms contained in the sample.

Extending the domain of values that ν may take into the negative region and defining f as an even function of ν, g as an odd function, one easily verifies that the dispersion formulas of Lorentz's classical theory and also those of modern quantum mechanics are equivalent to the formula

$$f(\nu) = \frac{1}{\pi} \fint_{-\infty}^{+\infty} \frac{g(\nu')}{\nu - \nu'} d\nu', \tag{2}$$

where the sign $\fint$ indicates the 'principal' value of the integral.

This formula can easily be applied to atoms showing continuous absorption regions and is equivalent to the formulas proposed for these cases by R. de Laer Kronig and by Mark and Kallmann. There is hardly any doubt that this general formula may be derived from quantum mechanics, if one duly takes into account the absorption of radiation, basing oneself on Dirac's theory, for example.

From a mathematical point of view, formula (2) gives us the means to construct an analytic function of a complex variable ν that is holomorphic below the real axis and whose real part takes the values $g(\nu')$ on this axis. If one considers ν as a real variable, the integral equation (2) has the solution

$$g(\nu) = -\frac{1}{\pi} \fint_{-\infty}^{+\infty} \frac{f(\nu')}{\nu - \nu'} d\nu', \tag{3}$$

which shows that the imaginary part of the refractive index depends on the real part in nearly the same way as the real part depends on the imaginary part. The fact that the analytic function f of the complex variable ν, defined by (2) for the lower half of the complex plane, has no singularity in this half-plane, means that dispersion phenomena, when one studies them by means of waves whose amplitude grows in an exponential manner (ν complex), can never give rise to singular behaviour for the atoms.

MR COMPTON. – The measurements of refractive indices of X-rays made by Doan agree better with the Drude-Lorentz formula than with the expression derived by Kronig based on the quantum theory of dispersion.

MR DE BROGLIE. – I should like to draw attention to recent experiments carried out by Messrs J. Thibaud and A. Soltan,[a] which touch on the questions raised by Mr Bragg. In these experiments Messrs Thibaud and Soltan measured, by the tangent grating method, the wavelength of a certain number of X-rays in the domain 20 to 70 Å. Some of these wavelengths had already been determined by Mr Dauvillier using diffraction by fatty-acid gratings. Now, comparing the results of Dauvillier with those of Thibaud and Soltan, one notices that there is a systematic discrepancy between them that increases with wavelength. Thus for the K_α line of boron, Thibaud and Soltan find 68 Å, while Dauvillier had found 73.5 Å, that is, a difference of 5.5 Å. This systematic discrepancy appears to be due to the increase of the refractive index with wavelength. The index does not actually play a role in the tangent grating method, while it distorts in a systematic way the results obtained by crystalline diffraction when one uses the Bragg formula. Starting from the difference between their results and those of Mr Dauvillier, Messrs Thibaud and Soltan have calculated the value of the refractive index of fatty acids around 70 Å and found

$$\delta = 1 - \mu = 10^{-2}$$

thereabouts. This agrees well with a law of the form $\delta = K\lambda^2$; since in the ordinary X-ray domain the wavelengths are about 100 times smaller, δ is of order 10^{-6}. One could object that, according to the Drude-Lorentz law, the presence of K discontinuities of oxygen, nitrogen and carbon between 30 and 45 Å should perturb the law in λ^2. But in the X-ray domain the validity of Drude's law is doubtful, and if one uses in its place the formula proposed by Kallmann and Mark[b] the agreement with the experimental results is very good. Let us note finally that the existence of an index appreciably different from 1 can contribute to explaining why large-wavelength lines, obtained with a fatty-acid grating, are broad and spread out.

MR LORENTZ makes a remark concerning the refractive index of a crystal for Röntgen rays and the deviations from the Bragg law. It is clear that, according

[a] *C. R. Acad. Sc.*, **185** (1927), 642.
[b] *Ann. d. Phys.*, **82** (1927), 585.

to the classical theory, the index must be less than unity, because the electrons contained in the atoms have eigenfrequencies smaller than the frequency of the rays, which gives rise to a speed of propagation greater than c. But in order to speak of this speed, one must adopt the macroscopic point of view, abstracting away the molecular discontinuity. Now, if one wishes to explain Laue's phenomenon in all its details, one must consider, for example, the action of the vibrations excited in the particles of a crystallographic layer on a particle of a neighbouring layer. This gives rise to series that one cannot replace by integrals. It is for this reason that I found some difficulty in the explanation of deviations from the Bragg law.[80]

MR DEBYE. – Ewald has tried to do similar calculations.

MR LORENTZ. – It is very interesting to note that with Röntgen rays one finds, in the vicinity of an absorption edge, phenomena similar to those that in classical optics are produced close to an absorption band. There is, however, a profound difference between the two cases, the absorption edge not corresponding to a frequency that really exists in the particles.

Notes on the translation

1 Here and in a few other places, the French adds (or omits) inverted commas.
2 [réflexion des radiations des raies]
3 [ne fait prévoir aucun écart]
4 The French edition adds 'en couches' [in layers].
5 Typescript: 'have often been . . . it has led'; French version: 'a souvent été . . . elles ont conduit'.
6 [ont eu confiance dans les méthodes quantitatives]
7 [portions]
8 [sous une forme plus primitive]
9 [il soit possible de les trouver]
10 [logiquement]
11 [thermodynamique]
12 Word omitted in the French version.
13 The French edition adds 'Sir'.
14 [les données obtenues par réflexion]
15 Here following the French edition; the typescript reads 'total radiation'.
16 Emphasis omitted in the French edition.
17 [servirent à comparer]
18 The French omits 'of reflection'.
19 [déjà observée]
20 [il proposa d'employer, pour l'interprétation des mesures, la méthode de Fourier]
21 [rapport]
22 [ou]
23 [Mc Innes]
24 [potassium]
25 The typescript has a spurious comma after 'Bergen'.
26 [à l'aide de cristaux]
27 [les deux constantes électroniques]
28 Here and in some other instances, the French renders 'single' as 'simple'.
29 ['facteur atomique']
30 Not printed as a quotation in the French edition.
31 choisies simplement
32 The French adds 'de ses expériences'.
33 The French adds 'dans ces conditions'.
34 This is indeed a quotation from p. 272 from the lecture by W. H. Bragg. The typescript has a comma instead of the closing quotation mark, while the French edition omits the opening quotation mark. The typescript has a spurious denominator 'a' instead of 'd' in the second and third terms (but tacitly corrects another typo in the original).
35 [dans son livre sur 'les rayons X et les électrons']
36 The French adds 'on peut procéder de la façon inverse, c'est-à dire'.
37 'Beryl.' omitted in French edition.
38 The French edition uses superscripts throughout.
39 The French edition omits the overbar in the caption.
40 [Wyckhoff]
41 Word omitted in French edition.
42 [frappante]
43 Again, the French edition omits the overbar in the caption.
44 [son livre sur les rayons X et les électrons]
45 [bien que les figures (*sic*) sur lesquelles la nouvelle courbe se base soient plus exactes]
46 Here and in several other places, the French adds 'Mlle'.
47 The French reads 'une énergie au zéro absolu (énergie de structure)'.
48 Word missing in the French edition.
49 Bracket printed as a footnote in the French edition.

50 [tout près de ce point]
51 [irrégularités]
52 [pour]
53 [brillante]
54 [λ $\mathrm{Sr}_{\mathrm{K}\alpha} = 0.871$Å]
55 The French omits ‘B. J. B. figures for NaCl’.
56 [faites d’après 1921 figures]
57 [figures]
58 Again, in the French, the false friend ‘figures’.
59 [déduits]
60 [CaFl, LiFl, NaFl]
61 The French edition omits ‘& Compton’ and has ‘Modèle de Pauling et Schrödinger’.
62 [Une fois que nous disposerons de données précises]
63 The French translates as if the comma were after ‘of Thomas’ model’ rather than before.
64 [M. Lorentz]
65 The typescript reads ‘much smaller in the ordinary X-ray region’, but given the context the text should be amended as shown (as also done in the French version).
66 [parfaitement d’accord]
67 Reference omitted in the French edition.
68 [$\frac{Z}{A}$ la valeur moyenne du rapport du nombre atomique au poids atomique pour ses divers constituants (pour tous les atomes légers ce rapport est à peu près égal à 0.5]
69 [rayons]
70 Typescript: ‘of the K adsorption edge’; French version: ‘de la discontinuité K’.
71 Typescript: ‘adsorption edge’; French: ‘bord d’absorption’.
72 The French adds ‘dans la pyrite’.
73 French edition: ‘Mac Innes’.
74 Typescript and French edition both have ‘532’.
75 Both typescript and French edition give this reference as ‘B. Davis and C. C. Hatley’. The typescript has ‘291’.
76 Authors added in the French edition.
77 Here and in the following displayed formula, the published version has square brackets instead of absolute bars.
78 Arrow missing on $\vec{r_k}$ in the published volume.
79 The original text mistakenly states that both fractions are ‘not modified’.
80 The mixing of first and third person, here and in a few similar instances throughout the discussions, is as in the published text.

Disagreements between experiment and the electromagnetic theory of radiation[a]

BY MR ARTHUR H. COMPTON

INTRODUCTION

Professor W. L. Bragg has just discussed a whole series of radiation phenomena in which the electromagnetic theory is confirmed. He has even dwelt on some of the limiting cases, such as the reflection of X-rays by crystals, in which the electromagnetic theory of radiation gives us, at least approximately, a correct interpretation of the facts, although there are reasons to doubt that its predictions are truly exact. I have been left the task of pleading the opposing cause to that of the electromagnetic theory of radiation, seen from the experimental viewpoint.

I have to declare from the outset that in playing this role of the accuser I have no intention of diminishing the importance of the electromagnetic theory as applied to a great variety of problems.[b] It is, however, only by acquainting ourselves with the real or apparent[1] failures of this powerful theory that we can hope to develop a more complete theory of radiation which will describe the facts as we know them.

The more serious difficulties which present themselves in connection with the theory that radiation consists of electromagnetic waves, propagated through space

[a] An English version of this report (Compton 1928) was published in the *Journal of the Franklin Institute*. The French version appears to be essentially a translation of the English paper with some additions. Whenever there are no discrepancies, we reproduce Compton's own English (we have corrected some obvious typos and harmonised some of the spelling). Interesting variants are footnoted. Other discrepancies between the two versions are reported in the endnotes (*eds.*).

[b] The opening has been translated from the French edition. The English version has the following different opening (*eds.*):

During the last few years it has become increasingly evident that the classical electromagnetic theory of radiation is incapable of accounting for certain large classes of phenomena, especially those concerned with the interaction between radiation and matter. It is not that we question the wave character of light – the striking successes of this conception in explaining polarisation and interference of light can leave no doubt that radiation has the characteristics of waves; but it is equally true that certain other properties of radiation are not easily interpreted in terms of waves. The power of the electromagnetic theory as applied to a great variety of problems of radiation is too well known to require emphasis.

in accord with the demands of Maxwell's equations, may be classified conveniently under five heads:[2]

(1) Is there an ether? If there are oscillations, there must be a medium in which these oscillations are produced. Assuming the existence of such a medium, however, one encounters great difficulties.
(2) How are the waves produced? The classical electrodynamics requires as a source of an electromagnetic wave an oscillator of the same frequency as that of the waves it radiates. Our studies of spectra,[3] however, make it appear impossible that an atom should contain oscillators of the same frequencies as the emitted rays.
(3) The photoelectric effect. This phenomenon is wholly anomalous when viewed from the standpoint of waves.
(4) The scattering of X-rays, and the recoil electrons, phenomena in which we find gradually increasing departures from the predictions of the classical wave theory as the frequency increases.
(5) Experiments on individual interactions between quanta of radiation and electrons. If the results of the experiments of this type are reliable, they seem to show definitely that individual quanta of radiation, of energy $h\nu$, proceed in definite directions.

The photon hypothesis.[4] – In order to exhibit more clearly the difficulties with the classical theory of radiation, it will be helpful to keep in mind the suggestion that[5] light consists of corpuscles. We need not think of these two views as necessarily alternative. It may well be that the two conceptions are complementary. Perhaps the corpuscle is related to the wave in somewhat the same manner that the molecule is related to matter in bulk; or there may be a guiding wave which directs the corpuscles which carry the energy. In any case, the phenomena which we have just mentioned suggest the hypothesis that radiation is divisible into units possessing energy $h\nu$, and which proceed in definite directions with momentum $h\nu/c$. This is obviously similar to Newton's old conception of light corpuscles. It was revived in its present form by Professor Einstein,[6] it was defended under the name of the 'Neutron Theory' by Sir William [H.] Bragg, and has been given new life by the recent discoveries associated with the scattering of X-rays.

In referring to this unit of radiation I shall use the name 'photon', suggested recently by G. N. Lewis.[a] This word avoids any implication regarding the nature of the unit, as contained for example in the name 'needle ray'. As compared with the terms 'radiation quantum' and 'light quant',[7] this name has the advantages of brevity and of avoiding any implied dependence upon the much more general quantum mechanics or quantum theory of atomic structure.

[a] G. N. Lewis, *Nature*, [**118**], [874] (Dec. 18, 1926).

Virtual radiation. – Another conception of the nature of radiation which it will be desirable to compare with the experiments is Bohr, Kramers and Slater's important theory of virtual radiation.[a] According to this theory, an atom in an excited state is continually emitting virtual radiation, to which no energy characteristics are to be ascribed. The normal atoms have associated with them virtual oscillators, of the frequencies corresponding to jumps of the atom to all of the stationary states of higher energy. The virtual radiation may be thought of as being absorbed by these virtual oscillators, and any atom which has a virtual oscillator absorbing this virtual radiation has a certain probability of jumping suddenly to the higher state of energy corresponding to the frequency of the particular virtual oscillator. On the average, if the radiation is completely absorbed, the number of such jumps to levels of higher energy is equal to the number of emitting atoms which pass from higher to lower states. But there is no direct connection between the falling of one atom from a higher to a lower state and a corresponding rise of a second atom from a lower to a higher state. Thus on this view the energy of the emitting atoms and of the absorbing atoms is only statistically conserved.

THE PROBLEM OF THE ETHER[8]

The constancy of the speed of radiation of different wavelengths has long been considered as one of the most powerful arguments in favour of the wave theory of light. This constancy suggests that a perturbation is travelling through a fixed medium in space, the ether.

If experiments like those by Michelson and Morley's were to show the existence of a relative motion with respect to such a medium, this argument would be considerably strengthened. For then we could imagine light as having a speed determined with reference to a fixed axis in space. But, except for the recent and quite doubtful experiments by Miller,[b] no-one has ever detected such a relative motion. We thus find ourselves in the difficult position of having to imagine a medium in which perturbations travel with a definite speed, not with reference to a fixed system of axes, but with reference to each individual observer, whatever his motion. If we think of the complex properties a medium must have in order to transmit a perturbation in this way, we find that the medium differs so considerably from the simple ether from which we started that the analogy between a wave in such a medium and a pertubation travelling in an elastic medium is very distant. It is true that doubts have often been expressed as to the usefulness of retaining the notion of the ether. Nevertheless, if light is truly a wave motion, in the sense of

[a] N. Bohr, H. A. Kramers and J. C. Slater, *Phil. Mag.*, **47** (1924), 785; *Zeits. f. Phys.*, **24** (1924), 69.
[b] D. C. Miller, *Nat. Acad. Sci. Proc.*, **11** (1925), 306.

Maxwell, there must be a medium in order to transmit this motion, without which the notion of wave would have no meaning. This means that, instead of being a support for the wave theory, the concept of the ether has become an uncomfortable burden of which the wave theory has been unable to rid itself.

If, on the other hand, we accept the view suggested by the theory of relativity, in which for the motion of matter or energy there is a limiting speed relative to the observer, it is not surprising to find a form of energy that moves at this limiting speed. If we abandon the idea of an ether, it is simpler to suppose that this energy moves in the form of corpuscles rather than waves.

THE EMISSION OF RADIATION

When we trace a sound to its origin, we find it coming from an oscillator vibrating with the frequency of the sound itself. The same is true of electric waves, such as radio waves, where the source of the radiation is a stream of electrons oscillating back and forth in a wire. But when we trace a light ray or an X-ray back to its origin, we fail to find any oscillator which has the same frequency as the ray itself. The more complete our knowledge becomes of the origin of spectrum lines, the more clearly we see that if we are to assign any frequencies to the electrons within the atoms, these frequencies are not the frequencies of the emitted rays, but are the frequencies associated with the stationary states of the atom. This result cannot be reconciled with the electromagnetic theory of radiation, nor has any mechanism been suggested whereby radiation of one frequency can be excited by an oscillator of another frequency. The wave theory of radiation is thus powerless to suggest how the waves originate.

The origin of the radiation is considerably simpler when we consider it from the photon viewpoint. We find that an atom changes from a stationary state of one energy to a state of less energy, and associated with this change radiation is emitted. What is simpler than to suppose that the energy lost by the atom is radiated away as a single photon? It is on this view unnecessary to say anything regarding the frequency of the radiation. We are concerned only with the energy of the photon, its direction of emission, and its state of polarisation.

The problem of the emission of radiation takes an especially interesting form when we consider the production of the continuous X-ray spectrum.[a] Experiment shows that both the intensity and the average frequency of the X-rays emitted at angles less than 90 degrees with the cathode-ray stream are greater than at angles greater than 90 degrees. This is just what we should expect due to the Doppler effect if the X-rays are emitted by a radiator moving in the direction of the cathode rays.

[a] The difficulty here discussed was first emphasised by D. L. Webster, *Phys. Rev.*, **13** (1919), 303.

In order to account for the observed dissymmetry between the rays in the forward and backward directions, the particles emitting the radiation must be moving with a speed of the order of 25 per cent that of light. This means that the emitting particles must be free electrons, since it would require an impossibly large energy to set an atom into motion with such a speed.

But it will be recalled that the continuous X-ray spectrum has a sharp upper limit. Such a sharp limit is, however, possible on the wave theory only in case the rays come in trains of waves of considerable length, so that the interference between the waves in different parts of the train can be complete at small glancing angles of reflection from the crystal. This implies that the oscillator which emits the rays must vibrate back and forth with constant frequency a large number of times while the ray is being emitted. Such an oscillation might be imagined for an electron within an atom; but it is impossible for an electron moving through an irregular assemblage of atoms with a speed comparable with that of light.

Thus the Doppler effect in the primary X-rays demands that the rays shall be emitted by rapidly moving electrons, while the sharp limit to the continuous spectrum requires that the rays be emitted by an electron bound within an atom.

The only possible escape from this dilemma on the wave theory is to suppose that the electron is itself capable of internal oscillation of such a character as to emit radiation. This would, however, introduce an undesirable complexity into our conception of the electron, and would ascribe the continuous X-rays to an origin entirely different from that of other known sources of radiation.

Here again the photon theory affords a simple solution. It is a consequence of Ehrenfest's adiabatic principle[a] that photons emitted by a moving radiator will show the same Doppler effect, with regard to both frequency and intensity, as does a beam of waves.[b] But if we suppose that photons are radiated by the moving cathode electrons, the energy of each photon will be the energy lost by the electron, and the limit of the X-ray spectrum is necessarily reached when the energy of the photon is equal to the initial energy of the electron, i.e., $h\nu = eV$. In this case, if we consider the initial state as an electron approaching an atom with large kinetic energy and the final state as the electron leaving the atom with a smaller kinetic energy, we see that the emission of the continuous X-ray spectrum is the same kind of event as the emission of any other type of radiation.

[a] The adiabatic principle consists in the following. Since for a quantised quantity there should be no quantum jumps induced by an infinitely slowly varying external force (in this case, one that gently accelerates a radiator), there is an analogy between these quantities and the classical adiabatic invariants. Ehrenfest (1917) accordingly formulated a principle identifying the classical quantities to be quantised as the adiabatic invariants of a system (*eds.*).

[b] Cf., e.g., A. H. Compton, *Phys. Rev.*, **21** (1923), 483.

Absorption of radiation. – According to the photon theory, absorption occurs when a photon meets an atom and imparts its energy to the atom. The atom is thereby raised to a stationary state of higher energy – precisely the reverse of the emission process.

On the wave theory, absorption is necessarily a continuous process, if we admit the conservation of energy, since on no part of the wave front is there enough energy available to change the atom suddenly from a state of low energy to a state of higher energy. What evidence we have is, however, strongly against the atom having for any considerable length of time an energy intermediate between two stationary states; and if such intermediate states cannot exist, the gradual absorption of radiation is not possible. Thus the absorption of energy from waves[9] is irreconcilable with the conception of stationary states.

We have seen that on the theory of virtual radiation the energy of the emitting atoms and of the absorbing atoms is only statistically conserved. There is according to this view therefore no difficulty with supposing that the absorbing atom suddenly jumps to a higher level of energy, even though it has not received from the radiation as much energy as is necessary to make the jump. It is thus possible through virtual oscillators and virtual radiation to reconcile the wave theory of radiation with the sudden absorption of energy, and hence to retain the idea of stationary states.

THE PHOTOELECTRIC EFFECT

It is well known that the photon hypothesis was introduced by Einstein to account for the photoelectric effect.[a] The assumption that light consists of discrete units which can be absorbed by atoms only as units, each giving rise to a photoelectron, accounted at once for the fact that the number of photoelectrons is proportional to the intensity of the light; and the assumption that the energy of the light unit is equal to $h\nu$, where h is Planck's constant, made it possible to predict the kinetic energy with which the photoelectrons should be ejected, as expressed by Einstein's well-known photoelectric equation,

$$mc^2\left(\frac{1}{\sqrt{1-\beta^2}}-1\right)=h\nu-w_p. \tag{1}$$

Seven years elapsed before experiments by Richardson and Compton[b] and by Hughes[c] showed that the energy of the emitted electrons was indeed proportional

[a] A. Einstein, *Ann. d. Phys.*, **17** (1905), [132].[10]
[b] O. W. Richardson and K. T. Compton, *Phil. Mag.*, **24** (1912), 575.
[c] A. L. Hughes, *Phil. Trans. A*, **212** (1912), 205.[11]

to the frequency less a constant,[12] and that the factor of proportionality was close to the value of h calculated from Planck's radiation formula. Millikan's more recent precision photoelectric experiments with the alkali metals[a] confirmed the identity of the constant h in the photoelectric equation with that in Planck's radiation formula. De Broglie's beautiful experiments[b] with the magnetic spectrograph showed that in the region of X-ray frequencies the same equation holds, if only we interpret the work function w_p as the work required to remove the electron from the pth energy level of the atom. Thibaud has made use of this result[c] in comparing the velocities of the photoelectrons ejected by γ-rays from different elements, and has thus shown that the photoelectric equation (1) holds with precision even for β-rays of the highest speed. Thus from light of frequency so low that it is barely able to eject photoelectrons from metals to γ-rays that eject photoelectrons with a speed almost as great as that of light, the photon theory expresses accurately the speed of the photoelectrons.

The direction in which the photoelectrons are emitted is no less instructive than is the velocity. Experiments using the cloud expansion method, performed[13] by C. T. R. Wilson[d] and others,[e] have shown that the most probable direction in which the photoelectron is ejected from an atom is nearly the direction of the electric vector of the incident wave, but with an appreciable forward component to its motion. There is, however, a very considerable variation in the direction of emission. For example, if we plot the number of photoelectrons ejected at different angles with the primary beam we find, according to Auger, the distribution shown in Fig. 1.

Each of these curves, taken at a different potential, represents the distribution of about 200 photoelectron tracks. It will be seen that as the potential on the X-ray tube increases, the average forward component of the photoelectron's motion also increases.

When polarised X-rays are used, there is a strong preponderance of the photoelectrons in or near the plane including the electric vector of the incident rays. Thus Fig. 2 shows the distribution found by Bubb of the direction of the photoelectrons ejected from moist air when traversed by X-rays that have been polarised by scattering at right angles from a block of paraffin. Because of multiple scattering in the paraffin, the scattered rays are not completely polarised, and this is probably sufficient to account for the fact that some photoelectrons appear to start

[a] R. A. Millikan, *Phys. Rev.*, **7** (1916), 355.
[b] M. de Broglie, *Jour. de Phys.*, **2** (1921), 265.
[c] J. Thibaud, *C. R.*, **179** (1924), 165, 1053 and 1322.
[d] C. T. R. Wilson, *Proc. Roy. Soc. A*, **104** (1923), 1.
[e] A. H. Compton, *Bull. Natl. Res. Coun.*, No. 20 (1922), 25; F. W. Bubb, *Phys. Rev.*, **23** (1924), 137; P. Auger, *C. R.*, **178** (1924), 1535; D. H. Loughridge, *Phys. Rev.*, **26** (1925), 697; F. Kirchner, *Zeits. f. Phys.*, **27** (1926), 385.

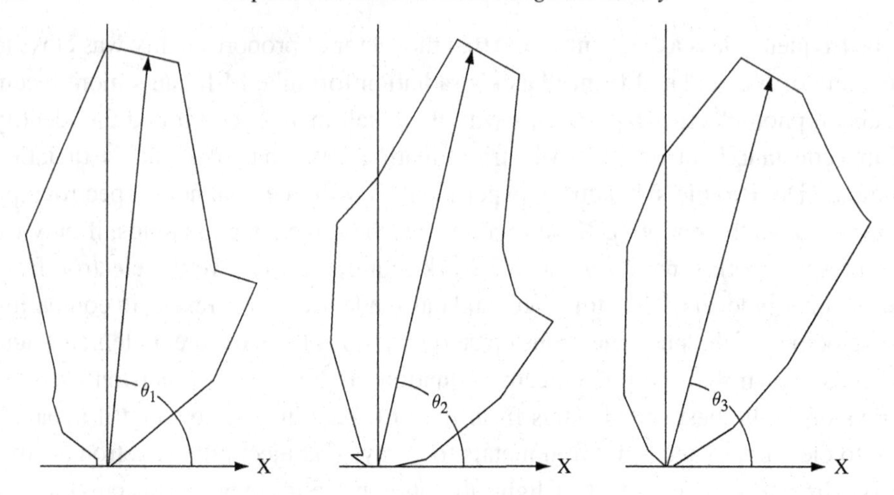

Fig. 1. Longitudinal distribution of photoelectrons for X-rays of three different effective wavelengths, according to Auger.

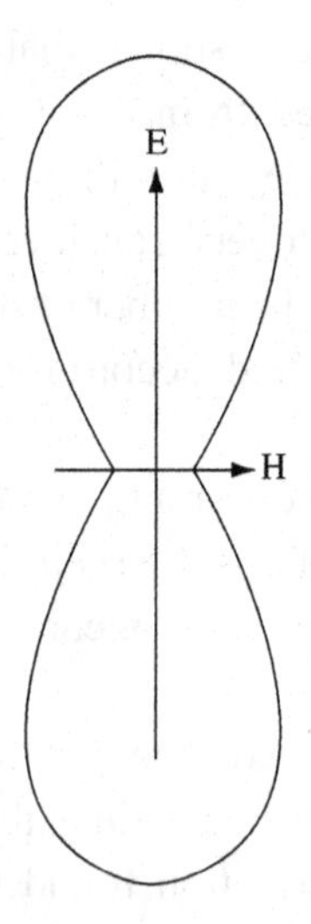

Fig. 2. Lateral distribution of photoelectrons for incompletely polarised X-rays, according to Bubb.

at right angles with the electric vector. This effect with X-rays is doubtless similar in character to the selective photoelectric effect discovered many years ago by Pohl and Pringsheim, in which the number of electrons ejected by light from the liquid surface of sodium-potassium alloy is greater when the electric vector is in a plane perpendicular to the surface than when parallel to the surface.

Recent experiments have shown that the direction in which the photoelectrons are ejected by X-rays is at least very nearly independent of the material from which the electrons come.[a]

Can electromagnetic waves produce photoelectrons? – Before discussing the production of photoelectrons from the standpoint of radiation quanta, let us see what success meets the attempt[15] to account for them on the basis of electromagnetic waves. The fact that they are emitted approximately in the direction of the electric vector would suggest that the photoelectrons are ejected by the direct action of the electric field of the incident rays. If this were the case, however, we should expect the speed of the ejected electrons to be greater for greater intensity of radiation, whereas experiment shows that for the same wavelength intense sunlight ejects an electron no faster than does the feeble light from a star. Furthermore, the energy available from the electromagnetic wave is wholly inadequate. Thus in a recent experiment performed by Joffe and Dobronrawov,[b] X-rays were produced by the impact on a target of 10^4 to 10^5 electrons per second. Since on the electromagnetic theory an X-ray pulse is of the order of 10^3 waves in length or 10^{-16} seconds in duration, the X-ray pulses must have followed each other at widely separated intervals. It was found, however, that photoelectrons were occasionally ejected from a bismuth particle which subtended a solid angle not greater than 10^{-5}. It is clearly impossible that all the energy of an X-ray pulse which has spread out in a spherical wave should spend itself on this bismuth particle. Thus on the wave theory the ejection of the photoelectron, which has almost as much energy as the original cathode electron, could not have been accomplished by a single[16] pulse. It cannot therefore be the direct action of the electric vector of the wave, taken in the usual sense,[17] which has ejected the electron.

We may assume, on the other hand, that the energy is gradually absorbed in the bismuth particle of Joffe's experiment until an amount $h\nu$ has accumulated, which is then spent in ejecting the photoelectron. We have already called attention to the fact that this gradual absorption hypothesis implies the existence of stationary states in the atom having infinitesimal gradations of energy, whereas the evidence is very strong that atoms cannot endure except in certain definitely defined stationary states. But new difficulties also arise. Why do the photoelectrons tend to start in the direction of the electric field of the incident wave? If we suppose that it is the gradual absorption of energy from a wave which liberates the electron, why does there exist a tendency for the electron to start with a large component of its motion in a forward direction?[18] The forward impulse due to the radiation pressure

[a] E. A. Owen, *Proc. Phys. Soc.*, **30** (1918), 133; Auger, Kirchner, Loughridge, *loc. cit.*[14]
[b] A. Joffe and N. Dobronrawov, *Zeits. f. Phys.*, **34** (1925), 889.

as[19] the energy is gradually absorbed will be transferred to the atom and not left with [the] absorbing electron. The accumulation hypothesis is thus difficult to defend.

Photons and photoelectrons. – On the photon theory it is possible to account in a simple manner for most of the properties of the photoelectrons. We have seen how Einstein was able to predict accurately the velocity of the photoelectrons, assuming only that energy is conserved when a photon acts on an electron. In order to account for the direction of emission we must ascribe to the photon some of the properties of an electromagnetic pulse. Bubb introduced the suggestion[a] that we ascribe to the photon a vector property similar to the electric vector of an electromagnetic wave, so that when the photon traverses an atom the electrons and the nucleus receive impulses in opposite directions perpendicular to the direction of propagation. Associated with this electric vector, we should also expect to find a magnetic vector. Thus if an electron is set in motion by the electric vector of the photon at right angles to the direction of propagation, the magnetic vector of the photon will act on the moving electron in the direction of propagation. This is strictly analogous to the radiation pressure exerted by an electromagnetic wave on an electron which it traverses, and means that the forward momentum of the absorbed photon is transferred to the photoelectron.

In the simplest case, where we neglect the initial momentum of the electron in its orbital motion in the atom, the angle between the direction of the incident ray and the direction of ejection is found from these assumptions to be

$$\theta = \tan^{-1}\sqrt{2/\alpha}, \tag{2}$$

where $\alpha = \gamma/\lambda$, and $\gamma = h/mc = 0.0242$ Å. The quantity α is small compared with unity, except for very hard X-rays and γ-rays. Thus for light, equation (2) predicts the expulsion of photoelectrons at nearly 90 degrees. This is in accord with the rather uncertain data which have been obtained with visible and ultra-violet light.[b]

The only really significant test of this result is in its application to X-ray photoelectrons. In Fig. 1 are drawn the lines θ_1, θ_2 and θ_3 for the three curves, at the angles calculated by Auger from equation (2). It will be seen that they fall very satisfactorily in the direction of maximum emission of the photoelectrons. Similar

[a] F. W. Bubb, *Phys. Rev.*, **23** (1924), 137.
[b] Cf. A. Partsch and W. Hallwachs, *Ann. d. Phys.*, **41** (1913), 247.

results have been obtained by other investigators.[a] This may be taken as proof that a photon imparts not only its energy, but also its momentum to the photoelectrons.[b]

Honesty[c] obliges me to point out a difficulty that arises in this explanation of the motion of the photoelectrons. It is the failure of the attempts made to account properly for the fact that the photoelectrons are emitted over a wide range of angles instead of in a definite direction, as would be suggested by the calculation just outlined. The most interesting of these attempts is that of Bubb,[d] who takes into account the momentum of the electron immediately before the absorption of the photon. Bubb finds a dispersion of the directions of emission of the photoelectrons of the correct order of magnitude, but which is larger when the electron issues from a heavy atom than when it issues from a light one. We have seen, however, that experiment has shown this dispersion of the directions of emission to be notably independent of the element from which the photoelectron originates.

Whatever may be the cause of the dispersion in the directions of motion of the photoelectrons,[21] it will readily be seen that if the time during which the photon exerts a force on the electron is comparable with the natural period of the electron[22] in the atom, the impulse imparted to the electron will be transferred in part to the positive nucleus about which the electron is moving. The fact that the photoelectrons are ejected with a forward component equal, within the limits of experimental error, to the momentum of the incident photon[23] means that no appreciable part of the photon's momentum is spent on the remainder of the atom.

[a] W. Bothe, *Zeits. f. Phys.*, **26** (1925), 59; F. Kirchner, *Zeits. f. Phys.*, **27** (1926), 385.[20]

[b] The English version includes here the following footnote. Cf. also the comments by Bragg on p. 325 and the ensuing discussion (*eds.*).

Since this was written, experiments by Loughridge ([D. H. Loughridge], *Phys. Rev.*, **30** (1927), [488]) have been published which show a forward component to the photoelectron's motion which seems to be greater than that predicted by equation (2). Williams, in experiments as yet unpublished, finds that the forward component is almost twice as great as that predicted by this theory. These results indicate that the mechanism of interaction between the photon and the atom must be more complex than here postulated. The fact that the forward momentum of the photoelectron is found to be of the same order of magnitude as that of the incident photon, however, suggests that the momentum of the photon is acquired by the photoelectron, while an additional forward impulse is imparted by the atom. Thus these more recent experiments also support the view that the photoelectron acquires both the energy and the momentum of the photon.

[c] This paragraph is present only in the French edition. The corresponding one in the English edition reads:

If the angular momentum of the atomic system from which the photoelectron is ejected is to be conserved when acted upon by the radiation, the electron cannot be ejected exactly in the direction of θ, but must receive an impulse in a direction determined by the position of the electron in the atom at the instant it is traversed by the photon.* Thus we should probably consider the electric vector of the X-ray wave as defining merely the most probable direction in which the impulse should be imparted to the electron. This is doubtless the chief reason why the photoelectrons are emitted over a wide range of angles instead of in a definite direction, as would be suggested by the calculation just outlined.

With the footnote: *Cf. A. H. Compton, *Phys. Rev.*, [**31**] (1928), [59] (*eds.*).

[d] F. W. Bubb, *Phil. Mag.*, **49** (1925), 824.

This can only be the case if the time of action of the photon on the electron is short compared with the time of revolution of the electron in its orbit.[a]

The photoelectric effect and virtual radiation. – It is to be noted that none of these properties of the photoelectron is inconsistent with the virtual radiation theory of Bohr, Kramers and Slater. The difficulties which applied to the classical wave theory do not apply here, since the energy and momentum are conserved only statistically. There is nothing in this theory, however, which would enable us to predict anything regarding the motion of the photoelectrons. The degree of success that has attended the application of the photon hypothesis to the motion of these electrons has come directly from the application of the conservation principles to the individual action of a photon on an electron. The power of these principles as applied to this case is surprising if the assumption is correct that they are only statistically valid.

PHENOMENA ASSOCIATED WITH THE SCATTERING OF X-RAYS

As is now well known, there is a group of phenomena associated with the scattering of X-rays for which the classical wave theory of radiation fails to account. These phenomena may be considered under the heads of: (1) The change of wavelength of X-rays due to scattering, (2) the intensity of scattered X-rays, and (3) the recoil electrons.

The earliest experiments on secondary X-rays and γ-rays[24] showed a difference in the penetrating power of the primary and the secondary rays. In the case of X-rays, Barkla and his collaborators[b] showed that the secondary rays from the heavy elements consisted largely of fluorescent radiations characteristic of the radiator, and that it was the presence of these softer rays which was chiefly responsible for the greater absorption of the secondary rays. When later experiments[c] showed a measurable difference in penetration even for light elements such as carbon, from which no fluorescent K or L radiation appears, it was natural to ascribe[d] this difference to a new type of fluorescent radiation, similar to the K and L types, but of shorter wavelength. Careful absorption measurements[e] failed, however, to reveal any critical absorption limit for these assumed 'J' radiations similar to those corresponding to the K and L radiations. Moreover,

[a] The English edition includes the further sentence: 'Such a short duration of interaction is a natural consequence of the photon conception of radiation, but is quite contrary to the consequences of the electromagnetic theory' (*eds.*).

[b] C. [G.] Barkla and C. A. Sadler, *Phil. Mag.*, **16**, 550 (1908).[25]

[c] C. A. Sadler and P. Mesham, *Phil. Mag.*, **24** (1912), 138; J. Laub, *Ann. d. Phys.*, **46** (1915), 785.

[d] [C. G.] Barkla and [M. P.] White, *Phil. Mag.*, **34** (1917), 270; J. Laub, *Ann. d. Phys.*, **46** (1915), 785, *et al.*

[e] E.g., [F. K.] Richtmyer and [K.] Grant, *Phys. Rev.*, **15** (1920), 547.

direct spectroscopic observations[a] failed to reveal the existence of any spectrum lines[26] under conditions for which the supposed J-rays should appear. It thus became evident that the softening of the secondary X-rays from the lighter elements was due to a different kind of process than the softening of the secondary rays from heavy elements where fluorescent X-rays are present.

A series of skilfully devised absorption experiments performed by J. A. Gray[b] showed, on the other hand, that both in the case of γ-rays and in that of X-rays an increase in wavelength accompanies the scattering of the rays of light elements.

It was at this stage that the first spectroscopic investigations of the secondary X-rays from light elements were made.[c] According to the usual electron theory of scattering it is obvious that the scattered rays will be of the same frequency as the forced oscillations of the electrons which emit them, and hence will be identical in frequency with the primary waves which set the electrons in motion. Instead of showing scattered rays of the same wavelength as the primary rays, however, these spectra revealed lines in the secondary rays corresponding to those in the primary beam, but with each line displaced slightly toward the longer wavelengths.

This result might have been predicted from Gray's absorption measurements; but the spectrum measurements had the advantage of affording a quantitative measurement of the change in wavelength, which gave a basis for its theoretical interpretation.

The spectroscopic experiments which have shown this change in wavelength are too well known[d] to require discussion. The interpretation of the wavelength change in terms of photons being deflected by individual[27] electrons and imparting a part of their energy to the scattering electrons is also very familiar. For purposes of discussion, however, let us recall that when we consider the interaction of a single photon with a single electron the principles of the conservation of energy and momentum lead us[e] to the result that the change in wavelength of the deflected photon is

$$\delta\lambda = \frac{h}{mc}(1 - \cos\varphi), \tag{3}$$

where φ is the angle through which the photon is deflected. The electron at the same time recoils from the photon at an angle of θ given by,[28]

$$\cot\theta = -(1+\alpha)\tan\frac{1}{2}\varphi\ ; \tag{4}$$

[a] E.g., [W.] Duane and [T.] Shimizu, *Phys. Rev.*, **13** (1919), [289]; *ibid.*, **14** (1919), 389.
[b] J. A. Gray, *Phil. Mag.*, **26** (1913), 611; *Jour. Frank. Inst.*, [**190**], 643 (Nov. 1920).
[c] A. H. Compton, *Bull. Natl. Res. Coun.*, No. 20 ([October] 1922), [18]; *Phys. Rev.*, **22** (1923), 409.
[d] Cf., e.g., A. H. Compton, *Phys. Rev.*, **22** (1923), 409; P. A. Ross, *Proc. Nat. Acad.*, **10** (1924), 304.
[e] A. H. Compton, *Phys. Rev.*, [**21**] (1923), 483; P. Debye, *Phys. Zeits.*, **24** (1923), 161.

and the kinetic energy of the recoiling electron is,

$$E_{\text{kin}} = h\nu \frac{2\alpha \cos^2 \theta}{(1+\alpha)^2 - \alpha^2 \cos^2 \theta}. \tag{5}$$

The experiments show in the spectrum of the scattered rays two lines corresponding to each line of the primary ray. One of these lines is of precisely the same wavelength as the primary ray, and the second line, though somewhat broadened, has its centre of gravity displaced by the amount predicted by equation (3). According to experiments by Kallmann and Mark[a] and by Sharp,[b] this agreement between the theoretical[29] and the observed shift is precise within a small fraction of 1 per cent.

The recoil electrons. – From the quantitative agreement between the theoretical and the observed wavelengths of the scattered rays, the recoil electrons predicted by the photon theory of scattering were looked for with some confidence.[30] When this theory was proposed, there was no direct evidence for the existence of such electrons, though indirect evidence suggested that the secondary β-rays ejected from matter by hard γ-rays are mostly of this type. Within a few months of their prediction, however, C. T. R. Wilson[c] and W. Bothe[d] independently announced their discovery. The recoil electrons show as short tracks, pointed in the direction of the primary X-ray beam, mixed among the much longer tracks due to the photoelectrons ejected by the X-rays.

Perhaps the most convincing reason for associating these short tracks with the scattered X-rays comes from a study of their number. Each photoelectron in a cloud photograph represents a quantum of truly absorbed X-ray energy. If the short tracks are due to recoil electrons, each one should represent the scattering of a photon. Thus the ratio N_r/N_p of the number of short tracks to the number of long tracks should be the same as the ratio σ/τ of the scattered to the truly absorbed energy[31] when the X-rays pass through air. The latter ratio is known from absorption measurements, and the former ratio can be determined by counting the tracks on the photographs. The satisfactory agreement between the two ratios[e] for X-rays of different wavelengths means that on the average there is about one quantum of energy scattered for each short track that is produced.

This result is in itself contrary to the predictions of the classical wave theory, since on this basis all the energy spent on a free electron (except the

[a] H. Kallmann and H. Mark, *Naturwiss.*, **13** (1925), 297.
[b] H. M. Sharp, *Phys. Rev.*, **26** (1925), 691.
[c] C. T. R. Wilson, *Proc. Roy. Soc.* [A], **104** (1923), 1.
[d] W. Bothe, *Zeits. f. Phys.*, **16** (1923), 319.
[e] A. H. Compton and A. W. Simon, *Phys. Rev.*, **25** (1925), 306; J. M. Nuttall and E. J. Williams, *Manchester Memoirs*, **70** (1926), 1.

insignificant effect of radiation pressure) should reappear as scattered X-rays. In these experiments, on the contrary, 5 or 10 per cent as much energy appears in the motion of the recoil electrons as appears in the scattered X-rays.

That these short tracks associated with the scattered X-rays correspond to the recoil electrons predicted by the photon theory of scattering becomes clear from a study of their energies. The energy of the electron which produces a track can be calculated from the range of the track. The ranges of tracks which start in different directions have been studied[a] using primary X-rays of different wavelengths, with the result that equation (5)[33] has been satisfactorily verified.

In view of the fact that electrons of this type were unknown at the time the photon theory of scattering was presented, their existence, and the close agreement with the predictions as to their number, direction and velocity, supply strong evidence in favour of the fundamental hypotheses of the theory.

Interpretation of these experiments. – It is impossible to account for scattered rays of altered frequency, and for the existence of the recoil electrons, if we assume that X-rays consist of electromagnetic waves in the usual sense. Yet some progress has been made on the basis of semi-classical theories. It is an interesting fact that the wavelength of the scattered ray according to equation (3)[34] varies with the angle just as one would expect from a Doppler effect if the rays are scattered from an electron moving in the direction of the primary beam. Moreover, the velocity that must be assigned to the electron in order to give the proper magnitude to the change of wavelength is that which the electron would acquire by radiation pressure if it should absorb a quantum of the incident rays. Several writers[b] have therefore assumed that an electron takes from the incident beam a whole quantum of the incident radiation, and then emits this energy as a spherical wave while moving forward[36] with high velocity.

This conception that the radiation occurs in spherical waves, and that the scattering electron can nevertheless acquire suddenly the impulses from a whole quantum of incident radiation is inconsistent with the principle of energy conservation. But there is the more serious experimental difficulty that this theory predicts recoil electrons all moving in the same direction and with the same velocity. The experiments show, on the other hand, a variety of directions and velocities, with the velocity and direction correlated as demanded by the photon hypothesis. Moreover, the maximum range of the recoil electrons, though in

[a] Compton and Simon, *loc. cit.*[32]

[b] C. R. Bauer, *C. R.*, **177** (1923), 1211; C. T. R. Wilson, *Proc. Roy. Soc.* [*A*], **104** (1923), 1; K. Fosterling, *Phys. Zeits.*, **25** (1924), 313; O. Halpern, *Zeits. f. Phys.*, **30** (1924), 153.[35]

agreement with the predictions of the photon theory, is found to be about four times as great as that predicted by the semi-classical theory.

There is nothing in these experiments, as far as we have described them, which is inconsistent with the idea of virtual oscillators continually scattering virtual radiation. In order to account for the change of wavelength on this view, Bohr, Kramers and Slater assumed that the virtual oscillators scatter as if moving in the direction of the primary beam, accounting for the change of wavelength as a Doppler effect. They then supposed that occasionally an electron, under the stimulation of the primary virtual rays, will suddenly move forward with a momentum large compared with the impulse received from the radiation pressure. Though we have seen that not all of the recoil electrons move directly forward, but in a variety of different directions, the theory could easily be extended to include the type of motion that is actually observed.

The only objection that one can raise against this virtual radiation theory in connection with the scattering phenomena as viewed on a large scale, is that it is difficult to see how such a theory could by itself predict the change of wavelength and the motion of the recoil electrons. These phenomena are directly predictable if the conservation of energy and momentum are assumed to apply to the individual actions of radiation on electrons; but this is precisely where the virtual radiation theory denies the validity of the conservation principles.

We may conclude that the photon theory predicts quantitatively and in detail the change of wavelength of the scattered X-rays and the characteristics of the recoil electrons. The virtual radiation theory is probably not inconsistent with these experiments, but is incapable of predicting the results. The classical theory, however, is altogether helpless to deal with these phenomena.

The origin of the unmodified line. – The unmodified line is probably due to X-rays which are scattered by electrons so firmly held within the atom that they are not ejected by the impulse from the deflected photons. This view is adequate to account for the major characteristics of the unmodified rays, though as yet no quantitatively satisfactory theory of their origin has been published.[a] It is probable that a detailed account of these rays will involve definite assumptions regarding the nature and the duration of the interaction between a photon and an electron; but it is doubtful whether such investigations will add new evidence as to the existence of the photons themselves.

[a] Cf., however, G. E. M. Jauncey, *Phys. Rev.*, **25** (1925), 314 and *ibid.*, 723; G. Wentzel, *Zeits. f. Phys.*, **43** (1927), 14, 779; I. Waller, *Nature*, [**120**, 155] (July 30, 1927).[37] [The footnote in the English edition continues with the sentence: 'It is possible that the theories of the latter authors may be satisfactory, but they have not yet been stated in a form suitable for quantitative test' (*eds.*).]

A similar situation holds regarding the intensity of the scattered X-rays. Historically it was the fact that the classical electromagnetic theory is unable to account for the low intensity of the scattered X-rays which called attention to the importance of the problem of scattering. But the solutions which have been offered by Breit,[a] Dirac[b] and others[c] of this intensity problem as distinguished from that of the change of wavelength, seem to introduce no new concepts regarding the nature of radiation or of the scattering process. Let us therefore turn our attention to the experiments that have been performed on the individual process of interaction between photons and electrons.

INTERACTIONS BETWEEN RADIATION AND SINGLE ELECTRONS[39]

The most significant of the experiments which show departures from the predictions of the classical wave theory are those that study the action of radiation on individual atoms or on individual electrons. Two methods have been found suitable for performing these experiments, Geiger's point counters, and Wilson's cloud expansion photographs.

(1) *Test for coincidences with fluorescent X-rays.* – Bothe has performed an experiment[d] in which fluorescent K radiation from a thin copper foil is excited by a beam of incident X-rays. The emitted rays are so feeble that only about five quanta of energy are radiated per second. Two point counters are mounted, one on either side of the copper foil in each of which an average of one photoelectron is produced and recorded for about twenty quanta radiated by the foil. If we assume that the fluorescent radiation is emitted in quanta of energy, but proceed[s] in spherical waves in all directions, there should thus be about 1 chance in 20 that the recording of a photoelectron in one chamber should be simultaneous with the recording of a photoelectron in the other.

The experiments showed no coincidences other than those which were explicable by such sources as high-speed β-particles which traverse both counting chambers.

This result is in accord with the photon hypothesis,[e] according to which coincidences should not occur. It is, nevertheless, equally in accord with the virtual radiation hypothesis, if one assumes that the virtual oscillators in the copper

[a] G. Breit, *Phys. Rev.*, **27** (1926), 242.

[b] P. A. M. Dirac, *Proc. Roy. Soc. A*, [**111**] (1926), [405].

[c] W. Gordon, *Zeits. f. Phys.*, **40** (1926), 117; E. Schrödinger, *Ann. d. Phys.*, **82** (1927), 257; O. Klein, *Zeits. f. Phys.*, **41** (1927), 407; G. Wentzel, *Zeits. f. Phys.*, **43** (1927), 1, 779.[38]

[d] W. Bothe, *Zeits. f. Phys.*, **37** (1926), 547.

[e] The English edition continues: 'For if a photon of fluorescent radiation produces a β-ray in one counting chamber it cannot traverse the second chamber. Coincidences should therefore not occur' (*eds.*).

continuously emit virtual fluorescent radiation, so that the photoelectrons should be observed in the counting chambers at arbitrary intervals.[a]

But the experiment is important in the sense that it refutes the often suggested idea that a quantum of radiation energy is suddenly emitted in the form of a spherical wave when an atom passes from one stationary state to another.

(2) *The composite photoelectric effect.*[40] – Wilson[b] and Auger[c] have noticed in their cloud expansion photographs that when X-rays eject photoelectrons from heavy atoms, it often occurs that two or more electrons are ejected simultaneously from the same atom. Auger has deduced from studying the ranges of these electrons that, when this occurs, the total energy of all the emitted electrons is no larger than that of a quantum of the incident radiation. When two electrons are emitted simultaneously it is usually the case that the energy of one of them is

$$E_{\text{kin}} = h\nu - h\nu_K,$$

which according to the photon theory means that this electron is due to the absorption of an incident photon accompanied by the ejection of an electron from the K energy level. The second electron has in general the energy

$$E_{\text{kin}} = h\nu_K - h\nu_L.$$

This electron can be explained as the result of the absorption by an L electron of the Kα-ray emitted when another L electron occupies the place left vacant in the K orbit by the primary photoelectron. It is established that all the electrons that are observed in the composite photoelectric effect have to be interpreted in the same way. Their interpretation according to the photon theory thus meets with no difficulties.

With regard to the virtual radiation theory, we can take two points of view: first, under the influence of the excitation produced by the primary virtual radiation, virtual fluorescent K radiation is emitted by virtual oscillators associated with all the atoms traversed by the primary beam. In this view, the probability that this

[a] At this point in the English version Compton is much more critical of the BKS theory (*eds.*):

According to the virtual radiation hypothesis, however, coincidences should have been observed. For on this view the fluorescent K radiation is emitted by virtual oscillators associated with atoms in which there is a vacancy in the K shell. That is, the copper foil can emit fluorescent K radiation only during the short interval of time following the expulsion of a photoelectron from the K shell, until the shell is again occupied by another electron. This time interval is so short (of the order of 10^{-15} sec.) as to be sensibly instantaneous on the scale of Bothe's experiments. Since on this view the virtual fluorescent radiation is emitted in spherical waves, the counting chambers on both sides of the foil should be simultaneously affected, and coincident pulses in the two chambers should frequently occur. The results of the experiment are thus contrary to the predictions of the virtual radiation hypothesis.

[b] C. T. R. Wilson, *Proc. Roy. Soc. A*, **104** (1923), 1.

[c] P. Auger, *Journ. d. Phys.*, **6** (1926), 183.

virtual fluorescent radiation will cause the ejection of a photoelectron from the same atom as the one that has emitted the primary photoelectron is so small that such an event will almost never occur; second, we can alternatively assume that a virtual oscillator emitting virtual K radiation is associated only with an atom in which there is a vacant place in the K shell. In this case, since the virtual radiation proceeds from the atom that has emitted the primary photoelectron, we could expect with extremely large probability that it should excite a photoelectron from the L shell of its own atom, thus accounting for the composite photoelectric effect. But in this view the virtual fluorescent radiation is emitted only during a very short interval after the ejection of the primary photoelectron, in which case Bothe's fluorescence experiment, described above, should have shown some coincidences.

One sees thus that the virtual radiation hypothesis is irreconcilable both with the composite photoelectric effect and with the absence of coincidences in Bothe's fluorescence experiment. The photon hypothesis, instead, is in complete accord with both these experimental facts.

(3) *Bothe and Geiger's coincidence experiments.*[41] – We have seen that according to Bohr, Kramers and Slater's theory, virtual radiation[42] is being continually scattered by matter traversed by X-rays, but only occasionally is a recoil electron emitted. This is in sharp contrast with the photon theory, according to which a recoil electron appears every time a photon is scattered. A crucial test between the two points of view is afforded by an experiment devised and brilliantly performed[43] by Bothe and Geiger.[a] X-rays were passed through hydrogen gas, and the resulting recoil electrons and scattered rays were detected by means of two different point counters placed on opposite sides of the column of gas. The chamber for counting the recoil electrons was left open, but a sheet of thin platinum prevented the recoil electrons from entering the chamber for counting the scattered rays. Of course not every photon entering the second counter could be noticed, for its detection depends upon the production of a β-ray. It was found that there were about ten recoil electrons for every scattered photon that recorded itself.

The impulses from the counting chambers were recorded on a moving photographic film. In observations over a total period of over five hours, sixty-six such coincidences were observed. Bothe and Geiger calculate that according to the statistics of the virtual radiation theory the chance was only 1 in 400 000 that so many coincidences should have occurred. This result therefore is in accord with the predictions of the photon theory, but is directly contrary to the statistical view of the scattering process.

[a] W. Bothe and H. Geiger, *Zeits. f. Phys.*, **26** (1924), 44; **32** (1925), 639.

(4) *Directional emission of scattered X-rays.* – Additional information regarding the nature of scattered X-rays has been obtained by studying the relation between the direction of ejection of the recoil electron and the direction in which the associated photon proceeds. According to the photon theory, we have a definite relation (equation (4)) between the angle at which the photon is scattered and the angle at which the recoil electron is ejected. But according to any form of spreading wave theory, including that of Bohr, Kramers and Slater, the scattered rays may produce effects in any direction whatever, and there should be no correlation between the directions in which the recoil electrons proceed and the directions in which the secondary β-rays are ejected by the scattered X-rays.

A test to see whether such a relation exists has been made,[a] using Wilson's cloud apparatus, in the manner shown diagrammatically in Fig. 3. Each recoil electron

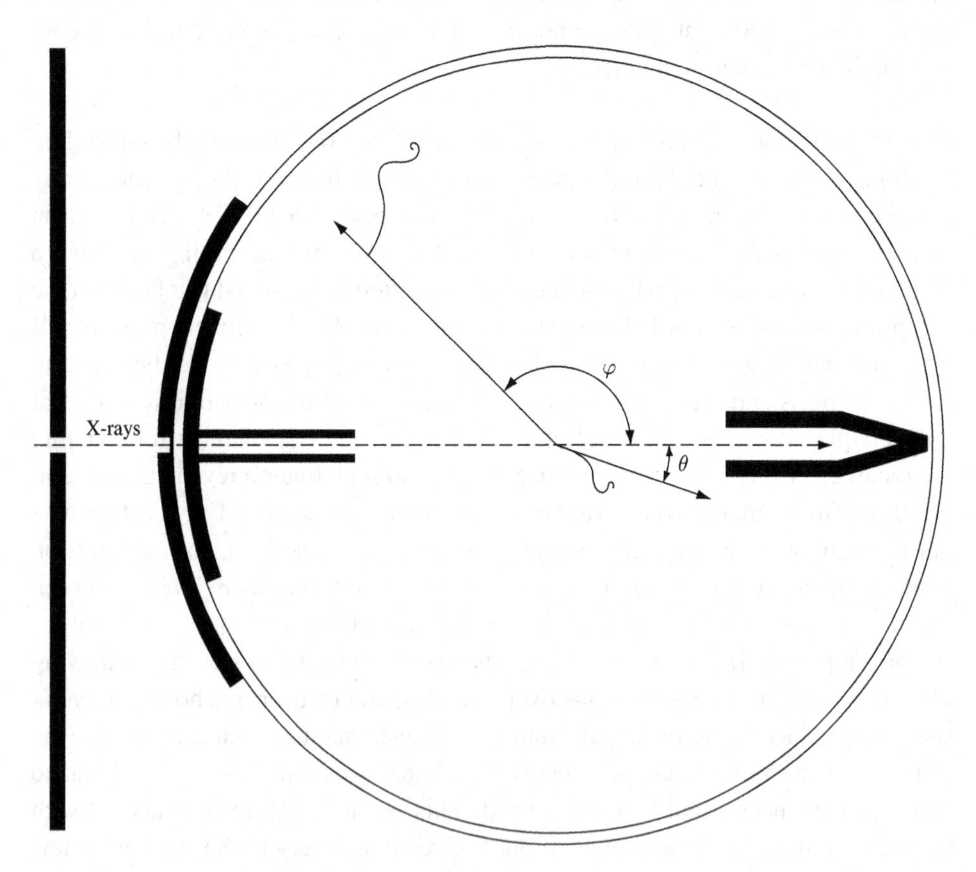

Fig. 3. If the X-rays excite a recoil electron at an angle θ, the photon theory predicts a secondary β-particle at an angle φ.

[a] A. H. Compton and A. W. Simon, *Phys. Rev.*, **26** (1925), 289.

produces a visible track, and occasionally a secondary track is produced by the scattered X-ray. When but one recoil electron appears on the same plate with the track due to the scattered rays, it is possible to tell at once whether the angles satisfy equation (4). If two or three recoil tracks appear,[44] the measurements on each track can be approximately[45] weighted.

Out of 850 plates taken in the final series of readings, thirty-eight show both recoil tracks and secondary β-ray tracks. On eighteen of these plates the observed angle φ[46] is within 20 degrees of the angle calculated from the measured value of θ, while the other twenty tracks are distributed at random angles. This ratio 18:20 is about that to be expected for the ratio of the rays scattered by the part of the air from which the recoil tracks could be measured to the stray rays from various sources. There is only about 1 chance in 250 that so many secondary β-rays should have appeared at the theoretical angle.

If this experiment is reliable, it means that there is scattered X-ray energy associated with each recoil electron sufficient to produce a β-ray, and proceeding in a direction determined at the moment of ejection of the recoil electron. In other words, the scattered X-rays proceed in photons, that is[47] in directed quanta of radiant energy.

This result, like that of Bothe and Geiger, is irreconcilable with Bohr, Kramers and Slater's hypothesis of the statistical production of recoil and photoelectrons. On the other hand, both of these experiments are in complete accord with the predictions of the photon theory.

Reliability of Experimental Evidence

While all of the experiments that we have considered are difficult to reconcile with the classical theory that radiation consists of electromagnetic waves, only those dealing with the individual scattering process[48] afford crucial tests between the photon theory and the statistical theory of virtual radiation. It becomes of especial importance, therefore, to consider the errors to which these experiments are subject.

When two point counters are set side by side, it is very easy to obtain coincidences from extraneous sources. Thus, for example, the apparatus must be electrically shielded so perfectly that a spark on the high-tension outfit that operates the X-ray tube may not produce coincident impulses in the two counters. Then there are high-speed α- and β-rays, due to radium emanation in the air and other radioactive impurities, which may pass through both chambers and produce spurious coincidences. The method which Bothe and Geiger used to detect the coincidences, of[49] recording on a photographic film the time of each pulse, makes it possible to estimate reliably[50] the probability that the coincidences are due to chance. Moreover, it is possible by auxiliary tests to determine whether spurious coincidences are occurring – for example, by operating the outfit as usual, except that the X-rays are absorbed by a sheet of lead. It is especially worthy of note that in the fluorescence experiment the photon theory predicted absence of coincidences, while in the scattering experiment it predicted their presence. It is thus difficult to see how both of these counter experiments can have been seriously affected by systematic errors.

In the cloud expansion experiment the effect of stray radiation is to hide the effect sought for, rather than to introduce a spurious effect. It is possible that due to radioactive contamination and to stray scattered X-rays β-particles may appear in different parts of the chamber, but it will be only a matter of chance if these β-particles appear in the position predicted from the direction of ejection of the recoil electrons. It was in fact only by taking great care to reduce such stray radiations to a minimum that the directional relations were clearly observed in the photographs. It would seem that the only form of consistent error that could vitiate the result of this experiment would be the psychological[51] one of misjudging the angles at which the β-particles appear. It hardly seems possible, however, that errors in the measurement of these angles could be large enough to account for the strong apparent tendency for the angles to fit with the theoretical formula.

It is perhaps worth mentioning further that the initial publications of the two experiments on the individual scattering process were made simultaneously, which means that both sets of experimenters had independently reached a conclusion opposed to the statistical theory of the production of the β-rays.

Nevertheless,[52] given the difficulty of the experiments and the importance of the conclusions to which they have led, it is highly desirable that both experiments should be repeated by physicists from other laboratories.

SUMMARY

The classical theory that radiation consists of electromagnetic waves propagated in all directions through space[53] is intimately connected to the idea of the ether, which is difficult to conceive. It affords no adequate picture of the manner in which radiation is emitted or absorbed. It is inconsistent with the experiments on the photoelectric effect, and is entirely helpless to account for the change of wavelength of scattered radiation or the production of recoil electrons.

The theory of virtual oscillators and virtual radiation which are associated statistically with sudden jumps of atomic energy and the emission of photoelectrons and recoil electrons, does not seem to be inconsistent with any of these phenomena as viewed on a macroscopic scale. This theory, however,[54] retains the difficulties inherent in the conception of the ether and seems powerless to predict the characteristics of the photoelectrons and the recoil electrons. It[55] is further difficult to reconcile with the composite photoelectric effect and is also contrary to Bothe's and Bothe and Geiger's coincidence experiments and to the ray track experiments relating the directions of ejection of a recoil electron and of emission of the associated scattered X-ray.

The photon theory avoids the difficulties associated with the conception of the ether.[56] The production and absorption of radiation is very simply connected with the modern idea of stationary states. It supplies a straightforward explanation of the major characteristics of the photoelectric effect, and it accounts in the simplest possible manner for the change of wavelength accompanying scattering and the existence of recoil electrons. Moreover, it predicts accurately the results of the experiments with individual radiation quanta, where the statistical theory fails.

Unless the four[57] experiments on the individual events[58] are subject to improbably large experimental errors, the conclusion is, I believe, unescapable that radiation consists of directed quanta of energy, i.e., of photons, and that energy and momentum are conserved when these photons interact with electrons or atoms.

Let me say again that this result does not mean that there is no truth in the concept of waves of radiation. The conclusion is rather that energy is not transmitted by such waves. The power of the wave concept in problems of interference, refraction, etc., is too well known to require emphasis. Whether the waves serve to guide the photons, or whether there is some other relation between photons and waves is another and a difficult question.

Discussion of Mr Compton's report

MR LORENTZ. – I would like to make two comments. First on the question of the ether. Mr Compton considers it an advantage of the photon theory that it allows us to do without the hypothesis of an ether which leads to great difficulties. I must say that these difficulties do not seem so great to me and that in my opinion the theory of relativity does not necessarily rule out the concept of a universal medium. Indeed, Maxwell's equations are compatible with relativity, and one can well imagine a medium for which these equations hold. One can even, as Maxwell and other physicists have done with some success, construct a mechanical model of such a medium. One would have to add only the hypothesis of the permeability of ponderable matter by the ether to have all that is required. Of course, in making these remarks, I should not wish to return in any way to these mechanical models, from which physics has turned away for good reasons. One can be satisfied with the concept of a medium that can pass freely through matter and to which Maxwell's equations can be applied.

In the second place: it is quite certain that, in the phenomena of light, there must yet be something other than the photons. For instance, in a diffraction experiment performed with very weak light, it can happen that the number of photons present at a given instant between the diffracting screen and the plane on which one observes the distribution of light, is very limited. The average number can even be smaller than one, which means that there are instants when no photon is present in the space under consideration.

This clearly shows that the diffraction phenomena cannot be produced by some novel action among the photons. There must be something that guides them in their progress and it is natural to seek this something in the electromagnetic field as determined by the classical theory. This notion of electromagnetic field, with its waves and vibrations would bring us back, in Mr Compton's view, to the notion of ether.

MR COMPTON. – It seems, indeed, difficult to avoid the idea of waves in the discussion of optical phenomena. According to Maxwell's theory the electric and magnetic properties of space lead to the idea of waves as directly as did the elastic ether imagined by Fresnel. Why the space having such magnetic properties should bear the name of ether is perhaps simply a matter of words. The fact that these properties of space immediately lead to the wave equation with velocity c is a much more solid basis for the hypothesis of the existence of waves than the old elastic ether. That *something* (E and H) propagates like a wave with velocity c seems evident. However, experiments of the kind we have just discussed show, if they are correct, that the *energy* of the bundle of X-rays propagates in the form of *particles* and not in the form of extended waves. So then, not even the electromagnetic ether appears to be satisfactory.

MR BRAGG. – In his report Mr Compton has discussed the average momentum component of the electrons in the direction of motion of the photon, and he has informed us of the conclusion, at which several experimenters have arrived, that this forward average component is equal to the momentum of the light quantum whose energy has been absorbed and is found again in that of the photoelectron.

I would like to report in this connection some results obtained by Mr Williams.[a] Monochromatic X-rays, with wavelength lying between 0.5 Å and 0.7 Å, enter a Wilson cloud chamber containing oxygen or nitrogen. The trajectories of the photoelectrons are observed through a stereoscope and their initial directions are measured. Since the speed of the photoelectrons is exactly known (the ionisation energy being weak by comparison to the quantity $h\nu$), a measurement of the initial direction is equivalent to a measurement of momentum in the forward direction. Williams finds that the average momentum component in this direction is in all cases markedly larger than the quantity $\frac{h\nu}{c}$ or $\frac{h}{\lambda}$. These results can be summarised by a comparison with the scheme proposed by Perrin and Auger ([P. Auger and F. Perrin], *Journ. d. Phys.* [6th series, vol. **8**] (February 1927), [93]). They are in perfect agreement with the $\cos^2\theta$ law, provided one assumes that the magnetic impulse T_m is equal to $1.8\frac{h\nu}{c}$ and not just $\frac{h\nu}{c}$ as these authors assume. One should not attach any particular importance to this number 1.8, because the range of the examined wavelengths is too small. I mention it only to show that it is possible that the simple law proposed by Mr Compton might not be exact.

I would like to point out that this method of measuring the forward component of the momentum is more precise than an attempt made to establish results about the most probable direction of emission.

MR WILSON says that his own observations, discussed in his Memoir of 1923[b] (but which do not pretend to be very precise) seem to show that in fact the forward momentum component of the photoelectrons is, on average, much larger than what one would derive from the idea that the absorbed quantum yields all of its momentum to the expelled electron.

MR RICHARDSON. – When they are expelled by certain X-rays, the electrons have a momentum in the direction of propagation of the rays equal to $1.8\frac{h\nu}{c}$. If I have understood Mr Bragg correctly, this result is not the effect of some

[a] Cf. the relevant footnote on p. 311 (*eds.*).
[b] Referenced in footnote on p. 307 (*eds.*).

specific elementary process [action], but the average result for a great number of observations in which the electrons were expelled in different directions. Whether or not the laws of energy and momentum conservation apply to an elementary process, it is certain that they apply to the average result for a great number of these processes. Therefore, the process [processus] we are talking about must be governed by the equations for momentum and energy. If for simplicity we ignore the refinements introduced by relativity, these equations are

$$\frac{h\nu}{c} = m\overline{v} + M\overline{V}$$

and

$$h\nu = \frac{1}{2}m\overline{v}^2 + \frac{1}{2}M\overline{V}^2,$$

where m and M are the masses, v and V the velocities of the electrons and of the positive residue; the overbars express that these are averages. The experiments show that the average value of mv is $1.8\frac{h\nu}{c}$ and not $\frac{h\nu}{c}$. This means that $M\overline{V}$ is not zero, so that we cannot ignore this term in the equation. If we consider, for instance, the photoelectric effect on a hydrogen atom, we have to take the collision energy of the hydrogen nucleus into account in the energy equation.

MR LORENTZ. – The term $\frac{1}{2}M\overline{V}^2$ will however be much smaller than $\frac{1}{2}m\overline{v}^2$?[59]

MR RICHARDSON. – It is approximately its 1850th part: that cannot always be considered negligible.

MR BORN thinks that he is speaking also for several other members in asking Mr Compton to explain why one should expect that the momentum imparted to the electron be equal to $\frac{h\nu}{c}$.

MR COMPTON. – When radiation of energy $h\nu$ is absorbed by an atom – which one surely has to assume in order to account for the kinetic energy of the photoelectron – the momentum imparted to the atom by this radiation is $\frac{h\nu}{c}$. According to the classical electron theory, when an atom composed of a negative charge $-e$ of mass m and a positive charge $+e$ of mass M absorbs energy from an electromagnetic wave, the momenta imparted to the two elementary charges [électrons] are inversely proportional to their masses. This depends on the fact that the forward momentum is due to the magnetic vector, which acts with a force proportional to the velocity and consequently more strongly on the charge having the smaller mass.[60] Effectively, the momentum is thus received by the charge with the smaller mass.

MR DEBYE. – Is the reason why you think that the rest of the atom does not receive any of the forward momentum purely theoretical?

MR COMPTON. – The photographs of the trajectories of the photoelectrons show, in accordance with Auger's prediction, that the forward component of the momentum of the photoelectron is, on average, the same as that of the photon. That means, clearly, that on average the rest of the atom does not receive any momentum.

MR DIRAC. – I have examined the motion of an electron placed in an arbitrary force field according to the classical theory, when it is subject to incident radiation, and I have shown in a completely general way that at every instant the fraction of the rate of change [vitesse de variation] of the forward momentum of the electron due to the incident radiation is equal to $\frac{1}{c}$ times the fraction of the rate of change of the energy due to the incident radiation. The nucleus and the other electrons of the atom produce changes of momentum and of energy that at each instant are simply added to those produced by the incident radiation. Since the radiation must modify the electron's orbit, it must also change the fraction of the rate of change of the momentum and of the energy that comes from the nucleus and the other electrons, so that it would be necessary to integrate the motion in order to determine the total change produced by the incident radiation in the energy and the momentum.

MR BORN. – I would like to mention here a paper by Wentzel,[a] which contains a rigorous treatment of the scattering of light by atoms according to quantum mechanics. In it, the author considers also the influence of the magnetic force, which allows him to obtain the quantum analogue of the classical light pressure. It is only in the limiting case of very short wavelengths that one finds that the momentum of light $\frac{h\nu}{c}$ is completely transmitted to the electron; in the case of large wavelengths an influence of the binding forces appears.

MR EHRENFEST. – One can show by a very simple example where the surplus of forward momentum, which we have just discussed, can have its origin. Take a box whose inner walls reflect light completely, but diffusedly, and assume that on the bottom there is a little hole. Through the latter I shine a ray of light into the box which comes and goes inside the box and pushes away its lid and bottom. The lid then has a surplus of forward momentum.

[a] Born is presumably referring to Wentzel's second paper on the photoelectric effect (Wentzel 1927). Compare Mehra and Rechenberg (1987, pp. 835ff.) (*eds.*).

MR BOHR.[a] – With regard to the question of waves or photons discussed by Mr Compton, I would like to make a few remarks, without pre-empting the general discussion. The radiation experiments have indeed revealed features that are not easy to reconcile within a classical picture. This difficulty arises particularly in the Compton effect itself. Several aspects of this phenomenon can be described very simply with the aid of photons, but we must not forget that the change of frequency that takes place is measured using instruments whose functioning is interpreted according to the wave theory. There seems to be a logical contradiction here, since the description of the incident wave as well as that of the scattered wave require that these waves be finitely extended [limitées] in space and time, while the change in energy and in momentum of the electron is considered as an instantaneous phenomenon at a given point in spacetime. It is precisely because of such difficulties that Messrs Kramers, Slater and myself were led to think that one should completely reject the idea of the existence of photons and assume that the laws of conservation of energy and momentum are true only in a statistical way.

The well-known experiments by Geiger and Bothe and by Compton and Simon, however, have shown that this point of view is not admissible and that the conservation laws are valid for the individual processes, in accordance with the concept of photons. But the dilemma before which we are placed regarding the nature of light is only a typical example of the difficulties that one encounters when one wishes to interpret the atomic phenomena using classical concepts. The logical difficulties with a description in space and time have since been removed in large part by the fact that it has been realised that one encounters a similar paradox with respect to the nature of material particles. According to the fundamental ideas of Mr de Broglie, which have found such perfect confirmation in the experiments of Davisson and Germer, the concept of waves is as indispensable in the interpretation of the properties of material particles as in the case of light. We know thereby that it is equally necessary to attribute to the wave field a finite extension in space and in time, if one wishes to define the energy and the momentum of the electron, just as one has to assume a similar finite extension in the case of the light quantum in order to be able to talk about frequency and wavelength.

Therefore, in the case of the scattering process, in order to describe the two changes affecting the electron and the light we must work with four wave fields (two for the electron, before and after the phenomenon, and two for the quantum of light, incident and scattered), finite in extension, which meet in the same region of spacetime.[b] In such a representation all possibility of incompatibility with a

[a] This discussion contribution by Bohr is reprinted and translated also in vol. 5 of Bohr's *Collected Works* (Bohr 1984, pp. 207–12). Compare also the transcript in the Appendix (*eds.*).
[b] Compare also the discussion contributions below, by Pauli, Schrödinger and others (*eds.*).

description in space and time disappears. I hope the general discussion will give me the opportunity to enter more deeply into the details of this question, which is intimately tied to the general problem of quantum theory.

MR BRILLOUIN. – I have had the opportunity to discuss Mr Compton's report with Mr Auger,[a] and wish to make a few comments on this topic. A purely corpuscular description of radiation is not sufficient to understand the peculiarities of the phenomena; to assume that energy is transported by photons $h\nu$ is not enough to account for all the effects of radiation. It is essential to complete our information by giving the direction of the electric field; we cannot do without this field, whose role in the wave description is well known.

I shall recall in this context a simple argument, recently given by Auger and F. Perrin, and which illustrates this remark clearly. Let us consider the emission of electrons by an atom subject to radiation, and let us examine the distribution of the directions of emission. This distribution has usually been observed in a plane containing the light ray and the direction of the electric field (the incident radiation is assumed to be polarised); let φ be the angle formed by the direction of emission of the photoelectrons and the electric field h; as long as the incident radiation is not too hard, the distribution of the photoelectrons is symmetric around the electric field; one can then show that the probability law necessarily takes the form $A\cos^2\varphi$. Indeed, instead of observing the distribution in the plane of incidence (Fig. 1), let us examine it in the plane of the wave; the same distribution law will still be valid; and it is the only one that would allow us to obtain, through

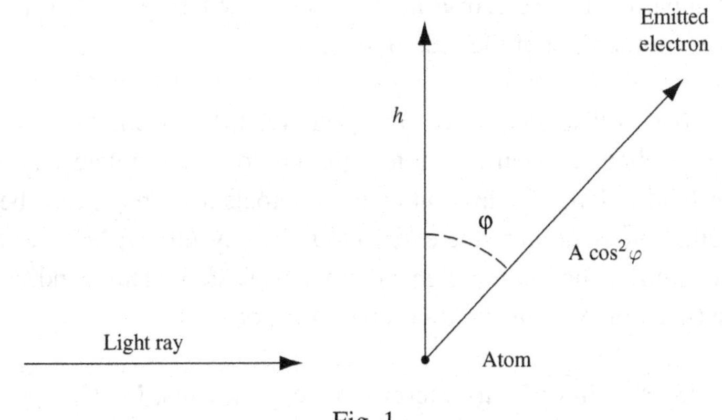

Fig. 1.

[a] As noted in Section 1.4, this and other reports had been circulated among the participants before the conference (*eds.*).

the superposition of two waves polarised at right angles, an entirely symmetric distribution

$$A\cos^2\varphi + A\sin^2\varphi = A.$$

Now, from the point of view of waves, one must necessarily obtain this result, a beam [rayonnement] of natural light having no privileged direction in the plane of the wave. These symmetry considerations, which any theory of radiation must respect, provide a substantial difficulty for the structural theories of the photon (Bubb's quantum vector, for instance).

Summing up, the discontinuity of the radiation manifests itself just in the most elementary way, through the laws of conservation of energy and momentum, but the detailed analysis of the phenomena is interpreted more naturally from the continuous point of view. For the problem of emission of the photoelectrons, a complete theory has been given by Wentzel, by means of wave mechanics.[a] He finds the $A\cos^2\varphi$ law of F. Perrin and Auger for radiation of low penetration; when the radiation is harder, Wentzel obtains a more complex law, in which the electrons tend to be emitted in larger numbers in the forward direction. His theory, however, seems incomplete with regard to this point since, if I am not mistaken, he has assumed the immobility of the atomic nucleus; now, nothing tells us *a priori* how the momentum $\frac{h\nu}{c}$ of the photon is going to be distributed between the nucleus and the emitted electron.

MR LORENTZ. – Allow me to point out that according to the old electron theory, when one has a nucleus and an electron on which a beam of polarised light falls, the initial angular momentum of the system is always conserved. The angular momentum imparted to the electron-nucleus system will be provided at the expense of the angular momentum of the radiation field.

MR COMPTON. – The conception of the photon differs from the classical theory in that, when a photoelectron is emitted, the photon is completely absorbed and no radiation field is left. The motion of the photoelectron must thus be such that the final angular momentum of the electron-nucleus system will be the same as the initial momentum of the photon-electron-nucleus system. This condition restricts the possible trajectories of the emitted photoelectron.

MR KRAMERS. – In order to interpret his experiments, Mr Compton needs to know how the absorption μ is divided between a component τ, due to the 'true'

[a] This is presumably Wentzel's treatment from his first paper on the photoelectric effect (Wentzel 1926). Compare again Mehra and Rechenberg (1987, pp. 835ff.) (*eds.*).

absorption, and a component σ due to the scattering. We do not know with certainty that, if μ can be written in the form $C\lambda^a + D$, the constant D truly represents the scattering for large wavelengths, where $C\lambda^a$ is no longer small compared to D. In general, thus, specific measurements of σ are necessary. Did you have sufficient information regarding the values of σ and τ in your experiments?

MR COMPTON. – The most important case in which it is necessary to distinguish between the true absorption τ and the absorption σ due to the scattering, is that of carbon. For this case, Hewlett[a] has measured σ directly for the wavelength 0.71 Å and the total absorption μ over a large range of wavelengths. The difference between μ and σ for the wavelength 0.71 Å corresponds to τ for this wavelength. According to Owen's formula this τ is proportional to λ^3; we can thus calculate τ for all wavelengths. The difference between this value of τ and the measured value of μ corresponds to the value of σ for the wavelengths considered. Since τ is relatively small in the case of carbon, especially for small wavelengths, this procedure yields a value for σ that cannot be very imprecise.

MR BRAGG. – When one consults the original literature on this subject, one is struck by how much the X-ray absorption measurements leave to be desired, both with regard to precision as well as with regard to the extent of the scale of wavelengths for which they have been performed.

MR PAULI. – How large is the broadening of the modified rays?

MR COMPTON. – The experiments have shown clearly that the modified ray is broader than the unmodified ray. In the typical case of the ray λ 0.7 Å scattered by carbon, the broadening is of order 0.005 angström. Unfortunately, the experiments concerning this point are far from being satisfactory, and this number should be considered only as a rough approximation.

MR PAULI. – The broadening of the modified ray can be interpreted theoretically in two ways, which to tell the truth reduce to the same according to quantum mechanics. First, the electron, in a given stationary state of the atom, has a certain velocity distribution with regard to magnitude and direction. That gives rise to a broadening of the frequency of the scattered rays through the Doppler effect, a broadening whose order of magnitude is $\frac{\Delta\lambda}{\lambda} = \frac{v}{c}$, where v denotes the average velocity of the electron in the atom.

[a] [C. W. Hewlett, *Phys. Rev.*, **17** (1921), 284.]

In order to convey the second means of explanation, I would like to sketch briefly the meaning of the Compton effect in wave mechanics.[a] This meaning is based first of all on the wave equation

$$\sum_\alpha \frac{\partial^2 \psi}{\partial x_\alpha^2} - \frac{4\pi i}{h}\frac{e}{c}\sum_\alpha \varphi_\alpha \frac{\partial \psi}{\partial x_\alpha} - \frac{4\pi^2}{h^2}\left(\frac{e^2}{c^2}\sum_\alpha \varphi_\alpha^2 + m_0^2 c^2\right)\psi = 0$$

and further on the expression

$$i S_\alpha = \psi \frac{\partial \psi^*}{\partial x_\alpha} - \psi^* \frac{\partial \psi}{\partial x_\alpha} + \frac{4\pi i}{h}\frac{e}{c}\varphi_\alpha \psi \psi^*,$$

in which ψ is Schrödinger's function, ψ^* the complex conjugate value and φ_α the four-potential of the electromagnetic field. Given S_α, one calculates the radiation from *classical* electrodynamics. If now in the wave equation one replaces φ_α by the potential of an incident plane wave, the terms that are proportional to the amplitude of this wave can be considered infinitely small in the first order, and one can apply the approximation methods of perturbation theory. This now is a point where one needs to be especially careful. It is all-important to know what one will take as the unperturbed field ψ, which must correspond to a solution of the wave equation for the free particle (corresponding to $\varphi_\alpha = 0$).[61] One finds that in order to agree with the observations, it is necessary to take *two* infinitely extended monochromatic wave trains as being already present in the unperturbed solution, of which one corresponds to the initial state, the other to the final state of the Compton process. In my opinion this assumption, on which the theories of the Compton effect by Schrödinger, Gordon and Klein are based, is unsatisfactory and this defect is corrected only by Dirac's quantum electrodynamics.[b] But if one makes this assumption, the current distribution of the unperturbed solution corresponds to that of an infinitely extended diffraction grating that moves with a constant speed, and the action of the radiation on this grating leads to a sharp modified ray.

If one considers a bound electron in an atom, one has to replace one component of the solution ψ in the unperturbed charge and current distribution by the eigenfunction of the atom in the stationary state considered, and the other component by a solution corresponding to the final state of the Compton process (belonging to the continuous spectrum of the atom), which at great distance from the atom behaves more or less as a plane wave. One thus has a moving grating that first of all depends only on the finite extension of the atom and in the second place has components no longer moving with the same speed at all. This gives rise to a lack of sharpness of the shifted ray of the scattered radiation.

[a] For a modern discussion, see Björken and Drell (1964, chapter 9) (*eds.*).
[b] Compare below Schrödinger's contribution and the ensuing discussion (*eds.*).

But one can show that, from the point of view of quantum mechanics, this explanation for the lack of sharpness of the shifted ray is just another form of the explanation given in the first instance and which relies on the different directions of the initial velocities of the electrons in the atom. For according to quantum mechanics if

$$\psi = f(x, y, z)e^{2\pi i \nu t}$$

is the eigenfunction corresponding to a given stationary state of the atom, the function[62]

$$\varphi(p_x, p_y, p_z) = \iiint f(x, y, z)e^{-\frac{2\pi i}{\lambda}(p_x x + p_y y + p_z z)} dx dy dz,$$

which one obtains by decomposing f in plane waves according to Fourier can be interpreted in the sense that $|\varphi(p)|^2 dp_x dp_y dp_z$ denotes the probability that in the given stationary state the components of the momentum of the electron lie between p_x, p_y, p_z and $p_x + dp_x$, etc. Now, if through the resulting velocity distribution of the electrons in the atom one calculates the broadening of the shifted line according to the first point of view, for light of sufficiently short wavelength with respect to which the electron can be considered free in the atom (and it is only under these conditions that the procedure is legitimate), one finds exactly the same result as with the other method described.[a]

MR COMPTON. – Jauncey has calculated the broadening of the modified ray using essentially the method that Mr Pauli has just described. Jauncey assumed, however, that the velocities of the electron are the ones given by Bohr's theory of orbital motions. The broadening thus obtained is larger than that found experimentally.

MRS CURIE. – In his very interesting report, Professor Compton has dwelt on emphasising the reasons that lead one to adopt the theory of a collision between a quantum and a free electron. Along the same line of thought, I think it is useful to point out the following two views:

First, the existence of collision electrons seems to play a fundamental role in the biological effects produced on living tissues by very high-frequency radiation, such as the most penetrating γ-rays emitted by radioelements. If one assumes that the biological effect may be attributed to the ionisation produced in the cells subjected to radiation, this effect cannot depend directly on the γ-rays, but is due to the emission of secondary β-rays that accompanies the passage of the γ-rays through

[a] Pauli was possibly the first to introduce the probability interpretation of the wave function in momentum space, in a letter to Heisenberg of 19 October 1926 (Pauli, 1979, pp. 347–8). Cf. the footnote on p. 105 above (*eds.*).

matter. Before the discovery of the collision electrons, only a single mechanism was known for the production of these secondary rays, that consisting in the total absorption of a quantum of radiation by the atom, with the emission of a photoelectron. The absorption coefficient τ relating to this process varies with the wavelength λ of the primary γ-radiation, as well as with the density [ρ] of the absorbing matter and the atomic number N of the atoms composing it, according to the well-known relation of Bragg and Peirce $\frac{\tau}{\rho} = \Lambda N^3 \lambda^3$, where Λ is a coefficient that has a constant value for frequencies higher than that of the K discontinuity. If this relation valid in the domain of X-rays can be applied to high-frequency γ-rays, the resulting value of $\frac{\tau}{\rho}$ for the light elements is so weak that the emission of photoelectrons appears unable to explain the biological effects of radiation on the living tissues traversed.[a]

The issue appears altogether different if one takes into consideration the emission of collision electrons in these tissues, following Compton's theory. For a collimated primary beam of γ-rays, the fraction of electromagnetic energy converted into kinetic energy of the electrons per unit mass of the absorbing matter is given by the coefficient

$$\frac{\sigma_a}{\rho} = \frac{\alpha}{(1 - 2\alpha)^2} \frac{\sigma_0}{\rho},$$

where $\frac{\sigma_0}{\rho}$ is the scattering coefficient per unit mass valid for medium frequency X-rays, according to the theory of J. J. Thomson, and is close to 0.2, while α is Compton's parameter $\alpha = \frac{h\nu}{mc^2}$ (h Planck's constant, ν primary frequency, m rest mass of the electron, c speed of light). Taking $\alpha = 1.2$, a value suitable for an important group of γ-rays (equivalent potential 610 kilovolts), one finds $\frac{\sigma_a}{\rho} = 0.02$, that is, 2 per cent of the primary energy is converted to energy of the electron per unit mass of absorbing matter, whence a possibility of interpreting the observed biological effects. To this direct production of collision electrons along the trajectory of the primary beam is added, in an extended medium, a supplementary production, from the fact that to each of these electrons corresponds a scattered quantum, with a smaller value than the primary quantum, and that this scattered quantum can in turn be subject to the Compton effect in the medium through which it propagates, with production of a new collision electron and of an even smaller quantum. This process, indefinitely repeatable and called the 'multiple Compton effect' seems in fact to have been observed by certain authors.[b] Not only is the number of collision electrons thereby multiplied, but, further, the primary

[a] It is true that several authors have recently contested the legitimacy of extending the absorption law of Bragg and Peirce to X-rays.

[b] [B.] Rajewsky, *Fortschritte auf dem Gebiet der Roentgenstrahlung*, **35** (1926), 262.

quantum, reduced by successive collisions takes on values for which the absorption with emission of photoelectrons becomes more and more probable.

These facts have an important repercussion on the technique of X-ray therapy. Certain authors had, in fact, denied the usefulness of producing very high-voltage apparatus providing X-rays of very high frequency and very high penetrating power, whose use is otherwise convenient owing to the uniformity of irradiation they allow one to attain. If these rays had been devoid of efficacy, one would have had to give up on their use. Such is not the case if one adopts the point of view of the Compton effect, and it is then legitimate to direct the technique towards the use of high voltages.

Another interesting point of view to examine is that of the emission of β-rays by radioactive bodies. Professor Compton has pointed out that among the β-rays of secondary origin, some could be collision electrons produced by the scattering of the primary γ-rays on the electrons contained in the matter they traverse.

It is in an effect of this type that Thibaud thinks one may find the explanation for the appearance of the magnetic spectra of the secondary γ-rays. These spectra are composed of lines that may be attributed to groups of photoelectrons of the same speed, each of which is emitted by absorption in a thin metallic envelope of a group of homogeneous γ-rays emitted by a radioelement contained in this envelope. Each line of photoelectric origin is accompanied by a band beginning at the line itself and extending towards the region of low velocities. Thibaud thinks that this band could be due to photoelectrons expelled from the screen by those γ-rays that, in this same screen, had suffered the Compton effect with reduction of frequency. This interpretation appears plausible; however, in order to prove it, it would be necessary to study the structure of the band and find in the same spectrum the band that may be attributed to the collision electrons corresponding to the scattered γ-rays.

An analogous problem arises regarding the emission of β-rays by radioactive bodies with negligible thickness, so as to eliminate, as far as possible, the secondary effects due to the supports and envelopes. One then observes a magnetic spectrum that may be attributed to the radioelement alone and consisting either of a continuous band, or of the superposition of a continuous spectrum and a line spectrum. The latter has received a satisfactory interpretation in some recent papers (L. Meitner, Ellis, Thibaud, etc.).

A line is due to a group of photoelectrons with the same speed expelled from the levels of the radioactive atoms by a group of homogeneous γ-rays produced in their nuclei. This effect is called 'internal conversion', since one assumes that the quantum emitted by an atomic nucleus is reabsorbed in the electron cloud [enveloppe électronique] of the same atom. The great majority of observed lines find their explanation in this hypothesis.

The interpretation of the continuous spectrum appears to present more difficulties. Some authors attribute it only to the primary β-rays, while others consider the possibility of a secondary origin and invoke the Compton effect as a possible cause of its production (L. Meitner). This would be an 'internal' Compton effect, such that a γ-ray emitted from the nucleus of an atom would experience a collision with one of the weakly bound electrons at the periphery of the same atom. If that were the case, the velocity distribution of the emitted collision electrons would not be arbitrary, but would have to conform to the predictions of Compton's theory.

I have closely examined this problem, which has a very complex appearance.[a] Each group of homogeneous γ-rays is accompanied by scattered γ-rays, so that in the diffraction spectrum of the γ-rays, each line should experience a broadening of 0.0485 Å units. The experiments on the diffraction of γ-rays are difficult and not very numerous; so far the broadening effect has not been reported.

Each homogeneous group of γ-rays must correspond to a group of collision electrons, whose velocity varies continuously from zero to an upper limit derived from Compton's theory and which in the magnetic spectrum corresponds to a band bounded sharply on the side of the large velocities. The same group of γ-rays may correspond to further groups of photoelectrons expelled from the different levels K, L, etc. of the atom through internal absorption of the scattered γ-rays. For each group of photoelectrons, the velocity of emission lies between two well-defined limits. The upper limit corresponds to the surplus energy of the primary γ-rays with respect to the extraction work W characteristic of the given level; the lower limit corresponds to the surplus energy, with respect to the same work, of the γ-rays scattered in the direction opposite to that of the primary rays, and having experienced because of that the highest loss of frequency. In the magnetic spectrum, each group of photoelectrons will be represented by a band equally well bounded on the side of the large and of the small velocities, with the same difference between the extreme energies for each band.

It is easy to see that in the same magnetic spectrum the different bands corresponding to the same group of γ-rays may partially overlap, making it difficult to analyse the spectrum comparing the distribution of β-rays with that predicted by the theory. For substances emitting several groups of γ-rays, the difficulty must become considerable, unless there are large differences in their relative effectiveness in producing the desired effect. Let us also point out that the continuous spectrum due to the Compton effect may be superposed with a continuous spectrum independent of this effect (that may be attributed for instance to the primary β-rays).

[a] [M.] Curie, *Le Journal de Physique et le Radium*, **7** (1926), 97.

Examination of the experimental data available so far does not yet allow one to draw conclusions convincingly. Most of the spectra are very complex, and their precise study with respect to the energy distributions of the β-rays will require very detailed work. In certain simple spectra such as that of the β-rays of RaD, one observes lines of photoelectric origin that may be attributed to a single group of monochromatic γ-rays. These lines form the upper edge of bands extending towards low velocities and probably arising from photoelectrons produced by the scattered γ-rays. In certain magnetic spectra obtained from the β-rays of mesothorium 2 in the region of low velocities, one notices in the continuous spectrum a gap that might correspond, for the group of primary γ-rays with 58 kilovolts, to the separation between the band due to the collision electrons and that due to the photoelectrons of the scattered γ-rays.[a]

MR SCHRÖDINGER, at the invitation of Mr Ehrenfest, draws on the blackboard in coloured chalk the system of four wave trains by which he has tried to represent the Compton effect in an anschaulich way [d'une façon intuitive][b] (*Ann. d. Phys.* 4th series, vol. **82** (1927), 257).[c]

MR BOHR. – The simultaneous consideration of two systems of waves has not the aim of giving a causal theory in the classical sense, but one can show that it leads to a symbolic analogy. This has been studied in particular by Klein. Furthermore, it has been possible to treat the problem in more depth through the way Dirac has formulated Schrödinger's theory. We find here an even more advanced renunciation of Anschaulichkeit [intuitivité], a fact very characteristic of the symbolic methods in quantum theory.

MR LORENTZ. – Mr Schrödinger has shown how one can explain the Compton effect in wave mechanics. In this explanation one considers the waves associated with the electron (e) and the photon (ph), before (1) and after (2) the encounter. It is natural to think that, of these four systems of waves e_1, ph_1, e_2 and ph_2, the latter two are produced by the encounter. But they are not *determined* by e_1 and ph_1, because one can for example choose arbitrarily the direction of e_2. Thus, for the problem to be well-defined, it is not sufficient to know e_1 and ph_1; another piece of data is necessary, just as in the case of the collision of two elastic balls one must know not only their initial velocities but also a parameter that determines the greater or lesser eccentricity of the collision, for instance the angle between the

[a] D. K. Yovanovitch and A. Prola, *Comptes Rendus*, **183** (1926), 878.
[b] For discussions of the notion of Anschaulichkeit, see Sections 3.4.7, 4.6 and 8.3 (*eds.*).
[c] Schrödinger (1928, p. x) later remarked on a mistake in the figure as published in the original paper, pointed out to him by Ehrenfest (*eds.*).

relative velocity and the common normal at the moment of the encounter. Perhaps one could introduce into the explanation given by Mr Schrödinger something that would play the role of this accessory parameter.

MR BORN. – I think it is easy to understand why three of the four waves have to be given in order for the process to be determined; it suffices to consider analogous circumstances in the classical theory. If the motions of the two particles approaching each other are given, the effect of the collision is not yet determined; it can be made determinate by giving the position of closest approach or an equivalent piece of data. But in wave mechanics such microscopic data are not available. That is why it is necessary to prescribe the motion of one of the particles after the collision, if one wants the motion of the second particle after the collision to be determined. But there is nothing surprising in this, everything being exactly as in classical mechanics. The only difference is that in the old theory one introduces microscopic quantities, such as the radii of the atoms that collide, which are eliminated from subsequent calculations, while in the new theory one avoids the introduction of these quantities.

Notes on the translation

1 The words 'réels ou apparents' are present only in the French version.
2 The English version has only four headings (starting with '(1) How are the waves produced?'), and accordingly omits the next section, on 'The problem of the ether', and later references to the ether.
3 [d'après les résultats de l'étude des spectres]
4 The English edition distinguishes sections and subsections more systematically than the French edition, and in this and other small details of layout we shall mostly follow the former.
5 [rappeler qu'il existe une théorie dans laquelle]
6 The words 'le professeur' are present only in the French edition.
7 ['élément de radiation' ou 'quantum de lumière']
8 This section is present only in the French version.
9 [énergie ondulatoire]
10 The original footnote gives page '145'.
11 The English edition has '**213**'.
12 [à part une constante]
13 [perfectionnée]
14 The second part of the footnote is printed only in the English edition.
15 The French edition here includes the clause 'qui a été faite'.
16 Here and in several places in the following, the French edition has 'simple' where the English one has 'single'.
17 [l'action directe du vecteur électrique de l'onde, prise dans le sens ordinaire]
18 [dans la direction de propagation de l'onde]
19 [puisque]
20 This footnote is only present in the English edition.
21 The preceding clause is only present in the French edition.
22 [la période de l'électron dans son mouvement orbital]
23 In the English edition this reads: 'The fact that the photoelectrons receive the momentum of the incident photon'.
24 [sur les rayons X secondaires et les rayons γ]
25 This footnote appears only in the French edition.
26 [ne fournirent aucune preuve de l'existence d'un spectre de raies]
27 This word is missing in the French edition.
28 The English edition reads '$(1 + x)$'.
29 [prédit]
30 [on eut quelque confiance dans les électrons de recul]
31 The French edition uses ρ instead of τ in the text, but uses τ in the discussion (where ρ is used for matter density). The English edition uses t.
32 Footnote mark missing in the French edition.
33 The French edition gives (4).
34 The French edition gives (2).
35 This footnote is present only in the English edition.
36 This word is missing in the French edition.
37 The French edition reads 'J. Waller'.
38 The page numbers for Wentzel appear only in the French edition.
39 The English edition describes only three experiments, omitting the section on the composite photoelectric effect as well as references to it later.
40 This section is present only in the French edition.
41 This and the next section are of course numbered (2) and (3) in the English edition.
42 [rayonnement de fluorescence]
43 [une expérience cruciale entre les deux points de vue a été imaginée et brillamment réalisée]
44 The French edition includes also 'en même temps'.
45 [d'une façon appropriée]
46 The French edition has 'θ'.

47 The words ‘photons, c’est-à-dire’ are present only in the French edition.
48 [au phénomène de la diffusion par les électrons individuels]
49 [ou]
50 [avec certitude]
51 [physiologique]
52 This sentence is only printed in the French edition.
53 The English edition omits reference to the ether and continues directly with ‘affords no adequate picture’.
54 The English edition continues directly with: ‘seems powerless’.
55 The English edition continues: ‘is also contrary to’.
56 In the English edition this reads simply: ‘According to the photon theory, the production . . . ’.
57 In the English edition: ‘three’.
58 [processus]
59 Overbars have been added.
60 The French text reads ‘avec moins d’intensité sur la charge ayant la plus petite masse’. This is evidently an error: the (transverse) velocity of the charges stems from the electric field, which imparts the larger velocity to the charge with the smaller mass, which therefore experiences the larger magnetic force.
61 The printed text reads ‘$\varphi_\alpha \omega$’.
62 Brackets in the exponent added.

The new dynamics of quanta[a]

BY MR LOUIS DE BROGLIE

I. – PRINCIPAL POINTS OF VIEW[b]

1. *First works of Mr Louis de Broglie* [1]. – In his first works on the Dynamics of Quanta, the author of the present report started with the following idea: taking the existence of elementary corpuscles of matter and radiation as an experimental fact, these corpuscles are supposed to be endowed with a periodicity. In this way of seeing things, one no longer conceives of the 'material point' as a static entity pertaining to only a tiny region of space, but as the centre of a periodic phenomenon spread all around it.

Let us consider, then, a completely isolated material point and, in a system of reference attached to this point, let us attribute to the postulated periodic phenomenon the appearance of a stationary wave defined by the function

$$u(x_0, y_0, z_0, t_0) = f(x_0, y_0, z_0) \cos 2\pi \nu_0 t_0.$$

In another Galilean system x, y, z, t, the material point will have a rectilinear and uniform motion with velocity $v = \beta c$. Simple application of the Lorentz transformation shows that, as far as the phase is concerned, in the new system the periodic phenomenon has the appearance of a plane wave propagating in the direction of motion whose frequency and phase velocity are

$$\nu = \frac{\nu_0}{\sqrt{1-\beta^2}}, \quad V = \frac{c^2}{v} = \frac{c}{\beta}.$$

[a] Our translation of the title ('La nouvelle dynamique des quanta') reflects de Broglie's frequent use of the word 'quantum' to refer to a (pointlike) particle, an association that would be lost if the title were translated as, for example, 'The new quantum dynamics' (*eds.*).

[b] On beginning this exposition, it seems right to underline that Mr Marcel Brillouin was the true precursor of wave Mechanics, as one may realise by referring to the following works: *C. R.* **168** (1919), 1318; **169** (1919), 48; **171** (1920), 1000. – *Journ. Physique* **3** (1922), 65.

The appearance of this phase propagation with a speed superior to c, as an immediate consequence of the theory of Relativity, is quite striking.

There exists a noteworthy relation between v and V. The formulas giving ν and V allow us in fact to define a refractive index of the vacuum, for the waves of the material point of proper frequency ν_0, by the dispersion law

$$n = \frac{c}{V} = \sqrt{1 - \frac{\nu_0^2}{\nu^2}}.$$

One then easily shows that

$$\frac{1}{v} = \frac{1}{c}\frac{\partial(n\nu)}{\partial\nu},$$

that is, that the velocity v of the material point is equal to the group velocity corresponding to the dispersion law.

With the free material point being thus defined by wave quantities, the dynamical quantities must be related back to these. Now, since the frequency ν transforms like an energy, the obvious thing to do is to assume the quantum relation

$$W = h\nu,$$

a relation that is valid in all systems, and from which one derives the undulatory definition of the proper mass m_0

$$m_0 c^2 = h\nu_0.$$

Let us write the function representing the wave in the system x, y, z, t in the form

$$u(x, y, z, t) = f(x, y, z, t) \cos \frac{2\pi}{h}\varphi(x, y, z, t).$$

Denoting by W and p the energy and momentum, one easily shows that one has[a, b]

$$W = \frac{\partial\varphi}{\partial t}, \quad \vec{p} = -\overrightarrow{\text{grad}}\, \varphi.$$

The function φ is then none other than the Jacobi function.[c] One deduces from this that, in the case of uniform rectilinear motion, the principles of least action and of Fermat are identical.

To look for a generalisation of these results, let us now assume that the material point moving in a field derived from a potential function $F(x, y, z, t)$ is represented by the function

[a] These are the relativistic guidance equations of de Broglie's early pilot-wave theory of 1923–4, for the special case of a free particle (*eds.*).

[b] The vector '$\overrightarrow{\text{grad}}\, \varphi$' is the vector whose components are $\partial\varphi/\partial x$, $\partial\varphi/\partial y$, $\partial\varphi/\partial z$.

[c] Usually called the Hamilton–Jacobi function (*eds.*).

$$u(x, y, z, t) = f(x, y, z, t) \cos \frac{2\pi}{h} \varphi(x, y, z, t),$$

where φ is the Jacobi function of the old Dynamics. This assimilation of the phase into the Jacobi function then leads us to assume the following two relations, which establish a general link between mechanical quantities and wave quantities:

$$W = h\nu = \frac{\partial \varphi}{\partial t}, \quad \vec{p} = \frac{h\nu}{V} = -\overrightarrow{\text{grad}}\, \varphi.$$

One then deduces that, for the waves of the new Mechanics, the space occupied by the field has a refractive index

$$n = \sqrt{\left(1 - \frac{F}{h\nu}\right)^2 - \frac{\nu_0^2}{\nu^2}}.$$

Hamilton's equations show in addition that, here again, the velocity of the moving body is equal to the group velocity.[a]

These conceptions lead to an interpretation of the stability conditions introduced by quantum theory. If, indeed, one considers a closed trajectory, the phase must be a single-valued function along this curve, and as a result one is led to write the Planck condition[1]

$$\oint (p \cdot dl) = k \cdot h \quad (k \text{ integer}).$$

The Sommerfeld conditions for quasi-periodic motions may also be derived. The phenomena of quantum stability thus appear to be analogous to phenomena of resonance, and the appearance of whole numbers here becomes as natural as in the theory of vibrating strings or plates. Nevertheless, as we shall see, the interpretation that has just been recalled still constitutes only a first approximation.

The application of the new conceptions to corpuscles of light leads to difficulties if one considers their proper mass to be finite. One avoids these difficulties by assuming that the properties of the corpuscles of light are deduced from those of ordinary material points by letting the proper mass tend to zero. The two speeds v and V then both tend to c, and in the limit one obtains the two fundamental relations of the theory of light quanta

$$h\nu = W, \quad \frac{h\nu}{c} = p,$$

with the aid of which one can account for Doppler effects, radiation pressure, the photoelectric effect and the Compton effect.

[a] In the case of motion of a point charge in a magnetic field, space behaves like an anisotropic medium (see *Thesis*, p. 39).

The new wave conception of Mechanics leads to a new statistical Mechanics, which allows us to unify the kinetic theory of gases and the theory of blackbody radiation into a single doctrine. This statistics coincides with that proposed independently by Mr Bose [2]; Mr Einstein [3] has shown its scope and clarified its significance. Since then, numerous papers [4] have developed it in various directions.

Let us add a few remarks. First, the author of this report has always assumed that the material point occupies a well-defined position in space. As a result, the amplitude f should contain a singularity or at the very least have abnormally high values in a very small region. But, in fact, the form of the amplitude plays no role in the results reviewed above. Only the phase intervenes: hence the name *phase waves* originally given to the waves of the new Mechanics.

On the other hand, the author, after having reduced the old forms of Dynamics to geometrical Optics, realised clearly that this was only a first stage. The existence of diffraction phenomena appeared to him to require the construction of a new Mechanics 'which would be to the old Mechanics (including that of Einstein) what wave Optics is to geometrical Optics'.[a] It is Mr Schrödinger who has had the merit of definitively constructing the new doctrine.

2. *The work of Mr E. Schrödinger* [5]. – Mr Schrödinger's fundamental idea seems to have been the following: the new Mechanics must begin from wave equations, these equations being constructed in such a way that in each case the phase of their sinusoidal solutions should be a solution of the Jacobi equation in the approximation of geometrical Optics.

Instead of considering waves whose amplitude contains a singularity, Mr Schrödinger systematically looks at waves of classical type, that is to say, waves whose amplitude is a continuous function. For him, the waves of the new Mechanics are therefore represented by functions Ψ that one can always write in the canonical form

$$\Psi = a \cos \frac{2\pi}{h} \varphi,$$

a being a continuous function and φ being *in the first approximation* a solution of the Jacobi equation. We may understand the words 'in the first approximation' in two different ways: first, if the conditions that legitimate the use of geometrical Optics are realised, the phase φ will obey the equation called the equation of geometrical Optics, and this equation will have to be identical to that of Jacobi; second, one must equally recover the Jacobi equation if one makes Planck's

[a] *Revue Générale des Sciences*, 30 November 1924, p. 633.

constant tend to zero, because we know in advance that the old Dynamics must then become valid.

Let us first consider the case of the motion of a single material point in a static field derived from the potential function $F(x, y, z)$. In his first Memoir Schrödinger shows that the wave equation, at least in the approximation of Newton's Mechanics, is in this case

$$\Delta\Psi + \frac{8\pi^2 m_0}{h^2}(E - F)\Psi = 0.$$

It is also just this equation that one arrives at beginning from the dispersion law noted in the first section.

Having obtained this equation, Mr Schrödinger used it to study the quantisation of motion at the atomic scale (hydrogen atom, Planck oscillator, etc.). He made the following fundamental observation: in the problems considered in micromechanics, the approximations of geometrical Optics are no longer valid at all. As a result, the interpretation of the quantum conditions proposed by L. de Broglie shows only that the Bohr-Sommerfeld formulas correspond to the approximation of the old Dynamics. To resolve the problem of quantisation rigorously, one must therefore consider the atom as the seat of stationary waves satisfying certain conditions. Schrödinger assumed, as is very natural, that the wave functions must be finite, single-valued and continuous over all space. These conditions define a set of fundamental functions (Eigenfunktionen) for the amplitude, which represent the various stable states of the atomic system being considered. The results obtained have proven that this new quantisation method, to which Messrs Léon Brillouin, G. Wentzel and Kramers [6] have made important contributions, is the correct one.

For Mr Schrödinger, one must look at continuous waves, that is to say, waves whose amplitude does not have any singularities. How can one then represent the 'material point'? Relying on the equality of the velocity of the moving body and the group velocity, Schrödinger sees the material point as a group of waves (Wellenpaket[2]) of closely neighbouring frequencies propagating in directions contained within the interior of a very narrow cone. The material point would then not be really pointlike; it would occupy a region of space that would be at least of the order of magnitude of its wavelength. Since, in intra-atomic phenomena, the domain where motion takes place has dimensions of the order of the wavelengths, there the material point would no longer be defined at all; for Mr Schrödinger, the electron in the atom is in some sense 'smeared out' ['fondu'], and one can no longer speak of its position or velocity. This manner of conceiving of material points seems to us to raise many difficulties; if, for example, the quantum of ultraviolet light occupies a volume whose dimensions are of the order of its wavelength, it is quite difficult to conceive that this quantum could be absorbed by an atom of dimensions a thousand times smaller.

Having established the wave equation for a material point in a static field, Mr Schrödinger then turned to the Dynamics of many-body systems [la Dynamique des systèmes]. Still limiting himself to the Newtonian[a] approximation, and inspired by Hamilton's ideas, he arrived at the following statement: Given an isolated system whose potential energy is $F(q_1, q_2, \ldots, q_n)$, the kinetic energy is a homogeneous quadratic form in the momenta p_k and one may write

$$2T = \sum_{kl} m^{kl} p_k p_l,$$

the m^{kl} being functions of the q. If m denotes the determinant $|m^{kl}|$ and if E is the constant of energy in the classical sense, then according to Schrödinger one must begin with the wave equation

$$m^{+\frac{1}{2}} \sum_{kl} \frac{\partial}{\partial q_k} \left[m^{-\frac{1}{2}} m^{kl} \frac{\partial \Psi}{\partial q^l} \right] + \frac{8\pi^2}{h^2}(E - F)\Psi = 0,$$

which describes the propagation of a wave in the configuration space constructed by means of the variables q. Setting

$$\Psi = a \cos \frac{2\pi}{h} \varphi,$$

and letting h tend to zero, in the limit one indeed recovers the Jacobi equation

$$\frac{1}{2} \sum_{kl} m^{kl} \frac{\partial \varphi}{\partial q^k} \frac{\partial \varphi}{\partial q^l} + F = E.$$

To quantise an atomic system, one will here again determine the fundamental functions of the corresponding wave equation.

We cannot recall here the successes obtained by this method (papers by Messrs Schrödinger, Fues,[3] Manneback [7], etc.), but we must insist on the difficulties of a conceptual type that it raises. Indeed let us consider, for simplicity, a system of N material points each possessing three degrees of freedom. The configuration space is in an essential way formed by means of the coordinates of the points and yet Mr Schrödinger assumes that in atomic systems material points no longer have a clearly defined position. It seems a little paradoxical to construct a configuration space with the coordinates of points that do not exist. Furthermore, if the propagation of a wave in space has a clear physical meaning, it is not the same as the propagation of a wave in the abstract configuration space, for which the number of dimensions is determined by the number of degrees of freedom of

[a] That is, non-relativistic (*eds.*).

the system. We shall therefore have to return later to the exact meaning of the Schrödinger equation for many-body systems.

By a transformation of admirable ingenuity, Mr Schrödinger has shown that the quantum Mechanics invented by Mr Heisenberg and developed by Messrs Born, Jordan, Pauli, etc., can be translated into the language of wave Mechanics. By comparison with Heisenberg's matrix elements, he was able to derive the expression for the mean charge density of the atom from the functions Ψ, an expression to which we shall return later.

The Schrödinger equations are not relativistic. For the case of a single material point, various authors [8] have given a more general wave equation that is in accord with the principle of Relativity. Let e be the electric charge of the point, $\mathcal{V}$ and $\vec{A}$ the two electromagnetic potentials. The equation that the wave Ψ, written in complex form, must satisfy is[a]

$$\Box\Psi + \frac{4\pi i}{h}\frac{e}{c}\left[\frac{\mathcal{V}}{c}\frac{\partial\Psi}{\partial t} + \sum_{xyz} A_x \frac{\partial\Psi}{\partial x}\right] - \frac{4\pi^2}{h^2}\left[m_0^2c^2 - \frac{e^2}{c^2}(\mathcal{V}^2 - A^2)\right]\Psi = 0.$$

As Mr O. Klein [9] and then the author [10] have shown, the theory of the Universe with five dimensions allows one to give the wave equation a more elegant form in which the imaginary terms, whose presence is somewhat shocking for the physicist, have disappeared.

We must also make a special mention of the beautiful Memoirs in which Mr De Donder [11] has connected the formulas of wave Mechanics to his general theory of Einsteinian Gravity.

3. *The ideas of Mr Born* [12]. – Mr Born was struck by the fact that the continuous wave functions Ψ do not allow us to say *where* the particle whose motion one is studying is and, rejecting the concept of the Wellenpaket, he considers the waves Ψ as giving only a statistical representation of the phenomena. Mr Born seems even to abandon the idea of the determinism of individual physical phenomena: the Quantum Dynamics, he wrote in his letter to *Nature*, 'would then be a singular fusion of mechanics and statistics . . .. A knowledge of Ψ enables us to follow the course of a physical process in so far as it is quantum mechanically determinate: not in a causal sense, but in a statistical one'.[4]

These conceptions were developed in a mathematical form by their author, in Memoirs of fundamental interest. Here, by way of example, is how he treats the collision of an electron and an atom. He writes the Schrödinger equation for the electron-atom system, and he remarks that before the collision, the wave Ψ must be

[a] This is the complex, time-dependent Klein–Gordon equation in an external electromagnetic field (*eds.*).

expressed by the product of the fundamental function[a] representing the initial state of the atom and the plane wave function corresponding to the uniform rectilinear motion of the electron. During the collision, there is an interaction between the electron and the atom, an interaction that appears in the wave equation as the mutual potential energy term. Starting from the initial form of Ψ, Mr Born derives by methods of successive approximation its final form after the collision, in the case of an elastic collision, which does not modify the internal state of the atom, as well as in the case of an inelastic collision, where the atom passes from one stable state to another taking energy from or yielding it to the electron. According to Mr Born, the final form of Ψ determines the probability that the collision may produce this or that result.

The ideas of Mr Born seem to us to contain a great deal of truth, and the considerations that shall now be developed show a great analogy with them.

II. – PROBABLE MEANING OF THE CONTINUOUS WAVES Ψ [13]

4. *Case of a single material point in a static field.* – The body of experimental discoveries made over forty years seems to require the idea that matter and radiation possess an atomic structure. Nevertheless, classical optics has with immense success described the propagation of light by means of the concept of continuous waves and, since the work of Mr Schrödinger, also in wave Mechanics one always considers continuous waves which, not showing any singularities, do not allow us to define the material point. If one does not wish to adopt the hypothesis of the 'Wellenpaket', whose development seems to raise difficulties, how can one reconcile the existence of pointlike elements of energy with the success of theories that consider the waves Ψ? What link must one establish between the corpuscles and the waves? These are the chief questions that arise in the present state of wave Mechanics.

To try to answer this, let us begin by considering the case of a single corpuscle carrying a charge e and moving in an electromagnetic field[b] defined by the potentials $\mathcal{V}$ and $\vec{A}$. Let us suppose first that the motion is one for which the old Mechanics (in relativistic form) is sufficient. If we write the wave Ψ in the canonical form

$$\Psi = a \cos \frac{2\pi}{h} \varphi,$$

[a] That is, eigenfunction (*eds.*).

[b] Here we leave aside the case where there also exists a gravitational field. Besides, the considerations that follow extend without difficulty to that case.

the function φ is then, as we have seen, the Jacobi function, and the velocity of the corpuscle is defined by the formula of Einsteinian Dynamics

$$\vec{v} = -c^2 \frac{\overrightarrow{\text{grad}}\,\varphi + \frac{e}{c}\vec{A}}{\frac{\partial\varphi}{\partial t} - e\mathcal{V}}. \tag{I}$$

We propose to assume by induction that this formula is still valid when the old Mechanics is no longer sufficient, that is to say when φ is no longer a solution of the Jacobi equation.[a] If one accepts this hypothesis, which appears justified by its[5] consequences, the formula (I) completely determines the motion of the corpuscle *as soon as one is given its position at an initial instant*. In other words, the function φ, just like the Jacobi function of which it is the generalisation, determines a whole class of motions, and to know which of these motions is actually described it suffices to know the initial position.

Let us now consider a whole cloud of corpuscles, identical and without interaction, whose motions, determined by (I), correspond to the same function φ but differ in the initial positions. Simple reasoning shows that if the density of the cloud at the initial moment is equal to

$$Ka^2\left(\frac{\partial\varphi}{\partial t} - e\mathcal{V}\right),$$

where K is a constant, it will subsequently remain constantly given by this expression. We can state this result in another form. Let us suppose there be only a single corpuscle whose initial position we ignore; from the preceding, the *probability for its presence* [sa *probabilité de présence*] at a given instant in a volume $d\tau$ of space will be

$$\pi\, d\tau = Ka^2\left(\frac{\partial\varphi}{\partial t} - e\mathcal{V}\right) d\tau. \tag{II}$$

In brief, in our hypotheses, each wave Ψ determines a 'class of motions', and each one of these motions is governed by equation (I) when one knows the initial position of the corpuscle. If one ignores this initial position, the formula (II) gives the probability for the presence of the corpuscle in the element of volume $d\tau$ at the instant t. The wave Ψ then appears as both a *pilot wave* (Führungsfeld of Mr Born) and a *probability wave*. Since the motion of the corpuscle seems to us to be strictly determined by equation (I), it does not seem to us that there is any reason to renounce believing in the determinism of individual physical phenomena,[b] and

[a] Mr De Donder assumes equation (I) as we do, but denoting by φ not the phase of the wave, but the classical Jacobi function. As a result his theory and ours diverge as soon as one leaves the domain where the old relativistic Mechanics is sufficient.

[b] Here, that is, of the motion of individual corpuscles.

it is in this that our conceptions, which are very similar in other respects to those of Mr Born, appear nevertheless to differ from them markedly.

Let us remark that, if one limits oneself to the Newtonian approximation, in (I) and (II) one can replace: $\frac{\partial \varphi}{\partial t} - e\mathcal{V}$ by $m_0 c^2$, and one obtains the simplified forms

$$\vec{v} = -\frac{1}{m_0}\left(\overrightarrow{\text{grad}\,\varphi} + \frac{e}{c}\vec{A}\right), \qquad \text{(I}'\text{)}$$

$$\pi = \text{const} \cdot a^2. \qquad \text{(II}'\text{)}$$

There is one case where the application of the preceding ideas is done in a remarkably clear form: when the initial motion of the corpuscles is uniform and rectilinear in a region free of all fields. In this region, the cloud of corpuscles we have just imagined may be represented by the homogeneous plane wave[6]

$$\Psi = a\cos\frac{2\pi}{h}W\left(t - \frac{vx}{c^2}\right);$$

here a is a constant, and this means that a corpuscle has the same probability to be at any point of the cloud. The question of knowing how this homogeneous plane wave will behave when penetrating a region where a field is present is analogous to that of determining the form of an initially plane light wave that penetrates a refracting medium. In his Memoir 'Quantenmechanik der Stossvorgänge', Mr Born has given a general method of successive approximation to solve this problem, and Mr Wentzel [14] has shown that one can thus recover the Rutherford formula for the deflection of β-rays by a charged centre.

We shall present yet another observation on the Dynamics of the material point such as results from equation (I): for the material point one can always write the equations of the Dynamics of Relativity even when the approximation of the old mechanics is not valid, on condition that one attributes to the body a variable proper mass M_0 given by the formula

$$M_0 = \sqrt{m_0^2 - \frac{h^2}{4\pi^2 c^2}\frac{\Box a}{a}}.$$

5. *The interpretation of interference.* – The new Dynamics allows us to interpret the phenomena of wave Optics in exactly the way that was foreseen, a long time ago now, by Mr Einstein.[a] In the case of light, the wave Ψ is indeed the light wave of the classical theories.[b,c] If we consider the propagation of light in a region strewn

[a] Cf. Chapter 9 (*eds.*).

[b] We then consider Ψ as the 'light variable' without at all specifying the physical meaning of this quantity.

[c] By 'classical theories' de Broglie seems to mean scalar wave optics. In the general discussion (p. 460), de Broglie states that the physical nature of Ψ for photons is unknown (*eds.*).

with fixed obstacles, the propagation of the wave Ψ will depend on the nature and arrangement of these obstacles, but the frequency $\frac{1}{h}\frac{\partial\varphi}{\partial t}$ will not vary (no Doppler effect). The formulas (I) and (II) will then take the form

$$\vec{v} = -\frac{c^2}{h\nu}\overrightarrow{\text{grad}\,\varphi}; \quad \pi = \text{const}\cdot a^2.$$

The second of these formulas shows immediately that the bright and dark fringes predicted by the new theory will coincide with those predicted by the old. To record the fringes, for example by photography, one can do an experiment of short duration with intense irradiation, or an experiment of long duration with feeble irradiation (Taylor's experiment); in the first case one takes a mean in space, in the second case a mean in time, but if the light quanta do not act on each other the statistical result must evidently be the same.

Mr Bothe [15] believed he could deduce, from certain experiments on the Compton effect in a field of interference, the inexactitude of the first formula written above, the one giving the velocity of the quantum, but in our opinion this conclusion can be contested.

6. *The energy-momentum tensor of the waves* Ψ. – In one of his Memoirs [16], Mr Schrödinger gave the expression for the energy-momentum tensor in the interior of a wave Ψ.[a] Following the ideas expounded here, the wave Ψ represents the motion of a cloud of corpuscles; examining the expression given by Schrödinger and taking into account the relations (I) and (II), one then perceives that it decomposes into one part giving the energy and momentum of the particles, and another that can be interpreted as representing a state of stress existing in the wave around the particles. These stresses are zero in the states of motion consistent with the old Dynamics; they characterise the new states predicted by wave Mechanics, which thus appear as 'constrained states' of the material point and are intimately related to the variability of the proper mass M_0. Mr De Donder has also drawn attention to this fact, and he was led to denote the amplitude of the waves that he considered by the name of 'internal stress potential'.

The existence of these stresses allows one to explain how a mirror reflecting a beam of light suffers a radiation pressure, even though according to equation (I), because of interference, the corpuscles of light do not 'strike' its surface.[b]

7. *The dynamics of many-body systems.* – We must now examine how these conceptions may serve to interpret the wave equation proposed by Schrödinger for the Dynamics of many-body systems. We have pointed out above the two

[a] Cf. Schrödinger's report, Section II (*eds.*).
[b] Cf. Brillouin's example in the discussion at the end of de Broglie's lecture (*eds.*).

difficulties that this equation raises. The first, relating to the meaning of the variables that serve to construct the configuration space, disappears if one assumes that the material points always have a quite definite position. The second difficulty remains. It appears to us certain that if one wants to *physically* represent the evolution of a system of N corpuscles, one must consider the propagation of N waves in space, each of the N propagations being determined by the action of the $N-1$ corpuscles connected to the other waves.[a] Nevertheless, if one focusses one's attention only on the corpuscles, one can represent their states by a point in configuration space, and one can try to relate the motion of this representative point to the propagation of a fictitious wave Ψ in configuration space. It appears to us very probable that the wave[b]

$$\Psi = a(q_1, q_2, \ldots, q_n) \cos \frac{2\pi}{h} \varphi(t, q_1, \ldots, q_n),$$

a solution of the Schrödinger equation, is only a fictitious wave which, in the *Newtonian approximation*, plays for the representative point of the system in configuration space the same role of pilot wave and of probability wave that the wave Ψ plays in ordinary space in the case of a single material point.

Let us suppose the system to be formed of N points having for rectangular coordinates

$$x_1^1, x_2^1, x_3^1, \ldots, x_1^N, x_2^N, x_3^N.$$

In the configuration space formed by means of these coordinates, the representative point of the system has for [velocity] components along the axis x_i^k

$$v_{x_i^k} = -\frac{1}{m_k} \frac{\partial \varphi}{\partial x_i^k},$$

m_k being the mass of the kth corpuscle. This is the relation that replaces (I′) for many-body systems. From this, one deduces that the probability for the presence of the representative point in the element of volume $d\tau$ of configuration space is

$$\pi \, d\tau = \text{const} \cdot a^2 \, d\tau.$$

This new relation replaces relation (II′) for many-body systems. It fully accords, it seems to us, with the results obtained by Mr Born for the collision of an electron and an atom, and by Mr Fermi [17] for the collision of an electron and a rotator.[c]

[a] Cf. Section 2.4 (*eds.*).

[b] The amplitude a is time-independent because de Broglie is assuming the time-independent Schrödinger equation. Later in his report, de Broglie applies his dynamics to a non-stationary wave function as well, for the case of an atomic transition (*eds.*).

[c] Cf. the remarks by Born and Brillouin in the discussion at the end of de Broglie's lecture, and the de Broglie–Pauli encounter in the general discussion at the end of the conference (pp. 462ff.) (*eds.*).

Contrary to what happens for a single material point, it does not appear easy to find a wave Ψ that would define the motion of the system taking Relativity into account.

8. *The waves Ψ in micromechanics.* – Many authors think it is illusory to wonder what the position or the velocity of an electron in the atom is at a given instant. We are, on the contrary, inclined to believe that it is possible to attribute to the corpuscles a position and a velocity even in atomic systems, in a way that gives a precise meaning to the variables of configuration space.

This leads to conclusions that deserve to be emphasised. Let us consider a hydrogen atom in one of its stable states. According to Schrödinger, in spherical coordinates[a] the corresponding function Ψ_n is of the form

$$\Psi_n = F(r, \theta)[A \cos m\alpha + B \sin m\alpha] \begin{matrix} \sin \\ \cos \end{matrix} \frac{2\pi}{h} W_n t \quad (m \text{ integer})$$

with

$$W_n = m_0 c^2 - \frac{2\pi^2 m_0 e^4}{n^2 h^2}.$$

If we then apply our formula (I′), we conclude that the electron is motionless in the atom, a conclusion which would evidently be inadmissible in the old Mechanics. However, the examination of various questions and notably of the Zeeman effect has led us to believe that, in its stable states, the H atom must rather be represented by the function

$$\Psi_n = F(r, \theta) \cos \frac{2\pi}{h} \left(W_n t - \frac{mh}{2\pi} \alpha \right),$$

which, being a linear combination of expressions of the type written above, is equally acceptable.[b] If this is true the electron will have, from (I′), a uniform circular motion of speed

$$v = \frac{1}{m_0 r} \frac{mh}{2\pi}.$$

It will then be motionless only in states where $m = 0$.

Generally speaking, the states of the atom at a given instant can always be represented by a function

$$\Psi = \sum_n c_n \Psi_n,$$

[a] r, radius vector; θ, latitude; α, longitude.

[b] In his memoir, 'Les moments de rotation et le magnétisme dans la mécanique ondulatoire' (*Journal de Physique* **8** (1927), 74), Mr Léon Brillouin has implicitly assumed the hypothesis that we formulate in the text.

the Ψ_n being Schrödinger's Eigenfunktionen. In particular, the state of transition $i \to j$ during which the atom emits the frequency ν_{ij} would be given by (this appears to be in keeping with Schrödinger's ideas)

$$\Psi = c_i \Psi_i + c_j \Psi_j,$$

c_i and c_j being two functions of time that change very slowly compared with the trigonometric factors of the Ψ_n, the first varying from 1 to 0 and the second from 0 to 1 during the transition. Writing the function Ψ in the canonical form $a \cos \frac{2\pi}{h}\varphi$, which is always possible, formula (I′) will give the velocity of the electron during the transition, if one assumes the initial position to be given. So it does not seem to be impossible to arrive in this way at a visual representation of the transition.[a]

Let us now consider an ensemble of hydrogen atoms that are all in the same state represented by the same function

$$\Psi = \sum_n c_n \Psi_n.$$

The position of the electron in each atom is unknown to us, but if, in our imagination, we superpose all these atoms, we obtain a *mean atom* where the probability for the presence of one of the electrons in an element of volume $d\tau$ will be given by the formula (II),[b] K being determined by the fact that the total probability for all the possible positions must be equal to unity. The charge density ρ and the current density $\vec{J} = \rho \vec{v}$ in the mean atom are then, from (I) and (II),

$$\rho = Kea^2 \left(\frac{\partial \varphi}{\partial t} - eV \right),$$

$$\vec{J} = -Kec^2a^2 \left(\overrightarrow{\text{grad}\,\varphi} + \frac{e}{c}\vec{A} \right)$$

and these formulas coincide, apart from notation, with those of Messrs Gordon, Schrödinger and O. Klein [18].

Limiting ourselves to the Newtonian approximation, and for a moment denoting by Ψ the wave written in *complex* form, and by $\bar{\Psi}$ the conjugate function, it follows that

$$\rho = \text{const}\cdot a^2 = \text{const}\cdot \Psi\bar{\Psi}.$$

This is the formula to which Mr Schrödinger was led in reformulating the matrix theory; it shows that the electric dipole moment of the mean atom during the

[a] In this example, de Broglie is applying his dynamics to a case where the wave function Ψ has a time-dependent amplitude a (*eds.*).

[b] In this section on atomic physics ('micromechanics') de Broglie considers the non-relativistic approximation, using the limiting formula (I′) – except in this paragraph where he reverts to the relativistic formulas (I) and (II), for the purpose of comparison with the relativistic formulas for charge and current density obtained by other authors (*eds.*).

transition $i \to j$ contains a term of frequency ν_{ij}, and thus allows us to interpret Bohr's frequency relation.

Today it appears certain that one can predict the mean energy radiated by an atom by using the Maxwell-Lorentz equations, on condition that one introduces in these equations the mean quantities ρ and $\rho \overrightarrow{v}$ which have just been defined.[a] One can thus give the correspondence principle an entirely precise meaning, as Mr Debye [19] has in fact shown in the particular case of motion with one degree of freedom. It seems indeed that classical electromagnetism can from now on retain only a statistical value; this is an important fact, whose meaning one will have to try to explore more deeply.

To study the interaction of radiation with an ensemble of atoms, it is rather natural to consider a 'mean atom', immersed in a 'mean light' which one defines by a homogeneous plane wave of the vector potential. The density ρ of the mean atom is perturbed by the action of the light and one deduces from this the scattered radiation. This method, which gives good mean predictions, is related more or less directly to the theories of scattering by Messrs Schrödinger and Klein [20], to the theory of the Compton effect by Messrs Gordon and Schrödinger [21], and to the Memoirs of Mr Wentzel [22] on the photoelectric effect and the Compton effect, etc. The scope of this report does not permit us to dwell any further on this interesting work.

9. *Conclusions and remarks.* – So far we have considered the corpuscles as 'exterior' to the wave Ψ, their motion being only determined by the propagation of the wave. This is, no doubt, only a provisional point of view: a true theory of the atomic structure of matter and radiation should, it seems to us, *incorporate* the corpuscles in the wave phenomenon by considering singular solutions of the wave equations. One should then show that there exists a correspondence between the singular waves and the waves Ψ, such that the motion of the singularities is connected to the propagation of the waves Ψ by the relation (I).[b] In the case of no [external] field, this correspondence is easily established, but it is not so in the general case.

We have seen that the quantities ρ and $\rho \overrightarrow{v}$ appearing in the Maxwell-Lorentz equations must be calculated in terms of the functions Ψ, but that does not suffice to establish a deep link between the electromagnetic quantities and those of wave Mechanics. To establish this link,[c] one should probably begin with singular waves, for Mr Schrödinger has very rightly remarked that the

[a] Cf. Schrödinger's report, p. 412, and the ensuing discussion, and Section 4.5 (*eds.*).

[b] Cf. Section 2.3.1 (*eds.*).

[c] The few attempts made till now in this direction, notably by Mr Bateman (*Nature* **118** (1926), 839) and by the author (*Ondes et mouvements*, chap. VIII, and *C. R.* **184** (1927), 81) can hardly be regarded as satisfactory.

potentials appearing in the wave equations are those that result from the discontinuous structure of electricity and not those that could be deduced from the functions Ψ.

Finally, we point out that Messrs Uhlenbeck and Goudsmit's hypothesis of the magnetic electron, so necessary to explain a great number of phenomena, has not yet found its place in the scope of wave Mechanics.

III. – Experiments showing preliminary direct evidence for the new Dynamics of the electron

10. *Phenomena whose existence is suggested by the new conceptions.* – The ideas that have just been presented lead one to consider the motion of an electron as guided by the propagation of a certain wave. In many usual cases, the old Mechanics remains entirely adequate as a first approximation; but our new point of view, as Elsasser[7] [23] pointed out already in 1925, necessarily raises the following question: 'Could one not observe electron motions that the old Mechanics would be incapable of predicting, and which would therefore be characteristic of wave Mechanics? In other words, for electrons, could one not find the analogue of the phenomena of diffraction and interference?'[a]

These new phenomena, if they exist, must depend on the wavelength of the wave associated with the electron motion. For an electron of speed v, the fundamental formula

$$p = \frac{h\nu}{V}$$

gives

$$\lambda = \frac{V}{\nu} = \frac{h}{p} = \frac{h\sqrt{1-\beta^2}}{m_0 v}.$$

If β is not too close to 1, it suffices to write

$$\lambda = \frac{h}{m_0 v}.$$

[a] This is not a quotation: these words do not appear in the cited 1925 paper by Elsasser. Further, it was de Broglie who first suggested electron diffraction, in a paper of 1923 (see Section 2.2.1) (*eds.*).

Let $\mathcal{V}$ be the potential difference, expressed in volts, that is capable of imparting the speed v to the electron; numerically, for the wavelength in centimetres, one will have[a]

$$\lambda = \frac{7.28}{v} = \frac{12.25}{\sqrt{\mathcal{V}}} \times 10^{-8}.$$

To do precise experiments, it is necessary to use electrons of at least a few volts: from which one has an upper limit for λ of a few angstroms. One then sees that, even for slow electrons, the phenomena being sought are analogous to those shown by X-rays and not to those of ordinary light. As a result, it will be difficult to observe the diffraction of a beam of electrons by a small opening, and if one wishes to have some chance of obtaining diffraction by a grating, one must either consider those natural three-dimensional gratings, the crystals, or use ordinary gratings under a very grazing incidence, as has been done recently for X-rays. On making slow electrons pass through a crystalline powder or an amorphous substance, one could also hope to notice the appearance of rings analogous to those that have been obtained and interpreted in the X-ray domain by Messrs Hull, Debye and Scherrer, Debierne, Keesom and De Smedt, etc.

The exact theoretical prediction of the phenomena to be observed along these lines is still not very advanced. Let us consider the diffraction of a beam of electrons with the same velocity by a crystal; the wave Ψ will propagate following the general equation, in which one has to insert the potentials created by the atoms of the crystal considered as centres of force. One does not know the exact expression for these potentials but, because of the regular distribution of atoms in the crystal, one easily realises that the scattered amplitude will show maxima in the directions predicted by Mr von Laue's theory. Because of the role of pilot wave played by the wave Ψ, one must then observe a selective scattering of the electrons in these directions.

Using his methods, Mr Born has studied another problem: that of the collision of a narrow beam of electrons with an atom. According to him, the curve giving the number of electrons that have suffered an inelastic collision as a function of the scattering angle must show maxima and minima; in other words, these electrons will display rings on a screen placed normally to the continuation of the incident beam.

It would still be premature to speak of agreement between theory and experiment; nevertheless, we shall present experiments that have revealed phenomena showing at least broadly the predicted character.

[a] Here we have adopted the following values:

$$h = 6.55 \times 10^{-27} \text{ erg-seconds},$$
$$m_0 = 9 \times 10^{-28} \text{ gr},$$
$$e = 4.77 \times 10^{-10} \text{ e.s.u.}$$

11. *Experiments by Mr E.G. Dymond* [24]. – Without feeling obliged to follow the chronological order, we shall first present Mr Dymond's experiments:

A flask of purified helium contained an 'electron gun', which consisted of a brass tube containing an incandescent filament of tungsten and in whose end a slit was cut. This gun discharged a well-collimated beam of electrons into the gas, with a speed determined by the potential difference (50 to 400 volts) established between the filament and the wall of the tube. The wall of the flask had a slit through which the electrons could enter a chamber where the pressure was kept low by pumping and where, by curving their trajectories, a magnetic field brought them onto a Faraday cylinder.

Mr Dymond first kept the orientation of the gun fixed and measured the speed of the electrons thus scattered by a given angle. He noticed that most of the scattered electrons have the same energy as the primary electrons; they have therefore suffered an elastic collision. Quite a large number of electrons have a lower speed corresponding to an energy loss from about 20 to 55 volts: this shows that they made the He atom pass from the normal state 1^1S to the excited state 2^1S. One also observes a lower proportion of other values for the energy of the scattered electrons; we shall not discuss the interpretation that Mr Dymond has given them, because what interests us most here is the variation of the number of scattered particles with the scattering angle θ. To determine this number, Mr Dymond varied the orientation of the gun inside the flask, and for different scattering angles collected the electrons that suffered an energy loss equivalent to 20 to 55 volts; he constructed a series of curves of the angular distribution of these electrons for different values of the tension $\mathcal{V}$ applied to the electron gun. The angular distribution curve shows a very pronounced maximum for a low value of θ, and this maximum appears to approach $\theta = 0$ for increasing values of $\mathcal{V}$.

Another, less important, maximum appears towards $\theta = 50°$ for a primary energy of about a hundred volts, and then moves for increasing values of $\mathcal{V}$ towards increasing θ. Finally, a very sharp maximum appears for a primary energy of about 200 volts at $\theta = 30°$, and then seems independent of $\mathcal{V}$. These facts are summarised in the following table given by Dymond:[8]

$\mathcal{V}$ (volts)	Positions of the maxima (°)		
48.9	24	–	–
72.3	8	–	–
97.5	5	–	50
195	<2.5	30	59
294	<2.5	30	69
400	<2.5	30	70

The above results must very probably be interpreted with the aid of the new Mechanics and are to be related to Mr Born's predictions. Nevertheless, as Mr Dymond very rightly says, 'the theoretical side of the problem is however not yet sufficiently advanced to give detailed information on the phenomena to be expected, so that the results above reported cannot be said to substantiate the wave mechanics except in the most general way'.[9]

12. *Experiments by Messrs C. Davisson and L. H. Germer.* – In 1923, Messrs Davisson and Kunsman [25] published peculiar results on the scattering of electrons at low speed. They directed a beam of electrons, accelerated by a potential difference of less than 1000 volts, onto a block of platinum at an incidence of 45° and determined the distribution of scattered electrons by collecting them in a Faraday cylinder. For potentials above 200 volts, one observed a steady decrease in scattering for increasing values of the deviation angle, but for smaller voltages the curve of angular variation showed two maxima. By covering the platinum with a deposit of magnesium, one obtained a single small maximum for electrons of less than 150 volts. Messrs Davisson and Kunsman attributed the observed phenomena to the action of various layers of intra-atomic electrons on the incident electrons, but it seems rather, according to Elsasser's opinion, that the interpretation of these phenomena is a matter for the new Mechanics.

Resuming analogous experiments with Mr Germer [26], Mr Davisson obtained very important results this year, which appear to confirm the general predictions and even the formulas of Wave Mechanics.

The two American physicists sent homogeneous beams of electrons onto a crystal of nickel, cut following one of the 111 faces of the regular octahedron (nickel is a cubic crystal). The incidence being normal, the phenomenon necessarily had to show the ternary symmetry around the direction of the incident beam. In a cubic crystal cut in this manner, the face of entry is cut obliquely by three series of 111 planes, three series of 100 planes, and six series of 110 planes. If one takes as positive orientation of the normals to these series of planes the one forming an acute angle with the face of entry, then these normals, together with the direction of incidence, determine distinguished azimuths, which Messrs Davisson and Germer call azimuths (111), (100), (110), and for which they studied the scattering; because of the ternary symmetry, it evidently suffices to explore a single azimuth of each type.

Let us place ourselves at one of the distinguished azimuths and let us consider only the distribution of Ni atoms on the face of entry of the crystal, which we assume to be perfect. These atoms form lines perpendicular to the azimuth being considered and whose equidistance d is known from crystallographic data. The different directions of scattering being identified in the azimuthal plane by the angle

θ of co-latitude, the waves scattered by the atoms in the face of entry must be in phase in directions such that one has

$$\theta = \arcsin\left(\frac{n\lambda}{d}\right) = \arcsin\left(\frac{n}{d}\frac{12.25}{\sqrt{\mathcal{V}}}\cdot 10^{-8}\right) \qquad (n \text{ integer}).$$

One must then expect to observe maxima in these directions, for the scattering of the electrons by the crystal.

Now here is what Messrs Davisson and Germer observed. By gradually varying the voltage $\mathcal{V}$ that accelerates the electrons one observes, in the neighbourhood of *certain* values of $\mathcal{V}$, very distinct scattering maxima in directions whose co-latitude is accurately given by the above formula (provided one sets in general $n = 1$, and sometimes $n = 2$). There is direct numerical confirmation of the formulas of the new Dynamics; this is evidently a result of the highest importance.

However, the explanation of the phenomenon is not complete: one must explain why the scattering maxima are observed only in the neighbourhood of certain particular values of $\mathcal{V}$, and not for all values of $\mathcal{V}$. One interpretation naturally comes to mind: we assumed above that only the face of entry of the crystal played a role, but one can assume that the electron wave penetrates somewhat into the crystal and, further, in reality the face of entry will never be perfect and will be formed by several parallel 111 planes forming steps. In these conditions, it is not sufficient to consider the interference of the waves scattered by a single reticular plane at the surface, one must take into account the interference of the waves scattered by several parallel reticular planes. In order for there to be a strong scattering in a direction θ, θ and $\mathcal{V}$ must then satisfy not only the relation written above, but also another relation which is easy to find; the scattering must then be *selective*, that is to say, occur with [significant] intensity only for certain values of $\mathcal{V}$, as experiment shows. Of course, the theory that has just been outlined is a special case of Laue's general theory.

Unfortunately, as Messrs Davisson and Germer have themselves remarked, in order to obtain an exact prediction of the facts in this way, it is necessary to attribute to the separation of the 111 planes next to the face of entry a smaller value (of about 30%) than that provided by Crystallography and by direct measurements by means of X-rays. It is moreover not unreasonable to assume that the very superficial reticular planes have a spacing different from those of the deeper planes, and one can even try to connect this idea to our current conceptions concerning the equilibrium of crystalline gratings.

If one accepts the preceding hypothesis, the scattering must be produced by a very small number of reticular planes in the entirely superficial layer of the crystal; the concentration of electrons in preferred directions must then be much less pronounced than in the case of scattering by a whole unlimited spatial grating.

Is it nevertheless sufficient in order to explain the 'peaks' observed by Davisson and Germer? To this question, Mr Patterson has recently provided an affirmative answer, by showing that the involvement of just two superficial reticular planes already suffices to predict exactly the variations of the selective reflection of electrons observed in the neighbourhood of

$$\theta = 50^\circ, \quad \mathcal{V} = 54 \text{ volts}.$$

To conclude, we can do no better than quote the conclusion of Mr Patterson [27]: 'The agreement of these results with calculation seems to indicate that the phenomenon can be explained as a diffraction of waves in the *outermost layers* of the crystal surface. It also appears [...] that a complete analysis of the results of such experiments will give valuable information as to the conditions prevailing in the actual surface, and that a new method has been made available for the investigation of the structure of crystals in a region which has up to the present almost completely escaped observation'.[10]

13. *Experiments by Messrs G. P. Thomson and A. Reid* [28]. – Very recently, Messrs Thomson and Reid have made the following results known: if a narrow pencil of homogeneous cathode rays passes normally through a celluloid film, and is then received on a photographic plate placed parallel to the film at 10 cm behind it, one observes rings around the central spot. With rays of 13 000 volts, a photometric examination has revealed the existence of three rings. By gradually increasing the energy of the electrons, one sees the rings appear around 2500 volts, and they have been observed up to 16 500 volts. The radii of the rings decrease when the energy increases and, it seems, approximately in inverse proportion to the speed, that is, to our wavelength λ.

These observations are very interesting, and again confirm the new conceptions in broad outline. Is it a question here of an atomic phenomenon analogous to those observed by Dymond, or else of a phenomenon of mutual interference falling into one of the categories studied by Debye and Scherrer, Hull, Debierne, Keesom and De Smedt? We are unable to say, and we limit ourselves to remarking that here the electrons used are relatively fast; this is interesting from the experimental point of view, because it is much easier to study electrons of a few thousand volts than electrons of about a hundred volts.

Bibliography[a]

[1] Louis de Broglie, *C. R.*, **177** (1923), 507, 548 and 630; **179** (1924), 39, 676 and 1039. Doctoral thesis, November 1924 (Masson, publisher), *Annales de Physique* (10), **III** [(1925)], 22. *J. de Phys.* (6), **VII** [(1926)], 1.
[2] S. N. Bose, *Zts. f. Phys.*, **27** (1924), 384.
[3] A. Einstein, *Berl. Ber.* (1924), 261; (1925), [3].
[4] P. Jordan, *Zts. f. Phys.*, **33** (1925), 649.
E. Schrödinger, *Physik. Zts.*, **27** (1926), 95.
P. Dirac, *Proc. Roy. Soc. A*, **112** (1926), 661.
E. Fermi, *Zts. f. Phys.*, **36** (1926), 902.
L. S. Ornstein and H. A. Kramers, *Zts. f. Phys.*, **42** (1927), 481.
[5] E. Schrödinger, *Ann. der Phys.*, **79** (1926), 361, 489 and 734; **80** (1926), 437; **81** (1926), 109. *Naturwissensch.*, 14th year [(1926)], 664. *Phys. Rev.*, **28** (1926), 1051.
[6] L. Brillouin, *C. R.*, **183** (1926), 24 and 270. *J. de Phys.* [(6)], **VII** (1926), 353.
G. Wentzel, *Zts. f. Phys.*, **38** (1926), 518.
H. A. Kramers, *Zts. f. Phys.*, **39** (1926), 828.
[7] E. Fues, *Ann. der Phys.*, **80** (1926), 367; **81** (1926), 281.
C. Manneback, *Physik. Zts.*, **27** (1926), 563; **28** (1927), 72.
[8] L. de Broglie, *C. R.*, **183** (1926), 272. *J. de Phys.* (6), **VII** (1926), 332.
O. Klein, *Zts. f. Phys.*, **37** (1926), 895.
[V.] Fock, *Zts. f. Phys.*, **38** (1926), 242.
W. Gordon, *Zts. f. Phys.*, **40** (1926), 117.
L. Flamm, *Physik. Zts.*, **27** (1926), 600.
[9] O. Klein, *loc. cit.* in [8].
[10] L. de Broglie, *J. de Phys.* (6), **VIII** (1927), 65.
(See also L. Rosenfeld, *Acad. Roy. Belg.* (5), **13**, n. 6.)
[11] T. De Donder, *C. R.*, **182** (1926), 1512; **183** (1926), 22 (with Mr Van den Dungen); **183** (1926), 594; **184** (1927), 439 and 810. *Acad. Roy. Belg.*, sessions of 9 October 1926, of 8 January, 5 February, 5 March and 2 April 1927.
[12] M. Born, *Zts. f. Phys.*, **37** (1926), 863; **38** (1926), 803; **40** (1926), 167. *Nature*, **119** (1927), 354.
[13] L. de Broglie, *C. R.*, **183** (1926), 447, and **184** (1927), 273. *Nature*, **118** (1926), 441. *J. de Phys.* (6), **VIII** (1927), 225. *C. R.*, **185** (1927), 380.
(See also E. Madelung, *Zts. f. Phys.*, **40** (1926), 322.)
[14] G. Wentzel, *Zts. f. Phys.*, **40** (1926), 590.
[15] W. Bothe, *Zts. f. Phys.*, **41** (1927), 332.
[16] E. Schrödinger, *Ann. der Phys.*, **82** (1927), 265.
[17] E. Fermi, *Zts. f. Phys.*, **40** (1926), 399.
[18] W. Gordon and E. Schrödinger, *loc. cit.* in [8] and [16].
O. Klein, *Zts. f. Phys.*, **41** (1927), 407.
[19] P. Debye, *Physik. Zts.*, **27** (1927), 170.
[20] E. Schrödinger and O. Klein, *loc. cit.* in [5] and [18].
[21] W. Gordon, *loc. cit.* in [8].
E. Schrödinger, *Ann. der Phys.*, **82** (1927), 257.
[22] G. Wentzel, *Zts. f. Phys.*, **43** (1927), 1; **41** (1927), 828.
(See also G. Beck, *Zts. f. Phys.*, **41** (1927), 443; J. R. Oppenheimer, *Zts. f. Phys.*, **41** (1927), 268.)

[a] The style of the bibliography has been slightly modernised and uniformised with that used in the other reports (*eds.*).

[23] [W.] Elsasser, *Naturwissensch.*, **13** (1925), 711.
[24] E. G. Dymond, *Nature*, **118** (1926), 336. *Phys. Rev.*, **29** (1927), 433.
[25] C. Davisson and C. H. Kunsman, *Phys. Rev.*, **22** (1923), 242.
[26] C. Davisson and L. H. Germer, *Nature*, **119** (1927), 558.
[27] A. L. Patterson, *Nature*, **120** (1927), 46.
[28] C. P. Thomson and A. Reid, *Nature*, **119** (1927), 890.

Discussion of Mr de Broglie's report

MR LORENTZ. – I should like to see clearly how, in the first form of your theory, you recovered Sommerfeld's quantisation conditions. You obtained a single condition, applicable only to the case where the orbit is closed: the wave must, after travelling along the orbit, finish in phase when it comes back to the initial point. But in most cases the trajectory is not closed; this happens, for example, for the hydrogen atom when one takes relativity into account; the trajectory is then a rosette, and never comes back to its initial point.

How did you find the quantisation conditions applicable to these multiperiodic problems?

MR DE BROGLIE. – The difficulty is resolved by considering pseudo-periods, as I pointed out in my Thesis (chap. III, p. 41). When a system is multiperiodic, with partial periods $\tau_1, \tau_2, \ldots, \tau_n$, one can prove that one can find quasi-periods τ that are nearly exactly whole multiples of the partial periods:

$$\tau = m_1\tau_1 + \varepsilon_1 = m_2\tau_2 + \varepsilon_2 \ldots = m_n\tau_n + \varepsilon_n,$$

the $m_1, m_2, \ldots, m_n$ being integers and the $\varepsilon_1, \varepsilon_2, \ldots, \varepsilon_n$ as small as one likes. The trajectory then never comes back to its initial point, but at the end of a quasi-period τ it comes back as closely as one likes to the initial position. One will then be led to write that, at the end of a quasi-period, the wave finishes in phase; now, there is an infinite number of quasi-periods, corresponding to all kinds of systems of values of the integers $m_1, m_2, \ldots, m_n$. In order that the wave finishes in phase after any one of these quasi-periods, it is necessary that one have[11]

$$\int_{\tau_1} p_1 dq_1 = n_1 h, \quad \int_{\tau_2} p_2 dq_2 = n_2 h, \quad \ldots, \quad \int_{\tau_n} p_n dq_n = n_n h,$$

which gives exactly Sommerfeld's conditions.[a]

MR BORN. – The definition of the trajectory of a particle that Mr de Broglie has given seems to me to present difficulties in the case of a collision between an electron and an atom. In an elastic collision, the speed of the particle must be the same after the collision as before. I should like to ask Mr de Broglie if that follows from his formula.

MR DE BROGLIE. – That follows from it, indeed.

[a] Darrigol (1993, pp. 342–3, 364–5) shows that this derivation is faulty: the condition that the wave should finish in phase after any quasi-period does *not* imply the n separate conditions listed above.

MR BRILLOUIN. – It seems to me that no serious objection can be made to the point of view of L. de Broglie. Mr Born can doubt the real existence of the trajectories calculated by L. de Broglie, and assert that one will never be able to observe them, but he cannot prove to us that these trajectories do not exist. There is no contradiction between the point of view of L. de Broglie and that of other authors, since, in his report (§8, p. 354[12]) L. de Broglie shows us that his formulas[a] are in exact agreement with those of Gordon, at present accepted by all physicists.

MR PAULI. – I should like to make a small remark on what seems to me to be the mathematical basis of Mr de Broglie's viewpoint concerning particles in motion on definite trajectories. His conception is based on the principle of conservation of charge:

$$\frac{\partial \rho}{\partial t} + \frac{\partial s_1}{\partial x} + \frac{\partial s_2}{\partial y} + \frac{\partial s_3}{\partial z} = 0 \quad \text{or} \quad \sum_{k=1}^{4} \frac{\partial s_k}{\partial x_k} = 0, \tag{a}$$

which is a consequence of the wave equation, when one sets

$$is_k = \psi \frac{\partial \psi^*}{\partial x_k} - \psi^* \frac{\partial \psi}{\partial x_k} + \frac{4\pi i}{h} \frac{e}{c} \Phi_k \psi \psi^*.$$

Mr de Broglie introduces, in place of the complex function ψ, the two real functions a and φ defined by

$$\psi = a e^{\frac{2\pi i}{h}\varphi}, \quad \psi^* = a e^{-\frac{2\pi i}{h}\varphi}.$$

Substituting these expressions into the expression for s_k yields:

$$s_k = \frac{4\pi}{h} a^2 \left(\frac{\partial \varphi}{\partial x_k} + \frac{e}{c} \Phi_k \right).$$

From this follow the expressions given by Mr de Broglie for the velocity vector, defined by

$$v_1 = \frac{s_1}{\rho}, \quad v_2 = \frac{s_2}{\rho}, \quad v_3 = \frac{s_3}{\rho}. \tag{b}$$

Now if in a field theory there exists a conservation principle of the form (a), it is always formally possible to introduce a velocity vector (b), depending on space and time, and to imagine furthermore corpuscles that move following the current lines of this vector. Something similar was already proposed in optics by Slater; according to him, light quanta should always move following the lines of the Poynting vector. Mr de Broglie now introduces an analogous representation for material particles.

[a] That is, de Broglie's equations for the mean charge and current density to be used in semiclassical radiation theory (*eds.*).

In any case, I do not believe that this representation may be developed in a satisfactory manner; I intend to return to this during the general discussion.[a]

MR SCHRÖDINGER. – If I have properly understood Mr de Broglie, the *velocity* of the particles must have its analogue in a vector field composed of the three *spatial* components of the *current in a four-dimensional space*, after division of these by the component *with respect to time* (that is, the charge density). I should like simply to recall now that there exist still other vector quantities of a field, which can be made to correspond with the velocity of the particles, such as the components of the *momentum density* (see *Ann. d. Phys.*[13] **82**, 265). Which of the two analogies is the more convincing?

MR KRAMERS. – The fact that with independent particles in motion one cannot construct an energy-momentum tensor having the properties required by Maxwell's theory constitutes nevertheless a difficulty.

MR PAULI. – The quotient of the momentum by the energy density which Mr Schrödinger considers would in fact lead in a relativistic calculation to other particle trajectories than would the quotient of the densities of current and of charge.

MR LORENTZ. – In using his formulas for the velocity of the electron, has Mr de Broglie not calculated this velocity in particular cases, for example for the hydrogen atom?

MR DE BROGLIE. – When one applies the formula for the velocity to a wave function representing a stable state of the hydrogen atom according to Mr Schrödinger, one finds circular orbits. One does not recover the elliptical orbits of the old theory (see my report, §8).

MR EHRENFEST. – Can the speed of an electron in a stationary orbit be zero?

MR DE BROGLIE. – Yes, the speed of the electron can be zero.

MR SCHRÖDINGER. – Mr de Broglie says that in the case of the hydrogen atom his hypothesis leads to *circular* orbits. That is true for the particular solutions of the wave equation that one obtains when one separates the problem in polar coordinates in space; perhaps it is still true for the solutions that one obtains by

[a] See pp. 462ff. (*eds.*).

making use of parabolic or elliptical coordinates. But in the case of a degeneracy (as he considers it here) it is, in reality, not at all the particular solutions which have a significance, but only an *arbitrary* linear combination, with constant coefficients, of all the particular solutions belonging to the same eigenvalue, because there is no means of distinguishing between them, all linear combinations being equally justified in principle. In these conditions, much more complicated types of orbit will certainly appear. But I do not believe that in the atomic domain one may still speak of 'orbits'.

MR LORENTZ. – Does one know of such more complicated orbits?

MR SCHRÖDINGER. – No, one does not know of them; but I simply wanted to say that if one finds circular orbits, that is due to a fortuitous choice of particular solutions that one considers, and this choice cannot be motivated in a way that has no arbitrariness.

MR BRILLOUIN. – Perhaps it is not superfluous to give some examples that illustrate well the meaning of Mr L. de Broglie's formulas, and that allow one to follow the motion of the particles guided by the phase wave. If the wave is plane and propagates freely, the trajectories of the particles are the rays normal to the wave surface. Let us suppose that the wave is reflected by a plane mirror, and let θ be the angle of incidence; the wave motion in front of the mirror is given by a superposition of the incident wave

$$\psi_1 = a_1 \cos 2\pi \left(\frac{t}{T} - \frac{x \sin\theta - z \cos\theta}{\lambda} \right)$$

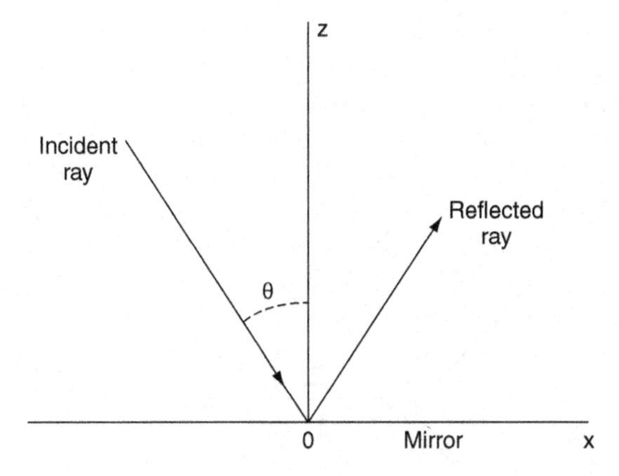

Fig. 1.

and the reflected wave

$$\psi_2 = a_1 \cos 2\pi \left(\frac{t}{T} - \frac{x \sin\theta + z\cos\theta}{\lambda} \right),$$

which gives

$$\psi = 2a_1 \cos \frac{2\pi z \cos\theta}{\lambda} \cos 2\pi \left(\frac{t}{T} - \frac{x \sin\theta}{\lambda} \right).$$

This wave is put in L. de Broglie's canonical form

$$\psi = a \cos \frac{2\pi}{h} \varphi$$

with

$$a = 2a_1 \cos \frac{2\pi z \cos\theta}{\lambda} \quad \text{and} \quad \varphi = h \left(\frac{t}{T} - \frac{x \sin\theta}{\lambda} \right).$$

Let us then apply L. de Broglie's formulas, in the simplified form given on page 351 (§5);[14] and let us suppose that it is a light wave guiding the photons; the velocity of these is

$$\vec{v} = -\frac{c^2}{h\nu} \overrightarrow{\text{grad}}\, \varphi.$$

We see that the projectiles move parallel to the mirror, with a speed $v_x = c \sin\theta$, less than c. Their energy remains equal to $h\nu$, because their mass has undergone a variation, according to the following formula (report by L. de Broglie, p. 350):[15]

$$M_0 = \sqrt{m_0^2 - \frac{h^2}{4\pi^2 c^2} \frac{\Box a}{a}} = \frac{h}{2\pi c} \sqrt{-\frac{\Box a}{a}} = \frac{h\nu}{c^2} \cos\theta.$$

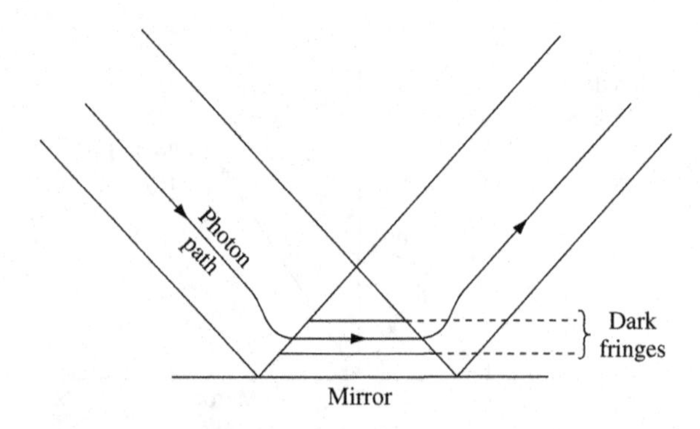

Fig. 2.

The mass of the photons, which is zero in the case where the wave propagates freely, is then assumed to take a non-zero value in the whole region where there is interference or deviation of the wave.

Let us draw a diagram for the case of a limited beam of light falling on a plane mirror; the interference is produced in the region of overlap of two beams. The trajectory of a photon will be as follows: at first a rectilinear path in the incident beam, then a bending at the edge of the interference zone, then a rectilinear path parallel to the mirror, with the photon travelling in a bright fringe and avoiding the dark interference fringes; then, when it comes out, the photon retreats following the direction of the reflected light beam.

No photon actually strikes the mirror, nevertheless the mirror suffers classical radiation pressure; it is in order to explain this fact that L. de Broglie assumes the existence of special stresses in the interference zone; these stresses, when added to the tensor of momentum flux transported by the photons, reproduce the classical Maxwell tensor; there is then no difference in the mechanical effects produced by the wave during its reflection by the mirror.

These remarks show how L. de Broglie's system of hypotheses preserves the classical formulas, and avoids a certain number of awkward paradoxes. One thus obtains, for example, the solution to a curious problem posed by G. N. Lewis (*Proc. Nat. Acad.* **12** (1926), 22 and 439), which was the subject of discussions between this author and R. C. Tolman and S. Smith (*Proc. Nat. Acad.* **12** (1926), 343 and 508).

Lewis assumed that the photons always follow the path of a light ray of geometrical optics, but that they choose, among the different rays, only those that lead from the luminous source to a bright fringe situated on an absorbing body. He then considered a source S whose light is reflected by two mirrors AA′ and BB′; the light beams overlap, producing interference [zones] in which one places a

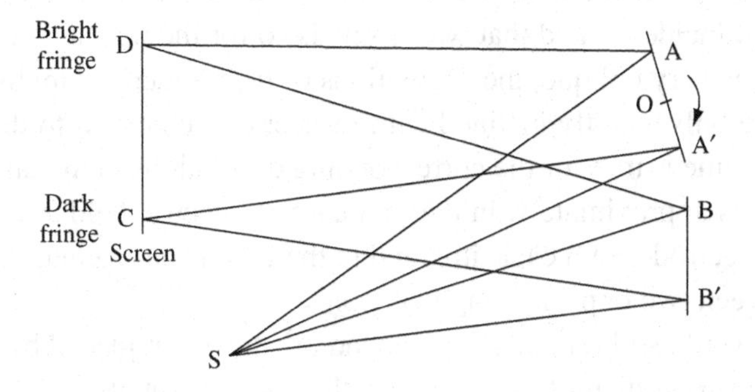

Fig. 3.

screen CD; the dimensions are assumed to be such that there is a bright fringe on one of the edges D of the screen and a dark fringe on the other edge C. Following the hypothesis of Lewis, the photons would follow only the paths SBD and SAD, which end at the bright fringe D; no photon will take the path SA′C or SB′C. All the photons come to strike the mirror AA′ on the edge A, so one could predict that this mirror would suffer a torque; if one made it movable around an axis O, it would tend to turn in the direction of the arrow.

This paradoxical conclusion is entirely avoided by L. de Broglie, since his system of hypotheses preserves the values of radiation pressure. This example shows clearly that there is a contradiction between the hypothesis of rectilinear paths for the photons (following the light rays) and the necessity of finding photons only where a bright interference fringe is produced, no photon going through the regions of dark fringes.

MR LORENTZ draws attention to a case where the classical theory and the photon hypothesis lead to different results concerning the ponderomotive forces produced by light. Let us consider reflection by the hypotenuse face of a glass prism, the angle of incidence being larger than the angle of total reflection. Let us place a[16] second prism behind the first, at a distance of the order of magnitude of the wavelength, or only a fraction of this length. Then, the reflection will no longer be total. The light waves that penetrate the layer of air reach the second prism before their intensity is too much weakened, and there give rise to a beam transmitted in the direction of the incident rays.

If, now, one calculates the Maxwell stresses on a plane situated in the layer of air and parallel to its surfaces, one finds that, if the angle of incidence exceeds a certain value (60° for example), there will be an *attraction* between the two prisms. Such an effect can never be produced by the motion of corpuscles, this motion always giving rise to a [positive] pressure as in the kinetic theory of gases.

What is more, in the classical theory one easily sees the origin of the 'negative pressure'. One can distinguish two cases, that where the electric oscillations are in the plane of incidence and that where this is so for the magnetic oscillations. If the incidence is very oblique, the oscillations of the incident beam that I have just mentioned are only slightly inclined with respect to the normal to the hypotenuse face, and the same is true for the corresponding oscillations in the layer of air.

One then has approximately, in the first case an electric field such as one finds between the electrodes of a capacitor, and in the second case a magnetic field such as exists between two opposite magnetic poles.

The effect would still remain if the second prism were replaced by a glass plate, but it must be very difficult to demonstrate this experimentally.

Notes on the translation

1 The integral sign is printed as '$\int_0$' in the original.
2 The French uses 'Wellenpacket' throughout.
3 Mis-spelt as 'Fuess'.
4 We follow the original English, which is a translation by Oppenheimer, from Born (1927, p. 355). De Broglie translates 'mechanics' as 'dynamique' and includes the words 'La Dynamique des Quanta' in the quotation, where Born has 'it' (referring to 'quantum mechanics').
5 The French reads 'ces' [these] rather than 'ses' [its].
6 'v' is misprinted as '$\mathcal{V}$'.
7 Consistently mis-spelt throughout the text as 'Elsaesser'.
8 For clarity, the presentation of the table has been slightly altered.
9 We have used the original English (Dymond 1927, p. 441).
10 We use here the original English text (Patterson 1927, p. 47). De Broglie changes 'of these results' to 'des résultats expérimentaux', omits the italics and translates 'valuable' by 'exacts'.
11 The last equality is misprinted as '$\int_{\tau_n} p_n dq = n_n h_n$'.
12 The original reads 'p. 18', which is presumably in reference to de Broglie's Gauthier-Villars 'preprint', in all likelihood circulated before the conference (preprints of other lectures were circulated as mimeographs). Cf. Chapter 1, p. 17.
13 '*Ann. de Phys.*' in the original.
14 This is 'p. 117' in the original.
15 Again, 'p. 117' in the original.
16 Misprinted as 'au' instead of 'un'.

Quantum mechanics[a]

BY MESSRS MAX BORN AND WERNER HEISENBERG

INTRODUCTION

Quantum mechanics is based on the intuition that the essential difference between atomic physics and classical physics is the occurrence of discontinuities (see in particular [1,4,58–63]).[b] Quantum mechanics should thus be considered a direct continuation of the quantum theory founded by Planck, Einstein and Bohr. Bohr in particular stressed repeatedly, already before the birth of quantum mechanics, that the discontinuities must lead to the introduction of new kinematical and mechanical concepts, so that indeed classical mechanics and its corresponding conceptual scheme should be abandoned [1,4]. Quantum mechanics tries to introduce the new concepts through a precise analysis of what is 'observable in principle'. In fact, this does not mean setting up the principle that a sharp division between 'observable' and 'unobservable' quantities is possible and necessary. As soon as a conceptual scheme is given, one can infer from the observations to other facts that are actually not observable directly, and the boundary between 'observable' and 'unobservable'[1] quantities becomes altogether indeterminate. But if the conceptual scheme itself is still unknown, it will be expedient to enquire only about the observations themselves, without drawing conclusions from them, because otherwise wrong concepts and prejudices taken over from before will block the way to recognising the physical relationships [Zusammenhänge]. At the same time the new conceptual scheme provides the anschaulich content of the new theory.[c] From a theory that is anschaulich in this sense, one can thus demand only that it is consistent in itself and that it allows one to predict unambiguously the

[a] Our translation follows the German typescript in AHQP-RDN, document M-0309. Discrepancies between the typescript and the published version are reported in the endnotes. The published version is reprinted in Heisenberg (1984, ser. B, vol. 2, pp. 58–99) (*eds.*).

[b] Numbers in square brackets refer to the bibliography at the end [of the report].

[c] For the notion of Anschaulichkeit, see the comments in Sections 3.4.7, 4.6 and 8.3 (*eds.*).

results for all experiments conceivable in its domain. Quantum mechanics is meant as a theory that is in this sense anschaulich and complete for the micromechanical processes [46].[2]

Two kinds of discontinuities are characteristic of atomic physics: the existence of corpuscles (electrons, light quanta) on the one hand, and the occurrence of discrete stationary states (discrete[3] energy values, momentum values etc.) on the other. Both kinds of discontinuities can be introduced in the classical theory only through artificial auxiliary assumptions. For quantum mechanics, the existence of discrete stationary states and energy values is just as natural as the existence of discrete eigenoscillations in a classical oscillation problem [4]. The existence of corpuscles will perhaps later turn out to be reducible just as easily to discrete stationary states of the wave processes (quantisation of the electromagnetic waves on the one hand, and of the de Broglie waves on the other) [4], [54].

The discontinuities, as the notion of 'transition probabilities' already shows, introduce a statistical element into atomic physics. This statistical element forms an *essential* part of the foundations of quantum mechanics (see in particular [4,30,38,39,46,60,61,62]);[4] according to the latter, for instance, in many cases the course of an experiment is determinable from the initial conditions only statistically, at least if in fixing the initial conditions one takes into account only the experiments conceivable in principle up to now. This consequence of quantum mechanics is empirically testable. Despite its statistical character, the theory nevertheless accounts for the apparently fully causal determination of macroscopic[5] processes. In particular, the principles of conservation of energy and momentum hold exactly also in quantum mechanics. There seems thus to be no empirical argument against accepting fundamental indeterminism for the microcosm.

I. – THE MATHEMATICAL METHODS OF QUANTUM MECHANICS[a]

The phenomenon for whose study the mathematical formalism of quantum mechanics was first developed is the spontaneous radiation of an excited atom. After innumerable attempts to explain the structure of the line spectra with classical mechanical models had proved inadequate, one returned to the direct description of the phenomenon on the basis of its simplest empirical laws (Heisenberg [1]). First among these is Ritz's combination principle, according to which the frequency of each spectral line of an atom appears as the difference of two terms $\nu_{ik} = T_i - T_k$; thus the set of all lines of the atom will be best described by specifying a quadratic array [Schema], and since each line possesses besides its frequency also

[a] Section 3.3 contains additional material on the less familiar aspects of the formalisms presented here (*eds.*).

an intensity and a phase, one will write in each position of the array an elementary oscillation function with complex amplitude:

$$\begin{pmatrix} q_{11}e^{2\pi i\nu_{11}t} & q_{12}e^{2\pi i\nu_{12}t} & \dots \\ q_{21}e^{2\pi i\nu_{21}t} & q_{22}e^{2\pi i\nu_{22}t} & \dots \\ \dots\dots\dots & \dots\dots\dots & \dots \end{pmatrix}. \tag{1}$$

This array is understood as representing a coordinate q as a function of time in a similar way as the totality of terms of the Fourier series

$$q(t) = \sum_n q_n e^{2\pi i\nu_n t}, \qquad \nu_n = n\nu_0$$

in the classical theory; except that now because of the two indices the sum no longer makes sense. The question arises of which expressions correspond to functions of the classical coordinate, for instance to the square q^2. Now, such arrays ordered by two indices occur as *matrices* in mathematics in the theory of quadratic forms and of linear transformations; the composition of two linear transformations,

$$x_k = \sum_l a_{kl}y_l, \qquad y_l = \sum_j b_{lj}z_j,$$

to form a new one,

$$x_k = \sum_j c_{kj}z_j,$$

then corresponds to the composition or multiplication of the matrices

$$ab = c, \quad \text{that is,} \quad \sum_l a_{kl}b_{lj} = c_{kj}. \tag{2}$$

This multiplication in general is *not* commutative. It is natural to apply this recipe to the array of the atomic oscillations (Born and Jordan [2], Dirac [3]); it is immediately evident that because of Ritz's formula $\nu_{ik} = T_i - T_k$ no new frequencies appear, just as in the classical theory in the multiplication of two Fourier series, and herein lies the first justification for the procedure. By repeated application of additions and multiplications one can define arbitrary matrix functions.

The analogy with the classical theory leads further to allowing as representatives of real quantities only those matrices that are 'Hermitian', that is, whose elements go over to the complex conjugate numbers under permutation of the indices. The discontinuous nature of the atomic processes here is put into the theory from the start as empirically established. However, this does not establish yet the connection with quantum theory and its characteristic constant h. This is also achieved, by carrying over the content[6] of the Bohr–Sommerfeld quantum conditions in a form given by Kuhn and Thomas, in which they are written as relations between the Fourier coefficients of the coordinates q and momenta p. In this way one obtains the matrix equation

$$pq - qp = \frac{h}{2\pi i} \cdot 1, \tag{3}$$

where 1 means the unit matrix. The matrix p thus does not 'commute' with q. For several degrees of freedom the commutation relation (3) holds for every pair of conjugate quantities, while the q_k commute with each other, the p_k with each other, and also the p_k with the non-corresponding q_k.

In order to construct the new mechanics (Born, Heisenberg and Jordan [4]), one carries over as far as possible the notions of the classical theory. It is possible to define the differentiation of a matrix with respect to time and that of a matrix function with respect to an argument matrix. One can thus carry over to the matrix theory the canonical equations

$$\frac{dq}{dt} = \frac{\partial H}{\partial p}, \qquad \frac{dp}{dt} = -\frac{\partial H}{\partial q},$$

where one should understand $H(p, q)$ as the same function of the matrices p, q that occurs in the classical theory as a function of the numbers p, q. (To be sure,[7] ambiguity can occur because of the non-commutativity of the multiplication; for example, p^2q is different from pqp.) This procedure was tested in simple examples (harmonic and anharmonic oscillator). Further, one can prove the theorem of conservation of energy, which for non-degenerate systems (all terms T_k different from each other, or: all frequencies ν_{ik} different from zero) here takes the form: for the solutions p, q of the canonical equations the Hamiltonian function $H(p, q)$ becomes a diagonal matrix W. It follows immediately that the elements of this diagonal matrix represent the terms T_n of Ritz's formula multiplied by h (Bohr's frequency condition). It is particularly important to realise that conversely the requirement

$$H(p, q) = W \qquad \text{(diagonal matrix)}$$

is a complete substitute for the canonical equations of motion, and leads to unambiguously determined solutions even if one allows for degeneracies (equality of terms, vanishing frequencies).

By a matrix with elements that are harmonic functions of time, one can of course represent only quantities (coordinates) that correspond to time-periodic quantities of the classical theory. Therefore cyclic coordinates (angles), which increase proportionally to time, cannot be treated at present.[a] Nevertheless, one

[a] This point is taken up again shortly after eq. (10). (*eds.*).

easily manages to subject rotating systems to the matrix method by representing the Cartesian components of the angular momentum with matrices [4].[8] One obtains thereby expressions for the energy[a] that differ characteristically from the corresponding classical ones; for instance the modulus[9] of the total angular momentum is not equal to $\frac{h}{2\pi}j$ $(j = 0, 1, 2, \ldots)$, but to $\frac{h}{2\pi}\sqrt{j(j+1)}$, in accordance with empirical rules that Landé and others had derived from the term splitting in the Zeeman effect.[b] Further, one obtains for the changes in the angular quantum numbers [Rotationsquantenzahlen] the correct selection rules and intensity formulas, as had already been arrived at earlier by correspondence arguments and confirmed by the Utrecht observations.[c]

Pauli [6], avoiding angular variables, even managed to work out the hydrogen atom with matrix mechanics, at least with regard to the energy values and some aspects of the intensities.

Asking for the most general coordinates for which the quantum mechanical laws are valid leads to the generalisation of the notions of canonical variables and canonical transformations known from the classical theory. Dirac [3] has noted that the content of the expressions such as $\frac{2\pi i}{h}(p_k q_l - q_l p_k) - \delta_{kl}$, which appear in the commutation relations of the type (3) corresponds[10] to that of the Poisson brackets, whose vanishing in classical mechanics characterises a system of variables as canonical. Therefore also in quantum mechanics one will denote as canonical every system of matrix pairs p, q that satisfy the commutation relations, and as a canonical transformation every transformation that leaves these relations invariant. One can write these with the help of an arbitrary matrix S in the form[d]

$$P = S^{-1}pS, \quad Q = S^{-1}qS, \tag{4}$$

and in a certain sense this is the most general canonical transformation. Then for an arbitrary function one has

$$f(P, Q) = S^{-1}f(p, q)S.$$

Now one can also carry over the main idea of the Hamilton-Jacobi theory [4]. Indeed, if the Hamiltonian function H is given as a function of any known canonical matrices p_0, q_0, then the solution of the mechanical problem defined by H reduces to finding a matrix S that satisfies the equation

$$S^{-1}H(p_0, q_0)S = W. \tag{5}$$

[a] Angular momentum is of course responsible for a characteristic splitting of the energy terms (*eds.*).
[b] For Landé's work on the anomalous Zeeman effect, see Mehra and Rechenberg (1982a, section IV.4, esp. pp. 467–76 and 482–5) (*eds.*).
[c] For the 'Utrecht observations' see Mehra and Rechenberg (1982a, section VI.6, pp. 647–8) and Mehra and Rechenberg (1982b, section III.4, esp. pp. 154–61) (*eds.*).
[d] Cf. p. 89 above (*eds.*).

This is an analogue of the Hamilton–Jacobi differential equation of classical mechanics.

Exactly as there, also here perturbation theory can be treated most clearly with the help of equation (5). If H is given as a power series in some small parameter

$$H = H_0 + \lambda H_1 + \lambda^2 H_2 + \cdots$$

and the mechanical problem is solved for $\lambda = 0$, that is, $H_0 = W_0$ is known as a diagonal matrix, then the solution to (5) can be obtained easily as a power series

$$S = 1 + \lambda S_1 + \lambda^2 S_2 + \cdots$$

by successive approximations. Among the numerous applications of this procedure, only the derivation of Kramers' dispersion formula shall be mentioned here, which results if one assumes that the light-emitting and the scattering systems are weakly coupled and if one calculates the perturbation on the latter ignoring the backreaction [4].[a]

The theory of the canonical transformations leads to a deeper conception, which later became essential in understanding the physical meaning of the formalism.

To each matrix $a = (a_{nm})$ one can associate a quadratic (more precisely: Hermitian) form[b]

$$\sum_{nm} a_{nm} \varphi_n \bar{\varphi}_m$$

of a sequence of variables $\varphi_1, \varphi_2 \ldots$, or also a linear transformation of the sequence of variables $\varphi_1, \varphi_2 \ldots$, into another one $\psi_1, \psi_2 \ldots$[11]

$$\psi_n = \sum_m a_{nm} \varphi_m, \tag{6}$$

where provisionally the meaning of the variables φ_n and ψ_n shall be left unspecified; we shall return to this.

A transformation (6) is called 'orthogonal' if it maps the identity form into itself

$$\sum_n \varphi_n \bar{\varphi}_n = \sum_n \psi_n \bar{\psi}_n. \tag{7}$$

Now these orthogonal transformations of the auxiliary variables φ_n immediately turn out to be essentially identical to the canonical transformations of the q and p matrices; the Hermitian character and the commutation relations are preserved.

[a] In other words, one considers just the scattering system under an external perturbation (Born, Heisenberg and Jordan [4], section 2.4, in particular eq. (32)). See also Mehra and Rechenberg (1982c, ch. III, esp. pp. 93–4 and 103–9) (*eds.*).

[b] $\bar{\varphi}$ denotes the complex number conjugate to φ.

Further, one can replace the matrix equation (5) by the equivalent requirement [4]: the form

$$\sum_{nm} H_{nm}(q_0, p_0)\varphi_n\bar{\varphi}_m$$

is to be transformed orthogonally into a sum of squares

$$\sum_n w_n\psi_n\bar{\psi}_n. \tag{8}$$

The fundamental problem of mechanics is thus none other than the principal axes problem for surfaces of second order in infinite-dimensional space, occurring everywhere in pure and applied mathematics and variously studied. As is well known, this is equivalent to asking for the values of the parameter W for which the linear equations

$$W\varphi_n = \sum_m H_{nm}\varphi_m \tag{9}$$

have a non-identically vanishing solution. The values $W = W_1, W_2, \ldots$ are called eigenvalues of the form H; they are the energy values (terms) of the mechanical system. To each eigenvalue W_n corresponds an eigensolution $\varphi_k = \varphi_{kn}$.[a] The set of these eigensolutions evidently again forms a matrix and it is easy to see that this is identical with the transformation matrix S appearing in (5).[b]

The eigenvalues, as is well known, are invariant under orthogonal transformations of the φ_k,[12] and since these correspond to the canonical substitutions of the p and q matrices, one recognises immediately the canonical invariance of the energy values W_n.

While the quantum theoretical matrices do not belong to the class of matrices (finite and bounded infinite[13]) investigated by the mathematicians (especially by Hilbert and his school), one can nevertheless carry over the main aspects of the known theory to the more general case. The precise formulation of these theorems[14] has been recently given by J. von Neumann [42] in a paper to which we shall have to return.[15]

The most important result that is achieved in this way is the theorem that a form cannot always be decomposed into a sum of squares (8), but that there also occur invariant integral components

$$\int W\psi(W)\overline{\psi(W)}dW, \tag{10}$$

[a] This is the notation used by Born and Heisenberg: the *n*th eigensolution is represented by an infinite vector with components labelled by k (*eds.*).

[b] This point is made more explicit after eq. (17). See also the relevant contributions by Dirac and by Kramers in the general discussion, pp. 445 and 448 (*eds.*).

where the sequence of variables $\psi_1, \psi_2, \ldots$ has to be complemented by the continuous distribution $\psi(W)$.

In this way the continuous spectra appear in the theory in the most natural way. But this implies by no means that in this domain the classical theory comes again into its own. Also here the characteristic discontinuities of quantum theory remain; also in the continuous spectrum a (spontaneous) state transition consists of a 'jump' of the system from a point W' to another one W'' with emission of a wave $q(W, W')e^{2\pi i \nu t}$ with the frequency $\nu = \frac{1}{h}(W' - W'')$.

The main defect of matrix mechanics consists in its clumsiness, even helplessness, in the treatment of non-periodic quantities, such as angular variables or coordinates that attain infinitely large values (e.g. hyperbolic trajectories). To overcome this difficulty two essentially different routes have been taken, the operator calculus of Born and Wiener [21], and the so-called[16] q-number theory of Dirac [7].

The latter starts from the idea that a great part of the matrix relations can be obtained without an explicit representation of the matrices, simply on the basis of the rules for operating with the matrix symbols. These depart from the rules for numbers only in that the multiplication is generally not commutative. Dirac therefore considers abstract quantities, which he calls q-numbers (as opposed to the ordinary c-numbers) and with which he operates according to the rules of the non-commutative algebra. It is therefore a kind of hypercomplex number system. The commutation relations are of course preserved. The theory acquires an extraordinary resemblance to the classical one; for instance, one can introduce angle and action variables w, J and expand any q-number into a Fourier series with respect to the w; the coefficients are functions of the J and turn out to be identical to the matrix elements if one replaces the J by integer multiples of h. By his method Dirac has achieved important results, for instance worked out the hydrogen atom independently of Pauli [7] and determined the intensity of radiation in the Compton effect [12]. A drawback of this formalism – apart from the quite tiresome dealing with the non-commutative algebra – is the necessity to replace at a certain point of the calculation certain q-numbers with ordinary numbers (e.g. $J = hn$), in order to obtain results comparable with experiment. Special 'quantum conditions' which had disappeared from matrix mechanics are thus needed again.

The operator calculus differs from the q-number method in that it does not introduce abstract hypercomplex numbers, but concrete, constructible mathematical objects that obey the same laws, namely operators or functions in the space of infinitely many variables. The method is by Eckart [22] and was then developed further by many others following on from Schrödinger's wave mechanics, especially by Dirac [38] and Jordan [39] and in an impeccable mathematical form by J. von Neumann [42]; it rests roughly on the following idea.

A sequence of variables $\varphi_1, \varphi_2, \ldots$ can be interpreted as a point in an infinite-dimensional space. If the sum of squares $\sum_n |\varphi_n|^2$ converges, then it represents a measure of distance, a Euclidean metric [Massbestimmung], in this space; this metric space of infinitely many dimensions is called for short a Hilbert space. The canonical transformations of matrix mechanics correspond thus to the rotations of the Hilbert space. Now, however, one can also fix a point in this space other than by the specification of discrete coordinates $\varphi_1, \varphi_2, \ldots$. Take for instance a complete, normalised orthogonal system of functions $f_1(q), f_2(q), \ldots$, that is one for which[17]

$$\int f_n(q)\overline{f_m(q)}dq = \delta_{nm} = \begin{cases} 1 & \text{for } n = m \\ 0 & \text{for } n \neq m \,; \end{cases} \tag{11}$$

the variable q can range here over an arbitrary, also multi-dimensional domain. If one then sets (Lanczos [23])[18]

$$\begin{cases} \varphi(q) = \sum_n \varphi_n f_n(q) \\ H(q', q'') = \sum_{nm} H_{nm} f_n(q')\overline{f_m(q'')} \,, \end{cases} \tag{12}$$

the linear equations (9)[19] turn into the integral equation[20]

$$W\varphi(q') = \int H(q', q'')\varphi(q'')dq''. \tag{13}$$

This relation established through (12) means thus nothing but a change of the coordinate system in the Hilbert space, given by the orthogonal transformation matrix $f_n(q)$ with one discrete and one continuous index.

One sees thus that the preference for 'discrete' coordinate systems in the original version of the matrix theory is by no means something essential. One can just as well use 'continuous matrices' such as $H(q', q'')$. Indeed, the specific representation of a point in the Hilbert space by projection onto certain orthogonal coordinate axes does not matter at all; rather, one can summarise equations (9) and (13) in the more general equation

$$W\varphi = H\varphi, \tag{14}$$

where H denotes a linear operator which transforms the point φ of the Hilbert space into another. The equation requires to find those points φ which under the operation H only suffer a displacement along the line joining them to the origin.[21] The points satisfying this condition determine an orthogonal system of axes, the principal axes frame of the operator H; the number of axes is finite or infinite, in the latter case distributed discretely or continuously, and the eigenvalues W are the lengths of the principal axes. The linear operators in the Hilbert space are thus the general concept that can serve to represent a physical quantity mathematically. The calculus with operators proceeds obviously according to the same rules as the one with Dirac's q-numbers; they[22] constitute a realisation of this abstract notion. So far we have analysed the situation with the example of the Hamiltonian function, but the same holds for any quantum mechanical quantity. Any coordinate q can be written, instead of as a matrix with discrete indices q_{nm}, also as a function of two continuous variables $q(q', q'')$ by projection onto an orthogonal system of functions, or, more generally, can also be considered as a linear operator in the Hilbert space; then it has eigenvalues that are invariant, and eigensolutions with respect to each orthogonal coordinate system. The same holds for a momentum p and every function of q and p, indeed for every quantum mechanical 'quantity'. While in the classical theory physical quantities are represented by variables that can take numerical values from an arbitrary value range, a physical quantity in quantum theory is represented by a linear operator and the stock of values that it can take by the eigenvalues of the corresponding principal axes problem in the Hilbert space.

In this view, Schrödinger's wave mechanics [24] appears formally as a special case. The simplest operator whose characteristic values are all the real numbers, is in fact the multiplication of a function $F(q)$ by the real number q; one writes it simply q. Then, however, the eigenfunctions are 'improper' functions; for according to (14) they must have the property of being everywhere zero except if $W = q$. Dirac [38] has introduced for the representation of such improper functions the 'unit function' $\delta(s)$, which should always be zero when $s \neq 0$, but for which nonetheless $\int_{-\infty}^{+\infty} \delta(s) ds = 1$ should hold. Then one can write down the (normalised) eigenfunctions

$$\varphi(q, W) = q\delta(W - q) \tag{15}$$

belonging to the operator q.

The conjugate to the operator q is the differential operator

$$p = \frac{h}{2\pi i} \frac{\partial}{\partial q} \, ; \tag{16}$$

indeed, the commutation relation (3) holds, which means just the trivial identity

$$(pq - qp)\mathcal{F}(q) = \frac{h}{2\pi i} \left\{ \frac{d}{dq}(q\mathcal{F}) - q\frac{d\mathcal{F}}{dq} \right\} = \frac{h}{2\pi i} \mathcal{F}(q).$$

If one now constructs a Hamiltonian function out of p, q (or out of several such conjugate pairs), then equation (14) becomes a differential equation for the quantity $\varphi(q)$:

$$H(q, \frac{h}{2\pi i}\frac{\partial}{\partial q})\varphi(q) = W\varphi(q). \tag{17}$$

This is Schrödinger's wave equation, which appears here as a special case of the operator theory. The most important point about this formulation of the quantum laws (apart from the great advantage of connecting to known mathematical methods) is the replacement of all 'quantum conditions', such as were still necessary in Dirac's theory of q-numbers, by the simple requirement that the eigenfunction $\varphi(q) = \varphi(q, W)$ should be everywhere finite in the domain of definition of the variables q; from this, in the event, a discontinuous spectrum of eigenvalues W_n (along with continuous ones) arises automatically. But Schrödinger's eigenfunction $\varphi(q, W)$ is actually nothing but the transformation matrix S of equation (5), which one can indeed also write in the form

$$HS = SW,$$

analogous to (17).

Dirac [38] has made this state of affairs even clearer by writing the operators q and p and thereby also H as integral operators, as in (13); then one has to set

$$\begin{aligned} q\mathcal{F}(q') &= \int q''\delta(q'-q'')\mathcal{F}(q'')dq'' &&= q'\mathcal{F}(q'), \\ p\mathcal{F}(q') &= \int \frac{h}{2\pi i}\delta'(q'-q'')\mathcal{F}(q'')dq'' &&= \frac{h}{2\pi i}\frac{d\mathcal{F}}{dq'}, \end{aligned} \tag{18}$$

where, however, the occurrence of the derivative of the singular function δ has to be taken into the bargain. Then Schrödinger's equation (17) takes the form (13).

The direct passage to the matrix representation in the strict sense takes place by inverting the formulas (12), in which one identifies the orthogonal system $f_n(q)$ with the eigenfunctions $\varphi(q, W_n)$ belonging to the discrete spectrum. If T is an arbitrary operator (constructed from q and $p = \frac{h}{2\pi i}\frac{\partial}{\partial q}$), define the corresponding matrix T_{nm} by the coefficients of the expansion

$$T\varphi_n(q) = \sum_m T_{nm}\varphi_m(q) \tag{19}$$

or

$$T_{nm} = \int \overline{\varphi_m(q)}T\varphi_n(q)dq\ ; \tag{19a}$$

then one easily sees that equation (17) is equivalent to (9).

The further development of the formal theory has taken place in close connection with its physical interpretation, to which we therefore turn first.

II. – PHYSICAL INTERPRETATION

The most noticeable defect of the original matrix mechanics consists in the fact that at first it appears to give information not about actual phenomena, but rather only about possible states and processes. It allows one to calculate the possible stationary states of a system; further it makes a statement about the nature of the harmonic oscillation that can manifest itself as a light wave in a quantum jump. But it says nothing about when a given state is present, or when a change is to be expected. The reason for this is clear: matrix mechanics deals only with closed periodic systems, and in these there are indeed no changes. In order to have true processes,[23] as long as one remains in the domain of matrix mechanics, one must direct one's attention to a *part* of the system; this is no longer closed and enters into interaction with the rest of the system. The question is what matrix mechanics can tell us about this.

Imagine, for instance, two systems 1 and 2 weakly coupled to each other (Heisenberg [35], Jordan [36]).[a] For the total system conservation of energy then holds; that is, H is a diagonal matrix. But for a subsystem, for instance 1, $H^{(1)}$ is not constant, the matrix has elements off the diagonal.[24] The energy exchange can now be interpreted in two ways: for one, the periodic elements of the matrix of $H^{(1)}$ (or of $H^{(2)}$) represent a slow beating, a continuous oscillation of the energy to and fro; but at the same time, one can also describe the process with the concepts of the discontinuum theory and say that system 1 performs quantum jumps and carries over the energy that is thereby freed to system 2 as quanta, and vice versa. But one can now show that these two apparently very different views do not contradict each other at all. This rests on a mathematical theorem that states the following:

Let $f(W_n^{(1)})$ be any function of the energy values $W_n^{(1)}$ of the isolated subsystem 1; if one forms the same function of the matrix $H^{(1)}$ that represents the energy of system 1 in the presence of the coupling to system 2, then $f(H^{(1)})$ is a matrix that does not consist only of diagonal elements $f(H^{(1)})_{nn}$. But these represent the time-averaged value of the quantity $f(H^{(1)})$. The effect of the coupling is thus measured by the difference[25]

$$\delta\overline{f_n} = f(H^{(1)})_{nn} - f(W_n^{(1)}).$$

[a] The form of the result as given here is similar to that in Heisenberg [35]. For further details, see the discussion in Section 3.4.4 (*eds.*).

The first part of the said theorem now states that $\delta \overline{f_n}$ can be brought into the form[26]

$$\delta \overline{f_n} = \sum_m \{f(W_m) - f(W_n)\}\Phi_{nm}. \tag{20}$$

This can be interpreted thus: the time average of the change in f due to the coupling is the arithmetic mean, with certain weightings Φ_{nm}, of all possible jumps of f for the isolated system.

These Φ_{nm} will have to be called 'transition probabilities'. The second part of the theorem determines the Φ_{nm} through the features of the coupling. Namely, if $p_1^0, q_1^0, p_2^0, q_2^0$ are coordinates satisfying the evolution equations of the uncoupled systems, for which therefore $H^{(1)}$ and $H^{(2)}$ on their own are diagonal matrices, one can then think of the energy, including the interaction, as expressed as a function of these quantities. Then the solution of the mechanical problem according to (5)[27] reduces to constructing a matrix S that satisfies the equation

$$S^{-1}H(p_1^0, q_1^0, p_2^0, q_2^0)S = W.$$

Denoting the states of system 1 by n_1, those of system 2 by n_2, a state of the total system is given by $n_1 n_2$,[28] and to each transition $n_1 n_2 \rightarrow m_1 m_2$ corresponds an element of S, $S_{n_1 n_2, m_1 m_2}$. Then the result is:[29]

$$\Phi_{n_1 m_1} = \sum_{n_2 m_2} |S_{n_1 n_2, m_1 m_2}|^2. \tag{21}$$

The squares of the elements of the S-matrix thus determine the transition probabilities. The individual sum term $|S_{n_1 n_2, m_1 m_2}|^2$ in (21) obviously means that component of the transition probability for the jump $n_1 \rightarrow m_1$ of system 1 that is induced by the jump $n_2 \rightarrow m_2$ of system 2.

By means of these results the contradiction between the two views from which we started is removed. Indeed, for the mean values, which alone may be observed, the conception of continuous beating always leads to the same result as the conception of quantum jumps.

If one asks the question *when* a quantum jump occurs, the theory provides no answer. At first it seemed as if there were a gap here which might be filled with further probing. But soon it became apparent that this is not so, rather, that it is a failure of principle, which is deeply anchored in the nature of the possibility of physical knowledge [physikalisches Erkenntnisvermögen].

One sees that quantum mechanics yields mean values correctly, but cannot predict the occurrence of an individual event. Thus determinism, held so far to be the foundation of the exact natural sciences, appears here to go no longer unchallenged. Each further advance in the interpretation of the formulas has shown that the system of quantum mechanical formulas can be interpreted consistently only from the point of view of a fundamental indeterminism, but also, at the same time, that the totality of empirically ascertainable facts can be reproduced by the system of the theory.

In fact, almost all observations in the field of atomic physics have a statistical character; they are countings, for instance of atoms in a certain state. While the determinateness of an individual process is assumed by classical physics, practically it plays no role in fact, because the microcoordinates that determine exactly an atomic process can never all be given; therefore by averaging they are eliminated from the formulas, which thereby become statistical statements. It has become apparent that quantum mechanics represents a merging of mechanics and statistics, in which the unobservable microcoordinates are eliminated.

The clumsiness of the matrix theory in the description of processes developing in time can be avoided by making use of the more general formalisms[30] we have described above. In the general equation (14) one can easily introduce time explicitly by invoking the theorem of classical mechanics that energy W and time t behave as canonically conjugate quantities; in quantum mechanics it corresponds to having a commutation relation

$$Wt - tW = \frac{h}{2\pi i}.$$

Thus for W one can posit the operator $\frac{h}{2\pi i}\frac{\partial}{\partial t}$. Equation (14) then reads

$$\frac{h}{2\pi i}\frac{\partial \varphi}{\partial t} = H\varphi, \tag{22}$$

and here one can consider H as depending explicitly on time. A special case of this is the equation

$$\left\{H(q, \frac{h}{2\pi i}\frac{\partial}{\partial q}) - \frac{h}{2\pi i}\frac{\partial}{\partial t}\right\}\varphi(q) = 0, \tag{22a}$$

given by Schrödinger [24],[31] which stands to (17) in the same relation as (22) to (14), as well as the form:

$$\frac{h}{2\pi i}\frac{\partial \varphi(q')}{\partial t} = \int H(q', q'')\varphi(q'')dq'', \tag{22b}$$

much used by Dirac, which relates to the integral formula (13). Essentially, the introduction of time as a numerical variable reduces to thinking of the system under consideration as coupled to another one and neglecting the reaction on the latter. But this formalism is very convenient and leads to a further development of the statistical view,[a] namely, if one considers the case where an explicitly

[a] See the discussion in Section 3.4 (*eds.*).

time-dependent perturbation $V(t)$ is added to a time-independent energy function H^0, so that one has the equation

$$\frac{h}{2\pi i}\frac{\partial\varphi}{\partial t} = \{H^0 + V(t)\}\,\varphi \tag{23}$$

(Dirac [37], Born [34]).[32] Now if φ_n^0 are the eigenfunctions of the operator H^0, which for the sake of simplicity we assume to be discrete, the desired quantity φ can be expanded in terms of these:

$$\varphi(t) = \sum_n c_n(t)\varphi_n^0. \tag{24}$$

The $c_n(t)$ are then the coordinates of φ in the Hilbert space with respect to the orthogonal system φ_n^0; they can be calculated from the differential equation (23), if their initial values $c_n(0)$ are given. The result can be expressed as:

$$c_n(t) = \sum_m S_{nm}(t)c_m(0), \tag{25}$$

where $S_{nm}(t)$ is an orthogonal matrix depending[33] on t and determined by $V(t)$.

The temporal process is thus represented by a rotation of the Hilbert space or by a canonical transformation (4) with the time-dependent matrix S.

Now how is one to interpret this?

From the point of view of Bohr's theory a system can always be in only *one* quantum state. To each of these belongs an eigensolution φ_n^0 of the unperturbed system. If now one wishes to calculate what happens to a system that is initially in a certain state, say the kth, one has to choose $\varphi = \varphi_k^0$ as the initial condition for equation (23), i.e. $c_n(0) = 0$ for $n \neq k$, and $c_k(0) = 1$. But then, after the perturbation is over, $c_n(t)$ will have become equal to $S_{nk}(t)$, and the solution consists of a superposition of eigensolutions. According to Bohr's principles it makes no sense to say a system is simultaneously in several states. The only possible interpretation seems to be statistical: the superposition of several eigensolutions expresses that through the perturbation the initial state can go over to any other quantum state, and it is clear that as measure for the transition probability one has to take the quantity

$$\Phi_{nk} = |S_{nk}(t)|^2\,;$$

because then one obtains again equation (20) for the average change of any state function.

This interpretation is supported by the fact that one establishes the validity of Ehrenfest's adiabatic theorem (Born [34]); one can show that under an infinitely slow action, one has

$$\Phi_{nn} \to 1, \qquad \Phi_{nk} \to 0 \qquad (n \neq k),$$

that is, the probability of a jump tends to zero.

But this assumption also leads immediately to an interpretation of the $c_n(t)$ themselves: the $|c_n(t)|^2$ must be the state probabilities [Zustandswahrscheinlichkeiten].

Here, however, one runs into a difficulty of principle that is of great importance, as soon as one starts from an initial state for which not all the $c_n(0)$ except one vanish. Physically, this case occurs if a system is given for which one does not know exactly the quantum state in which it is, but knows only the probability $|c_n(0)|^2$ for each quantum state. As a matter of fact, the phases [Arcus] of the complex quantities $c_n(0)$ still remain indefinite; if one sets $c_n(0) = |c_n(0)|e^{i\gamma_n}$, then the γ_n denote some phases whose meaning needs to be established. The probability distribution at the end of the perturbation according to (25) is then

$$|c_n(t)|^2 = \left|\sum_m S_{nm}(t)c_m(0)\right|^2 \tag{26}$$

and not

$$\sum_m |S_{nm}(t)|^2 |c_m(0)|^2, \tag{27}$$

as one might suppose from the usual probability calculus.

Formula (26), following Pauli, can be called the theorem of the *interference of probabilities*; its deeper meaning has become clear only through the wave mechanics of de Broglie and Schrödinger, which we shall presently discuss. Before this, however, it should be noted that this 'interference' does not represent a contradiction with the rules of the probability calculus, that is, with the assumption that the $|S_{nk}|^2$ are quite usual probabilities.[a] In fact, the composition rule (27) follows from the concept of probability for the problem treated here when and only when the relative number, that is, the probability $|c_n|^2$ of the atoms in the state n, has been *established* beforehand *experimentally*.[34] In this case the phases γ_n are unknown in principle,[35] so that (26) then naturally goes over to (27) [46].

It should be noted further that the formula (26) goes over to the expression (27) if the perturbation function proceeds totally irregularly as a function of time. That is for instance the case when the perturbation is produced by 'white light'.[b] Then, on average, the surplus terms in (26) drop out and one obtains (27). In this way it is easy to derive the Einstein coefficient B_{nm} for the probability per unit radiation of the quantum jumps induced by light absorption (Dirac [37], Born [30]). But, in general, according to (26) the knowledge of the probabilities $|c_n(0)|^2$ is by no

[a] The notation $|S_{nm}|^2$ would probably be clearer, at least according to the reading of this passage proposed in Section 3.4.6 (*eds.*).

[b] Compare also Born's discussion in Born (1926c [34]) (*eds.*).

means sufficient to calculate the course of the perturbation, rather one has to know also the phases γ_n.

This circumstance recalls vividly the behaviour of light in interference phenomena. The intensity of illumination on a screen is by no means always equal to the sum of the light intensities of the individual beams of rays that impinge on the screen, or, as one can well say, it is by no means equal to the sum of the light quanta that move in the individual beams; instead it depends essentially on the phases of the waves. Thus at this point an analogy between the quantum mechanics of corpuscles and the wave theory of light becomes apparent.

As a matter of fact this connection was found by de Broglie in a quite different way. It is not our purpose to discuss this. It is enough to formulate the result of de Broglie's considerations, and their further development by Schrödinger, and to put it in relation to quantum mechanics.

The dual nature of light – waves, light quanta – corresponds to the analogous dual nature of material particles; these also behave in a certain respect like waves. Schrödinger has set up the laws of propagation of these waves [24] and has arrived at equation [(17)],[36] here derived in a different way. His view, however, that these waves exhaust the essence of matter and that particles are nothing but wave packets, not only stands in contradiction with the principles of Bohr's empirically very well-founded theory, but also leads to impossible conclusions; here therefore it shall be left to one side. Instead we attribute a dual nature to matter also: its description requires both corpuscles (discontinuities) and waves (continuous processes). From the viewpoint of the statistical approach to quantum mechanics it is now clear why these can be reconciled: the waves are probability waves. Indeed, it is not the probabilities themselves, rather certain 'probability amplitudes' that propagate continuously and obey differential or integral equations, as in classical continuum physics; but additionally there are discontinuities, corpuscles whose frequency is governed by the square of these amplitudes.

The most definite support for this conception is given by collision phenomena for material particles (Born [30]). Already Einstein [16], when he deduced from de Broglie's daring theory the possibility of 'diffraction' of material particles,[a] tacitly assumed that it is the particle number that is determined by the intensity of the waves. The same occurs in the interpretation given by Elsasser [17] of the experiments by Davisson and Kunsman [18,19] on the reflection of electrons by crystals; also here one assumes directly that the *number* of electrons is a maximum in the diffraction maxima. The same holds for Dymond's [20] experiments on the diffraction of electrons by helium atoms.

[a] Note that the first prediction of such diffraction appears in fact to have been made by de Broglie in 1923; cf. Section 2.2.1 (*eds.*).

The application of wave mechanics to the calculation of collision processes takes a form quite analogous to the theory of diffraction of light by small particles. One has to find the solution to Schrödinger's wave equation (17) that goes over at infinity to a given incident plane wave; this solution behaves everywhere at infinity like an outgoing[37] spherical wave. The intensity of this spherical wave in any direction compared to the intensity of the incoming wave determines the relative number of particles deflected in this direction from a parallel ray. As a measure of the intensity one has to take a 'current vector'[38] which can be constructed from the solution $\varphi(q, W)$, and which is formed quite analogously to the Poynting vector of the electromagnetic theory of light, and which measures the number of particles crossing a unit surface in unit time.

In this way Wentzel [31] and Oppenheimer [32] have derived wave mechanically the famous Rutherford law for the scattering of α-particles by heavy nuclei.[a]

If one wishes to calculate the probabilities of excitation and ionisation of atoms [30], then one must introduce the coordinates of the atomic electrons as variables on an equal footing with those of the colliding electron. The waves then propagate no longer in three-dimensional space but in multi-dimensional configuration space. From this one sees that the quantum mechanical waves are indeed something quite different from the light waves of the classical theory.

If one constructs the current vector just defined for a solution of the generalised Schrödinger equation (22), which describes time evolution, one sees that the time derivative of the integral

$$\int |\varphi|^2 dq',$$

ranging over an arbitrary domain of the independent numerical variables q', can be transformed into the surface integral of the current vector over the boundary of that domain. From this it emerges that

$$|\varphi|^2$$

has to be interpreted as particle density or, better, as probability density. The solution φ itself is called 'probability amplitude'.

The amplitude $\varphi(q', W')$ belonging to a stationary state thus yields via $|\varphi(q', W')|^2$ the probability that for given energy W' the coordinate q' is in some given element dq'.[39] But this can be generalised immediately. In fact, $\varphi(q', W')$ is the projection of the principal axis W' of the operator H onto the principal axis q' of the operator q. One can therefore say in general (Jordan [39]): if two physical quantities are given by the operators q and Q and if one knows the principal axes

[a] Cf. Born (1969, appendix XX). The current vector, as defined there, is the usual $\mathbf{j} = \frac{h}{2\pi i}\frac{1}{2m}(\psi^*\nabla\psi - \psi\nabla\psi)$ (*eds.*).

of the former, for instance, according to magnitude and direction,[40] then from the equation

$$Q\varphi(q', Q') = Q'\varphi(q', Q')$$

one can determine the principal axes Q' of Q[41] and their projections $\varphi(q', Q')$ on the axes of q. Then $|\varphi(q', Q')|^2 dq'$ is the probability that for given Q' the value of q' falls in a given interval dq'.

If conversely one imagines the principal axes of Q as given, then those of q are obtained[42] through the inverse rotation; from this one easily recognises that $\overline{\varphi(Q', q')}$ is the corresponding amplitude,[43] so that $|\varphi(Q', q')|^2 dQ'$ means the probability, given q', to find the value of Q' in dQ'. If for instance one takes for Q the operator $p = \frac{h}{2\pi i}\frac{\partial}{\partial q}$, then one has the equation

$$\frac{h}{2\pi i}\frac{\partial\varphi}{\partial q'} = p'\varphi,$$

thus

$$\varphi = Ce^{\frac{2\pi i}{h}q'p'}. \tag{28}$$

This is therefore the probability amplitude for a pair of conjugate quantities. For the probability density one obtains $|\varphi|^2 = C$, that is, for given q' every value p' is equally probable.

This is an important result, since it allows one to retain the concept of 'conjugate quantity' even in the case where the differential definition fails, namely when the quantity q has only a discrete spectrum or even when it is only capable of taking finitely many values. The latter for instance is the case for angles with quantised direction [richtungsgequantelte Winkel],[a] say for the magnetic electron, or in the Stern–Gerlach experiments. One can then, as Jordan does, call by definition a quantity p conjugate to q, if the corresponding probability amplitude has the expression (28).

As the amplitudes are the elements of the rotation matrix of one orthogonal system into another, they are composed according to the matrix rule:

$$\varphi(q', Q') = \int \psi(q', \beta')\chi(\beta', Q')d\beta'; \tag{29}$$

in the case of discrete spectra, instead of the integral one has finite or infinite sums. This is the general formulation of the theorem of the interference of probabilities. As an application, let us look again at formula (24). Here $c_n(t)$ was the amplitude

[a] This was a standard term referring to the fact that in the presence of an external magnetic field, the projection of the angular momentum in the direction of the field has to be quantised (quantum number m). Therefore, the direction of the angular momentum with respect to the magnetic field can be said to be quantised. Cf. Born (1969, p. 121) (*eds.*).

for the probability that the system at time t has energy W_n; $\varphi_n^0(q')$ is the amplitude for the probability that for given energy W_n the coordinate q' has a given value. Thus

$$\varphi(q', t) = \sum_n c_n(t)\varphi_n^0(q')$$

expresses the amplitude for the probability that q' at time t has a given value.

Alongside the concept of the relative state probability $|\varphi(q', Q')|^2$, there also occurs the concept of the transition probability,[44] namely, every time one considers a system as depending on an external parameter, be it time or any property of a weakly coupled external system. Then the system of principal axes of any quantity becomes dependent on this parameter; it experiences a rotation, represented by an orthogonal transformation $S(q', q'')$, in which the parameter enters (as in formula (25)). The quantities $|S(q', q'')|^2$ are the 'transition probabilities';[45] in general, however, they are not independent, instead the 'transition amplitudes' are composed according to the interference rule.

III. – FORMULATION OF THE PRINCIPLES AND DELIMITATION OF THEIR SCOPE

After the general concepts of the theory have been developed through analysis of empirical findings, the dual task arises, first of giving a system of principles as simple as possible and connected directly to the observations, from which the entire theory can be deduced as from a mathematical system of axioms, and second of critically scrutinising experience to assure oneself that no observation conceivable by today's means stands in contradiction to the principles.

Jordan [39] has formulated such a 'system of axioms', which takes the following statements as fundamental:[46]

1) One requires for each pair of quantum mechanical quantities q, Q the existence of a probability amplitude $\varphi(q', Q')$, such that $|\varphi|^2$ gives the probability[47] that for given Q' the value of q' falls in a given infinitesimal interval.
2) Upon permutation of q and Q, the corresponding amplitude should be $\overline{\varphi(Q', q')}$.
3) The theorem (29) of the composition of probability amplitudes.
4) To each quantity q there should belong a canonically conjugate one p, defined by the amplitude (28). This is the only place where the quantum constant h appears.[48]

Finally one also takes as obvious that, if the quantities q and Q are identical, the amplitude $\varphi(q', q'')$ becomes equal to the 'unit matrix' $\delta(q' - q'')$, that is, always to zero, except when $q' = q''$. This assumption and the multiplication theorem 3) together characterise the amplitudes thus defined as the coefficients of an orthogonal transformation; one obtains the orthogonality conditions simply by stating that the composition of the amplitude belonging to q, Q with that belonging to Q, q must yield the identity.

One can then reduce all given quantities, including the operators, to amplitudes by writing them as integral operators as in formula (13). The non-commutative operator multiplication is then a consequence of the axioms and loses all the strangeness attached to it in the original matrix theory.

Dirac's method [38] is completely equivalent to Jordan's formulation, except in that he does not arrange the principles in axiomatic form.[a]

This theory now indeed summarises all of quantum mechanics in a system in which the simple concept of the calculable probability [berechenbare Wahrscheinlichkeit][49] for a given event plays the main role.[50] It also has some shortcomings, however. One formal shortcoming is the occurrence of improper functions, like the Dirac δ, which one needs for the representation of the unit matrix for continuous ranges of variables. More serious is the circumstance that the amplitudes are not directly measurable quantities, rather, only the squares of their moduli; the phase factors are indeed essential for how different phenomena are connected [für den Zusammenhang der verschiedenen Erscheinungen wesentlich], but are only indirectly determinable, exactly as phases in optics are deduced indirectly by combining measurements of intensity. It is, however, a tried and proven principle, particularly in quantum mechanics, that one should introduce as far as possible only directly observable quantities as fundamental concepts of a theory. This defect[51] is related mathematically to the fact that the definition of probability in terms of the amplitudes does not express the invariance under orthogonal transformations of the Hilbert space (canonical transformations).

These gaps in the theory have been filled by von Neumann [41,42]. There is[52] an invariant definition of the eigenvalue spectrum for arbitrary operators, and of the relative probabilities, without presupposing the existence of eigenfunctions or indeed using improper functions. Even though this theory has not yet been elaborated in all directions, one can however say with certainty that a mathematically irreproachable grounding of quantum mechanics is possible.

Now the second question has to be answered: is this theory in accord with the totality of our experience? In particular, given that the individual process is only statistically determined, how can the usual deterministic order be preserved in the composite macroscopic phenomena?[53]

The most important step in testing the new conceptual system in this direction consists in the determination of the boundaries within which the application of the old (classical) words and concepts is allowed, such as 'position, velocity,

[a] There are nevertheless some differences between the approaches of Dirac and Jordan. Cf. Darrigol (1992, pp. 343–4) (*eds.*).

momentum, energy of a particle (electron)' (Heisenberg [46]). It now turns out that all these quantities can be *individually* exactly measured and defined, as in the classical theory, but that for simultaneous measurements of canonically conjugate quantities (more generally: quantities whose operators do not commute) one cannot get below a characteristic limit of indeterminacy [Unbestimmtheit].[a] To determine this, according to Bohr [47][54] one can start quite generally from the empirically given dualism between waves and corpuscles. One has essentially the same phenomenon already in every diffraction of light by a slit. If a wave impinges perpendicularly on an (infinitely long) slit of width q_1, then the light distribution as a function of the deviation angle φ is given according to Kirchhoff by the square of the modulus of the quantity

$$a \int_{-\frac{q_1}{2}}^{+\frac{q_1}{2}} e^{\frac{2\pi i}{\lambda} \sin\varphi q} dq = 2a \frac{\sin\left(\frac{\pi q_1}{\lambda} \sin\varphi\right)}{\frac{\pi q_1}{\lambda} \sin\varphi},$$

and thus ranges over a domain whose order of magnitude is given by[55] $\sin\varphi_1 = \frac{\lambda}{q_1}$ and gets ever larger with decreasing slit width q_1. If one considers this process from the point of view of the corpuscular theory, and if the association given by de Broglie of frequency and wavelength with energy and momentum of the light quantum is valid,

$$h\nu = W, \quad \frac{h}{\lambda} = P,$$

then the momentum component perpendicular to the direction of the slit is

$$p = P \sin\varphi = \frac{h}{\lambda} \sin\varphi.$$

One sees thus that after the passage through the slit the light quanta have a distribution whose amplitude is given by

$$e^{\frac{2\pi i}{\lambda} \sin\varphi \cdot q} = e^{\frac{2\pi i}{h} p \cdot q},$$

precisely as quantum mechanics requires for two canonically conjugate variables; further, the width of the domain of the variable p that contains the greatest number of light quanta is

$$p_1 = P \sin\varphi_1 = \frac{P\lambda}{q_1} = \frac{h}{q_1}.$$

[a] Here and in the following, the choice of translation reflects the characteristic terminology of the original. Born and Heisenberg use the terms 'Unbestimmtheit' (indeterminacy) and 'Ungenauigkeit' (imprecision), while the standard German terms today are 'Unbestimmtheit' and 'Unschärfe' (unsharpness) (*eds.*).

By general considerations of this kind one arrives at the insight that the imprecisions (average errors) of two canonically conjugate variables p and q always stand in the relation

$$p_1 q_1 \geq h. \tag{30}$$

The narrowing of the range of one variable, which forms the essence of a measurement, widens unfailingly the range of the other. The same follows immediately from the mathematical formalism of quantum mechanics on the basis of formula (28). The actual meaning of Planck's constant h is thus that it is the universal measure of the indeterminacy that enters the laws of nature through the dualism of waves and corpuscles.

That quantum mechanics is a mixture of strictly mechanical and statistical principles can be considered a consequence of this indeterminacy. Indeed, in the classical theory one may fix the state of a mechanical system by, for instance, measuring the initial values of p and q at a certain instant. In quantum mechanics such a measurement of the initial state is possible only with the accuracy (30). Thus the values of p and q are known also at later times only statistically.

The relation between the old and the new theory can therefore be described thus:

In classical mechanics one assumes the possibility of determining exactly the initial state; the further development is then determined by the laws themselves.

In quantum mechanics, because of the imprecision relation, the result of each measurement can be expressed by the choice of appropriate initial values for probability functions; the quantum mechanical laws determine the change (wave-like propagation) of these probability functions. The result of future experiments however remains in general indeterminate and only the expectation[56] of the result is statistically constrained. Each new experiment replaces the probability functions valid until now with new ones, which correspond to the result of the observation; it separates the physical quantities into known and unknown (more precisely and less precisely known) quantities in a way characteristic of the experiment.

That in this view certain laws, like the principles of conservation of energy and momentum, are strictly valid, follows from the fact that they are relations between commuting quantities (all quantities of the kind q or all quantities of the kind p).[a]

The transition from micro- to macromechanics results naturally from the imprecision relation because of the smallness of Planck's constant h. The fact of the

[a] A similar but more explicit phrasing is used by Born (1926e, lecture 15): assuming that $H(p,q) = H(p) + H(q)$, the time derivatives not only of H but also of all components of momentum and of angular momentum have the form $f(q) + g(p)$ with suitable functions f and g. Born states that since all q commute with one another and all p commute with one another, the expressions $f(q) + g(p)$ will vanish under the same circumstances as in classical mechanics (*eds.*).

spreading,[57] the 'melting away' of a 'wave or probability packet' is crucial to this. For some simple mechanical systems (free electron in a magnetic or electric field (Kennard [50]), harmonic oscillator (Schrödinger [25])), the quantum mechanical spreading of the wave packet agrees with the spreading of the system trajectories that would occur in the classical theory if the initial conditions were known only with the precision restriction (30). Here the purely classical treatment of α and β particles, for instance in the discussion of Wilson's photographs, immediately finds its justification. But in general the statistical laws of the spreading of a 'packet' for the classical and the quantum theory are different; one has particularly extreme examples of this in the cases of 'diffraction' or 'interference' of material rays, as in the already mentioned experiments of Davisson, Kunsman and Germer [18,19] on the reflection of electrons by metallic crystals.

That the totality of experience can be fitted into the system of this theory can of course be established only by calculation and discussion of all the experimentally accessible cases. Individual experimental setups,[58] in which the suspicion of a contradiction with the precision limit (30) might arise, have been discussed [46,47]; every time the reason for the impossibility of fixing exactly all determining data could be exhibited intuitively [anschaulich aufgewiesen].

There remains only to survey the most important consequences of the theory and their experimental verification.

IV. – Applications of quantum mechanics

In this section we shall briefly discuss those applications of quantum mechanics that stand in close relation to questions of principle. Here the Uhlenbeck–Goudsmit theory of the magnetic electron shall be mentioned first. Its formulation and the treatment of the anomalous Zeeman effects with the matrix calculus raise no difficulties [11]; the treatment with the method of eigenoscillations succeeds only with the help of the general Dirac–Jordan theory (Pauli [45]). Here, two three-dimensional wave functions are associated with each electron. It becomes natural thereby to look for an analogy between matter waves and polarised light waves, which in fact can be carried through to a certain extent (Darwin [49], Jordan [53]). What is common to both phenomena is that the number of terms is finite, so the representative matrix is also finite (two arrangements [Einstellungen] for the electron, two directions of polarisation for light). Here the definition of the conjugate quantity by means of differentiation thus fails; one must resort to Jordan's definition by means of the probability amplitudes (formula (28)).

From among the other applications, the quantum mechanics of many-body problems shall be mentioned [28,29,40]. In a system that contains a number of similar particles [gleicher Partikel],[59] there occurs between them a kind of 'resonance' and from that results a decomposition of the system of terms into subsystems that do not combine (Heisenberg, Dirac [28,37]). Wigner has systematically investigated this phenomenon by resorting to group theoretic methods, and has set up the totality of the non-combining systems of terms [40]; Hund has managed to derive the majority of these results by comparatively elementary means [48]. A special role is played by the 'symmetric' and 'antisymmetric' subsystems of terms; in the former every eigenfunction remains unchanged under permutation of arbitrary similar particles, in the latter it changes sign under permutation of any two particles. In applying this theory to the spectra of atoms with several electrons it turns out that the Pauli equivalence rule[a] allows only the antisymmetric subsystem.[60] On the basis of this insight one can establish quantum mechanically the systematics of the line spectra and of the electron grouping throughout the whole periodic system of elements.

If one has a large number of similar particles, which are to be given a statistical treatment (gas theory), one obtains different statistics depending on whether one chooses the corresponding wave function according to the one or the other subsystem. The symmetric system is characterised by the fact that no new state arises under permutation of the particles from[61] a state described by a symmetric eigenfunction; thus all permutations that belong to the same set of quantum numbers (lie in the same 'cell') together always have the weight 1. This corresponds to the Bose–Einstein statistics [56,16]. In the antisymmetric term system two quantum numbers may never become equal, because otherwise the eigenfunction vanishes; a set of quantum numbers corresponds therefore either to no proper function at all or at most to one, thus the weight of a state is 0 or 1. This is the Fermi–Dirac statistics [57,37].

Bose–Einstein statistics holds for light quanta, as emerges from the validity of Planck's radiation formula. Fermi–Dirac statistics certainly holds for (negative) electrons, as emerges from the above-mentioned systematics of the spectra on the basis of Pauli's equivalence rule, and with great likelihood also for the positive elementary particles (protons); one can infer this from observations of band spectra [28,43] and in particular from the specific heat of hydrogen at low temperatures [55].[62] The assumption of Fermi–Dirac statistics for the positive and negative elementary particles of matter has the consequence that Bose–Einstein statistics holds for all neutral structures, e.g. molecules (symmetry of the eigenfunctions under permutation of an even number[63] of particles of matter). Within quantum mechanics, in which a many-body problem is treated in configuration space, the new statistics of Bose–Einstein and Fermi–Dirac has a perfectly legitimate place,

[a] That is, the Pauli principle applied to electrons that are 'equivalent' in the sense of having the same quantum numbers n and l; cf. Born (1969, p. 178) (*eds.*).

unlike in the classical theory, where an arbitrary modification of the usual statistics is impossible; nevertheless the restrictions made on the form of the eigenfunctions appear as an arbitrary additional assumption. In particular, the example of light quanta indicates that the new statistics is related in an essential way to the wave-like properties of matter and light. If one decomposes the electromagnetic oscillations of a cavity into spatial harmonic components, each of these behaves like a harmonic oscillator as regards time evolution; it now turns out that under quantisation of this system of oscillators a solution results that behaves exactly like a system of light quanta obeying Bose–Einstein statistics [4]. Dirac has used this fact for a consistent treatment of electrodynamical problems [51,52], to which we shall return briefly.

The corpuscular structure of light thus appears here as quantisation of light waves, similarly to how vice versa the wave nature of matter manifests itself in the 'quantisation' of the corpuscular motion. Jordan has shown [54] that one can proceed analogously with electrons; one has then to decompose the Schrödinger function of a cavity into fundamental and harmonics and to quantise each of these as a harmonic oscillator, in such a way in fact that Fermi–Dirac statistics is obtained. The new quantum numbers, which express the 'weights' in the usual many-body theory, have thus only the values 0 and 1. Therefore one has again here a case of finite matrices, which can be treated only with Jordan's general theory. The existence of electrons thus plays the same role in the formal elaboration of the theory as that of light quanta; both are discontinuities no different in kind from the stationary states of a quantised system. However, if the material particles stand in interaction with each other, the development of this idea might run into difficulties of a deep nature.

The results of Dirac's investigations [51,52] of quantum electrodynamics consist above all in a rigorous derivation of Einstein's transition probabilities for spontaneous emission.[a] Here the electromagnetic field (resolved into quantised harmonic oscillations) and the atom are considered as a coupled system and quantum mechanics is applied in the form of the integral equation (13). The interaction energy appearing therein is obtained by carrying over classical formulas. In this connection, the nature of absorption and scattering of light by atoms is clarified. Finally, Dirac [52] has managed to derive a dispersion formula with damping term; this includes also the quantum mechanical interpretation of Wien's experiments on the decay in luminescence of canal rays.[b] His method consists in considering the process of the scattering of light by atoms as a collision of light quanta. However, since one can indeed attribute energy and momentum to the light quantum but not easily a spatial position, there is a failure of the wave

[a] As opposed to the induced emission discussed on p. 387 (*eds.*).
[b] See above, p. 129 (*eds.*).

mechanical collision theory (Born [30]), in which one presupposes knowledge of the interaction between the collision partners as a function of the relative position. It is thus necessary to use the momenta as independent variables, and an operator equation of matrix character instead of Schrödinger's wave equation. Here one has a case where the use of the general points of view which we have emphasised in this report cannot be avoided. At the same time, the theory of Dirac reveals anew the deep analogy between electrons and light quanta.

Conclusion

By way of summary, we wish to emphasise that while we consider the last-mentioned enquiries, which relate to a quantum mechanical treatment of the electromagnetic field, as not yet completed [unabgeschlossen], we consider *quantum mechanics* to be a closed theory [geschlossene Theorie], whose fundamental physical and mathematical assumptions are no longer susceptible of any modification. Assumptions about the physical meaning of quantum mechanical quantities that contradict Jordan's or equivalent postulates will in our opinion also contradict experience. (Such contradictions can arise for example if the square of the modulus of the eigenfunction is interpreted as charge density.[a]) On the question of the 'validity of the law of causality' we have this opinion: as long as one takes into account only experiments that lie in the domain of our currently acquired physical and quantum mechanical experience, the assumption of indeterminism in principle, here taken as fundamental, agrees with experience. The further development of the theory of radiation will change nothing in this state of affairs, because the dualism between corpuscles and waves, which in quantum mechanics appears as part of a contradiction-free, closed theory [abgeschlossene Theorie], holds in quite a similar way for radiation. The relation between light quanta and electromagnetic waves must be just as statistical as that between de Broglie waves and electrons. The difficulties still standing at present in the way of a complete theory of radiation thus do *not* lie in the dualism between light quanta and waves – which is entirely intelligible – instead they appear only when one attempts to arrive at a relativistically invariant, closed formulation of the electromagnetic laws; all questions for which such a formulation is unnecessary can be treated by Dirac's method [51,52]. However, the first steps also towards overcoming these relativistic difficulties have already been made.

[a] See Schrödinger's report, especially his Section I, and Section 4.4 above (*eds.*).

Bibliography[a]

[1] W. Heisenberg, Über quanten[theoretische] Umdeutung kinematischer und mechanischer Beziehungen, *Z. f. Phys.*, **33** (1925), 879.
[2] M. Born and P. Jordan, Zur Quantenmechanik, I, *Z. f. Phys.*, **34** (1925), 858.
[3] P. Dirac, The fundamental equations of quantum mechanics, *Proc. Roy. Soc. A*, **109** (1925), 642.
[4] M. Born, W. Heisenberg and P. Jordan, Zur Quantenmechanik, II, *Z. f. Phys.*, **35** (1926), 557.
[5] N. Bohr, Atomtheorie und Mechanik, *Naturwiss.*, **14** (1926), 1.
[6] W. Pauli, Über das Wasserstoffspektrum vom Standpunkt der neuen Quantenmechanik, *Z. f. Phys.*, **36** (1926), 336.
[7] P. Dirac, Quantum mechanics and a preliminary investigation of the hydrogen atom, *Proc. Roy. Soc. A*, **110** (1926), 561.
[8] G. E. Uhlenbeck and S. Goudsmit, Ersetzung der Hypothese vom unmechanischen Zwang durch eine Forderung bezüglich des inneren Verhaltens jedes einzelnen Elektrons, *Naturw.*, **13** (1925), 953.[64]
[9] L. H. Thomas, The motion of the spinning electron, *Nature*, **117** (1926), 514.
[10] J. Frenkel, Die Elektrodynamik des rotierenden Elektrons, *Z. f. Phys.*, **37** (1926), 243.
[11] W. Heisenberg and P. Jordan, Anwendung der Quantenmechanik auf das Problem der anomalen Zeemaneffekte, *Z. f. Phys.*, **37** (1926), 263.
[12] P. Dirac, Relativity quantum mechanics with an application to Compton scattering, *Proc. Roy. Soc.* [*A*], **111** (1926), 405.
[13] P. Dirac, The elimination of the nodes in quantum mechanics, *Proc. Roy. Soc.* [*A*], **111** (1926), 281.
[14] P. Dirac, On quantum algebra, *Proc. Cambridge Phil. Soc.*, **23** (1926), 412.
[15] P. Dirac, The Compton effect in wave mechanics, *Proc. Cambridge Phil. Soc.*, **23** (1926), 500.
[16] A. Einstein, Quantentheorie des einatomigen idealen Gases, II, *Berl. Ber.* (1925), [3].[65]
[17] W. Elsasser, Bemerkungen zur Quanten[mechanik] freier Elektronen, *Naturw.*, **13** (1925), 711.
[18] C. Davisson and [C. H.] Kunsman, [The s]cattering of low speed electrons by [Platinum] and [Magnesium], *Phys. Rev.*, **22** (1923), [242].[66]
[19] C. Davisson and L. Germer, The scattering of electrons by a single crystal of Nickel, *Nature*, **119** (1927), 558.
[20] E. G. Dymond, On electron scattering in Helium, *Phys. Rev.*, **29** (1927), 433.
[21] M. Born and N. Wiener, Eine neue Formulierung der Quantengesetze für periodische und [nichtperiodische] Vorgänge, *Z. f. Phys.*, **36** (1926), 174.
[22] C. Eckart, Operator calculus and the solution of the equation[s] of quantum dynamics, *Phys. Rev.*, **28** (1926), 711.
[23] K. Lanczos, Über eine feldmässige Darstellung der neuen Quantenmechanik, *Z. f. Phys.*, **35** (1926), 812.
[24] E. Schrödinger, Quantisierung als Eigenwertproblem, I to IV, *Ann. d. Phys.*, **79** (1926), 361; *ibid.*, 489; *ibid.*, **80** (1926), 437; *ibid.*, **81** (1926), 109.

[a] The style of the bibliography has been both modernised and uniformised. Amendments and fuller details (when missing) are given in square brackets, mostly without commentary. Amendments in the French edition of mistakes in the typescript (wrong initials, spelling of names etc.) are taken over also mostly without commentary. Mistakes occurring only in the French edition are endnoted (*eds.*).

[25] E. Schrödinger, Der stetige Übergang von der Mikro- zur Makromechanik, *Naturw.*, **14** (1926), 664.
[26] E. Schrödinger, Über das Verhältnis der Heisenberg-Born-Jordanschen Quantenmechanik zu der meinen, *Ann. d. Phys.*, **79** (1926), 734.
[27] P. Jordan, Bemerkung über einen Zusammenhang zwischen Duanes Quantentheorie der Interferenz und den de Broglieschen Wellen, *Z. f. Phys.*, **37** (1926), 376.
[28] W. Heisenberg, Mehrkörperproblem und Resonanz in der Quantenmechanik, I and II, *Z. f. Phys.*, **38** (1926), 411; *ibid.* **41** (1927), 239.
[29] W. Heisenberg, Über die Spektren von Atomsystemen mit zwei Elektronen, *Z. f. Phys.*, **39** (1926), 499.[67]
[30] M. Born, Zur Quantenmechanik der Stossvorgänge, *Z. f. Phys.*, **37** (1926), 863; [Quantenmechanik der Stossvorgänge], *ibid.*, **38** (1926), 803.
[31] G. Wentzel, Zwei Bemerkungen über die Zerstreuung korpuskularer Strahlen als Beugungserscheinung, *Z. f. Phys.*, **40** (1926), 590.
[32] J. R. Oppenheimer, Bemerkung zur Zerstreuung der α-Teilchen, *Z. f. Phys.*, **43** (1927), 413.
[33] W. Elsasser, Diss. Göttingen, [Zur Theorie der Stossprozesse bei Wasserstoff], *Z. f. Phys.*, [**45** (1927), 522].[68]
[34] M. Born, Das Adiabatenprinzip in der Quantenmechanik, *Z. f. Phys.*, **40** ([1926]), 167.[69]
[35] W. Heisenberg, Schwankungserscheinungen und Quantenmechanik, *Z. f. Phys.*, **40** (1926), 501.
[36] P. Jordan, Über quantenmechanische Darstellung von Quantensprüngen, *Z. f. Phys.*, **40** ([1927]), 661.
[37] P. Dirac, On the theory of quantum mechanics, *Proc. Roy. Soc. A*, **112** (1926), 661.
[38] P. Dirac, The physical interpretation of quantum dynamics, *Proc. Roy. Soc. A*, **113** ([1927]), 621.
[39] P. Jordan, Über eine neue Begründung der Quantenmechanik, *Z. f. Phys.*, **40** ([1927]), [809]; Second Part, *ibid.*, **44** (1927), 1.
[40] E. Wigner, Über nichtkombinierende Terme in der neueren Quantentheorie, First Part, *Z. f. Phys.*, **40** (1926), 492; Second Part, *ibid.*, **40** (1927), 883.
[41] D. Hilbert, [J.]. v. Neumann and L. Nordheim, [Über die Grundlagen der Quantenmechanik,] *Mathem. Ann.*, **98** [1928], 1.
[42] [J.] v. Neumann, [Mathematische Begründung der Quantenmechanik,] *Gött. Nachr.*, 20 May 1927, [1].
[43] F. Hund, Zur Deutung der Molekelspektr[en], *Z. f. Phys.*, **40** ([1927]), 742; *ibid.*, **42** (1927), 93; *ibid.*, **43** (1927), 805.
[44] W. Pauli, Über Gasentartung und Paramagnetismus, *Z. f. Phys.*, **41** (1927), 81.
[45] W. Pauli, Zur Quantenmechanik des magnetischen Elektrons, *Z. f. Phys.*, **43** (1927), 601.
[46] W. Heisenberg, Über den anschaulichen Inhalt der quantentheoretischen Kinematik und Mechanik, *Z. f. Phys.*, **43** (1927), 172.[70]
[47] N. Bohr, Über den begrifflichen Aufbau der Quantentheorie, forthcoming [im Erscheinen].
[48] F. Hund, Symmetriecharaktere von Termen bei Systemen mit gleichen Partikeln in der Quantenmechanik, *Z. f. Phys.*, **43** (1927), 788.
[49] [C. G.] Darwin, The electron as a vector wave, *Nature*, **119** (1927), 282.
[50] E. Kennard, Zur Quantenmechanik einfacher Bewegungstypen, *Z. f. Phys.*, **44** (1927), 326.

[51] P. Dirac, The quantum theory of emission and absorption of radiation, *Proc. Roy. Soc. A*, **114** (1927), 243.
[52] P. Dirac, The quantum theory of dispersion, *Proc. Roy. Soc. A*, **114** (1927), 710.
[53] P. Jordan, Über die Polarisation der Lichtquanten, *Z. f. Phys.*, **44** (1927), 292.
[54] P. Jordan, Zur Quantenmechanik der Gasentartung, *Z. f. Phys.*, forthcoming [im Erscheinen] [**44** (1927), 473].
[55] D. Dennison, A note on the specific heat of [the] Hydrogen [molecule], *Proc. Roy. Soc. A*, [**115**] (1927), 483.

On statistics also:
[56] N. S. Bose, Plancks Gesetz und Lichtquantenhypothese, *Z. f. Phys.*, **26** (1924), 178.
[57] E. Fermi, [Sulla quantizzazione del gas perfetto monatomico], *Lincei Rend.*, **3** (1926), 145.

General surveys:
[58] M. Born (*Theorie des Atombaus*?) [*Probleme der Atomdynamik*], Lectures given at the Massachusetts Institute of Technology (Springer, [1926]).[71]
[59] W. Heisenberg, [Über q]uantentheoretische Kinematik und Mechanik, *Mathem. Ann.*, **95** (1926), 683.
[60] W. Heisenberg, Quantenmechanik, *Naturw.*, **14** (1926), [989].
[61] M. Born, Quantenmechanik und Statistik, *Naturw.*, **15** (1927), 238.[72]
[62] P. Jordan, Kausalität und Statistik in der modernen Physik, *Naturw.*, **15** (1927), 105.
[63] P. Jordan, Die Entwicklung der neuen Quantenmechanik, *Naturw.*, **15** (1927), 614, 636.[73]

Discussion of Messrs Born and Heisenberg's report[a]

MR DIRAC. – I should like to point out the exact nature of the correspondence between the matrix mechanics and the classical mechanics. In classical mechanics one can work out a problem by two methods: (1) by taking all the variables to be numbers and working out the motion, e.g. by Newton's laws, which means one is calculating the motion resulting from one particular set of numerical values for the initial coordinates and momenta, and (2) by considering the variables to be functions of the J's (action-angle variables)[74] and using the general transformation theory of dynamics and thus determining simultaneously the motion resulting from all possible initial conditions.[75] The matrix theory corresponds to this second classical method. It gives information about all the states of the system simultaneously. A difference between the matrix method and the second classical method arises since in the latter one requires to treat simultaneously only states having nearly the same J's (one uses, for instance, the operators $\frac{\partial}{\partial J}$), while in the matrix theory one must treat simultaneously states whose J's differ by finite amounts.

To get results comparable with experiment when one uses the second classical method,[76] one must substitute numerical values for the J's in the functions of the J's obtained from the general treatment. One has to do the same in the matrix theory. This gives rise to a difficulty since the results of the general treatment are now matrix elements, each referring in general to two different sets of J's. It is only the diagonal elements, for which these two sets of J's coincide, that have a direct physical interpretation.

MR LORENTZ. – I was very surprised to see that the matrices satisfy equations of motion. In theory that is very beautiful, but to me it is a great mystery, which, I hope, will be clarified. I am told that by all these considerations one has come to construct matrices that represent what one can observe in the atom, for instance the frequencies of the emitted radiation. Nevertheless, the fact that the coordinates, the potential energy, and so on, are now represented by matrices indicates that the quantities have lost their original meaning and that one has made a huge step in the direction of abstraction.

Allow me to draw attention to another point that has struck me. Let us consider the elements of the matrices representing the coordinates of a particle in an atom, a hydrogen atom for instance, and satisfying the equations of motion. One can then change the phase of each element of the matrices without these ceasing to satisfy

[a] The two discussion contributions by Dirac follow his manuscript in AHQP, microfilm 36, Section 10. Deviations in the French edition (which may or may not be due to Dirac) are reported in endnote, as well as interesting variants or cancellations in the manuscript, and punctuation has been slightly altered (*eds.*).

the equations of motion; one can, for instance, change the time. But one can go even further and change the phases, not arbitrarily, but by multiplying each element by a factor of the form $e^{i(\delta_m - \delta_n)}$, and this is quite different from a change of time origin.[a]

Now these matrix elements ought to represent emitted radiation. If the emitted radiation were what is at the basis of all this, one could expect to be able to change all phases in an arbitrary way. The above-mentioned fact then leads us very naturally to the idea that it is not the radiation that is the fundamental thing: it leads us to think that behind the emitted oscillations are hidden some true oscillations, of which the emitted oscillations are difference oscillations.

In this way then, in the end there would be oscillations of which the emitted oscillations are differences, as in Schrödinger's theory,[b] and it seems to me that this is contained in the matrices. This circumstance indicates the existence of a simpler wave substrate.

MR BORN. – Mr Lorentz is surprised that the matrices satisfy the equations of motion; with regard to this I would like to note the analogy with complex numbers. Also here we have a case where in an extension of the number system the formal laws are preserved almost completely. Matrices are some kind of hypercomplex numbers, which are distinguished from the ordinary numbers by the fact that the law of commutativity no longer holds.

MR DIRAC. – The arbitrary phases occurring in the matrix method correspond exactly[77] to the arbitrary phases in the second classical method, where the variables are functions of the J's and w's (action and angle variables). There are arbitrary[78] phases in the w's, which may have different values for each different set of values for the J's. This is completely analogous[79] to the matrix theory, in which each arbitrary phase is associated with a row and column, and therefore with a set of values for the J's.

MR BORN. – The phases α_n which Mr Lorentz has just mentioned are associated with the different energy levels, quite like in classical mechanics. I do not think there is anything mysterious hiding behind this.

MR BOHR. – The issue of the meaning of the arbitrary phases, raised by Mr Lorentz, is of very great importance, I think, in the discussion of the consistency of the methods of quantum theory. Although the concept of phase is indispensable in the calculations, it hardly enters the interpretation of the observations.

[a] This corresponds of course to the choice of a phase factor $e^{i\delta_n}$ for each stationary state. This point (among others) had been raised in the correspondence between Lorentz and Ehrenfest in the months preceding the conference. See Lorentz to Ehrenfest, 4 July 1927, AHQP-EHR-23 (in Dutch) (*eds.*).

[b] See Section 4.4 (*eds.*).

Notes on the translation

1 Here and in a number of places in the following, the French edition omits quotation marks present in the German typescript. They are tacitly restored in this edition.
2 The French edition gives '[47]'.
3 [diskrete] – [déterminées]
4 The French edition omits '[60]'.
5 [makroskopische] – [microscopiques]
6 [durch sinngemässe Übertragung] – [par une extension logique]
7 In the French edition, the parenthetical remark is given as a footnote.
8 The typescript does not give the reference number, only the brackets. The French edition omits the reference entirely. The mentioned results are to be found in Section 4.1 of Born, Heisenberg and Jordan (1926 [4]).
9 Word omitted in the French edition.
10 [sind sinngemässe Übertragungen] – [sont des extensions logiques]
11 Misprint in the French edition: summation index 'n' in the equation.
12 [orthogonale Transformationen der φ_k] – [transformations orthogonales φ_k]
13 [beschränkte unendliche] – [partiellement infinies]
14 [Sätze] – [principes]
15 [noch zurückzukommen haben] – [n'avons pas à revenir]
16 Word omitted in the French edition.
17 The overbar is missing in the original typescript (only here), but is included in the French edition.
18 The typescript reads: 'Lanczos [ ]', the reference number is added in the French edition.
19 The typescript consistently gives this reference as '(q)', the French edition as '(9)'.
20 Equation number missing in the French edition.
21 [eine Verschiebung längs ihrer Verbindungslinie mit dem Nullpunkt] – [un déplacement de leur droite de jonction avec l'origine]
22 [sie] – [ces règles]
23 [Vorgänge] – [phénomènes]
24 Both the manuscript and the French edition read 'H_1' and 'H_2' in this paragraph and two paragraphs later, and '$H^{(1)}$' in the intervening paragraph. We have uniformised the notation.
25 The French edition consistently reads '$\overline{\delta f_n}$'.
26 The right-hand side of this equation reads '$\sum_m \{f(W_n) - f(W_m)\}\Phi_{nm}$' in both the typescript and the French edition, but it should be as shown (see above, p. 104).
27 The French edition gives '(2)'.
28 Both the typescript and the French edition read (only here) 'n_1, n_2'.
29 Both the typescript and the French edition read 'Φ_{nm}'.
30 Singular in the French edition.
31 The typescript includes the square brackets but no reference number. The French edition omits the reference entirely.
32 Only brackets in the typescript, references omitted in the French edition.
33 [abhängige] – [indépendante]
34 The French edition reads '$(c_{nk})^2$' instead of '$|c_n|^2$' and 'nk' instead of 'n'.
35 The French edition reads 'p_{nk}' instead of 'γ_n'.
36 Both the typescript and the French edition give '(11)', but this should evidently be either '(17)' or '(22a)'.
37 The adjective is omitted in the French edition.
38 ['Strahlvektor'] – ['vecteur radiant']
39 The absolute square is missing in the German typescript, but is added in the French edition.
40 [des einen, etwa, nach Grösse und Richtung] – [de l'un, par exemple en grandeur et en direction]
41 [von Q] – [et Q]
42 The French edition has a prime on 'q'.
43 The overbar is missing in the German typescript, but is added in the French edition.
44 Throughout this paragraph, the French edition translates 'Übergang' as 'transformation' instead of 'transition'.

45 '*S*' missing in the French edition.
46 [das folgende Sätze zugrunde legt] – [qui est à la base des théorèmes suivants]. Note that 'Satz' can indeed mean both 'statement' and 'theorem'.
47 The French edition omits absolute bars.
48 The '*h*' is present in the French edition but not in the typescript.
49 In the typescript, this is typed over an (illegible) previous alternative. Jordan in his habilitation lecture (1927f [62]) uses the term 'angebbare Wahrscheinlichkeit' ('assignable probability' in Oppenheimer's translation (Jordan 1927g)).
50 [in dem der einfache Begriff der berechenbaren Wahrscheinlichkeit für ein bestimmtes Ereignis die Hauptrolle spielt] – [dans lequel la simple notion de la probabilité calculable joue le rôle principale pour un événement déterminé]
51 [Überstand] – [défaut]. The word 'Überstand' may be characterising the phases as some kind of surplus structure, but it is quite likely a mistyping of 'Übelstand', which can indeed be translated as 'defect', as in the French version.
52 [Es gibt] – [Cet auteur donne]
53 [Wie kann insbesondere bei der nur statistischen Bestimmtheit des Einzelvorgangs in den zusammengesetzten makroskopischen Erscheinungen die gewohnte deterministische Ordnung aufrecht erhalten werden?] – [En particulier comment, vu la détermination uniquement statistique des processus individuels dans les phénomènes macroscopiques compliqués, l'ordre déterministe auquel nous sommes accoutumés peut-il être conservé?]
54 This reference is to a supposedly forthcoming 'Über den begrifflichen Aufbau der Quantentheorie'. Yet, no such published or unpublished work by Bohr is extant. Some pages titled 'Zur Frage des begrifflichen Aufbaus der Quantentheorie' are contained in the folder 'Como lecture II' in the Niels Bohr archive, microfilmed in AHQP-BMSS-11, Section 4. See also Bohr (1985, p. 478). We wish to thank Felicity Pors, of the Niels Bohr archive, for correspondence on this point.
55 The French edition incorrectly reads '$\sin\varphi_1 = \frac{q_1}{\lambda}$'.
56 [Erwartung] – [attente]
57 [Ausbreitung] – [extension]
58 [einzelne Versuchsanordnungen] – [des essais isolés]
59 Again, the terminology has changed both in German and in English. The term 'similar particles' for 'identical particles' is used for instance by Dirac (1927a [37]).
60 [nur das antisymmetrische Teilsystem zulässt] – [ne permet pas le système antisymétrique]
61 [aus] – [dans]
62 The French edition gives '[56]'.
63 Both the German version followed here and the French version ('a whole number of particles') seem rather infelicitous.
64 Both the typescript and the French edition add '(Magnetelektron)'. The French edition reads '*Nature*'.
65 Both the typescript and the French edition read 'p. 5'.
66 Typescript and published volume read 'Pt' and 'Mg', as well as '243'.
67 In the French edition: '409'.
68 This is indeed the (abridged) published version of Elsasser's Göttingen dissertation.
69 In the French edition: 'Das Adiabatenprinzip in den Quanten', as well as '1927' (the latter as in the typescript).
70 In both the typescript and the French edition the title of the paper is given as 'Über den anschaulichen Inhalt der Quantenmechanik'.
71 Date given as '1927' in typescript and volume.
72 In the French edition: '288'.
73 The French edition omits '614'.
74 The manuscript includes also 'and w's' and 'and angle', both cancelled.
75 The French edition breaks up and rearranges this sentence.
76 The French edition omits the temporal clause.
77 The French edition reads 'trouvent une analogie'.
78 In the manuscript this replaces the cancelled word 'unknown'.
79 In the manuscript this replaces 'corresponds exactly'.

Wave mechanics[a]

By Mr E. Schrödinger

Introduction

Under this name at present two theories are being carried on, which are indeed closely related but not identical. The first, which follows on directly from the famous doctoral thesis by L. de Broglie, concerns waves in three-dimensional space. Because of the strictly relativistic treatment that is adopted in this version from the outset, we shall refer to it as the *four-dimensional* wave mechanics. The other theory is more remote from Mr de Broglie's original ideas, insofar as it is based on a wave-like process in the space of *position coordinates* (q-space) of an arbitrary mechanical system.[1] We shall therefore call it the *multi-dimensional* wave mechanics. Of course this use of the q-space is to be seen only as a mathematical tool, as it is often applied also in the old mechanics; ultimately, in this version also, the process to be described is one in space and time. In truth, however, a complete unification of the two conceptions has not yet been achieved. Anything over and above the motion of a single electron could be treated so far only in the *multi*-dimensional version; also, this is the one that provides the mathematical solution to the problems posed by the Heisenberg–Born matrix mechanics. For these reasons I shall place it first, hoping in this way also to illustrate better the characteristic difficulties of the as such more beautiful four-dimensional version.[b]

[a] Our translation follows Schrödinger's German typescript in AHQP-RDN, document M-1354. Discrepancies between the typescript and the French edition are endnoted. Interspersed in the German text, Schrödinger provided his own summary of the paper (in French). We translate this in the footnotes. The French version of this report is also reprinted in Schrödinger (1984, vol. 3, pp. 302–23) (*eds.*).

[b] Summary of the introduction: Currently there are in fact *two* [theories of] wave mechanics, very closely related to each other but not identical, that is, the relativistic or *four-dimensional* theory, which concerns waves in ordinary space, and the *multi-dimensional* theory, which originally concerns waves in the configuration space of an arbitrary system. The former, until now, is able to deal only with the case of a single electron, while the latter, which provides the solution to the matrix problems of Heisenberg–Born, comes up against the difficulty of being put in relativistic form. We start with the latter.

I. – Multi-dimensional theory

Given a system whose configuration is described by the generalised position coordinates $q_1, q_2, \ldots, q_n$, classical mechanics considers its task as being that of determining the q_k as functions of time, that is, of exhibiting *all* systems of functions $q_1(t), q_2(t), \ldots, q_n(t)$ that correspond to a dynamically possible motion of the system. Instead, according to wave mechanics the solution to the problem of motion is not given by a system of n functions of the single variable t, but by a *single* function ψ of the n variables $q_1, q_2, \ldots, q_n$ and perhaps of time (see below). This is determined by a *partial* differential equation with $q_1, q_2, \ldots, q_n$ (and perhaps t) as *independent* variables. This change of role of the q_k, which from dependent become independent variables, appears to be the crucial point. More later on the meaning of the function ψ, which is still controversial. We first describe how it is determined, thus what corresponds to the equations of motion of the old mechanics.

First let the system be a *conservative* one. We start from its Hamiltonian function

$$H = T + V,$$

that is, from the total energy expressed as a function of the q_k and the canonically-conjugate momenta p_k. We take H to be a *homogeneous* quadratic function of the q_k and of unity and replace in it each p_k by $\frac{h}{2\pi}\frac{\partial\psi}{\partial q_k}$ and unity by ψ. We call the function of the q_k, $\frac{\partial\psi}{\partial q_k}$ and ψ thus obtained L (because in wave mechanics it plays the role of a Lagrange function). Thus

$$L = T\left(q_k, \frac{h}{2\pi}\frac{\partial\psi}{\partial q_k}\right) + V\psi^2. \tag{1}$$

Now we determine $\psi(q_1, q_2, \ldots, q_n)$ by the requirement that under variation of ψ,

$$\delta\int L d\tau = 0 \quad \text{with} \quad \int \psi^2 d\tau = 1. \tag{2}$$

The integration is to be performed over the whole of q-space (on whose perhaps infinitely distant boundary, $\partial\psi$ must disappear). However, $d\tau$ is not simply the product of the dq_k, rather the 'rationally measured' volume element in q-space:

$$d\tau = dq_1 dq_2 \ldots dq_n \left| \pm \frac{\partial^2 T}{\partial p_1 \ldots \partial p_k} \right|^{-\frac{1}{2}} \tag{3}$$

(it is the volume element of a *Riemannian* q-space, whose *metric*, as for instance also in Hertz's mechanics,[a] is determined by the kinetic energy). – Performing

[a] For Schrödinger's interest in Hertz's work on mechanics, see Mehra and Rechenberg (1987, pp. 522–32) (*eds.*).

the variation, taking the normalisation constraint with the multiplier [Factor] $-E$, yields the Euler equation

$$\Delta\psi + \frac{8\pi^2}{h^2}(E - V)\psi = 0 \tag{4}$$

(Δ stands for the analogue of the Laplace operator in the generalised Riemannian sense). As is well known,

$$\int L d\tau = E$$

for a function that satisfies the Euler equation (4) and the constraint in (2).

Now, it turns out that equation (4) in general does not have, for every E-value, a solution ψ that is single-valued and always finite and continuous together with its first and second derivatives; instead, in all special cases examined so far, this is the case precisely for the E-values that Bohr's theory would describe as stationary energy levels of the system (in the case of discrepancies, the recalculated values explain the facts of experience *better* than the old ones). The word 'stationary' used by Bohr is thus given a very pregnant meaning by the variation problem (4).

We shall refer to these values as eigenvalues, E_k, and to the corresponding solutions ψ_k as eigenfunctions.[a] We shall number the eigenvalues always in increasing order and shall number repeatedly those with multiple eigensolutions. The ψ_k form a normalised complete orthogonal system in the q-space, with respect to which every well-behaved function of the q_k can be expanded in a series. Of course this does not mean that every well-behaved function solves the homogeneous equation (4) and thus the variation problem, because (4) is indeed an equation *system*, each single eigensolution ψ_k satisfying a different element of the system, namely the one with $E = E_k$.[b]

One can take the view that one should be content in principle with what has been said so far and its very diverse special applications. The single stationary states of Bohr's theory would then in a way be described by the eigenfunctions ψ_k, which *do not contain time at all.*[c] One would find that one can derive much more from them that is worth knowing, in particular, one can form from them,

[a] As a rule, in certain domains of the energy axis[2] the eigenvalue spectrum is continuous, so that the index k is replaced by a continuous parameter. In the notation we shall generally not take this into account.

[b] Summary of the above: Wave mechanics demands that events in a mechanical system that is in motion be described not by giving n generalised coordinates $q_1, q_2 \ldots q_n$ as functions of the time t, but by giving a *single* function [ψ] of the n variables $q_1, q_2 \ldots q_n$ and maybe of the time t. The system of equations of motion of classical mechanics corresponds in wave mechanics to a *single* partial differential equation, eq. (4), which can be obtained by a certain variational procedure. E is a Lagrange multiplier, V is the potential energy, a function of the coordinates; h is Planck's constant, Δ denotes the Laplacian in q-space, generalised in the sense of Riemann. One finds in specific cases that finite and continuous solutions, 'eigenfunctions' ψ_k of eq. (4), exist only for certain 'eigenvalues' E_k of E. The set of these functions forms a complete orthogonal system in the coordinate space. The eigenvalues are precisely the 'stationary energy levels' of Bohr's theory.

[c] Cf. Section 8.1 (*eds.*).

by fixed general rules, quantities that can be aptly taken to be *jump probabilities* between the single stationary states. Indeed, it can be shown for instance that the integral

$$\int q_i \psi_k \psi_{k'} d\tau, \tag{5}$$

extended to the whole of q-space, yields precisely the matrix element bearing the indices k and k' of the 'matrix q' in the Heisenberg–Born theory; similarly, the elements of all matrices occurring there can be calculated from the wave mechanical eigenfunctions.

The theory as it stands, restricted to conservative systems, could treat already even the *interaction* between two or more systems, by considering these as one single system, with the addition of a suitable term in the potential energy depending on the coordinates of all subsystems. Even the interaction of a material system with the radiation field is not out of reach, if one imagines the system together with certain ether oscillators (eigenoscillations of a cavity) as a single conservative system, positing suitable interaction terms.

On this view, the *time variable* would play absolutely no role in an isolated system – a possibility to which N. Campbell (*Phil. Mag.*, [**1**] (1926), [1106]) has recently pointed. Limiting our attention to an isolated system, we would not perceive the passage of time in it any more than we can notice its possible progress in space, an assimilation of time to the spatial coordinates that is very much in the spirit of relativity. What we would notice would be merely a sequence of discontinuous transitions, so to speak a cinematic image, but without the possibility of comparing the time intervals between the transitions. Only secondarily, and in fact with increasing precision the more extended the system, would a *statistical* definition of time result from *counting* the transitions taking place (Campbell's 'radioactive clock'). Of course then one cannot understand the jump probability in the usual way as the probability of a transition calculated relative to unit time. Rather, a *single* jump probability is then utterly meaningless; only with *two* possibilities for jumps, the probability that the one may happen *before* the other is equal to *its* jump probability divided by the sum of the two.

I consider this view the only one that would make it possible to hold on to 'quantum jumps' in a coherent way. Either all changes in nature are discontinuous or not a single one. The first view may have many attractions; for the time being however, it still poses great difficulties. If one does not wish to be so radical and give up in principle the use of the *time variable* also for the single atomistic system, then it is very natural to assume that it is contained hidden also in equation (4). One will conjecture that equation system (4) is the *amplitude* equation of an *oscillation* equation, from which time has been eliminated by setting[a]

$$\psi \sim e^{2\pi i \nu t}. \tag{6}$$

E must then be proportional to a power of ν, and it is natural to set $E = h\nu$. Then the following is the oscillation equation that leads to (4) with the ansatz (6):[b]

$$\Delta\psi - \frac{8\pi^2}{h^2} V\psi - \frac{4\pi i}{h}\frac{\partial\psi}{\partial t} = 0. \tag{7}$$

Now *this* is satisfied not just by a single[3]

$$\psi_k e^{2\pi i \nu_k t} \quad (\nu_k = \frac{E_k}{h}),$$

but by an arbitrary linear combination

$$\psi = \sum_{k=1}^{\infty} c_k \psi_k e^{2\pi i \nu_k t} \tag{8}$$

with arbitrary (even complex) constants c_k. If one considers *this* ψ as the description[4] of a certain sequence of phenomena in the system, then this is now given by a (complex) function of the $q_1, q_2, \ldots, q_n$ *and* of time, a function which can even be given arbitrarily at $t = 0$ (because of the completeness[5] and orthogonality of the ψ_k); the oscillation equation (7), or its solution (8) with suitably chosen c_k, then governs the temporal development. Bohr's stationary states correspond to the eigenoscillations of the structure (*one* $c_k = 1$, all others $= 0$).

There now seems to be no obstacle to assuming that equation (7) is valid immediately also for non-conservative systems (that is, V may contain time explicitly). Then, however,[6] the solution no longer has the simple form (8). A particularly interesting application hereof is the perturbation of an atomic system by an electric alternating field. This leads to a theory of *dispersion*, but we must forgo here a more detailed description of the same. – From (7) there *always* follows

$$\frac{d}{dt}\int d\tau\,\psi\psi^* = 0. \tag{9}$$

(An asterisk shall always denote the complex conjugate.[7]) Instead of the earlier normalisation condition (2), one can thus require

$$\int d\tau\,\psi\psi^* = 1, \tag{10}$$

[a] Schrödinger introduces the time-dependent equation in his fourth paper on quantisation (1926g). There (p. 112), Schrödinger leaves the sign of time undetermined, settling on the same convention as in (6) – the opposite of today's convention – on pp. 114–15. As late as Schrödinger (1926h, p. 1065), one reads that 'the most general solution of the wave-problem will be (the real part of) [eq. (27) of that paper]'. Instead the wave function is characterised as 'essentially complex' in Schrödinger (1927c, fn. 3 on p. 957) (*eds.*).

[b] Recall that Schrödinger does not in fact set $m = 1$, but absorbs the mass in the definition of Δ (*eds.*).

which in the conservative case, equation (8), means

$$\sum_{k=1}^{\infty} c_k c_k^* = 1.^{\mathrm{a}} \tag{11}$$

What does the ψ-function mean now, that is, *how does the system described by it really look like in three dimensions*? Many physicists today are of the opinion that it does not describe[8] the occurrences in an individual system,[9] but only the processes in an ensemble of very many like constituted systems that do not sensibly influence one another[10] and are all under the very same conditions. I shall skip this point of view, since others are presenting it.[b] I myself have so far found useful the following perhaps somewhat naive but quite concrete idea [dafür recht greifbare Vorstellung]. The classical system of material points does not really exist, instead there exists something that continuously fills the entire space and of which one would obtain a 'snapshot' if one dragged the classical system, with the camera shutter open, through *all* its configurations, the representative point in q-space spending in each volume element $d\tau$ a time that is proportional to the *instantaneous* value of $\psi\psi^*$. (The value of $\psi\psi^*$ for only *one* value of the argument t is thus in question.) Otherwise stated: the real system is a superposition of the classical one in all its possible states, using $\psi\psi^*$ as 'weight function'.

The systems to which the theory is applied consist classically of several[12] charged point masses. In the interpretation just discussed[13] the charge of every single one of these is distributed continuously across space, the individual point mass with charge e yielding to the charge in the three-dimensional volume element $dx\,dy\,dz$ the contribution[14]

$$e \int' \psi\psi^* d\tau. \tag{12}$$

[a] Summary of the above: Even limiting oneself to what has been said up to now, it would be possible to derive much of interest from these results, for instance the transition probabilities, formula (5) yielding precisely the matrix element $q_i(k, k')$ for the same mechanical problem formulated according to the Heisenberg–Born theory. Although we have restricted ourselves so far to conservative systems, it would be possible to treat in this way also the mutual action between several systems and even between a material system and the radiation field; one would only have to add all relevant systems to the system under consideration. *Time* does not appear at all in our considerations and one could imagine that the only events that occur are sudden transitions from one quantum state of the total system to another quantum state, as Mr N. Campbell has recently thought. Time would be defined only statistically by counting the quantum jumps (Mr Campbell's 'radioactive clock'). – Another, less radical, point of view is to assume that time is hidden already in the family of equations (4) parametrised by E, this family being the *amplitude* equation of an *oscillation* equation, from which time has been eliminated by the ansatz (6). Assuming $h\nu = E$ one arrives at eq. (7), which, because it no longer contains the frequency ν, is solved by the *series* (8), where the c_k are arbitrary, generally complex, constants. Now ψ is a function of the $q_1, q_2 \ldots q_n$ *as well as* of time t and, by a suitable choice of the c_k, it can be adjusted to an arbitrary initial state. Nothing prevents us now from making the time appear also in the function V – this is the theory of non-conservative systems, one of whose most important applications is the theory of dispersion. – The important relation (9), which follows from eq. (7), allows one in all cases to normalise ψ according to eq. (10).

[b] See the report by Messrs Born and Heisenberg.[11]

The prime on the integration sign means: one has to integrate only over the part of the q-space corresponding to a position of the distinguished point mass within $dxdydz$. – Since $\psi\psi^*$ in general depends on time, these charges fluctuate; only in the special case of a conservative system oscillating with a single eigenoscillation are they distributed permanently, so to speak statically.

It must now be emphasised that by the claim that there *are*[15] these charge densities (and the current densities arising from their fluctuation), we can mean at best *half* of what classical electrodynamics would mean by that. Classically, charge and current densities are (1) *application points*, (2) *source points* of the electromagnetic field. As *application points* they are completely out of the question here; the assumption that these charges and currents act, say, according to the Coulomb or Biot–Savart law directly on one another, or are directly affected in such a way by external fields, this assumption is either superfluous or wrong (N.B. *de facto* wrong), because the changes in the ψ function and thereby in the charges are indeed to be determined through the oscillation equation (7) – thus we must not think of them as determined also in another way, by forces acting on them. An external electric field is to be taken account of in (7) in the potential function V, an external magnetic field in a similar way to be discussed below, – this is the way their application to the charge distribution is expressed in the present theory.

Instead, our spatially distributed charges prove themselves excellently as *source points* of the field, at least for the external action of the system, in particular with respect to its *radiation*. Considered as source points in the sense of the usual electrodynamics, they yield largely[16] correct information about its frequency, intensity and polarisation.[a] In most cases, the charge is in practice confined to a region that is small compared to the wavelengths of the emitted light. The radiation is then determined by the resulting *dipole moment* [*elektrisches Moment*] of the charge distribution. According to the principles determined above, this is calculated from the classical dipole moment of an arbitrary configuration by performing an average using $\psi\psi^*$

$$M_{\text{qu}} = \int M_{\text{cl}}\psi\psi^* d\tau. \tag{13}$$

A glance at (8) shows that in M_{qu} the *differences* of the ν_k will appear as emission frequencies; since the ν_k are the spectroscopic term values, our picture provides an *understanding*[17] *of Bohr's frequency condition*. The integrals that appear as amplitudes of the different partial oscillations of the dipole moment represent according to the remarks on (5) the elements of Born and Heisenberg's 'dipole moment matrix'. By evaluating these integrals one obtained the correct polarisations and intensities of the emitted light in many special cases, in particular

[a] See the discussion after the report, as well as Section 4.4 (*eds.*).

intensity zero in all cases where a line allowed by the frequency condition is missing according to experience (*understanding*[18] *of the selection principle*). – Even though all these results, if one so wishes, can be detached from the picture of the fluctuating charges and be represented in a more abstract form, yet they put quite beyond doubt that the picture is tremendously useful for one who has the need for Anschaulichkeit![19,a]

In no way should one claim that the provisional attempt of a classical-electrodynamic coupling of the field to the charges generating the field is already the last word on this issue. There are internal[20] reasons for doubting this. First, there is a serious difficulty in the question of the *reaction* of the emitted radiation on the emitting system, which is not yet expressed by the wave equation (7), according to which also such wave forms of the system that continuously emit radiation could and would in fact always persist unabated. Further, one should consider the following. We always observe the radiation emitted by an atom only through its action on another atom or molecule. Now, from the wave mechanical standpoint we can consider two charged point masses that belong to the *same* atom, neither as acting directly on each other in their pointlike form (standpoint of classical mechanics), nor are we allowed to think this of their 'smeared out' wave mechanical charge distributions (the wrong move taunted above). Rather, we have to take account of their classical potential energy, considered as a function in q-space, in the coefficient V of the wave equation (7). But then, when we have two *different* atoms, it will surely not be correct in principle to insert the fields generated by the *spread-out* charges of the first at the position of the second in the wave equation for the latter. And yet we do this when we calculate the radiation of an atom in the way described above and now treat wave mechanically the behaviour of another atom in this radiation field. I say this way of calculating the interaction between the charges of different atoms can be at most approximate, but not correct in principle. For *within one* system it is certainly wrong. But if we bring the two atoms closer together, then the distinction between the charges of one and those of the other gradually disappears, it is actually never a distinction

[a] Summary of the above: The physical meaning of the function ψ appears to be that the system of charged point particles imagined by classical mechanics does not in fact exist, but that there is a continuous distribution of electric charge, whose density can be calculated at each point of ordinary space using ψ or rather $\psi\psi^*$, the square of the absolute value of ψ. According to this idea, the quantum (or: real) system is a superposition of all the possible configurations of the classical system, the real function $\psi\psi^*$ in q-space occurring as 'weighting function'. Since $\psi\psi^*$ in general contains time, fluctuations of charge must occur. What we mean by the *existence* of these continuous and fluctuating charges is not at all that they should act on each other according to the Coulomb or Biot–Savart law – the *motion* of these charges is already completely governed by eq. (7). But what we mean is that they are the *sources* of the electric fields and magnetic fields proceeding from the atom, above all the sources of the observed radiation. In many a case one has obtained wonderful agreement with experiment by calculating the radiation of these fluctuating charges using classical electrodynamics. In particular, they yield a complete and general explanation of Bohr's 'frequency condition' and of the spectral 'polarisation and selection rules'.

of principle.[21] – The coherent wave mechanical route would surely be to combine both the emitting and the receiving system into a single one and to describe them through a *single* wave equation with appropriate coupling terms, however large the distance between emitter and receiver may be. Then one could be completely silent about the processes in the radiation field. But what would be the correct coupling terms? Of course not the usual Coulomb potentials, as soon as the distance is equal to several wavelengths![22] (One realises from here that without important amendments the *entire* theory in reality can only be applied to very *small* systems.) Perhaps one should use the *retarded* potentials. But these are not functions in the (common) q-space, instead they are something much more complicated. Evidently we encounter here the provisional limits of the theory and must be happy to possess in the procedure depicted above an approximate treatment that appears to be very useful.[a]

II. – FOUR-DIMENSIONAL THEORY

If one applies the *multi*-dimensional version of wave mechanics to a single electron of mass m and charge e moving in a space with the electrostatic potential φ and to be described by the three rectangular coordinates x, y, z, then the wave equation (7) becomes

$$\frac{1}{m}\left(\frac{\partial^2\psi}{\partial x^2}+\frac{\partial^2\psi}{\partial y^2}+\frac{\partial^2\psi}{\partial z^2}\right)-\frac{8\pi^2}{h^2}e\varphi\psi-\frac{4\pi i}{h}\frac{\partial\psi}{\partial t}=0. \tag{14}$$

(N.B. The factor $\frac{1}{m}$ derives from the fact that, given the way of determining the metric of the q-space through the kinetic energy, $x\sqrt{m}$, $y\sqrt{m}$, $z\sqrt{m}$ should be used as coordinates rather than x, y, z.[23]) It now turns out that the present equation is nothing else but the ordinary three-dimensional wave equation for de Broglie's 'phase waves' of the electron, except that the equation in the above form is shortened or truncated in a way that one can call 'neglecting the influence of relativity'.

[a] Summary of the above: However, there are reasons to believe that our fluctuating and purely classically radiating charges do not provide the last word on this question. Since we observe the radiation of an atom only by its effect on another atom or molecule (which we shall thus also treat quite naturally by the methods of wave mechanics), our procedure reduces to substituting into the wave equation of *one* system the potentials that would be produced according to the classical laws by the extended charges of another system. This way of accounting for the mutual action of the charges belonging to two different systems cannot be absolutely correct, since for the charges belonging to the same system it is not. The correct method of calculating the influence of a radiating atom on another atom would be perhaps to treat them as *one* total system according to the methods of wave mechanics. But that does not seem at all possible, since the retarded potentials, which should no doubt occur, are not simply functions of the configuration of the systems, but something much more complicated. Evidently, at present these are the limits of the method!

In fact, in the electrostatic field de Broglie gives the following expression[a] for the wave velocity u of his phase waves, depending on the potential φ (i.e. on position) and on the frequency ν:[24]

$$u = c\frac{h\nu}{\sqrt{(h\nu - e\varphi)^2 - h^2\nu_0^2}} \qquad \left(\nu_0 = \frac{mc^2}{h}\right). \tag{15}$$

If one inserts this into the ordinary three-dimensional wave equation

$$\frac{\partial^2\psi}{\partial x^2} + \frac{\partial^2\psi}{\partial y^2} + \frac{\partial^2\psi}{\partial z^2} - \frac{1}{u^2}\frac{\partial^2\psi}{\partial t^2} = 0,$$

and uses (6) to eliminate the frequency ν from the equation, one has[25] ($\Delta = \frac{\partial^2}{\partial x^2} + \frac{\partial^2}{\partial y^2} + \frac{\partial^2}{\partial z^2}$)

$$\Delta\psi - \frac{1}{c^2}\frac{\partial^2\psi}{\partial t^2} + \frac{4\pi i e\varphi}{hc^2}\frac{\partial\psi}{\partial t} + \frac{4\pi^2}{c^2}\left(\frac{e^2\varphi^2}{h^2} - \nu_0^2\right)\psi = 0. \tag{16}$$

Now if one considers that in the case of 'slow electron motion' (a) the occurring frequencies are always very nearly equal to the rest frequency ν_0, so that in order of magnitude the derivative with respect to time in (16) is equal to a multiplication by $2\pi i\nu_0$, and that (b) $\frac{e\varphi}{h}$ in this case[26] is always small with respect to ν_0; and if one then sets in equation (16)

$$\psi = e^{2\pi i\nu_0 t}\tilde{\psi}, \tag{17}$$

and disregarding squares of small quantities, one obtains for $\tilde{\psi}$ exactly equation (14) derived from the multi-dimensional version of wave mechanics. As claimed, this is thus indeed the 'classical approximation' of the wave equation holding for de Broglie's phase waves.[b] The transformation (17) here shows us that, considered from de Broglie's point of view, the multi-dimensional theory is committed to a so to speak truncated view of the frequency, in that it subtracts once and for all from all frequencies the rest frequency ν_0. (N.B. In calculating the charge density from $\psi\psi^*$,[27] the additional factor is of course irrelevant since it has modulus[28] 1.[29])[c]

Let us now keep to the form (16) of the wave equation. It still requires an important generalisation. In order to be truly relativistic it must be invariant with

[a] Cf. the formula for the refractive index on p. 343 of de Broglie's report (*eds.*).

[b] That is, the non-relativistic approximation (*eds.*).

[c] Summary of the above: The three-dimensional wave equation, eq. (14), obtained by applying the multi-dimensional theory to a single electron in an electrostatic potential field φ, is none other than the non-relativistic approximation of the wave equation that results from Mr L. [d]e Broglie's ideas for his 'phase waves'. The latter, eq. (16), is obtained by substituting into the ordinary wave equation expression (15), which Mr [d]e Broglie has derived for the phase velocity u as a function of the frequency ν and of the potential φ (that is, of the coordinates x, y, z, on which φ will depend) and by eliminating from the resulting formula the frequency ν by means of (6).

respect to Lorentz transformations. But if we perform such a transformation on our electric field, hitherto assumed to be static, then it loses this feature and a magnetic field appears by itself next to it. In this way one derives almost unavoidably the form of the wave equation in an *arbitrary* electromagnetic field. The result can be put in the following transparent form, which makes the complete equivalence [Gleichberechtigung] of time and the three spatial coordinates fully explicit:

$$\left[\left(\frac{\partial}{\partial x}+\frac{2\pi ie}{hc}a_x\right)^2+\left(\frac{\partial}{\partial y}+\frac{2\pi ie}{hc}a_y\right)^2 + \left(\frac{\partial}{\partial z}+\frac{2\pi ie}{hc}a_z\right)^2+\left(\frac{1}{ic}\frac{\partial}{\partial t}+\frac{2\pi ie}{hc}i\varphi\right)^2-\frac{4\pi^2\nu_0^2}{c^2}\right]\psi=0. \tag{18}$$

(N.B. a is the vector potential.[30] In evaluating the squares one has to take account of the *order* of the factors, since one is dealing with operators, and further of Maxwell's relation:

$$\frac{\partial a_x}{\partial x}+\frac{\partial a_y}{\partial y}+\frac{\partial a_z}{\partial z}+\frac{1}{ic}\frac{\partial(i\varphi)}{\partial t}=0.) \tag{19}$$

This wave equation is of very manifold interest. First, as shown by Gordon,[a] it can be derived in a way very similar to what we have seen above for the amplitude equation of conservative systems, from a variational principle, which now obtains in four dimensions, and where time plays a perfectly symmetrical role with respect to the three spatial coordinates. Further: if one adds to the *Lagrange function* of Gordon's variational principle the well-known Lagrange function of the Maxwell field *in vacuo* (that is, the half-difference of the squares of the magnetic and the electric field strengths) and varies in the spacetime integral of the new Lagrange function thus obtained not only ψ, but also the potential components φ, a_x, a_y, a_z, one obtains as the *five* Euler equations along with the wave equation (18) also the four retarded potential equations for φ, a_x, a_y, a_z.[b] (One could also say: Maxwell's *second* quadruple of equations, while the first, as is well-known, holds identically in the potentials.[32]) It contains as *charge and current density* quadratic forms in ψ and its first derivatives[33] that agree completely with the rule which we had given in the *multi*-dimensional theory for calculating the true charge distribution from the ψ-function. Second, one can further define[c] a *stress-energy-momentum tensor* of the *charges*, whose ten components are also quadratic forms of ψ and its first derivatives, and which together with the well-known Maxwell tensor obeys the

[a] W. Gordon, *Zeitschr. f. Phys.*, **40** (1926), 117.
[b] E. Schrödinger, *Ann. d. Phys.*, **82** (1927), [265].[31]
[c] E. Schrödinger, *loc. cit.*[34]

laws of conservation of energy and of momentum (that is, the sum of the two tensors has a vanishing divergence).[a]

But I shall not bother you here with the rather complex mathematical development of these issues, since the view still contains a serious inconsistency. Indeed, according to it, it would be the *same* potential components φ, a_x, a_y, a_z which *on the one hand* act to modify the wave equation (18) (one could say: they act *on* the charges *as movers*[35]) and which *on the other hand* are determined in turn, via the retarded potential equations, *by* these same charges, which occur as *sources* in the latter equations. (That is: the wave equation (18) determines the ψ function, from the latter one derives the charge and current densities, which as sources determine the potential components.) – In reality, however, one operates *otherwise* in the application of the wave equation (18) to the hydrogen electron, and one *must* operate otherwise to obtain the correct result: one substitutes in the wave equation (18) the *already given* potentials of the nucleus and of possible external fields (Stark and Zeeman effect). From the solution for ψ thus obtained one derives the fluctuating charge densities discussed above, which one in fact[36] has to use for the determination from sources of the *emitted radiation*; but one must *not* add *a posteriori* to the field of the nucleus and the possible external fields also the fields produced by these charges at the position of the atom itself in equation (18) – [37] something totally wrong would result.

Clearly this is a painful lacuna. The pure field theory is not enough, it has to be supplemented by performing a kind of *individualisation* of the charge densities coming from the single point charges of the classical model, where however each single 'individual' may be spread over the whole of space, so that they overlap. In the wave equation for the single individual one would have to take into account only the fields produced by the other individuals but not its self-field. These remarks, however, are only meant to characterise the general nature of the required supplement, not to constitute a programme to be taken completely literally.[b]

[a] Summary of the above: In order to generalise equation (16) so that it may apply to an arbitrary electromagnetic field, one subjects it to a Lorentz transformation, which automatically makes a magnetic field appear. One arrives at eq. (18), in which time enters in a perfectly symmetrical way with the spatial coordinates. Gordon has shown that this equation derives from a four-dimensional variational principle. By adding to Gordon's Lagrangian the well-known Lagrangian of the free field and by varying along with ψ also the four components of the potential, one derives from a single variational principle besides eq. (18) also the laws of electromagnetism with certain homogeneous quadratic functions of ψ and its first derivatives as charge and current densities. These agree well with what was said in the previous section regarding the calculation of the fluctuating charges using the ψ function. – One finds a definition of the stress-energy-momentum tensor, which, added to Maxwell's tensor, satisfies the conservation laws.

[b] Summary of the above: However, these last developments run into a great difficulty. From their direct application would follow the logical necessity of taking into account in the wave equation, for instance in the case of the hydrogen atom, not only the potential arising from the nucleus, but also the potentials arising from the fluctuating charges; which, apart from the enormous mathematical complications that would arise, would give completely wrong results. The field theory ('Feldtheorie') appears thus inadequate; it should be supplemented by a kind of individualisation of the electrons, despite these being extended over the whole of space.

We wish to present also the remarkable special result yielded by the relativistic form (18) of the wave equation for the hydrogen atom. One would at first expect and hope to find the well-known Sommerfeld formula for the fine structure of terms. Indeed one *does* obtain a fine structure and one *does* obtain Sommerfeld's formula, however the result contradicts experience, because it is exactly what one would find in the Bohr–Sommerfeld theory, if one were to posit the radial as well as the azimuthal quantum number as *half-integers* [halbzahlig], that is, half of an odd integer. – Today this result is not as disquieting as when it was first encountered.[a] In fact, it is well-known that the extension of Bohr's theory through the Uhlenbeck–Goudsmit electron spin [Elektronendrall], required by many other facts of experience, has to be supplemented in turn by the move to secondary quantum 'half'-numbers ['halbe' Nebenquantenzahlen] in order to obtain good results. How the spin is represented in wave mechanics is still uncertain. Very promising suggestions[b] point in the direction that instead of the *scalar* ψ a *vector* should be introduced. We cannot discuss here this latest turn in the theory.[c]

III. – THE MANY-ELECTRON PROBLEM

The attempts[d] to derive numerical results by means of approximation methods for the atom with *several* electrons, whose amplitude equation (4) or wave equation (7) cannot be solved directly, have led to the remarkable result that actually, despite the multi-dimensionality of the original equation, in this procedure one always needs to calculate only with the well-known three-dimensional eigenfunctions of hydrogen; indeed one has to calculate certain *three*-dimensional *charge distributions* that result from the hydrogen eigenfunctions according to the principles presented above, and one has to calculate according to the principles of classical electrostatics the self-potentials and interaction potentials of these charge distributions; these *constants* then enter as coefficients in a system of equations that in a simple way determines *in principle* the behaviour of the many-electron atom. Herein, I think, lies a hint that with the furthering of our understanding 'in the end everything will

[a] E. Schrödinger, *Ann. d. Phys.*, **79** (1926), [361], p. 372.
[b] C. G. Darwin, *Nature*, **119** (1927), 282, *Proc. Roy. Soc. A*, **116** (1927), 227.
[c] Summary of the above: For the hydrogen atom the relativistic equation (18) yields a result that, although disagreeing with experience, is rather remarkable, that is: one obtains the same fine structure as the one that would result from the Bohr–Sommerfeld theory by assuming the radial and azimuthal quantum numbers to be 'integral and a half', that is, half an odd integer. The theory has evidently to be completed by taking into account what in Bohr's theory is called the spin of the electron. In wave mechanics this is perhaps expressed (C. G. Darwin) by a polarisation of the ψ waves, this quantity having to be modified from a scalar to a vector.
[d] See in particular A. Unsöld, *Ann. d. Phys.*, **82** (1927), 355.

indeed become intelligible in three dimensions again'.[38] For this reason I want to elaborate a little on what has just been said.

Let

$$\psi_k(x, y, z) \quad \text{and} \quad E_k\,; \quad (k = 1, 2, 3, \ldots)$$

be the normalised eigenfunctions (for simplicity assumed as *real*) and corresponding eigenvalues of the *one*-electron atom with Z-fold positive nucleus, which for brevity we shall call the hydrogen problem. They satisfy the three-dimensional amplitude equation (compare equation (4)):

$$\left\{ \begin{array}{c} \dfrac{1}{m}\left(\dfrac{\partial^2\psi}{\partial x^2} + \dfrac{\partial^2\psi}{\partial y^2} + \dfrac{\partial^2\psi}{\partial z^2}\right) + \dfrac{8\pi^2}{h^2}\left(E + \dfrac{Ze^2}{r}\right)\psi = 0 \\ (r = \sqrt{x^2 + y^2 + z^2}). \end{array} \right. \tag{20}$$

If only one eigenoscillation is present, one has the *static* charge distribution[39]

$$\rho_{kk} = -e\psi_k^2. \tag{21}$$

If one imagines *two* being excited with maximal strength, one adds to $\rho_{kk} + \rho_{ll}$ a charge distribution oscillating with frequency $|E_k - E_l|/h$, whose *amplitude* distribution is given by

$$2\rho_{lk} = -2e\psi_k\psi_l. \tag{22}$$

The spatial integral of ρ_{kl} vanishes when $k \neq l$ (because of the orthogonality of the ψ_k) and it is $-e$ for $k = l$. The charge distribution resulting from the presence of two eigenoscillations together has thus at every instant the sum zero. – One can now form the electrostatic potential energies

$$p_{k,l;k',l'} = \int \ldots \int dx\,dy\,dz\,dx'\,dy'\,dz'\,\frac{\rho_{kl}(x, y, z)\rho_{k'l'}(x', y', z')}{r'}, \tag{23}$$

where $r' = \sqrt{(x - x')^2 + (y - y')^2 + (z - z')^2}$ and the indices k, l, k', l' may exhibit arbitrary degeneracies (to be sure, in the case $k = k', l = l'$, p is *twice* the potential self-energy of the charge distribution ρ_{kl}; but that is of no importance). *It is the constants p that control also the many-electron atom.*

Let us sketch this. Let the classical model now consist of n electrons and a Z-fold positively charged nucleus at the origin. We shall use the wave equation in the form (7). It becomes $3n$-dimensional,[40] say thus

$$\frac{1}{m}(\Delta_1 + \Delta_2 + \ldots + \Delta_n)\psi - \frac{8\pi^2}{h^2}(V_n + V_e)\psi - \frac{4\pi i}{h}\frac{\partial \psi}{\partial t} = 0. \quad (24)$$

Here

$$\Delta_\sigma = \frac{\partial^2}{\partial x_\sigma^2} + \frac{\partial^2}{\partial y_\sigma^2} + \frac{\partial^2}{\partial z_\sigma^2}; \quad \sigma = 1, 2, 3, \ldots, n. \quad (25)$$

We have considered the potential energy function as decomposed in two parts, $V_n + V_e$; V_n should correspond to the interaction of all n electrons with the nucleus, V_e to their interaction with one another, therefore[a]

$$V_n = -Ze^2 \sum_{\sigma=1}^{n} \frac{1}{r_\sigma}, \quad (26)$$

$$V_e = +e^2 \sum_{(\sigma,\tau)}{}' \frac{1}{r_{\sigma\tau}} \quad (27)$$

$$\left[r_\sigma = \sqrt{x_\sigma^2 + y_\sigma^2 + z_\sigma^2}, \quad r_{\sigma\tau} = \sqrt{(x_\sigma - x_\tau)^2 + (y_\sigma - y_\tau)^2 + (z_\sigma - z_\tau)^2} \right].$$

As the starting point for an approximation procedure we choose now the eigensolutions of equation (24) with $V_e = 0$, that is with the interaction between the electrons disregarded. The eigenfunctions are then *products* of hydrogen eigenfunctions, and the eigenvalues are *sums* of the corresponding eigenvalues of hydrogen. As a matter of fact, one easily shows that[41]

$$\psi_{k_1 \ldots k_n} = \psi_{k_1}(x_1\, y_1\, z_1) \ldots \psi_{k_n}(x_n\, y_n\, z_n) e^{\frac{2\pi i t}{h}(E_{k_1} + \ldots + E_{k_n})} \quad (28)$$

always satisfies equation (24) (with $V_e = 0$). And if one takes all possible sequences of numbers [Zahlenkombinationen] for the $k_1, k_2, \ldots, k_n$, then these products of ψ_k form a complete orthogonal system in the $3n$-dimensional q-space – one has thus integrated the approximate equation completely.

[a] Analogously to eq. (12), the prime on the summation sign should be interpreted as meaning that the sum is to be taken over all pairs with $\sigma \neq \tau$ (*eds.*).

One now aims to solve the full[42] equation (24) (with $V_e \neq 0$) by *expansion* with respect to this complete orthogonal system, that is one makes *this* ansatz:

$$\psi = \sum_{k_1=1}^{\infty} \cdots \sum_{k_n=1}^{\infty} a_{k_1 \ldots k_n} \psi_{k_1 \ldots k_n}. \tag{29}$$

But of course the coefficients a cannot be *constants*, otherwise the above sum would again be only a solution of the truncated equation with $V_e = 0$. It turns out, however, that it is enough to consider the a as functions of *time* alone ('method of the variation of constants').[a] Substituting (29) into (24) one finds that the following conditions on the time dependence of the a must hold:[43,44]

$$\frac{da_{k_1 \ldots k_n}}{dt} = \frac{2\pi i}{h} \sum_{l_1=1}^{\infty} \cdots \sum_{l_n=1}^{\infty} v_{k_1 \ldots k_n, l_1 \ldots l_n} a_{l_1 \ldots l_n} e^{\frac{2\pi i t}{h}(E_{l_1 \ldots l_n} - E_{k_1 \ldots k_n})}$$
$$[k_1 \ldots k_n = 1, 2, 3 \ldots]. \tag{30}$$

Here we have set for brevity

$$E_{k_1} + \ldots + E_{k_n} = E_{k_1 \ldots k_n}. \tag{31}$$

The v are *constants*, indeed they are *prima facie* $3n$-tuple integrals ranging over the whole of q-space. (Additional explanation:[45] Where do these $3n$-tuple integrals come from? They derive from the fact that after substituting (29) into (24) one replaces the latter equation by the mathematically equivalent condition that its left-hand side shall be orthogonal to all functions of the complete orthogonal system in R_{3n}. The system (30) expresses *this* condition.) Writing this out one has

$$v_{k_1 \ldots k_n, l_1 \ldots l_n} = \int^{3n\text{-fold}} \cdots \int dx_1 \ldots dz_n V_e \psi_{k_1}(x_1, y_1, z_1) \ldots \psi_{k_n}(x_n, y_n, z_n) \psi_{l_1}(x_1, y_1, z_1) \ldots \psi_{l_n}(x_n, y_n, z_n). \tag{32}$$

If one now considers the simple structure of V_e given in (27), one recognises that the v can be reduced to sextuple integrals, in fact each of them is a finite sum of some of the Coulomb potential energies defined in (23). Indeed, if in the finite sum representing V_e, we focus on an individual term, for example $e^2/r_{\sigma\tau}$, this contains only the six variables $x_\sigma, \ldots, z_\tau$. One can thus immediately perform in (32) precisely $3n - 6$ integrations on this term, yielding (because of the orthogonality and normalisation of the ψ_k) the factor 1, if $k_\rho = l_\rho$ for all indices ρ that coincide

[a] P. A. M. Dirac, *Proc. Roy. Soc. A*, **112** (1926), [661] p. 674.

neither with σ nor with τ, and yielding instead the factor 0 if even just for a single ρ different from σ and τ one has: $k_\rho \neq l_\rho$. (One sees thus that very many terms disappear.) For the non-vanishing terms, it is easy to see that they coincide with one of the p defined in (23). QED[a]

Let us now have a somewhat closer look at the equation system (30), whose coefficients, as we have just seen, have such a relatively simple structure, and which determines the varying amplitudes of our ansatz[46] (29) as functions of time. We can allow ourselves to introduce a somewhat simpler symbolic notation, by letting the *string* of indices $k_1, k_2, \ldots, k_n$ be represented by the *single* index k, and similarly $l_1, l_2, \ldots, l_n$ by l. One then has

$$\frac{da_k}{dt} = \frac{2\pi i}{h} \sum_{l=1}^{\infty} v_{kl} a_l e^{\frac{2\pi i t}{h}(E_l - E_k)}. \tag{33}$$

(One must not confuse, however, E_l, E_k with the *single*[47] eigenvalues of the hydrogen problem, which were earlier denoted in the same way.[48]) This is now a system of *infinitely many* differential equations, which we cannot solve directly: so, practically nothing seems to have been gained. In turn, however, we have as yet also *neglected nothing*: with *exact* solutions a_k of (33), (29) would be an *exact* solution of (24). This is precisely where I want to place the main emphasis, greater than on the practical implementation of the approximation procedure, which shall be sketched below only for the sake of completeness. *In principle* the equations (33) determine the solution of the many-electron problem exactly;[49] – and they no longer contain anything multi-dimensional; their coefficients are simple Coulomb energies of charge distributions that already occur in the hydrogen problem. Further, the equations (33) determine the solutions of the many-electron problem according to (29) as a combination of products of the hydrogen eigenfunctions. While these products (denoted above by $\psi_{k_1 k_2 \ldots k_n}$)[50] are still functions on the $3n$-dimensional q-space, any two of them yield in the

[a] Summary of the above: Calling ψ_k and E_k the eigenfunctions and eigenvalues of the problem for one electron, charge $-e$, in the field of a nucleus $+Ze$ (hydrogen problem), let us form the charge distributions (21) and (22), the former corresponding to the existence of a single normal mode, the latter to the cooperation of two of them. Taken as charge densities in ordinary electrostatics, each of these would have a certain potential energy and there would even be a certain mutual potential energy between two of them, assumed to coexist. These are the constants $p_{kl;k'l'}$ in (23). – With these givens, let us attack the problem of the n-electron atom. Dividing the potential energy in the wave equation (24) for this problem into two terms and neglecting at first the term V_e, due to the mutual action between the electrons, the eigensolutions would be given by (28), that is, by the products of n hydrogen functions. From these products, taken in all combinations, form the series (29), which will yield the *exact* solution of equation (24), provided that the coefficients $a_{k_1 k_2 \ldots k_n}$ are functions of time satisfying the equations (30); (see the abbreviation (31)). The coefficients v in (30) are *constants*, defined originally by the $3n$-tuple integrals (32), which however, thanks to the simple form of V_e (see (27)), reduce to sextuple integrals, namely precisely to the constants $p_{k,l;k',l'}$ (see (23)).

calculation of the three-dimensional charge distributions in the many-electron atom, as is easily seen, a charge distribution which if it is not identically zero corresponds again to a hydrogen distribution (denoted above by ρ_{kk} or ρ_{kl}).

These considerations are the analogue of the construction of the higher atoms from hydrogen trajectories in Bohr's theory. They reinforce the hope that by delving more deeply one will be able to interpret and *understand* the results of the multi-dimensional theory in three dimensions.[a]

Now, as far as the approximation method is concerned, it consists in fact of considering the contribution V_e made to the potential energy function V by the interaction of the electrons with one another, to be as far as possible *small* as compared to the action of the nucleus. The v_{kl} are then considered small compared to the eigenvalue differences $E_l - E_k$, except if $E_l = E_k$. The a_l will then vary *slowly* by comparison to the powers of e appearing on the right-hand side of equation (33), *as long as the latter are not equal to* 1, and all those terms on the right-hand side for which this is not the case will yield only small fluctuations of short period of the a_k and can be neglected in the approximation.[51] Thereby, first, the sums on the right become *finite*, because in fact always only a finite number of eigenvalues coincide. Second, the infinitely many equations separate into groups; each group contains only a finite number of a_l and can be integrated very easily.[52] This is the first step of the approximation procedure, which in theory can be continued indefinitely, but becomes more and more cumbersome. We shall not enter into details.

One can also transform the untruncated system of differential equations (33) at a *single* stroke into a system of ordinary linear equations (with infinitely many unknowns!) by setting

$$a_l = c_l e^{\frac{2\pi i t}{h}(E-E_l)}, \tag{34}$$

where the quantity E and the quantities c_l are unknown *constants*. Substituting into (33) one finds

$$(E - E_k)c_k = \sum_{l=1}^{\infty} v_{kl} c_l \,; \quad (k = 1, 2, 3, \ldots). \tag{35}$$

[a] Summary of the above: Although the system of eqs. (30) (abbreviated to (33)) does not admit a direct solution, the number of equations as well as the number of unknown functions being infinite, it seems to me very interesting that the solution to the multi-dimensional problem is provided in principle by a system of equations whose coefficients have such simple meanings in three dimensions. Further, one realises that the charge distribution that corresponds to the solution (29) of the n-electron problem turns out to be the superposition of the distributions ρ_{kk} and ρ_{kl} that occur already in the hydrogen problem. The hope of interpreting and of *understanding* the multi-dimensional theory in three dimensions is thus strengthened.

This equation system coincides with the Heisenberg–Born 'principal axes problem'. If the v_{kl} are very small quantities, then, if not *all* c_l are to be very small, E must be close to *one* of the E_l, let us say to E_k.

In the first approximation then only c_k, and all those c_l for which $E_l = E_k$, are different from zero. The problem thus separates in the first approximation into a denumerable set of *finite* principal axes problems.[a]

[a] Summary of the above: One can embark on the solution of the system of equations (33) by an approximation method. Positing (34), the *constants* E and c_l have to satisfy the system (35) of ordinary linear homogeneous equations, whose number as well as that of the unknown constants, however, is infinite. It is only by assuming all coefficients v_{kl} to be *small* that one can conclude that E has to be very close to one of the values E_l, for instance E_k, and that [c_l] approximately vanishes, unless E_l is equal to E_k. Since there is only a finite number of E_l that coincide with E_k, the problem reduces in the first approximation to a problem of a finite number of 'principal axes', or rather to an infinity of such finite problems. – As a matter of fact, the equations (35) coincide with the problem of an infinite number of principal axes, to which reduces the Heisenberg–Born mechanics.

Discussion of Mr Schrödinger's report

MR SCHRÖDINGER. – It would seem that my description in terms of a snapshot was not very fortunate, since it has been misunderstood. Perhaps the following explanation is clearer. The interpretation of Born is well-known, who takes $\psi\psi^* d\tau$ to be the probability for the system being in the volume element $d\tau$ of the configuration space. Distribute a very large number N of systems in the configuration space, taking the above probability as 'frequency function'. Imagine these systems as superposed in real space, dividing however by N the charge of each point mass in each system. In this way, in the limiting case where $N = \infty$ one obtains the wave mechanical picture of the system.

MR BOHR. – You have said that from the charge distribution $\psi\psi^* d\tau$ and the classical laws you obtain the frequency and intensity of light, but do the remarks about difficulties you made later indicate that what you had obtained was not correct?

MR SCHRÖDINGER. – The difficulty I mentioned is the following. If one expands the general solution as a series with respect to the eigenfunctions

$$\psi = \sum_k c_k \psi_k$$

and if one calculates the intensity of the radiation resulting from ψ_k and ψ_l together, one finds that it becomes proportional to $c_k^2 c_l^2$. However, according to the old theory, only the square of the amplitude corresponding to the 'initial level' should appear here; that of the 'final level' should be replaced by 1.

MR BOHR. – Has Dirac not found the solution to the difficulty?

MR SCHRÖDINGER. – Dirac's results are certainly very interesting and point the way toward a solution, if they do not contain it already. Only, we should first come to an understanding in physical terms [nous devrions d'abord nous entendre en langage physique]. I find it still impossible, for the time being, to see an answer to a physical question in the assertion that certain quantities obey a non-commutative algebra, especially when these quantities are meant to represent numbers of atoms. The relation between the continuous spatial densities, described earlier, and the observed intensities and polarisations of the spectral rays is [too natural][a] for me to deny all meaning to these densities only because some difficulties appear that are not yet resolved.

[a] The French here reads 'trop peu naturelle', which has the exact opposite meaning. The context would seem, however, to justify the amendment (*eds.*).

MR BORN. – It seems to me that interpreting the quantity $\psi\psi^*$ as a charge density leads to difficulties in the case of quadrupole moments. The latter in fact need to be taken into account in order to obtain the radiation, not only for theoretical reasons, but also for experimental reasons.

For brevity let us set

$$e^2\psi\psi^* = e^2|\psi|^2 = \Psi$$

and let us consider, for example, the case of two particles; Ψ becomes a function of x_1 and x_2, where for brevity x_1 stands for all the coordinates of the first particle; x_2 has a similar meaning. The electric density is then, according to Schrödinger,

$$\rho(x) = \int \Psi(x, x_2)dx_2 + \int \Psi(x_1, x)dx_1.$$

In wave mechanics the quadrupole moment

$$\iint x_1x_2\Psi(x_1, x_2)dx_1dx_2$$

cannot, as far as I can tell, be expressed using the function $\rho(x)$. I would like to know how one can, in this case, reduce the radiation of the quadrupole to the motion of a charge distribution $\rho(x)$ in the usual three-dimensional space.

MR SCHRÖDINGER. – I can assure you that the calculation of the dipole moments is perfecly correct and rigorous and that this objection by Mr Born is unfounded. Does the agreement between wave mechanics and matrix mechanics extend to the possible radiation of a quadrupole? That is a question I have not examined. Besides, we do not possess observations on this point that could allow us to use a possible disagreement between the two approaches to decide between them.

MR FOWLER asks for explanations regarding the method for solving the equations in the case of the many-electron problem.

MR DE DONDER. – Equation (24) of Mr Schrödinger's report can be extended to the case in which the n charged particles are *different* and where the external actions as well as the interactions can be described, in spacetime, by a gravitational field [champ gravifique].[a] The quantum equation thus obtained is the sum of the quantum equations for the n particles taken separately, each of the equations being divided by the (rest) mass of the corresponding particle. Thus, for instance,

[a] T. De Donder, L'équation fondamentale de la Chimie quantique, *Comptes Rendus Acad. Sci. Paris*, session of 10 October 1927, pp. 698–700. See esp. eq. (10).

the quantum equation for the nucleus will not enter if one assumes, as a first approximation, that the mass of the nucleus is infinitely large with respect to that of an electron.

When there is interaction, the problem is much more complex. One can, as Mr Schrödinger indicates, consider the action of the nucleus as an *external* action acting on the electrons of the cloud [couronne], and the (electrostatic) actions *between* the electrons in this cloud as a *perturbation*; but that is only a first approximation. In order to account for relativistic and electromagnetic effects I have assumed that the molecular systems have an *additive* character.[a] One can thus recover, as a special case, the above-mentioned method of quantisation by Schrödinger.

MR BORN. – In Göttingen we have embarked on a systematic calculation of the matrix elements that appear in perturbation theory, with the aim of collecting them in tables up to the principal quantum number 10. Part of these calculations, which are very extended, has already been done. My coworker Mr Biemüller has used them to calculate the lower terms of the helium atom according to the usual perturbation method up to perturbations of the second order.[b] The agreement of the ground term with the empirical value, despite the defects of the procedure, is hardly worse than in the recently published paper by Kellner [*Zeitschr. f. Phys.*, **44** (1927), 91], who has applied a more precise method (Ritz procedure).

MR LORENTZ. – Do you see the outcome of this long labour as satisfactory?

MR BORN. – The calculation has not attained yet the precision of the measurements. The calculations we have done applying the ordinary perturbation method [méthode des perturbations ordinaires] consist of a series expansion with respect to the inverse of the nuclear charge Z, of the form

$$E = Z\left(a + \frac{b}{Z} + \frac{c}{Z^2} + \cdots\right).$$

The three terms shown have been calculated. Nevertheless, in the case of helium ($Z = 2$) the precision is not yet as good as in the calculations done by Kellner using Ritz's approximation method.

MR LORENTZ. – But you hope however to improve your results.

[a] For more details, one can consult our note: 'L'équation de quantification des molécules comprenant n particules électrisées', published after this meeting, in the *Bull. Ac. R. Belg., Cl. des Sciences*, session of 5 November 1927.
[b] Cf. Hylleraas (1928) (*eds.*).

MR BORN. – Yes, only the convergence of the series is very slow.

MR HEISENBERG. – On the subject of this approximation method, Mr Schrödinger says at the end of his report that the discussion he has given reinforces the hope that when our knowledge will be deeper it will be possible to explain and to understand in three dimensions the results provided by the multi-dimensional theory. I see nothing in Mr Schrödinger's calculations that would justify this hope. What Mr Schrödinger does in his very beautiful approximation method, is to replace the n-dimensional differential equations by an infinity of linear equations. That reduces the problem, as Mr Schrödinger himself states, to a problem with ordinary matrices, in which the coefficients can be interpreted in three-dimensional space. The equations are thus 'three-dimensional' exactly in the same sense as in the usual matrix theory. It thus seems to me that, in the classical sense, we are just as far from understanding the theory in three dimensions as we are in the matrix theory.

MR SCHRÖDINGER. – I would not know how to express more precisely my hope of a possible formulation in a three-dimensional space. Besides, I do not believe that one would obtain simpler calculational methods in this way, and it is probable that one will always do calculations using the multi-dimensional wave equation. But then one will be able to grasp its physical meaning better. I am not precisely searching for a three-dimensional partial differential equation. Such a simple formulation is surely impossible. If I am not satisfied with the current state of the problem, it is because I do not understand yet the physical meaning of its solution.

What Mr Heisenberg has said is mathematically unexceptionable, but the point in question is that of the physical interpretation. This is indispensable for the further development of the theory. Now, this development is necessary. For one must agree that all current ways of formulating the results of the new quantum mechanics only correspond to the classical mechanics of actions at a distance. As soon as light crossing times become relevant in the system, the new mechanics fails, because the classical potential energy function no longer exists.

Allow me, to show that my hope of achieving a three-dimensional conception is not quite utopian, to recall what Mr Fowler has told us on the topic of Mr Hartree's approximation method.[a] It is true that this method abstracts from what one calls the 'exchange terms' (which correspond, for instance, to the distance between the ortho and para terms of neutral helium). But, abstracting from that, it already achieves the three-dimensional aim I tend to. Should one declare *a priori* impossible that

[a] See the discussion after Bragg's report, p. 291 (*eds.*).

Hartree's method might be modified or developed in such a way as to take into account the exchange terms while working with a satisfactory three-dimensional model?

MR BORN. – Regarding the question of knowing whether it is possible to describe a many-electron problem by a field equation in three dimensions, I would like to point out the following. The number of quantum numbers of an atom rises by three with each additional electron; it is thus equal to $3n$ for n electrons. It seems doubtful that there should be an ordinary, three-dimensional eigenvalue problem, whose eigenvalues have a range of size ∞^{3n} [dont la valeur caractéristique ait une multitude de ∞^{3n} dimensions].[a] Instead, it follows from recent papers by Dirac and Jordan[b] that one can build on a three-dimensional oscillation equation if one considers the eigenfunction itself not as an ordinary number, but as one of Dirac's q-numbers, that is, if one quantises again its amplitude as a function of time. An n-quanta oscillation with this amplitude then yields together with the three spatial quantum numbers the necessary range [multitude] of quantum numbers. From this point of view the number of electrons in a system appears itself as a quantum number, that is, the electrons themselves appear as discontinuities of the same nature as the stationary states.

MR SCHRÖDINGER. – Precisely the structure of the periodic system is already contained in the physics [mécanique] of the three-dimensional hydrogen problem. The degrees of degeneracy 1, 4, 9, 16, etc., multiplied by 2, yield precisely the periodic numbers [nombres de périodes]. The factor 2 that I have just mentioned derives from the spin. From the point of view of wave mechanics, the apparently mysterious 'Pauli action' of the first two electrons on the third (which they prevent from also following an orbit with quantum number 1) means strictly speaking nothing other than the non-existence of a third eigenfunction with principal quantum number 1. This non-existence is precisely a property of the three-dimensional model, or of the three-dimensional equation. The multi-dimensional equation has too many eigenfunctions; it is this [elle] that makes the 'Pauli exclusion' (Pauliverbot) necessary to eliminate this defect.[c]

[a] The French text here appears to make little sense, but Born is possibly referring to the dimension of the space of solutions (*eds.*).

[b] Cf. Section IV of Born and Heisenberg's report (*eds.*).

[c] The French text refers to the four-dimensional equation ('l'équation à quatre dimensions') as having too many solutions. This reading could be correct, in the sense that the exclusion principle was first introduced in the context of the relativistic (four-dimensional) Bohr–Sommerfeld theory of the atom, but the above reading seems much more natural in context. Note that Schrödinger throughout his report uses 'vierdimensional' and 'vieldimensional', which could be easily confused, for 'four-dimensional' and 'many-dimensional', respectively (*eds.*).

Notes on the translation

1 Here and in the following, the French edition omits some italics, which are quite characteristic of Schrödinger's writing style and which we tacitly restore.
2 [Energiegerade] – [série des énergies]
3 Bracket added in the French edition.
4 [als Beschreibung] – [comme la définition]
5 [Vollständigkeit] – [perfection]
6 [freilich] – [évidemment]
7 Printed as a footnote in the French edition.
8 [sie nicht.... beschreibe] – [qu'ils ne décrivent pas]
9 [Einzelsystem] – [système déterminé]
10 This clause is omitted in the French edition.
11 [von anderer Seite vertreten] – [défendue par d'autres]. Footnote only in the French edition.
12 [aus einer Anzahl] – [d'un grand nombre]
13 [Durch die eben besprochene Deutung] – [Ainsi que nous venons de le voir]
14 The equation number is missing in the French edition, and the following sentence is printed as a footnote.
15 [es *gebe*] – [sont données]
16 [weitgehend] – [tout à fait]
17 [*Verständnis*] – [*interprétation*]
18 [*Verständnis*] – [signification]
19 No exclamation mark in the French edition.
20 [innere] – [intimes]
21 eine prinzipielle] – [essentielle]
22 No exclamation mark in the French edition.
23 Printed as a footnote in the French edition.
24 In the French edition this equation number is given to the following equation (unnumbered in the typescript).
25 Bracket printed as a footnote in the French edition, with the addition: 'Δ stands for the Laplacian'.
26 $[\frac{e\phi}{h}] - [e\frac{e\varphi}{h}]$
27 [die Ladungsdichte aus $\psi\psi^*$] – [la densité de charge $\psi\psi^*$]
28 [Betrag] – [valeur]
29 Bracket printed as a footnote in the French edition.
30 The rest of the bracket is printed as a footnote in the French edition.
31 Both typescript and French edition give '365' as page number.
32 The French edition adds this to the footnote.
33 [treten darin in ψ und seinen ersten Ableitungen quadratische Formen auf] – [y figurent dans ψ et ses premières dérivées des formes quadratiques]
34 Footnote only in the French edition.
35 [*bewegend*] – [*par le mouvement*]
36 [allerdings] – [certainement]
37 [(18)] – [(8)]
38 The French edition omits the inverted commas.
39 The equation number is missing in the printed volume.
40 [$3n$-dimensional] – [tridimensionelle]
41 Misprint in the French edition: the E_k are not in the exponent.
42 [komplet[t]e] – [complexe]
43 [dass für die Abhängigkeit der a von der Zeit folgende Forderungen bestehen] – [que pour que a dépende du temps les conditions suivantes doivent être satisfaites]
44 Two misprints in the French edition: the E_k are not in the exponent, and the k_i run to n.
45 Printed as a footnote in the French edition.
46 [unseres Ansatzes] – [de notre expression fondamentale]
47 [den *einzelnen*] – [les *diverses*]

48 Bracket printed as footnote in the French edition.
49 [bestimmen die Lösung exact] – [déterminent la solution]
50 Misprint in the French edition: '$\psi(k_1, k_2, \ldots, k_n)$'.
51 Both the typescript and the French edition read 'c_l' and 'c_k' instead of 'a_l' and 'a_k'.
52 Again, both the typescript and the French edition read 'c_k' instead of 'a_k'.

General discussion of the new ideas presented[a]

CAUSALITY, DETERMINISM, PROBABILITY

MR LORENTZ. – I should like to draw attention to the difficulties one encounters in the old theories.

We wish to make a representation of the phenomena, to form an image of them in our minds. Until now, we have always wanted to form these images by means of the ordinary notions of time and space. These notions are perhaps innate; in any case, they have developed from our personal experience, by our daily observations. For me, these notions are clear and I confess that I should be unable to imagine physics without these notions. The image that I wish to form of phenomena must be absolutely sharp and definite, and it seems to me that we can form such an image only in the framework of space and time.

For me, an electron is a corpuscle that, at a given instant, is present at a definite point in space, and if I had the idea that at a following moment the corpuscle is present somewhere else, I must think of its trajectory, which is a line in space. And if the electron encounters an atom and penetrates it, and after several incidents leaves the atom, I make up a theory in which the electron preserves its individuality; that is to say, I imagine a line following which the electron passes through the atom. Obviously, such a theory may be very difficult to develop, but *a priori* it does not seem to me impossible.

I imagine that, in the new theory, one still has electrons. It is of course possible that in the new theory, once it is well-developed, one will have to suppose that the electrons undergo transformations. I happily concede that the electron may dissolve

[a] As mentioned in Section 1.6, the Bohr archives contain material by Verschaffelt and by Kramers on the general discussion, in particular relating to Bohr's contributions. (They also contain a copy of the galley proofs of the general discussion,[1] dated 1 June 1928.) We reproduce the greater part of this material in the appendices. For Bohr's discussion contributions in what follows, we have used material from Bohr's *Collected Works* (Bohr 1985) and notes taken by Richardson[2] (also mentioned in Section 1.6). For the discussion contributions by Dirac, we have followed his manuscript version when available.[3] See our notes below for further details (*eds.*).

into a cloud. But then I would try to discover on which occasion this transformation occurs. If one wished to forbid me such an enquiry by invoking a principle, that would trouble me very much. It seems to me that one may always hope one will do later that which we cannot yet do at the moment. Even if one abandons the old ideas, one may always preserve the old terminology [dénominations]. I should like to preserve this ideal of the past, to describe everything that happens in the world with distinct images. I am ready to accept other theories, on condition that one is able to re-express them in terms of clear and distinct images.

For my part, despite not having yet become familiar with the new ideas that I now hear expressed,[a] I could visualise these ideas thus. Let us take the case of an electron that encounters an atom; let us suppose that the electron leaves the atom and that at the same time there is emission of a light quantum. One must consider, in the first place, the systems of waves that correspond to the electron and to the atom before the collision. After the collision, we will have new systems of waves. These systems of waves can be described by a function ψ defined in a space with a large number of dimensions and satisfying a differential equation. The new wave mechanics will work with this equation and will determine the function ψ before and after the collision.

Now, there are phenomena that teach us that there is something else in addition to the waves, namely corpuscles; one can, for example, perform an experiment with a Faraday cylinder; one must then take into account the individuality of the electrons and also of the photons. I think I would find that, to explain the phenomena, it suffices to assume that the expression $\psi\psi^*$ gives the probability that the electrons and the photons exist in a given volume; that would suffice to explain the experiments. But the examples given by Mr Heisenberg teach me that I will have thus attained everything that experiment allows me to attain. However, I think that this notion of probability should be placed at the end, and as a conclusion, of theoretical considerations, and not as an *a priori* axiom, though I may well admit that this indeterminacy corresponds to experimental possibilities. I would always be able to keep my deterministic faith for the fundamental phenomena, of which I have not spoken. Could a deeper mind not be aware of the motions of these electrons? Could one not keep determinism by making it an article of faith? Must one necessarily elevate indeterminism to a principle?

MR BOHR expounds his point of view with respect to the problems of quantum theory.

[a] In fact, Lorentz had followed the recent developments rather closely. In particular, he had corresponded extensively with Ehrenfest and with Schrödinger, and had delivered seminars and lectures on wave mechanics and on matrix mechanics at Leiden, Cornell and Caltech. See Section 1.3 (*eds.*).

The original published proceedings add '(see the preceding article)'. In the proceedings, the article preceding the general discussion is a French translation of the German version of Bohr's Como lecture (Bohr 1928) (published in *Naturwissenschaften*). As described in Section 1.6, this article was included at Bohr's request, to replace his remarks made at this point in the general discussion. (In our translation of the proceedings, we have omitted this well-known article.)

Kalckar's introduction to volume 6 of Bohr's *Collected Works* (Bohr 1985) describes the corresponding notes (by Kramers and by Verschaffelt) in the Bohr archives as too incomplete to warrant reproduction, providing instead a short summary and comparison with the printed versions of the Como lecture.[a] We concur that for the most part the notes do not make easy reading, and for this reason we have reproduced them in the appendices. On the other hand, in order to give an idea of Bohr's contributions here in the main text, below we reproduce the relevant parts of notes on the general discussion taken by Richardson,[4] and comment on their relation to Bohr's paper translated in the proceedings.

The first part of Richardson's notes relating to Bohr reads as follows:

$$E = h\nu \qquad\qquad e^{i2\pi(\tau_x x + \tau_y y + \tau_z z - \nu t)}$$

$$p = h\tau$$

Int[er]f[eren]ce. $\quad ?h \to \infty$ [?]

$$\left.\begin{array}{lcl} \Delta x \Delta \tau_x & \sim & 1 \\ \Delta t \Delta \nu & \sim & 1 \end{array}\right\} \qquad \begin{array}{lcl} \Delta x \Delta p_x & \sim & h \\ \Delta t \Delta E & \sim & h \end{array}$$

This corresponds to part of section 2 of Bohr's paper translated in the proceedings. There Bohr introduces the concepts of energy and momentum for plane waves, and the idea that waves of limited extent in spacetime are obtained through the 'interference' (that is, superposition) of different plane waves, the resulting waves satisfying (at best) the given relations. (As a consequence, a group of waves has no well-defined phase, a point Bohr takes up again below.)

Richardson's notes then continue as shown in Fig. A. The γ-ray microscope is discussed in section 3 of Bohr (1928) (the section on measurement, which also discusses momentum measurements based on the Doppler effect). Bohr appears to have inserted this discussion as the physical illustration of the uncertainty-type relations above.

The next part of Richardson's notes returns to section 2 of the paper, and is reproduced in Fig. B. This corresponds in fact to the subsequent paragraphs of section 2, in which Bohr proposes his resolution of the perceived paradoxes related to the scattering of radiation by free electrons (note the extended – as opposed to pointlike – region of scattering in the

[a] 'The notes cover the wave-corpuscle aspects of light and matter (corresponding to the first sections of the printed lecture). The γ-ray microscope is analysed, although the notes are somewhat incomplete here (as in many other places), and the rôle of the finite wave trains is discussed in connection with the momentum measurement through the Doppler effect (as in the printed versions). After some questions . . . Bohr continues by discussing the significance of the phase and comments on the Stern–Gerlach experiment and the inobservability of the phase in a stationary state . . . ' (Bohr 1985, p. 37).

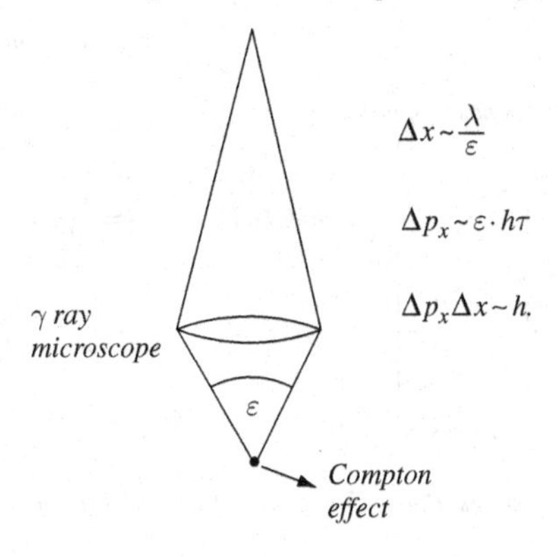

Fig. A.

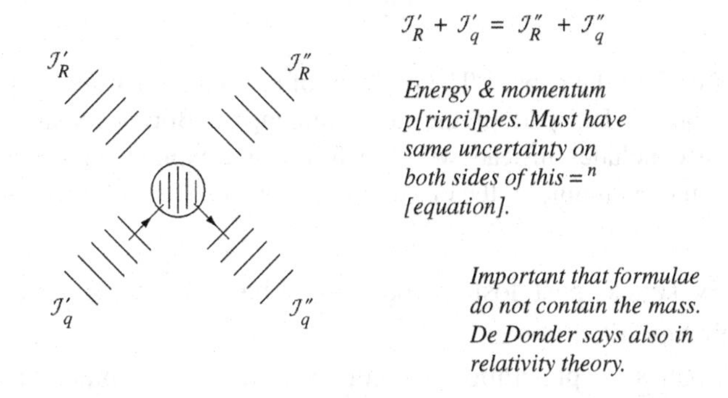

Fig. B.

diagram, and see Bohr's contribution to the discussion of Compton's report, pp. 328–9) as well as of the perceived paradoxes related to collisions (cf. Section 3.4.2).

The next part of Richardson's notes, shown in Fig. C, instead relates to part of section 6 of Bohr's paper (sections 4 and 5 of the paper are, respectively, a review of the correspondence principle and of matrix mechanics, and a discussion and critique of wave mechanics). In section 6 of the published paper, Bohr raises the following puzzle. According to Bohr, in any observation that distinguishes between different stationary states one has to disregard the past history of the atom, but, paradoxically, the theory assigns a phase to a stationary state. However, since the system will not be strictly isolated, one will work with a group of waves, which (as mentioned in section 2) has no well-defined phase. Bohr then illustrates this with the Stern-Gerlach experiment. The condition for distinguishability of the eigenstates of the hydrogen atom is that the angular spreading of the beam should be greater than that given by diffraction at the slit ($\varepsilon > \alpha$), which translates into the time-energy uncertainty relation. As Bohr mentions, Heisenberg (1927) uses this as an illustration of the uncertainty relation, while Bohr uses it as an illustration of how knowledge of the phase is lost.

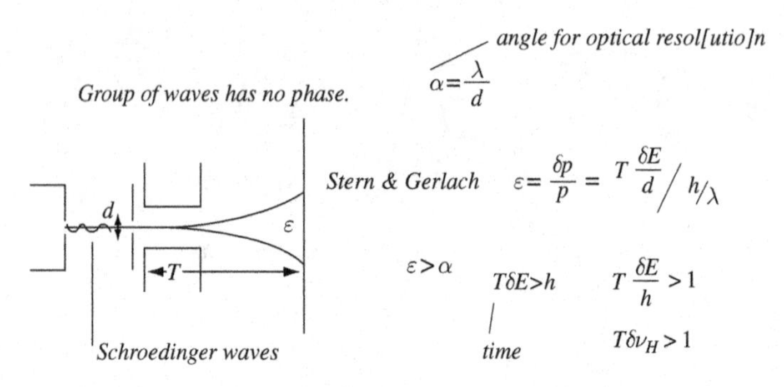

In stationary states. Phase is lost. (of <de B.>{Schrödinger} waves).

$Vt + \phi_1$
$Vt + \phi_2 + T\Delta\nu$

Fig. C.

The final section 7 of the paper ('The problem of elementary particles') has no parallel in Richardson's notes. The part of the notes relating to Bohr's remarks at this point in the discussion concludes instead with the following, explicitly labelled 'Bohr', and corresponding to the beginning of the discussion on the next day (cf. the Appendix):

1. [blank]

2. Stationary states, past lost ∵ [because] phase indetermination – Stern & Gerlach's Exp[erimen]t.

3. Schroedinger's ψ, prob[abilit]y of electron at a given place at a given time [?], uncertainty $\Delta\nu\Delta t \sim 1$

$$\nu[?] \times \frac{v}{c}$$

MR BRILLOUIN. – Mr Bohr insists on the uncertainty of simultaneous measurements of position and momentum; his point of view is closely connected to the notion of *cells in phase space* introduced by Planck a very long time ago. Planck assumed that if the representative point of a system is in a cell (of size $\Delta p \Delta q = h$) one cannot distinguish it from another point in the same cell. The examples brought by Mr Bohr aptly make precise the physical meaning of this quite abstract notion.

MR DE DONDER. – The considerations that Mr Bohr has just developed are, I think, in close relation with the following fact: in the Einsteinian Gravitation[a] of a continuous system or of a pointlike system, there appear not the masses and

[a] T. De Donder, *Théorie des champs gravifiques* (*Mémorial des sciences mathématiques*, part 14, Paris, 1926). See esp. equations (184), (184′) and (188), (188′). One can also consult our lectures: *The Mathematical Theory of Relativity* (Massachusetts Institute of Technology), Cambridge, Mass., 1927. See esp. equations (23), (24) and (28), (29).

charges of the particles, but entities $\tau^{(m)}$ and $\tau^{(e)}$ in *four* dimensions; note that these *generalised* masses and charges, localised in spacetime, *are conserved* along their worldlines.

MR BORN. – Mr Einstein has considered the following problem: A radioactive sample emits α-particles in all directions; these are made visible by the method of the Wilson cloud [chamber]. Now, if one associates a spherical wave with each emission process, how can one understand that the track of each α particle appears as a (very nearly) straight line? In other words: how can the corpuscular character of the phenomenon be reconciled here with the representation by waves?

To do this, one must appeal to the notion of 'reduction of the probability packet' developed by Heisenberg.[a] The description of the emission by a spherical wave is valid only for as long as one does not observe ionisation; as soon as such ionisation is shown by the appearance of cloud droplets, in order to describe what happens afterwards one must 'reduce' the wave packet in the immediate vicinity of the drops. One thus obtains a wave packet in the form of a ray, which corresponds to the corpuscular character of the phenomenon.

Mr Pauli[b] has asked me if it is not possible to describe the process without the reduction of wave packets, by resorting to a multi-dimensional space, whose number of dimensions is three times the number of all the particles present (α-particles and atoms hit by the radiation).

This is in fact possible and can even be represented in a very anschaulich manner [d'une manière fort intuitive] by means of an appropriate simplification, but this does not lead us further as regards the fundamental questions. Nevertheless, I should like to present this case here as an example of the multi-dimensional treatment of such problems. I assume, for simplicity, that there are only two atoms that may be hit. One then has to distinguish two cases: either the two atoms 1 and 2 lie on the same ray starting from the origin (the place where the preparation is), or they do not lie on the same ray. If we represent by ε the probability that an atom will be hit, we have the following probability diagram:[c]

I. The points 1 and 2 are located on the same ray starting from the origin.

Number of particles hit	Probability
0	$1-\varepsilon$
1	0
2	ε

[a] Born is referring here in particular to Heisenberg's uncertainty paper (Heisenberg 1927) (*eds.*).
[b] Cf. Pauli's letter to Bohr, 17 October 1927, discussed in Section 6.2.1 (*eds.*).
[c] In the following tables, the probability for the number of particles hit to equal 1 should be read as the probability for *each case* in which the number of particles hit equals 1 (*eds.*).

II. The points 1 and 2 are not on the same ray.

Number of particles hit	Probability
0	$1-2\varepsilon$
1	ε
2	0

This is how one should express the probability of events in the case of rectilinear propagation.

To make possible a graphical representation of the phenomenon, we will simplify it further by assuming that all the motions take place following only a single straight line, the axis x. We must then distinguish the two cases where the atoms lie on the same side and on either side of the origin. The corresponding probabilities are the following:

I. The points 1 and 2 are located on the same side.

Number of particles hit	Probability
0	$\frac{1}{2}$
1	0
2	$\frac{1}{2}$

II. The points 1 and 2 are located on different sides.

Number of particles hit	Probability
0	0
1	$\frac{1}{2}$
2	0

Now, these relations can be represented by the motion of a wave packet in a space with three dimensions x_0, x_1, x_2. To the initial state there corresponds:

In case I, the point $x_0 = 0, \quad x_1 = a \quad x_2 = b$
In case II, the point $x_0 = 0, \quad x_1 = a \quad x_2 = -b$

where a and b are positive numbers. The wave packet at first fills the space surrounding these points and subsequently moves parallel to the axis x_0, dividing itself into two packets of the same size going in opposite directions. Collisions are produced when $x_0 = x_1$ or $x_0 = x_2$, that is to say, on two planes of which one, P_1, is parallel to the axis x_2 and cuts the plane x_0x_1 following the bisector of the positive quadrant, while the second, P_2, is parallel to the axis x_1 and cuts the plane

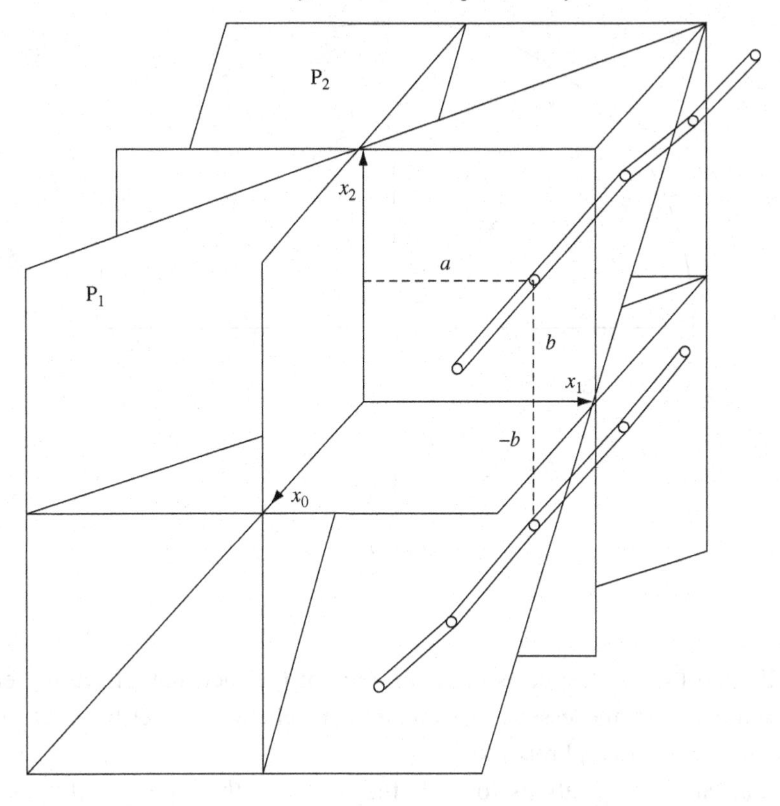

Fig. 1.

x_0x_2 following the bisector of the positive quadrant. As soon as the wave packet strikes the plane P_1, its trajectory receives a small kink in the direction x_1; as soon as it strikes P_2 the trajectory receives a kink in the direction x_2 (Fig. 1).

Now, one immediately sees in the figure that the upper part of the wave packet, which corresponds to case I, strikes the planes P_1, P_2 on the same side of the plane x_1x_2, while the lower part strikes them[5] on different sides. The figure then gives an anschaulich representation of the cases indicated in the above diagram. It allows us to recognise immediately whether, for a given size of wave packet, a given state, that is to say a given point x_0, x_1, x_2, can be hit or not.

To the 'reduction' of the wave packet corresponds the choice of one of the two directions of propagation $+x_0$, $-x_0$, which one must take as soon as it is established that one of the two points 1 and 2 is hit, that is to say, that the trajectory of the packet has received a kink.

This example serves only to make clear that a complete description of the processes taking place in a system composed of several molecules is possible only in a space of several dimensions.

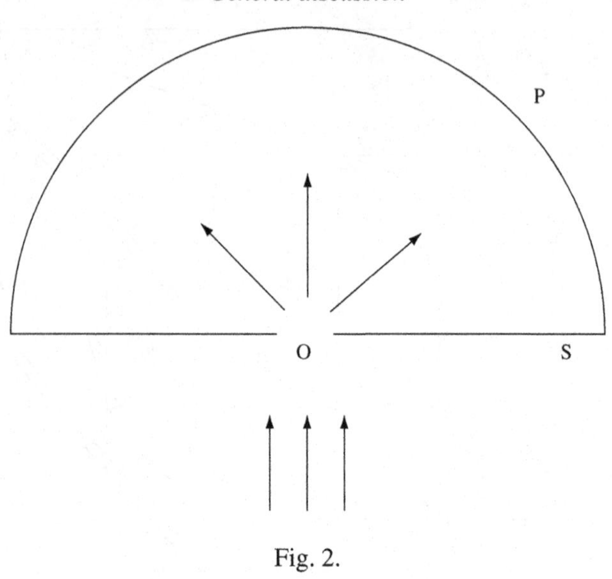

Fig. 2.

MR EINSTEIN.[a] – Despite being conscious of the fact that I have not entered deeply enough into the essence of quantum mechanics, nevertheless I want to present here some general remarks.[b]

One can take two positions towards the theory with respect to its postulated domain of validity, which I wish to characterise with the aid of a simple example.

Let S be a screen provided with a small opening O (Fig. 2), and P a hemispherical photographic film of large radius. Electrons impinge on S in the direction of the arrows. Some of these go through O, and because of the smallness of O and the speed of the particles, are dispersed uniformly over the directions of the hemisphere, and act on the film.

Both ways of conceiving the theory now have the following in common. There are de Broglie waves, which impinge approximately normally on S and are diffracted at O. Behind S there are spherical waves, which reach the screen P and whose intensity at P is responsible [massgebend] for what happens at P.[c]

We can now characterise the two points of view as follows.

[a] The extant manuscript in the Einstein archives[6] consists of the first four paragraphs only, which we have translated here (footnoting significant differences from the published French), ©Albert Einstein Archives, the Hebrew University of Jerusalem, Israel (*eds.*).

[b] The published French has: 'I must apologise for not having gone deeply into quantum mechanics. I should nevertheless want to make some general remarks' (*eds.*).

[c] In the published French, the German expression 'ist massgebend' is misrendered as 'gives the measure' [donne la mesure] instead of as 'is responsible'. This is of some significance for the interpretation of Einstein's remarks as a form of the later EPR argument; see Section 7.1 (*eds.*).

1. *Conception I.* – The de Broglie–Schrödinger waves do not correspond to a single electron, but to a cloud of electrons extended in space. The theory gives no information about individual processes, but only about the ensemble of an infinity of elementary processes.
2. *Conception II.* – The theory claims to be a complete theory of individual processes. Each particle directed towards the screen, as far as can be determined by its position and speed, is described by a packet of de Broglie–Schrödinger waves of short wavelength and small angular width. This wave packet is diffracted and, after diffraction, partly reaches the film P in a state of resolution [un état de résolution].

According to the first, purely statistical, point of view $|\psi|^2$ expresses the probability that there exists at the point considered *a particular* particle of the cloud, for example at a given point on the screen.

According to the second, $|\psi|^2$ expresses the probability that at a given instant *the same* particle is present at a given point (for example on the screen). Here, the theory refers to an individual process and claims to describe everything that is governed by laws.

The second conception goes further than the first, in the sense that all the information resulting from I results also from the theory by virtue of II, but the converse is not true.[a] It is only by virtue of II that the theory contains the consequence that the conservation laws are valid for the elementary process; it is only from II that the theory can derive the result of the experiment of Geiger and Bothe, and can explain the fact that in the Wilson [cloud] chamber the droplets stemming from an α-particle are situated very nearly on continuous lines.

But on the other hand, I have objections to make to conception II. The scattered wave directed towards P does not show any privileged direction. If $|\psi|^2$ were simply regarded as the probability that at a certain point a given particle is found at a given time, it could happen that *the same* elementary process produces an action *in two or several* places on the screen. But the interpretation, according to which $|\psi|^2$ expresses the probability that *this* particle is found at a given point, assumes an entirely peculiar mechanism of action at a distance, which prevents the wave continuously distributed in space from producing an action in *two* places on the screen.

In my opinion, one can remove this objection only in the following way, that one does not describe the process solely by the Schrödinger wave, but that at the same time one localises the particle during the propagation. I think that Mr de Broglie is right to search in this direction. If one works solely with the Schrödinger waves, interpretation II of $|\psi|^2$ implies to my mind a contradiction with the postulate of relativity.

[a] The French has 'I' and 'II' exchanged in this sentence, which is illogical (*eds.*).

I should also like to point out briefly two arguments which seem to me to speak against the point of view II. This [view] is essentially tied to a multi-dimensional representation (configuration space), since only this mode of representation makes possible the interpretation of $|\psi|^2$ peculiar to conception II. Now, it seems to me that objections of principle are opposed to this multi-dimensional representation. In this representation, indeed, two configurations of a system that are distinguished only by the permutation of two particles of the same species are represented by two different points (in configuration space), which is not in accord with the new results in statistics. Furthermore, the feature of forces of acting only at small *spatial* distances finds a less natural expression in configuration space than in the space of three or four dimensions.

MR BOHR.[a] – I feel myself in a very difficult position because I don't understand what precisely is the point which Einstein wants to [make]. No doubt it is my fault.

...

As regards general problem I feel its difficulties. I would put problem in other way. I do not know what quantum mechanics is. I think we are dealing with some mathematical methods which are adequate for description of our experiments. Using a rigorous wave theory we are claiming something which the theory cannot possibly give. [We must realise] that we are away from that state where we could hope of describing things on classical theories. Understand same view is held by Born and Heisenberg. I think that we actually just try to meet, as in all other theories, some requirements of nature, but difficulty is that we must use words which remind of older theories. The whole foundation for causal spacetime description is taken away by quantum theory, for it is based on assumption of observations without interference. ... excluding interference means exclusion of experiment and the whole meaning of space and time observation ... because we [have] interaction [between object and measuring instrument] and thereby we put us on a quite different standpoint than we thought we could take in classical theories. If we speak of observations we play with a statistical problem. There are certain features complementary to the wave pictures (existence of individuals). ...

...

The saying that spacetime is an abstraction might seem a philosophical triviality but nature reminds us that we are dealing with something of practical interest. Depends on how I consider theory. I may not have understood, but I think the

[a] These remarks by Bohr do not appear in the published French. We have reproduced them from Bohr's *Collected Works*, vol. 6 (Bohr 1985, p. 103), which contains a selection from Verschaffelt's notes (held in the Bohr archive). The tentative interpolations in square brackets are by the editor J. Kalckar. For the full version of this passage, see the Appendix (*eds.*).

whole thing lies [therein that the] theory is nothing else [but] a tool for meeting our requirements and I think it does.

MR LORENTZ. – To represent the motion of a system of n material points, one can of course make use of a space of 3 dimensions with n points or of a space of $3n$ dimensions where the systems will be represented by a single point. This must amount to exactly the same thing; there can be no fundamental difference. It is merely a question of knowing which of the two representations is the most suitable, which is the most convenient.

But I understand that there are cases where the matter is difficult. If one has a representation in a space of $3n$ dimensions, one will be able to return to a space of 3 dimensions only if one can reasonably separate the $3n$ coordinates into n groups of 3, each corresponding to a point, and I could imagine that there may be cases where that is neither natural nor simple. But, after all, it certainly seems to me that all this concerns the form rather than the substance of the theory.

MR PAULI. – I am wholly of the same opinion as Mr Bohr, when he says that the introduction of a space with several dimensions is only a technical means of formulating mathematically the laws of mutual action between several particles, actions which certainly do not allow themselves to be described simply, in the ordinary way, in space and time. It may perfectly well be that this technical means may one day be replaced by another, in the following fashion. By Dirac's method one can, for example, quantise the characteristic vibrations of a cavity filled with blackbody radiation, and introduce a function ψ depending on the amplitudes of these characteristic vibrations of unlimited number. One can similarly use, as do Jordan and Klein, the amplitudes of ordinary four-dimensional material waves as arguments of a multi-dimensional function ϕ. This gives, in the language of the corpuscular picture, the probability that at a given instant the numbers of particles of each species present, which have certain kinematical properties (given position or momentum), take certain values. This procedure also has the advantage that the defect of the ordinary multi-dimensional method, of which Mr Einstein has spoken and which appears when one permutes two particles of the same species, no longer exists. As Jordan and Klein have shown, making suitable assumptions concerning the equations that this function ϕ of the amplitudes of material waves in ordinary space must satisfy,[7] one arrives exactly at the same results as by basing oneself on Schrödinger's multi-dimensional theory.

To sum up, I wish then to say that Bohr's point of view, according to which the properties of physical objects of being defined and of being describable in space and time are complementary, seems to be more general than a special technical means. But, independently of such a means, one can, according to this idea, declare in any case that the mutual actions of several particles certainly cannot be described in the ordinary manner in space and time.

To make clear the state of things of which I have just spoken, allow me to give a special example. Imagine two hydrogen atoms in their ground state at a great distance from each other, and suppose one asks for their energy of mutual action. Each of the two atoms has a perfectly isotropic distribution of charge, is neutral as a whole, and does not yet emit radiation. According to the ordinary description of the mutual action of the atoms in space and time, one should then expect that such a mutual action does not exist when the distance between the two neutral spheres is so great that no notable interpenetration takes place between their charge clouds. But when one treats the same question by the multi-dimensional method, the result is quite different, and in accordance with experiment.

The classical analogy to this last result would be the following: Imagine inside each atom a classical oscillator whose moment p varies periodically. This moment produces a field at the location of the other atom whose periodically variable intensity is of order $\mathcal{E} \sim \frac{p}{r^3}$, where r is the distance between the two atoms. When two of these oscillators act on each other, a polarisation occurs with the following potential energy, corresponding to an attractive force between the atoms,

$$-\frac{1}{2}\alpha\mathcal{E}^2 \sim \frac{1}{2}\alpha p^2 \frac{1}{r^6},$$

where α represents the polarisability of the atom.

In speaking of these oscillators, I only wanted to point out a classical analogy with the effect that one obtains as a result of multi-dimensional wave mechanics. I had found this result by means of matrices, but Wang has derived it directly from the wave equation in several dimensions. In a paper by Heitler and London, which is likewise concerned with this problem, the authors have lost sight of the fact that, precisely for a large distance between the atoms, the contribution of polarisation effects to the energy of mutual action, a contribution which they have neglected, outweighs in order of magnitude the effects they have calculated.

MR DIRAC.[a] – I should like to express my ideas on a few questions.

The first is the one that has just been discussed and I have not much to add to this discussion. I shall just mention the explanation that the quantum theory would give of Bothe's experiment.[9] The difficulty arises from[10] the inadequacy of the 3-dimensional wave picture. This picture cannot distinguish between the case

[a] Here, we mostly follow the English version from Dirac's manuscript.[8] (The French translation may have been done from a typescript or fairer copy.) We generally follow the French paragraphing, and we uniformise Dirac's notation. Interesting variants, cancellations and additions will be noted, as will significant deviations from the published French (*eds.*).

when there is a probability p of a light-quant [*sic*] being in a certain small volume, and the case when there is a probability $\frac{1}{2}p$ of two light-quanta being in the volume, and no probability for only one. But the wave function in many-dimensional space does distinguish between these cases. The theory of Bothe's experiment in many-dimensional space would show that, while there is a certain probability for a light-quantum appearing in one or the other of the counting chambers, there is no probability of two appearing simultaneously.

At present the general theory of the wave function in many-dimensional space necessarily involves the abandonment of relativity.[11] One might, perhaps, be able to bring relativity into the general quantum theory in the way Pauli has mentioned of quantising 3-dimensional waves, but this would not lead to greater Anschaulichkeit[12] in the explanation of results such as Bothe's.

I shall now show how Schrödinger's expression for the electric density appears naturally in the matrix theory. This will show the exact signification of this density and the limitations which must be imposed on its use. Consider an electron moving in an arbitrary field, such as that of an H atom. Its coordinates x, y, z will be matrices. Divide the space up into a number of cells, and form that function of x, y, z that is equal to 1 when the electron is in a given cell and 0 otherwise. This function of the matrices x, y, z will also be a matrix.[a] There is one such matrix for each cell whose matrix elements will be functions of the coordinates a, b, c of the cell, so that it can be written $A(a, b, c)$.

Each of these matrices represents a quantity that if measured experimentally must have either the value 0 or 1. Hence each of these matrices has the characteristic values 0 and 1 and no others. If one takes the two matrices $A(a, b, c)$ and $A(a', b', c')$, one sees that they must commute,[13] since one can give a numerical value to both simultaneously; for example, if the electron is known to be in the cell a, b, c, it will certainly not be in the cell a', b', c', so that if one gives the numerical value 1 to $A(a, b, c)$, one must at the same time give the numerical value 0 to $A(a', b', c')$.

We can transform each of the matrices A into a diagonal matrix A^* by a transformation[14] of the type

$$A^* = BAB^{-1}.$$

Since all the matrices $A(a, b, c)$ commute,[15] they can be transformed simultaneously into diagonal matrices by a transformation of this type. The diagonal elements of each matrix $A^*(a, b, c)$ are its characteristic values, which are the same as the characteristic values of $A(a, b, c)$, that is, 0 and 1.

[a] The published version has: 'Divide the space up into a large number of small cells, and consider the function of three variables ξ, η, ζ that is equal to 1 when the point ξ, η, ζ is in a given cell and equal to 0 when the point is elsewhere. This function, applied to the matrices x, y, z, gives another matrix' (*eds.*).

Further, no two A^* matrices, such as $A^*(a, b, c)$ [and] $A^*(a', b', c')$, can both have 1 for the same diagonal element, as a simple argument shows that $A^*(a, b, c) + A^*(a', b', c')$ must also have only the characteristic values 0 and 1. We can without loss of generality assume that each A^* has just one diagonal element equal to 1 and all the others zero. By transforming back, by means of the formula

$$A(a, b, c) = B^{-1} A^*(a, b, c) B \,,$$

we now find that the matrix elements of $A(a, b, c)$ are of the form

$$A(a, b, c)_{mn} = B_m^{-1} B_n \,,$$

i.e. a function of the row multiplied by a function of the column.

It should be observed that the proof of this result is quite independent of equations of motion and quantum conditions. If we take these into account, we find that B_m^{-1} and B_n are apart from constants just Schrödinger's eigenfunctions $\bar{\psi}_m$ and ψ_n at the point a, b, c.

Thus Schrödinger's density function $\bar{\psi}_m(x, y, z)\psi_m(x, y, z)$ is a[16] diagonal element of the matrix A referring to a cell about the point x, y, z. The true quantum expression for the density is the whole matrix. Its diagonal elements give only the average density, and must not be used when the density is to be multiplied by a dynamical variable represented by a matrix.

I should now like to express my views on determinism and the nature of the numbers appearing in the calculations of the quantum theory, as they appear to me after thinking over Mr Bohr's remarks of yesterday.[17] In the classical theory one starts from certain numbers describing completely the initial state of the system, and deduces other numbers that describe completely the final state. This deterministic theory applies only to an isolated system.

But, as Professor Bohr has pointed out, an isolated system is by definition unobservable. One can observe the system only by disturbing it and observing its reaction to the disturbance. Now since physics is concerned only with observable quantities the deterministic classical theory is untenable.[18]

In the quantum theory one also begins with certain numbers and deduces others from them. Let us inquire into the distinguishing characteristics[19] of these two sets of numbers. The disturbances that an experimenter applies to a system to observe it are directly under his control, and are acts of free will by him. *It is only the numbers that describe these acts of free will that can be taken as initial numbers for a calculation in the quantum theory.* Other numbers describing the initial state of the system are inherently unobservable, and do[20] not appear in the quantum theoretical treatment.

Let us now consider the final numbers obtained as the result of an experiment. It is essential that the result of an experiment shall be a permanent record. The numbers that describe such a result must help to not only describe the state of the world at the instant the experiment is ended, but also help to describe the state of the world at any subsequent time. These numbers describe what is common to all the events in a certain chain of causally connected events, extending indefinitely into the future.

Take as an example a Wilson cloud expansion experiment. The causal chain here consists of the formation of drops of water round ions, the scattering of light by these drops of water, and the action of this light on a photographic plate, where it leaves a permanent record. The numbers that form the result of the experiment describe all of the events in this chain equally well and help to describe the state of the world at any time after the chain began.

One could perhaps extend the chain further into the past.[21] In the example one could, perhaps, ascribe the formation of the ions to a β-particle, so that the result of the experiment would be numbers describing the track of a β-particle. In general one tries with the help of theoretical considerations to extend the chain as far back into the past as possible, in order that the numbers obtained as the result of the experiment may apply as directly as possible to the process under investigation.[22]

This view of the nature of the results of experiments fits in very well with the new quantum mechanics. According to quantum mechanics the state of the world at any time is describable by a wave function ψ, which normally varies according to a causal law, so that its initial value determines its value at any later time. It may however happen that at a certain time t_1, ψ can be expanded in the form

$$\psi = \sum_n c_n \psi_n ,$$

where the ψ_n's are wave functions of such a nature that they cannot interfere with one another at any time subsequent to t_1. If such is the case, then the world at times later than t_1 will be described not by ψ but by one of the ψ_n's. The particular ψ_n that it shall be must be regarded as chosen by nature.[23] One may say that nature chooses which ψ_n it is to be, as the only information given by the theory is that the probability of any ψ_n being chosen is $|c_n|^2$.[24] The value of the suffix n that labels the particular ψ_n chosen may be the result of an experiment, and the result of an experiment must always be such a number. It is a number describing an irrevocable choice of nature, which must affect the whole of the future course of events.[a]

[a] The last two sentences appear differently in the published version: 'The choice, once made, is irrevocable and will affect the whole future state of the world. The value of n chosen by nature can be determined by experiment and *the results of all experiments* are numbers describing such choices of nature'.

As an example take the case of a simple collision problem. The wave packet representing the incident electron gets scattered in all directions. One must take for the wave function after the process not the whole scattered wave, but once again a wave packet moving in a definite direction. From the results of an experiment, by tracing back a chain of causally connected events one could determine in which direction the electron was scattered and one would thus infer that nature had chosen this direction. If, now, one arranged a mirror to reflect the electron wave scattered in one direction d_1 so as to make it interfere with the electron wave scattered in another direction d_2, one would not be able to distinguish between the case when the electron is scattered in the direction d_2 and when it is scattered in the direction d_1 and reflected back into d_2. One would then not be able to trace back the chain of causal events so far, and one would not be able to say that nature had chosen a direction as soon as the collision occurred, but only [that] at a later time nature chose where the electron should appear. The[25] interference between the ψ_n's compels nature to postpone her choice.

MR BOHR.[b] – Quite see that one must go into details of pictures, if one wants to control or illustrate general statements. I think still that you may simpler put it in my way. Just this distinction between observation and definition allows to let the quantum mechanics appear as generalisation. What does mean: get records which do not allow to work backwards. Even if we took all molecules in photographic plate one would have closed system. If we tell of a record we give up definition of plate. Whole point lies in that by observation we introduce something which does not allow to go on.

...

MR BORN. – I should like to point out, with regard to the considerations of Mr Dirac, that they seem closely related to the ideas expressed in a paper by my collaborator J.[26] von Neumann, which will appear shortly. The author of this paper shows that quantum mechanics can be built up using the ordinary probability calculus, starting from a small number of formal hypotheses; the probability amplitudes and the law of their composition do not really play a role there.

MR KRAMERS. – I think the most elegant way to arrive at the results of Mr Dirac's considerations is given to us by the methods he presented in his memoir

Dirac's notes contain a similar variant written in the margin: 'The value of n chosen by nature may be determined by experiment. The result of every experiment consists of numbers determining one of these choices of nature, and is permanent since such a choice is irrevocable and affects the whole future state of the world' (*eds.*).

[b] Again, these remarks do not appear in the published French, and we have reproduced them from Bohr's *Collected Works* (Bohr 1985, p. 105). For the full version of this passage, see the Appendix (*eds.*).

in the *Proc. Roy. Soc.*, ser. *A*, vol. **113**, p. 621. Let us consider a function of the coordinates q_1, q_2, q_3 of an electron, that is equal to 1 when the point considered is situated in the interior of a certain volume V of space and equal to zero for every exterior point, and let us represent by $\psi(q,\alpha)$ and $\overline{\psi}(\alpha,q)$ the transformation functions that allow us to transform a physical quantity F, whose form is known as a matrix (q',q''), into a matrix (α',α''), $\alpha_1,\alpha_2,\alpha_3$ being the first integrals of the equation of motion. The function f, written as a matrix (q',q''), will then take the form $f(q')\delta(q'-q'')$, where $\delta(q'-q'')$ represents Dirac's unit matrix. As a matrix (α',α''), f will then take the form

$$f(\alpha',\alpha'') = \int \bar{\psi}(\alpha',q')dq' f(q')\delta(q'-q'')dq''\psi(q'',\alpha'') = \int_V \bar{\psi}(\alpha',q')dq'\psi(q',\alpha''),$$

the integral having to be extended over the whole of the considered volume. The diagonal terms of $f(\alpha',\alpha'')$, which may be written in the form

$$f(\alpha) = \int \psi\bar{\psi}dq,$$

will directly represent, in accordance with Dirac's interpretation of the matrices, the probability that, for a state of the system characterised by given values of α, the coordinates of the electron are those of a point situated in the interior of V. As ψ is nothing other than the solution of Schrödinger's wave equation, we arrive at once at the interpretation of the expression $\psi\bar{\psi}$ under discussion.

MR HEISENBERG. – I do not agree with Mr Dirac when he says that, in the described experiment, nature makes a choice. Even if you place yourself very far away from your scattering material, and if you measure after a very long time, you are able to obtain interference by taking two mirrors. If nature had made a choice, it would be difficult to imagine how the interference is produced. Evidently, we say that this choice of nature can never be known before the decisive experiment has been done; for this reason, we can make no real objection to this choice, because the expression 'nature makes a choice' then implies no physical observation. I should rather say, as I did in my last paper, that the *observer himself* makes the choice,[a] because it is only at the moment when the observation is made that the 'choice' has

[a] From Heisenberg's publication record, it is clear that he is here referring to his uncertainty paper, which had appeared in May 1927. There we find the statement that 'all perceiving is a choice from a plenitude of possibilities' (Heisenberg 1927, p. 197). When Heisenberg says, in his above comment on Dirac, that the observer 'makes' the choice, he seems to mean this in the sense of the observer *bringing about* the choice (*eds.*).

become a physical reality and that the phase relationship in the waves, the power of interference, is destroyed.

MR LORENTZ. – There is then, it seems to me, a fundamental difference of opinion on the subject of the meaning of these choices made by nature.

To admit the possibility that nature makes a choice means, I think, that it is impossible for us to know in advance how phenomena will take place in the future. It is then indeterminism that you wish to erect as a principle. According to you there are events that we cannot predict, whereas until now we have always assumed the possibility of these predictions.

PHOTONS

MR KRAMERS. – During the discussion of Mr de Broglie's report, Mr Brillouin explained to us how radiation pressure is exerted in the case of interference and that one must assume an auxiliary stress. But how is radiation pressure exerted in the case where it is so weak that there is only one photon in the interference zone? And how does one obtain the auxiliary tensor in this case?

MR DE BROGLIE. – The proof of the existence of these stresses can be made only if one considers a cloud of photons.

MR KRAMERS. – And if there is only one photon, how can one account for the sudden change of momentum suffered by the reflecting object?

MR BRILLOUIN. – No theory currently gives the answer to Mr Kramers' question.

MR KRAMERS. – No doubt one would have to imagine a complicated mechanism, that cannot be derived from the electromagnetic theory of waves?

MR DE BROGLIE. – The dualist representation by corpuscles and associated waves does not constitute a definitive picture of the phenomena. It does not allow one to predict the pressures exerted on the different points of a mirror during the reflection of a single photon. It gives only the mean value of the pressure during the reflection of a cloud of photons.

MR KRAMERS. – What advantage do you see in giving a precise value to the velocity v of the photons?

MR DE BROGLIE. – This allows one to imagine the trajectory followed by the photons and to specify the meaning of these entities; one can thus consider the photon as a material point having a position and a velocity.

MR KRAMERS. – I do not very well see, for my part, what advantage there is, for the description of experiments, in making a picture where the photons travel along well-defined trajectories.

MR EINSTEIN. – During reflection on a mirror, Mr L. de Broglie assumes that the photons move parallel to the mirror with a speed $c \sin\theta$; but what happens if the incidence is normal? Do the photons then have zero speed, as required by the formula ($\theta = 0$)?

MR PICCARD. – Yes. In the case of reflection, one must assume that the component of the velocity of the photons parallel to the mirror is constant. In the interference zone, the component normal to the mirror disappears. The more the incidence increases, the more the photons are slowed down. One thus indeed arrives at stationary photons in the limiting case of normal incidence.[a]

MR LANGEVIN. – In this way then, in the interference zone, the photons no longer have the speed of light; they do not then always have the speed c?

MR DE BROGLIE. – No, in my theory the speed of photons is equal to c only outside any interference zone, when the radiation propagates freely in the vacuum. As soon as there are interference phenomena, the speed of the photons becomes smaller than c.

MR DE DONDER. – I should like to show how the research of Mr L. de Broglie is related to mine on some points.

By identifying the ten equations of the gravitational field and the four equations of the electromagnetic field with the fourteen equations of the wave mechanics of L. Rosenfeld, I have obtained[b] a *principle of correspondence* that clarifies and generalises that of O. Klein.[c]

In my principle of correspondence, there appear *the quantum current* and *the quantum tensor*. I will give the formulas for them later on; let it suffice to remark

[a] Note that here the wave train is tacitly assumed to be limited longitudinally. Cf. our discussion of the de Broglie–Pauli encounter, Section 10.2 (*eds.*).

[b] *Bull. Ac. Roy. de Belgique, Cl. des Sc.* (5) **XIII**, ns. 8–9, session of 2 August 1927, 504–9. See esp. equations (5) and (8).

[c] *Zeitschr. f. Phys.* **41**, n. 617 (1927). See esp. equations (18), p. 414.

now that the example of correspondence that Mr de Broglie has expounded is in harmony with my principle.

Mr L. Rosenfeld[a] has given another example. Here, the mass is *conserved* and, moreover, one resorts to the quantum current. We add that this model of quantisation is also included, as a particular case, in our principle of correspondence.

Mr Lorentz has remarked, with some surprise, that the continuity equation for charge is preserved in Mr de Broglie's example. Thanks to our principle of correspondence, and to Rosenfeld's compatibility[27] theorem, one can show that it will always be so for the total current (including the quantum current) and for the theorem of energy and momentum. The four equations that express this last theorem are satisfied by virtue of the two generalised quantum equations of de Broglie–Schrödinger.

One further small remark, to end with. Mr de Broglie said that *relativistic* systems do not exist yet. I have given the theory of *continuous* or *holonomic* systems.[b] But Mr de Broglie gives another meaning to the word *system*; he has in mind *interacting* systems, such as the Bohr atom, the system of three bodies, etc. I have remarked recently[c] that the quantisation of these systems should be done by means of a $(ds)^2$ taken in a *configuration* space with $4n$ dimensions, n denoting the number of particles. In a paper not yet published, I have studied particular systems called *additive*.

MR LORENTZ. – The stresses of which you speak and which you call quantum, are they those of Maxwell?

MR DE DONDER. – Our quantum stresses must contain the Maxwell stresses as a particular case; this results from the fact that our principle of correspondence is derived (in part, at least) from Maxwell's equations, and from the fact that these quantum stresses here formally play the same role as the stresses of electrostriction[d] in Einsteinian Gravity. Let us recall, on this subject, that our principle of correspondence is also derived from the fundamental equations of Einsteinian Gravity. Mr de Broglie has, by means of his calculations, thus recovered the stresses of radiation.

[a] L. Rosenfeld, 'L'univers à cinq dimensions et la mécanique ondulatoire (quatrième communication)', *Bull. Ac. Roy. Belg., Cl. des Sc.*, October 1927. See esp. paragraphs 4 and 5.

[b] *C. R. Acad. Sc. Paris*, 21 February 1927, and *Bull. Ac. Roy. Belgique, Cl. des Sc.*, 7 March 1927.

[c] *Bull. Ac. Roy. Belgique, Cl. des Sc.*, 2 August 1927. See esp. form. (22).

[d] For more details, see our Note: 'L'électrostriction déduite de la gravifique einsteinienne', *Bull. Ac. Roy. Belgique, Cl. des Sc.*, session of 9 October 1926, 673–8.

PHOTONS AND ELECTRONS

MR LANGEVIN makes a comparison between the old and modern statistics.

Formerly, one decomposed the phase space into cells, and one evaluated the number of representative points attributing an individuality to each constituent of the system.

It seems today that one must modify this method by suppressing the individuality of the constituents of the system, and substituting instead the individuality of the states of motion. By assuming that any number of constituents of the system can have the same state of motion, one obtains the statistics of Bose-Einstein.

One obtains a third statistics, that of Pauli-Fermi-Dirac,[28] by assuming that there can be only a single representative point in each cell of phase space.

The new type of representation seems more appropriate to the conception of photons and particles: since one attributes a complete identity of nature to them, it appears appropriate to not insist on their individuality, but to attribute an individuality to the states of motion.

In the report of Messrs Born and Heisenberg, I see that it results from quantum mechanics that the statistics of Bose-Einstein is suitable for molecules, that of Pauli-Dirac for electrons and protons. This means that for photons[29] and molecules there is superposition, while for protons and electrons there is impenetrability. Material particles are then distinguished from photons[30] by their impenetrability.

MR HEISENBERG. – There is no reason, in quantum mechanics, to prefer one statistics to another. One may always use different statistics, which can be considered as complete solutions of the problem of quantum mechanics. In the current state of the theory, the question of interaction has nothing to do with the question of statistics.

We feel nevertheless that Einstein-Bose statistics could be the more suitable for light quanta, Fermi-Dirac statistics for positive and negative electrons.[a] The statistics could be connected with the difference between radiation and matter, as Mr Bohr has pointed out. But it is difficult to establish a link between this question and the problem of interaction. I shall simply mention the difficulty created by electron spin.

MR KRAMERS reminds us of Dirac's research on statistics, which has shown that Bose-Einstein statistics can be expressed in an entirely different manner. The statistics of photons, for example, is obtained by considering a cavity filled with blackbody radiation as a system having an infinity of degrees of freedom. If one

[a] That is, for protons and electrons (*eds.*).

quantises this system according to the rules of quantum mechanics and applies Boltzmann statistics, one arrives at Planck's formula, which is equivalent to Bose-Einstein statistics applied to photons.

Jordan has shown that a formal modification of Dirac's method allows one to arrive equally at a statistical distribution that is equivalent to Fermi statistics. This method is suggested by Pauli's exclusion principle.

MR DIRAC[a] points out that this modification, considered from a general point of view, is quite artificial. Fermi statistics is not established on exactly the same basis as Einstein-Bose statistics, since the natural method of quantisation for waves leads precisely to the latter statistics for the particles associated with the waves. To obtain Fermi statistics, Jordan had to use an unusual method of quantisation for waves, chosen specially so as to give the desired result. There are mathematical errors in the work of Jordan that have not yet been redressed.

MR KRAMERS. – I willingly grant that Jordan's treatment does not seem as natural as the manner by which Mr Dirac quantises the solution of the Schrödinger equation. However, we do not yet understand why nature requires this quantisation, and we can hope that one day we will find the deeper reason for why it is necessary to quantise in one way in one case and in another way in the other.

MR BORN. – An essential difference between Debye's old theory, in which the characteristic vibrations of the blackbody cavity are treated like Planck oscillators, and the new theory is this, that both yield quite exactly Planck's radiation formula (for the mean density of radiation), but that the old theory leads to inexact values for the local fluctuations of radiation, while the new theory gives these values exactly.

MR HEISENBERG. – According to the experiments, protons and electrons both have an angular momentum and obey the laws of the statistics of Fermi-Dirac; these two points seem to be related. If one takes two particles together, if one asks, for example, which statistics one must apply to a gas made up of atoms of hydrogen, one finds that the statistics of Bose-Einstein is the right one, because by permuting two H atoms, we permute one positive electron and one negative electron,[b] so that we change the sign of the Schrödinger function *twice*. In other words, Bose–Einstein statistics is valid for all gases made up of neutral molecules, or more generally, composed of systems whose charge is an even multiple of e.

[a] On this criticism by Dirac, cf. Kragh (1990, pp. 128–30) (*eds.*).
[b] That is, we permute the two protons, and also the two electrons (*eds.*).

If the charge of the system is an odd multiple of e, the statistics of Fermi-Dirac applies to a collection of these systems.

The He nucleus does not rotate and a collection of He nuclei obeys the laws of Bose-Einstein statistics.

MR FOWLER asks if the fine details of the structure of the bands of helium agree better with the idea that we have only symmetric states of rotation of the nuclei of helium than with the idea that we have only antisymmetric states.

MR HEISENBERG. – In the bands of helium, the fact that each second line disappears teaches us that the He nucleus is not endowed with a spinning motion. But it is not yet possible to decide experimentally, on the basis of these bands, if the statistics of Bose-Einstein or that of Fermi-Dirac must be applied to the nucleus of He.

MR SCHRÖDINGER. – You have spoken of experimental evidence in favour of the hypothesis that the proton is endowed with a spinning motion just like the electron, and that protons obey the statistical law of Fermi-Dirac. What evidence are you alluding to?

MR HEISENBERG. – The experimental evidence is provided by the work of Dennison[a] on the specific heat of the hydrogen molecule, work which is based on Hund's research concerning the band spectra of hydrogen.

Hund found good agreement between his theoretical scheme and the experimental work of Dieke, Hopfield and Richardson, by means of the hypotheses mentioned by Mr Schrödinger. But for the specific heat, he found a curve very different from the experimental curve. The experimental curve of the specific heat seemed rather to speak in favour of Bose–Einstein statistics. But the difficulty was elucidated in the paper by Dennison, who showed that the systems of 'symmetric' and 'antisymmetric' terms (with regard to protons) do not combine in the time necessary to carry out the experiment. At low temperature, a transition takes place about every three months. The ratio of statistical weights of the systems of symmetric and antisymmetric terms is 1 : 3, as in the helium atom. But at low temperatures the specific heat must be calculated as if one had a mixture of two gases, an 'ortho' gas and a 'para' gas. If one wished to perform experiments on the specific heat with a gas of hydrogen, kept at low temperature for several months, the result would be totally different from the ordinary result.

[a] *Proc. Roy. Soc. A* **114** (1927), 483.

MR EHRENFEST wishes to formulate a question that has some relation to the recent experiments by Mr Langmuir on the disordered motion of electrons in the flow of electricity through a gas.

In the well-known Pauli exclusion (Pauliverbot), one introduces (at least in the language of the old quantum theory) a particular incompatibility relation between the quantum motions of the different particles of a single system, without speaking explicitly of the role possibly played by the forces acting between these particles. Now, suppose that through a small opening one allows particles that, so to speak, do not exert forces on each other, to pass from a large space into a small box bounded by quite rigid walls with a complicated shape, so that the particles encounter the opening and leave the box only at the end of a sufficiently long time. Before entering the box, if the particles have almost no motion relative to one another, the Pauli exclusion intervenes. After their exit, will they have very different energies, independently of the weakness of the mutual action between the particles? Or else what role do these forces play in the production of Pauli's incompatibility (choice of antisymmetric solutions of the wave equation)?

MR HEISENBERG. – The difficulty with Mr Ehrenfest's experiment is the following: the two electrons must have different energies. If the energy of interaction of the two electrons is very small, the time τ_1 required for the electrons to exchange an appreciable amount of energy is very long. But to find experimentally which state, symmetric or antisymmetric, the system of the two electrons in the box is in, we need a certain time τ_2 which is at least $\sim 1/\nu$, if $h\nu$ is the [energy] difference between the symmetric and antisymmetric states. Consequently, $\tau_1 \sim \tau_2$ and the difficulty disappears.

MR RICHARDSON. – The evidence for a nuclear spin is much more complete than Mr Heisenberg has just said. I have recently had occasion to classify a large number of lines in the visible bands of the spectrum of the H_2 molecule. One of the characteristic features of this spectrum is a rather pronounced alternation in the intensity of the successive lines. The intensities of the lines of this spectrum were recently measured by MacLennan, Grayson-Smith and Collins. Unfortunately, a large number of these lines overlap with each other, so that the intensity measurements must be accepted only with reservations.

But nevertheless, I think one can say, without fear of being mistaken, that all the bands that are sufficiently well-formed and sufficiently free of influences of the lines on each other (so that one can have confidence in the intensity measurements) have lines, generally numbered 1, 3, 5, ..., that are intrinsically three times more intense than the intermediate lines, generally numbered 2, 4, 6, ... By intrinsic intensity, I mean that which one obtains after having taken into account the effects on the intensity of temperature and quantum number (and also, of course, the effects of overlap with other lines, where it is possible to take this into account). In other words, I wish to say that the constant c of the intensity formula

$$\mathcal{J} = c\left(m+\frac{1}{2}\right)e^{\frac{-\left(m+\frac{1}{2}\right)^2 h^2}{8\pi K k T}},$$

where m is the number of the line and K the moment of inertia of the molecule, is three times bigger for the odd-numbered lines than for the even-numbered ones. This means that the ratio 3 : 1 applies, with an accuracy of about 5%, for at least five different vibration states of a three-electron state of excitation. It also applies to another state, which is probably 3^1P if the others are 3^3P. It is also shown, but in a less precise way, that it applies to two different vibration states of a state of excitation with four electrons.

At present, then, there is a great deal of experimental evidence that this nuclear spin persists through the different states of excitation of the hydrogen molecule.

MR LANGMUIR. – The question has often been raised of a similarity in the relation between light waves and photons on the one hand, and de Broglie waves and electrons on the other. How far can this analogy be developed? There are many remarkable parallels, but also I should like to see examined if there are no fundamental differences between these relations. Thus, for example, an electron is characterised by a constant charge. Is there a constant property of the photon that may be compared with the charge of the electron? The speed of the electron is variable; is that of the photon also? The electromagnetic theory of light has suggested a multitude of experiments, which have added considerably to our knowledge. The wave theory of the electron explains the beautiful results of Davisson and Germer. Can one hope that this theory will be as fertile in experimental suggestions as the wave theory of light has been?

MR EHRENFEST. – When one examines a system of plane waves of elliptically polarised light, placing oneself in differently moving coordinate systems, these waves show the same degree of ellipticity whatever system one places oneself in. Passing from the language of waves to that of photons, I should like to ask if one must attribute an elliptical polarisation (linear or circular in the limiting cases) to each photon? If the reply is affirmative, in view of the invariance of the degree of ellipticity in relativity, one must distinguish as many species of photons as there are degrees of ellipticity. That would yield, it seems to me, a new difference between the photon and the spinning electron. If, on the other hand, one wishes above all to retain the analogy with the electron, as far as I can see one comes up against two difficulties:

1. How then must one describe linearly polarised light in the language of photons? (It is instructive, in this respect, to consider the way in which the two linearly polarised components, emitted perpendicularly to the magnetic field by a flame showing the Zeeman effect, are absorbed by a second flame placed in a magnetic field with antiparallel orientation.)

   Mr Zeeman, to whom I posed the question, was kind enough to perform the experiment about a year ago, and he was able to notice that the absorption is the same in parallel and antiparallel fields, as one could have predicted, in fact, by considerations of continuity.
2. For electrons, which move always with a speed less than that of light, the universality of the spin may be expressed as follows, that one transforms the corresponding antisymmetric tensor into a system of coordinates carried with the electron in its translational motion ('at rest'). But photons always move with the speed of light!

MR COMPTON. – Can light be elliptically polarised when the photon has an angular momentum?

MR EHRENFEST. – Because the photons move with the speed of light, I do not really understand what it means when one says that each photon has a universal angular momentum just like an electron.

Allow me to remind you of yet another property of photons. When two photons move in directions that are not exactly the same, one can say quite arbitrarily that one of the photons is a radio-photon and the other a γ-ray photon, or inversely. That depends quite simply on the moving system of coordinates to which one refers the pair of photons.

MR LORENTZ. – Can you make them identical by such a transformation?

MR EHRENFEST. – Perfectly. If they move in different directions. One can then give them the same colour by adopting a suitable frame of reference. It is only in the case where their worldlines are exactly parallel that the ratio of their frequencies remains invariant.

MR PAULI. – The fact that the spinning electron can take two orientations in the field allowed by the quanta seems to invite us at first to compare it to the fact that there are, for a given direction of propagation of the light quanta, two characteristic vibrations of blackbody radiation, distinguished by their polarisation. Nevertheless there remain essential differences between the two cases. While in relativity one describes waves by a (real) sextuple vector $F_{ix} = -F_{xi}$, for the spinning electron one has proposed the following two modes of description for the associated de Broglie waves: 1. One describes these waves by two complex functions ψ_α, ψ_β

(and so by four real functions); but these functions transform in a way that is hardly intuitive during the change from one system of coordinates to another. That is the route I followed myself. Or else: 2. Following the example of Darwin, one introduces a *quadruple* vector with generally complex components (and so eight real functions in total). But this procedure has the inconvenience that the vector involves a redundancy [indétermination], because all the verifiable results depend on only *two* complex functions.

These two modes of description are mathematically equivalent, but independently of whether one decides in favour of one or the other, it seems to me that one cannot speak of a *simple* analogy between the polarisation of light waves and the polarisation of de Broglie waves associated with the spinning electron.

Another essential difference between electrons and light quanta is this, that between light quanta there does not exist direct (immediate) mutual action, whereas electrons, as a result of their carrying an electric charge, exert direct mutual actions on each other.

MR DIRAC.[a] – I should like to point out an important failure in the analogy between the spin of electrons and the polarisation of photons. In the present theory of the spinning electron one assumes that one can specify the direction of the spin axis of an electron at the same time as its position, or at the same time as its momentum. Thus the spin variable of an electron commutes[31] with its coordinate and with its momentum variables. The case is different for photons. One can specify a direction of polarisation for plane monochromatic light waves, representing photons of given momentum, so that the polarisation variable commutes with the momentum variables. On the other hand, if the position of a photon is specified, it means one has an electromagnetic disturbance confined to a very small volume,[32] and one cannot give a definite polarisation, i.e. a definite direction for the electric vector, to this disturbance. Thus the polarisation variable of a photon does not commute with its coordinates.

MR LORENTZ. – In these different theories, one deals with the probability $\psi\psi^*$. I should like to see quite clearly how this probability can exist when particles move in a well-defined manner following certain laws. In the case of electrons, this leads to the question of motions in the field ψ (de Broglie). But the same question arises for light quanta. Do photons allow us to recover all the classical properties of waves? Can one represent the energy, momentum and Poynting vector by photons? One sees immediately that, when one has an energy density and energy flow, if one wishes to explain this by photons then the number of photons per unit volume gives

[a] Again, here we follow Dirac's original English (*eds.*).

the density, and the number of photons per second that move across a unit surface gives the Poynting vector.

The photons will then have to move with a speed different from that of light. If one wished to assign always the same speed c to the photons, in some cases one would have to assume a superposition of several photon currents. Or else one would have to assume that the photons cannot be used to represent all the components of the energy-momentum tensor. Some of the terms must be continuous in the field. Or else the photons are smeared out [fondus].

A related question is to know whether the photons can have a speed different from that of light and whether they can even be at rest. That would altogether displease me. Could we speak of these photons and of their motion in a field of radiation?

MR DE BROGLIE. – When I tried to relate the motion of the photons to the propagation of the waves ψ of the new mechanics, I did not worry about putting this point of view in accord with the electromagnetic conception of light waves, and I considered only waves ψ of scalar character, which one has normally used until now.

MR LORENTZ. – One will need these waves for photons also. Are they of a different nature than light waves? It would please me less to have to introduce two types of waves.

MR DE BROGLIE. – At present one does not know at all the physical nature of the ψ-wave of the photons. Can one try to identify it with the electromagnetic wave? That is a question that remains open. In any case, one can provisionally try to develop a theory of photons by associating them with waves ψ.

MR LORENTZ. – Is the speed of the wave equal to that of light?

MR DE BROGLIE. – In my theory, the speed of photons is equal to c, except in interfering fields. In general, I find that one must assign to a moving corpuscle a proper mass M_0 given by the formula

$$M_0 = \sqrt{m_0^2 - \frac{h^2}{4\pi^2 c^2} \frac{\Box a}{a}}\,,$$

the function $\frac{\Box a}{a}$ being calculated at the point where the moving body is located at the given moment (a is the amplitude of the wave ψ). For photons, one has

$$m_0 = 0\,.$$

Thus, when a photon moves freely, that is to say, is associated with an ordinary plane wave, M_0 is zero and, to have a finite energy, the photon must have speed c. But, when there is interference, $\frac{\Box a}{a}$ becomes different from zero, M_0 is no longer zero and the photon, to maintain the same energy, must have a speed less than c, a speed that can even be zero.

MR LORENTZ. – The term $\frac{\Box a}{a}$ must be negative, otherwise the mass would become imaginary.

MR DE BROGLIE. – In the corpuscular conception of light, the existence of diffraction phenomena occurring at the edge of a screen requires us to assume that, in this case, the trajectory of the photons is curved. The supporters of the emission theory said that the edge of the screen exerts a force on the corpuscle. Now, if in the new mechanics as I develop it, one writes the Lagrange equations for the photon, one sees appear on the right-hand side of these equations a term proportional to the gradient of M_0.

This term represents a sort of force of a new kind, which exists only when the proper mass varies, that is to say, where there is interference. It is this force that will curve the trajectory of the photon when its wave ψ is diffracted by the edge of a screen.

Furthermore, for a cloud of photons the same Lagrange equations lead one to recover the internal stresses pointed out by Messrs Schrödinger and De Donder.[a] One finds, indeed, the relations

$$\frac{\partial}{\partial x^k}\left[T^{ik} + \Pi^{ik}\right] = 0\,,$$

where the tensor T^{ik} is the energy-momentum tensor of the corpuscles

$$T^{ik} = \rho_0 u^i u^k\,.$$

The tensor Π^{ik}, which depends on derivatives of the amplitude of the wave ψ and is zero when this amplitude is constant, represents stresses existing in the cloud of corpuscles, and these stresses allow us to recover the value of the radiation pressure in the case of reflection of light by a mirror.

The tensor $T^{ik} + \Pi^{ik}$ is certainly related to the Maxwell tensor but, to see clearly how, one would have to be able to clarify the relationship existing between the wave ψ of the photons and the electromagnetic light wave.

[a] Cf. Schrödinger (1927b) and De Donder's comments above (*eds.*).

MR PAULI.[a] – It seems to me that, concerning the statistical results of scattering experiments, the conception of Mr de Broglie is in full agreement with Born's theory in the case of elastic collisions, but that it is no longer so when one also considers inelastic collisions. I should like to illustrate this by the example of the rotator, which was already mentioned by Mr de Broglie himself. As Fermi[b] has shown, the treatment by wave mechanics of the problem of the collision of a particle that moves in the (x, y) plane and of a rotator situated in the same plane, may be made clear in the following manner.[c] One introduces a configuration space of three dimensions, of which two coordinates correspond to the x and y of the colliding particle, while as third coordinate one chooses the angle φ of the rotator. In the case where there is no mutual action between the rotator and the particle, the function ψ of the total system is given by[33]

$$\psi(x, y, \varphi) = Ae^{2\pi i\left[\frac{1}{\hbar}(p_x x + p_y y + p_\varphi \varphi) - \nu t\right]},$$

where one has put

$$p_\varphi = m\frac{h}{2\pi} \qquad (m = 0, 1, 2, \ldots).$$

In particular, the sinusoidal oscillation of the coordinate φ corresponds to a stationary state of the rotator. According to Born, the superposition of several partial waves of this type, corresponding to different values of m and by consequence of p_φ,[34] means that there is a probability different from zero for several stationary states of the rotator, while according to the point of view of Mr de Broglie, in this case the rotator no longer has a constant angular velocity and can also execute oscillations in certain circumstances.

Now, in the case of a finite energy of interaction between the colliding particle and the rotator, if we study the phenomenon of the collision by means of the wave equation in the space (x, y, φ), according to Fermi the result can be interpreted very simply. Indeed, since the energy of interaction depends on the angle φ in a periodic manner and vanishes at large distances from the rotator, that is to say from the axis φ, in the space (x, y, φ) we are dealing simply with a wave that falls on a grating and, in particular, on a grating that is unlimited in the direction of the axis φ. At large distances from the grating, waves come out only in fixed directions in configuration space, characterised by integral values of the difference $m' - m''$. Fermi has shown that the different spectral orders correspond simply to the different possible ways of transferring the energy of the colliding particle to

[a] Cf. Section 10.2 (*eds.*).
[b] *Zeitschr. f. Phys.* **40** (1926), 399.
[c] See Section 10.2 for a discussion of Fermi's argument (*eds.*).

the rotator, or conversely. Thus to each spectral order of the grating corresponds a given stationary state of the rotator after the collision.

It is, however, an essential point that, in the case where the rotator is in a stationary state before the collision, the incident wave is unlimited in the direction of the axis. For this reason, the different spectral orders of the grating will always be superposed at each point of configuration space. If we then calculate, according to the precepts of Mr de Broglie, the angular velocity of the rotator after the collision, we must find that this velocity is not constant. If one had assumed that the incident wave is limited[a] in the direction of the axis φ, it would have been the same before the collision. Mr de Broglie's point of view does not then seem to me compatible with the requirement of the postulate of the quantum theory, that the rotator is in a stationary state both before and after the collision.

To me this difficulty does not appear at all fortuitous or inherent in the particular example of the rotator; in my opinion, it is due directly to the condition assumed by Mr de Broglie, that in the individual collision process the behaviour of the particles should be completely determined and may at the same time be described completely by ordinary kinematics in spacetime. In Born's theory, agreement with the quantum postulate is realised thus, that the different partial waves in configuration space, of which the general solution of the wave equation after the collision is composed, are applicable [indiquées] *separately* in a statistical way. But this is no longer possible in a theory that, in principle, considers it possible to avoid the application of notions of probability to *individual* collision processes.

MR DE BROGLIE. – Fermi's problem is not of the same type as the one I treated earlier; indeed, he makes configuration space play a part, and not ordinary space.

The difficulty pointed out by Mr Pauli has an analogue in classical optics. One can speak of the beam diffracted by a grating in a given direction only if the grating and the incident wave are laterally limited, because otherwise all the diffracted beams will overlap and be bathed in the incident wave. In Fermi's problem, one must also assume the wave ψ to be limited laterally in configuration space.

MR LORENTZ. – The question is to know what a particle should do when it is immersed in two waves at the same time.

MR DE BROGLIE. – The whole question is to know if one has the right to assume the wave ψ to be limited laterally in configuration space. If one has this right, the velocity of the representative point of the system will have a constant value, and

[a] The French reads 'illimitée' [unlimited], which we interpret as a misprint. Pauli seems to be saying that if, on the other hand, the incident wave had been taken as limited, then *before* the collision the rotator could not have been in a stationary state and its angular velocity could not have been constant (*eds.*).

will correspond to a stationary state of the rotator, as soon as the waves diffracted by the φ-axis will have separated from the incident beam.

One can say that it is not possible to assume the incident beam to be limited laterally, because Fermi's configuration space is formed by the superposition of *identical* layers of height 2π in the direction of the φ-axis; in other words, two points of configuration space lying on the same parallel to the φ-axis and separated by a whole multiple of 2π represent the *same* state of the system. In my opinion, this proves above all the artificial character of configuration spaces, and in particular of that which one obtains here by rolling out along a line the cyclic variable φ.

MR DE DONDER. – In the course of the discussion of Mr L. de Broglie's report, we explained how we obtained our Principle of Correspondence; thanks to this principle, one will have[a]

$$\rho_{(e)}u^a + \Lambda^a = \sqrt{-g}K^2\frac{c}{e}\sum_n \frac{-h}{2i\pi}g^{an}\left(\psi\bar{\psi}_{.n} - \bar{\psi}\psi_{.n}\right) - 2\frac{e}{c}\Phi^a\psi\bar{\psi},$$

$$\rho_{(m)}u^a u^b + \Pi^{ab} = \sqrt{-g}\sum_\alpha\sum_\beta \gamma^{a\alpha}\gamma^{b\beta}\left(\psi_{.\alpha}\bar{\psi}_{.\beta} + \bar{\psi}_{.\alpha}\psi_{.\beta}\right) - \gamma^{ab}L$$

$$(a,\ b,\ n = 1, \ldots, 4;\ \alpha,\ \beta = 0, 1, \ldots, 4)\,.$$

The first relation represents *the total current* ($\equiv$ electronic current + quantum current) as a function of ψ and of the potentials g^{an}, Φ^a. Recall that one has set

$$L \equiv \sum_\alpha\sum_\beta \gamma^{\alpha\beta}\psi_{.\alpha}\overline{\psi_{.\beta}} + k^2\left(\mu^2 - \frac{1}{2\chi}\right)\psi\overline{\psi},$$

$$\gamma^{ab} \equiv g^{ab}, \quad \gamma^{0a} \equiv -\alpha\Phi^a, \quad \gamma^{00} \equiv \alpha^2\Phi^a\Phi_a - \frac{1}{\xi},$$

$$\xi\alpha^2 \equiv 2\chi, \quad \chi \equiv \frac{8\pi G}{c^2}, \qquad G \equiv 6.7 \times 10^{-8}\ \text{c.g.s.}$$

We have already mentioned the examples (or models) of correspondence found respectively by L. de Broglie and L. Rosenfeld. To be able to show clearly a new solution to the problem relating to *photons* that Mr L. de Broglie has just posed, I am going to display the formulas concerning the two above-mentioned models.[b]

[a] I adopt here L. Rosenfeld's notation, so as to facilitate the comparison with his formulas, given later.

[b] L. Rosenfeld, 'L'univers à cinq dimensions et la mécanique ondulatoire', *Bull. Ac. Roy. Belgique, Cl. des Sc.*, October 1927. See respectively the formulas (*38′), (*31), (*27), (21), (1), (8), (35), (28), (29), (*35).

Model of L. de Broglie.

Quantum current $\Lambda_a \equiv 0$.

Charge density $\rho_{(e)} = 2K^2A'^2\mu'$, where we have put

$$\mu'^2 = \mu^2 + \frac{\Box A'}{K^2A'},$$

which, retaining the charge e, reduces to substituting for the mass m_0 *the modified mass of L. de Broglie*:

$$M_0 \equiv \sqrt{m_0^2 + \frac{h^2}{4\pi^2c^2}\frac{\Box A'}{A'}}.$$

Model of L. Rosenfeld.

Quantum current $\Lambda_a = 2K^2A'^2C_{.a}$, where A' is the modulus of ψ and where the potential $C \equiv S' - S$. The function S satisfies the *classical* Jacobi equation; the function S' satisfies the *modified* Jacobi equation; one then has

$$\gamma^{\alpha\beta}S_{.\alpha}S_{.\beta} = \mu^2 - \frac{1}{2\chi},$$

$$\gamma^{\alpha\beta}S'_{.\alpha}S'_{.\beta} = \mu^2 - \frac{1}{2\chi} + \frac{\Box A'}{K^2A'}.$$

The quantum potential C produces *the difference* between *physical* quantisation and *geometrical* quantisation.

Recall that $\mu \equiv \frac{m_0c^2}{e}$, where m_0 and e are respectively the mass (at rest) and charge of the particle under consideration. We have also put

$$k \equiv iK \equiv i\frac{2\pi}{h}\frac{e}{c}.$$

Charge density $\rho_{(e)} = 2K^2A'^2\mu$. Here then *one retains*, at the same time, the mass m_0 and the charge e.

Let us respectively apply these formulas to the problem of the photon pointed out by Mr L. de Broglie. The *proper* mass m_0 of the photon is *zero*; in the model of Mr L. de Broglie, this mass must be replaced by the *modified* mass M_0; on the contrary, in the model of Mr L. Rosenfeld, one uses only the *proper* mass $m_0 \equiv 0$. In the two models, the charge density $\rho_{(e)}$ is zero. Finally, in the first model, the speed of the photon *must* vary; in contrast, in the second model, one can assume that this speed is always that of light. These conclusions obviously speak in favour of the model of L. Rosenfeld, and, in consequence, also in favour of the *physical* existence of our quantum current Λ^a $(a = 1, 2, 3, 4)$. This current will probably play a dominant role in still unexplained optical phenomena.[a]

MR LORENTZ. – Let us take an atom of hydrogen and let us form the Schrödinger function ψ.[35] We consider $\psi\psi^*$ as the probability for the presence of the electron in a volume element. Mr Born has mentioned all the trajectories in the classical theory: let us take them with all possible phases,[b] but let us now take the ψ corresponding to a single value W_n of energy and then let us form $\psi\psi^*$. Can one say that this product $\psi_n\psi_n^*$ represents the probability that the electrons move with the given energy W_n? We think that the electron cannot escape from a certain sphere. The atom is limited, whereas ψ extends to infinity. That is disagreeable.[36]

MR BORN. – The idea that $\psi\psi^*$ represents a probability density has great importance in applications. If, for example, in the classical theory an electron had two equilibrium positions separated by a considerable potential energy, then classically, for a sufficiently weak total energy only one oscillation could ever take place, around one of the two equilibrium positions. But according to quantum mechanics, each eigenfunction extends from one domain into the other; for this reason there always exists a probability that a particle, which at first vibrates in the neighbourhood of one of the equilibrium positions, jumps to the other. Hund has made important applications of this to molecular structure. This phenomenon probably also plays a role in the explanation of metallic conduction.

MR DE BROGLIE. – In the old theory of the motion of an electron in the hydrogen atom, an electron of total energy

$$W = \frac{m_0c^2}{\sqrt{1-\beta^2}} - \frac{e^2}{r}$$

cannot escape from a sphere of radius

$$R = -\frac{e^2}{W - m_0c^2}$$

because the value of the term $\frac{m_0c^2}{\sqrt{1-\beta^2}}$ has m_0c^2 as a lower limit.

[a] On this subject, Mr L. Brillouin has kindly drawn my attention to the experiments by Mr F. Wolfers: 'Sur un nouveau phénomène en optique: interférences par diffusion' (*Le Journal de Physique et le Radium* (VI) **6**, n. 11, November 1925, 354–68).

[b] The 'phases' of classical trajectories seem to be meant in the sense of action-angle variables (*eds.*).

In my conception one must take

$$W = \frac{M_0 c^2}{\sqrt{1-\beta^2}} - \frac{e^2}{r},$$

as the expression for the energy, where M_0 is the variable proper mass which I have already defined. Calculation shows that the proper mass M_0 diminishes when r increases, in such a way that an electron of energy W is no longer at all constrained to be in the interior of a sphere of radius R.

MR BORN. – Contrary to Mr Schrödinger's opinion, that it is nonsense to speak of the location and motion of an electron in the atom, Mr Bohr and I are of the opinion that this manner of speaking always has a meaning when one can specify an experiment allowing us to measure the coordinates and the velocities with a certain approximation.

Again in Richardson's notes on the general discussion (cf. p. 434), the following text together with Fig. D (both labelled 'Bohr'), and a similar figure with the shaded region labelled 'B', appear immediately after notes on De Donder's lengthy exposition just above, and clearly refer to remarks Bohr made on the topic being addressed here:

B[ohr] says it has no point to worry about the paradox that the electron in the atom is in a fixed path (ellipse or circle) and the probability that it should be found in a given place is given by the product $\psi\bar{\psi}$ which is a continuous function of space extending from zero to ∞. He says if we take a region such as B a long way from the atom in order to find if the electron is there we must illuminate it with long light waves and the frequency of these is so low that the electron is out of the region by reason of its motion in the stationary state before it has been illuminated long enough for the photoelectric act to occur. I am really not sure if this is right. But, anyway, it is no objection to pulling it out with an intense *static* electric field & this appears to be what is happening in the W experiments.

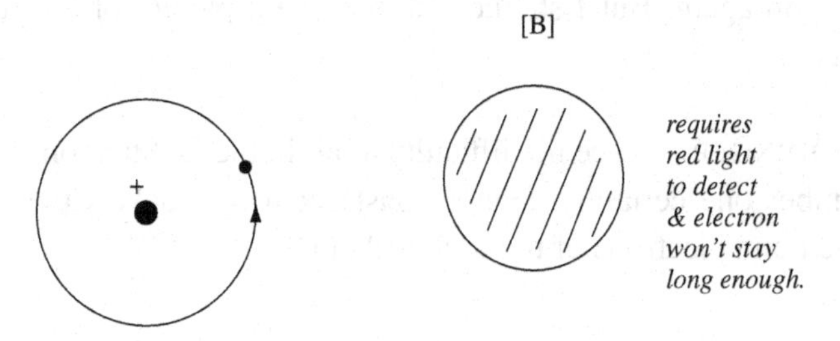

Fig. D.

MR PAULI. – One can indeed determine the location of the electron outside the sphere, but without modifying its energy to the point where an ionisation of the atom occurs.

MR LORENTZ. – I should like to make a remark on the subject of wave packets.[a] When Mr Schrödinger drew attention to the analogy between mechanics and optics, he suggested the idea of passing from corpuscular mechanics to wave mechanics by making a modification analogous to that which is made in the passage from geometrical optics to wave optics.[37] The wave packet gave a quite striking picture of the electron, but in the atom the electron had to be completely smeared out [fondu], the packet having the dimensions of the atom. When the dimensions of the wave packet become comparable to those of the trajectories of the classical theory, the material point would start to spread; having passed this stage, the electron will be completely smeared out.

The mathematical difficulty of constructing wave packets in the atom is due to the fact that we do not have at our disposal wavelengths sufficiently small or sufficiently close together. The frequencies of stable waves in the atom (eigenvalues) are more or less separated from each other; one cannot have frequencies very close together corresponding to states differing by very little, because the conditions at infinity would not be satisfied. To construct a packet, one must superpose waves of slightly different wavelengths; now, one can use only eigenfunctions ψ_n, which are sharply different from each other. In atoms, then, one cannot have wave packets. But there is a difficulty also for free electrons, because in reality a wave packet does not, in general, retain its shape in a lasting manner. Localised [limités] wave packets do not seem able to maintain themselves; spreading takes place. The picture of the electron given by a wave packet is therefore not satisfying, except perhaps during a short enough time.

What Mr Bohr does is this: after an observation he again localises [limite] the wave packet so as to make it represent what this observation has told us about the position and motion of the electron; a new period then starts during which the packet spreads again, until the moment when a new observation allows us to carry out the reduction again. But I should like to have a picture of all that during an unlimited time.

MR SCHRÖDINGER. – I see no difficulty at all in the fact that on orbits of small quantum number one certainly cannot construct wave packets that move in the manner of the point electrons of the old mechanics.

[a] Cf. also the discussion of the Lorentz–Schrödinger correspondence in Section 4.3 (*eds.*).

The fact that this is impossible is precisely the salient point of the wave mechanical view, the basis of the absolute powerlessness of the old mechanics in the domain of atomic dimensions. The original picture was this, that what moves is in reality not a point but a domain of excitation of finite dimensions, in particular at least of the order of magnitude of a few wavelengths. When such a domain of excitation propagates along a trajectory whose dimensions and radii of curvature are large compared with the dimensions of the domain itself, one can abstract away the details of its structure and consider only its progress along the trajectory. This progress takes place following exactly the laws of the old mechanics. But if the trajectory shrinks until it becomes of the order of magnitude of a few wavelengths, as is the case for orbits of small quantum number, all its points will be continually inside the domain of excitation and one can no longer reasonably speak of the propagation of an excitation along a trajectory, which implies that the old mechanics loses all meaning.

That is the original idea. One has since found that the naive identification of an electron, moving on a macroscopic orbit, with a wave packet encounters difficulties and so cannot be accepted to the letter. The main difficulty is this, that with certainty the wave packet spreads in all directions when it strikes an obstacle, an atom for example. We know today, from the interference experiments with cathode rays by Davisson and Germer, that this is part of the truth, while on the other hand the Wilson cloud chamber experiments have shown that there must be something that continues to describe a well-defined trajectory after the collision with the obstacle. I regard the compromise proposed from different sides, which consists of assuming a combination of waves and point electrons, as simply a provisional manner of resolving the difficulty.

MR BORN. – Also in the classical theory, the precision with which the future location of a particle can be predicted depends on the accuracy of the measurement of the initial location. It is then not in this that the manner of description of quantum mechanics, by wave packets, is different from classical mechanics. It is different because the laws of propagation of packets are slightly different in the two cases.

Notes on the translation

1. Microfilmed in AHQP-BMSS-11, section 5.
2. AHQP-36, section 10.
3. These notes are to be found in the Richardson collection in Houston, included with the copy of Born and Heisenberg's report (microfilmed in AHQP-RDN, document M-0309).
4. Included in AHQP-RDN, document M-0309.
5. French edition: 'les' is misprinted as 'le'.
6. AEA 16-617.00 (in German, with transcription and archival comments).
7. The French text has 'ψ' instead of 'ϕ', and 'doit satisfaire dans l'espace ordinaire' instead of the other way round. Note that ϕ is a functional of 'material' waves which themselves propagate in ordinary space.
8. AHQP-36, section 10.
9. The French adds: 'décrite par M. Compton'.
10. The French reads: 'provient uniquement de'.
11. Dirac's manuscript omits 'At present'.
12. The French reads 'intuitivité'.
13. Instead of 'commute' the French has 'permuter leurs valeurs'.
14. The French reads: 'transformation canonique'.
15. Instead of 'commute' the French has 'changent de valeur'.
16. Dirac's manuscript reads 'the'.
17. In Dirac's manuscript, the words 'determinism and' are cancelled and possibly reinstated. They appear in the French, which also omits 'of yesterday'.
18. <therefore unsatisfactory> <untenable>, the latter seems reinstated. The French has 'indéfendable'.
19. Instead of 'the distinguishing characteristics' the French has 'l'essence physique'.
20. <do>, {would} appears above the line, {can} below. The French reads 'ne figurent pas'.
21. This sentence does not appear in the French.
22. In the French, this sentence appears at the beginning of the paragraph.
23. This sentence does not appear in the French.
24. Dirac's manuscript has 'c_n^2'.
25. <Thus <a possibility> {the existence} of> .
26. The French has 'F.'.
27. Misprinted as 'comptabilité', despite being corrected in the galley proofs.
28. In the printed text, the word 'Dirac' is misplaced to later in the paragraph.
29. Misprinted as 'protons'.
30. Again misprinted as 'protons'.
31. The French renders 'commute' throughout with 'changer'.
32. The French adds: 'à un instant donné'.
33. 'h' misprinted as 'λ'.
34. 'p_φ' misprinted as 'φ'.
35. 'ψ' missing in the original, with a space instead.
36. The galley proofs add at the end: 'If one took the integral extended over the whole of this space, the exterior part would be comparable'.
37. The original mistakenly reads 'geometrical mechanics' and 'corpuscular optics'.

Appendix

This appendix reproduces most of the material contained in an envelope marked 'Solvaykongressen 1927', held in the Bohr Archive in Copenhagen. This material concerns the discussions at the Solvay conference, in particular Bohr's contributions and the general discussion. It has not been microfilmed, and only brief extracts have been published previously, in Bohr's *Collected Works* (Bohr 1985, pp. 103–5), where they are published together with some brief descriptions and comments (Bohr 1985, pp. 35–7, 100 and 478–9). The material consists of the following.

(1) In J.-É. Verschaffelt's hand (ten-and-a-half densely written pages), the transcripts of Bohr's major discussion contributions (after Compton's report and during the general discussion). These transcripts were presumably prepared on the basis of the (shorthand) notes taken during the conference, and they contain a large number of gaps. They are marked (in Danish) at the top: 'Verschaffelt – sent to Bohr', and they are evidently the material that Verschaffelt had supplied in order for Bohr to work out a final version of his contributions. The language is mainly English, with some German and French (and the annotation in Danish at the top).

(2) A typescript relating to the discussions after Compton's report (4 pp.), after Born and Heisenberg's report (2 pp.) and after Schrödinger's report (3 pp.), as well as to the general discussion (11 pp.), with some notes and insertions in Verschaffelt's hand. This, although also presumably prepared on the basis of the notes taken during the conference, is not a transcript, but appears to be a set of working notes or a skeleton outline of the discussions. The shorter contributions appear to be reproduced verbatim, and many have either disappeared from the published version or been moved or collated together. The longer contributions are represented only by the first few words or very brief descriptions. The discussion contributions are numbered, but there are gaps in the numbering, suggesting that some had already been discarded. There are also some notes (in Verschaffelt's hand) such as 'received already' or 'asked for'. While it may be unlikely that such

a document would have been widely circulated to the different contributors, it is quite likely that this copy might have been sent to Kramers, since he was helping Verschaffelt with some of the editing. The languages are a Babel of English, German and French.

(3) Six manuscript pages, in what appears to be Kramers' hand, of notes relating to the first session of the general discussion (Thursday morning). These appear to be part of Kramers' original notes taken during the conference, and they include in particular Bohr's speech at the beginning of the general discussion. Some actual timings are also given. One page must be missing from the middle of these notes, which otherwise are of very good quality. They are clearer and more concise than the transcripts, but for the most part one can easily identify parallel passages, except towards the end, where the correspondence is more difficult to establish. A numbering of the discussion contributions has evidently been added at a later date to try to match that in (2), and there are a few words translated into French between the lines and other brief annotations, suggesting an attempt to use these notes for producing a French version of Bohr's discussion contributions. The language is mainly English, with some French and Dutch.

(4) Two manuscript pages, the first one dated '30-2-1928' (*sic*), with two paragraphs that are incomplete and (at least partially) crossed-out, relating to two of Bohr's shorter discussion contributions (in English) and some other scattered annotations, mainly in Danish, the latter appearing to be in Kramers' hand. These two pages must date from when Bohr and Kramers were trying to prepare final versions of Bohr's contributions for inclusion in the volume.

Of this material, we reproduce (1), (2) and (3), in the original languages. A few cautionary remarks may be in order:

- We are reproducing this material for completeness and for possible use by future historians, without detailed analysis and comment (or indexing).
- The material in (2) probably gives us a more accurate record of who spoke and in what order. However, it clearly cannot be regarded as an accurate account of which topics were discussed or of how much time was spent on them: for the entries are very uneven, both in clarity and in degree of detail, with many of the longer published contributions being represented by only a short sentence.
- The entries in (2) are often cryptic, but usually make reasonable sense when compared with the published discussions.
- Most of these entries say nothing really substantial that is not already in the published discussions.
- A few of the entries (for example by Born, Bragg, Einstein and Planck) contain material that is not in the published discussions.
- This material provides useful information about the format, date and timing of the discussions.

- In a couple of instances, the material can be used to confirm uncertain readings in the published text.
- In (2), some of the entries are recorded as if they were quotations, others are written as summaries or paraphrase, sometimes in French when it would be reasonable to expect that the original was in English or German. Possibly, some of the paraphrase came from Lorentz, who acted as chairman and general interpreter.

For the text, we have generally tacitly taken into account any corrections, deletions or additions in the archival material, and we have uniformised the format and spelling with those used in the rest of the book.

We wish to thank very particularly Felicity Pors of the Niels Bohr Archive for her invaluable assistance.

The transcripts of Bohr's main contributions[a]

Discussion of Compton's report[b]

BOHR wishes to make a few remarks on Compton's paper; more general discussion must be postponed, but he wishes to speak now on an elementary point.

In how far gives experiment evidence about radiation. Features which in fact are not easily combined. There are properties of very different kinds which are not easily combined. To explain them Kramers, Slater and myself some years ago suggested views now shown to be wrong because there are points (e.g. scattering by crystals) which can only be described by waves. The difficulty is that Compton's interpretation of his effect used wave-theory, i.e. extension of the light in space and time; if we consider Compton's experiments, he does not find (?) to consider their corpuscular character; he uses instruments the action of which is interpreted by wave theory[.] We cannot describe frequency without ascribing some extension in space and time to waves. Yet we then ascribe to radiation properties of material particle[.] According to the theory a corpuscular collision is an event which took very little time in a very little space. It is very different to connect such waves with an event that happens at a point and a time. Velocity of electron suddenly changed in scattering phenomena. This led to formulation of theory that such a connection is only statistical. These were two quite different views. Now we know better, due to de Broglie: there is no such distinction between wave and electrons[.] De Broglie introduces the idea of groups of waves to represent corpuscles, such that four such groups meeting at one place in spacetime would give the scattering of radiation by an electron. In fact now also scattering process of an electron can be described as waves which may be so that energy and impulse of the particles are sufficiently defined. It is so difficult to unit[e] the idea of two agencies, one with extension in space and time, the other an event at a moment in space and time [ ]. Waves and electrons [   ] we may use such a group of waves, four groups of waves in the same spacetime extension; this explains the difficulties which have now disappeared. This gives the justification to certain extent; the recent developments remove these difficulties to a certain extent. There is no question now of deciding between the photon idea and the old idea. That is now Bohr–Kramers idea of virtual radiation. We know (de Broglie) that for an electron also we must suppose extension in space and time: groups of waves representing electrons also, four groups of waves, they may meet in same space and time region. We have not a simple choice between photon idea and wave idea (Compton lays stress on the simple choice). Space and time concepts are used in a different way when we consider waves on the one hand and the collision of corpuscles on the other hand[.] Consider wave side. Wave description is connected with solid bodies, for wave point of view is connected very much with our practical tools used in optical instruments[.] On the other hand particles can only be investigated by other particles that can move, so that we cannot find out so well where they are. Exchange of momentum cannot be studied by solid bodies, only by bodies allowed to move [   ] and this is just [   ] measured space and time not the same in the two. Energy and impulse want solid bodies which are in motion and change motion, and therefore are not exactly

[a] Tacit amendments to spelling and punctuation have been made when reasonably obvious. Other interpolations and amendments are denoted by square brackets. The sign [   ] denotes a clear blank space or dash or series of dashes in the typescript, the sign [ ] denotes a small space or an uncertain one (*eds.*).

[b] Marked in manuscript: 'Tuesday morning'. Overlay in pencil: 'Verschaffelt – Sendt til Bohr' (*eds.*).

defined in their position. We have to consider moving bodies and our ideas of space and time may have to be profoundly modified.

BOHR repeats: It is a point very much connected with problems of today that we all deal with 3 [?] difficulties in space and time description and we remark that in practical experiment the disagreement is not so strong as it might look because [ ] wave side and corpuscular side [ ] in spatial analysis in our instruments [ ] fixed frame [ ] we cannot fix a phenomenon within this same frame because impulses energies [ ] bring in bodies which are free to move [ ] we cannot know where they are in the fixed frame, or they are free and we do not know where they are. We have great difficulties in our ordinary spacetime description, but disagreement is not quite so strong from the practical point of view. In wave side our experiments are all made with respect to a fixed frame, while with corpuscles which are free we do not know where they are in the frame.

GENERAL DISCUSSION ON THURSDAY

After Lorentz's introductory speech:

BOHR. – I will try to explain as simply as I can certain attitudes to problem of quantum theory. I have nothing new. I only hope could help those simple points discussed in a way which could generate general discussion. [W]e come perhaps to some agreement. There are points in which I most heartily agree. I agree heartily with our president's explanation of his attitude. I feel very much that it is very difficult to start to speak on ultimate physical interpretation of words and symbols used in our calculations. We must first try to see what are our tools, see what our mathematical theorems are in themselves so very beautiful and how they suit requirements for accounting for experience. I think in that way...

But still I like to begin speaking about such points, for in the question of light in empty space and free particles we have already the means of speaking about what words mean and possibility of crucial discussions.

I should like to begin with [ ] and then consider some very simple problems connected with electrons, analyse what we mean by bound electrons and their inner attitude [?] [ ] very simply so as to see what we mean when we speak of stationary states. [ ] Then going a little further in details on problem which Heisenberg has brought for in order to discuss the difficulties in defining the very simplest things in space and time concepts. It all lies in the found[ation] of the wave theory, the whole difficulty is embodied in $E = h\nu$. This is an equation very difficult to say what it means, for the energy idea is so much connected with idea of particles and the ν with the idea of a wave and is not defined except for an infinite wave filling all space and to be compared with a free particle in space and time. The problem is just try to describe particles in space and time by waves. The expression for the phase wave of de Broglie is $e^{2\pi i(\tau_x x+\tau_y y+\tau_z z-\nu t)}$. If you have wave, τ_x is the number of waves per unit of length along x-axis. We will try to locate the particles by means of waves which cannot be done by interference with other waves [ ] and this very [ ] from theory of optical instruments [ ] is connected with probabilities in measuring wavelengths. We know from optical theory that if we have a certain wavefield it can be located because the waves interfere and annull outside. Location of wavefield is connected up with analysis in monochromatic parts. In order to have limited extension the condition is[:] outside a certain region Δx we must have different τ_x so that $\Delta x \Delta \tau_x \sim 1$, $\Delta \tau \Delta \nu \sim 1$.

Now from Einstein and de Broglie theory of light quanta $E = h\nu$, $P = h\tau$ and so we have at once $\Delta x \Delta p_x \sim h$, $\Delta t \Delta E \sim [h]$. [F]ormulae of the type of Heisenberg. They are the limitation of the possibilities of locating the wavefield in spacetime and determining the E and P in x, y, z, t from quantum relations. So far mathematics.

But if we really believe in properties of light and material particles these formulae play such large part that if we have a particle we can only find when it comes by its waves. Think first of experiments Davisson and Germer and recently of Thomson. We see that we hardly can deal with difficulties arising in connection with the measurement [ ] start from another such example given by Heisenberg. Consider now how to place an electron by means of a kind of microscope with X-rays. Just only ask what can we tell about position. If we see light we know that there was an electron. From the theory of optics how can we locate it? What is the greatest aperture, so that the smallest distance we can determine is $\Delta x \sim \frac{\lambda}{\varepsilon}$, ε = angular aperture of objective. So far we can say that the electron has been there. Now point is that if we will know something about [ ] we get into difficulties [ ] uncertainty [ ] of the velocity and momentum of the electron for we know from Compton's work there is a momentum imparted to electron which lies in the cone of the microscope but any direction. We only know that light comes in the room and we don't know at all from Compton effect in what exact direction electron gets impulse; we can say how large impulse is but this impulse will be only [ ] composed of [ ] that is not known, because we don't know whence the [ ] comes from in the cone. Consider impulse in x-direction, $\Delta p_x \sim \varepsilon h \tau$ and so $\Delta x \Delta p_x \sim h$. [ ] again. This is only an example for showing that we are in [ ] but directly connected with other formulae which express consequences of the formulae themselves [ ] essential relations $E = h\nu$, $P = h\tau$. Have we now another means for saying what that impulse is [ ] conservation of momentum [ ] Einstein recoil [ ] we could know influence [ ] we might say that if we could measure the [ ] gives [?] of the microscope [ ] we should know what [ ] left for the electrons [ ] then we might know full impulse [ ] already for the whole microscope [ ] fixed frame [ ] no possibility measuring impulse imparted to it the whole microscope connected up with [ ]. But this won't do, for either the [ ] is fixed [ ] and then we can argue on conservation of [ ] or else it is free [ ] it is a de Broglie wave itself and then we have the whole difficulty afresh [ ] not be able to know what is the place of the [ ] the whole very simple [ ] possibility of localising things [ ] we see this so simply [ ] we touch them [ ]. Suppose then we try to locate the electron with a sharp point (as we [ ] locate the microscope) [ ] just the same difficulty [ ] We can't do it because we can't place the pencil [ ] not able to say where it is [ ] have no start on it [ ] try to [ ] come nearer this problem [ ] experience embodied in this form.

Other example [ ] Heisenberg [ ] in experiment of Franck–Hertz [ ] use electrons give them amount of energy. [H]ow shall we know something about that energy.[a] These electrons fall through definite potentials [ ] but how do we know whence the electron comes. If we want to fix the time we need only close the hole for a short time [ ] but then the wave frequencies are [ ] so we cannot in any way ascribe to an

[a] Very simple sketch in the margin of electrons coming out of a hole (*eds.*).

electron coming out [ ]. You see how interwoven the whole problem is, but if you will want to know what energy the photoelectrons have you cannot know exact ν in a short time [ ] large number of waves. [ ] You cannot at all [ ] a long time [ ] is necessary.

Other example again: problem connected with Compton effect. The possibility of getting any consistency in quantum problem description lies in these wave particles relationships which due to de Broglie discovery. [ ] problem just the same as for material point.

Now I must mention a point. [ ] We deal only with free particles. [ ] These are pure abstraction, for you can never observe a free particle for observation means interaction [ ] Compton effect [ ] very simplest kind [ ] light is coming [ ] get change in impulse [ ]. We represent the electron and light by packets. I only point out small point don't go into many sides of the problem. We use in all the experiments light [ ] locate the electron by waves. [I]nner consistency [ ] The four wave packets meet in the same spacetime extension but point not well defined[.][a] For the limitation of spacetime extension connected up [ ] both for light and particles are the same [ ] We may say $(T_x) = h$ [ ] energy and time vector × position vector [ ] simple means limitation spacetime vector connected to impulse [ ] Now conservation equation (impulse energy vector) [ ]

But now in description of Compton effect we express that this vector for radiation and electron [ ] must before and after process. [ ] Some uncertainty in definition of energy and impulse vector [ ] we have the same uncertainty on both sides of equation.

[T]hen we have same extension in space on both sides. But in these various [ ] rest just a point of simple [ ] Therefore harmony in description of conservation and use of momentum relation.

Further:

BOHR. – We are rescued by a number of points which were not formulated explicitly in the theory. We completely avoided causality [ ] we cannot speak about a phase [ ] phase belonging to elementary wave [ ] but if we [ ] phase disappears because [ ] waves get out of phase [ ] and I think it is [ ] to remember consistency of quantum theory [ ] We must be very careful to avoid inconsistencies [ ] If we locate electron by light it is consistent so far that conservation theorems allow the light and particle meet before and after process in the same region. If the [ ] were not the same for both [ ] this would be. I am still postponing deep question of causality and go on to stationary states of atom.

Say a few words about important point of wave description[.] In the idea of a phase we embody the idea of an extended wave. If you have a group of waves to the group do not belong phases. There is a phase in an elementary wave. As soon as we limit the wave field, the phase disappears. In a group there is no idea of phase and the size of the group and lack of phase are essentially connected and allow us to escape contradiction in relating waves to particles. In its very definition [ ] implies that we have long train of waves, so long that waves get out of phase [ ] you cannot speak of a phase [ ] use quantum theory

[a] Figure in the margin similar to Richardson's Fig. B (p. 435) from the general discussion (*eds.*).

explicitly [ ] which has been discussed in latest paper of Heisenberg. Suppose from white light we take a narrow beam of red light by instruments. This implies we have a long train. But as regards the whole group we have no phase. Light and electrons but also motion of atoms can be used viz. in Stern-Gerlach experiment.[a] A beam of atoms or other beam comes from a hole in furnace and the beams [ ] by the inhomogeneous magnetic field [ ] we throw [ ] limited beam split [ ] atoms molecules pass [ ] properties of atoms analysed by splitting beam. [ ] It is possible tell something about most wonderful agreement with quantum theory.

But now problem of Heisenberg; he asks what is the new feature of wave theory given here. We must look in all such experiments what the wave theory tells us. We can't measure very small deflections; it must be larger than the natural diffraction of the bundle. This is very condition. Angle of diffraction given by $\alpha = \frac{\lambda}{d}$, $\frac{\text{wave length}}{\text{slit width}}$. The deflection is $\varepsilon = \frac{\delta p}{p}$, $\frac{\text{change momentum}}{\text{momentum itself}}$, not an uncertainty and $\frac{\delta p}{p} = \frac{T\frac{\delta E}{d}}{h/\lambda}$, T = time, E energy difference between A and B.[b] The condition $\varepsilon > \alpha$ is

$$T\delta E > h\,,$$

resembling much old formula. [D]o not believe it is actually the same, for we must distinguish between time and the uncertainties in time, e.g. uncertain [ ] in forming a certain point and this is not here in question. We get here $T\frac{\delta E}{h} > 1$ or, as $\frac{\delta E}{h}$ means difference of frequency $\delta\nu$ in Schrödinger wave of the [ ] in the field [ ], $T\delta\nu_H > 1$.

Now we say that we have a different natural frequency in the field. We cannot distinguish experimentally between parts A and B of beam, no possibility of discriminating between atoms and we see we do never get the difference in phase of the atoms going up and down. The condition $T\delta\nu_H > 1$ means that the condition of being able [ ] in Stern–Gerlach experiment / lost by that any knowledge of place which we may [ ] when the atoms enter the field is lost when they emerge [ ] discriminate between the various states. If we are able to give any meaning to [ ] that atom is in stationary state [ ] show that phase is lost.

BOHR.[c] – The problem in quantum theory is this. We have two sides of the question. We believe that the atom will be in one of the beams. [ ] Phase is necessarily lost when one can give a meaning to an atom being in a stationary state. As the phase is lost we may think of the atom in a stationary state and then other properties [ ] not before. If we have an atom we may illuminate it for a sec. say or for $\frac{1}{2}$sec. and so on. But a short or long train of waves is not the same and phase relations come in. But often [ ] vibration we cannot say that the atom is in a definite stationary state without contradiction. We can only put a meaning to idea of stationary state if we can really split the atoms up and then the past is destroyed. (If we split up a wave into a number of parts we must take the proper

[a] Sketch of a Stern–Gerlach experiment in the margin (*eds.*).

[b] Simple sketch in the margin of the two beams A and B (*eds.*).

[c] Here, the page numbering in the manuscript starts afresh, with the title 'General discussion (continued)'; the material still belongs to the session on Thursday morning, as can be seen by comparing with the other notes (*eds.*).

phase relationship into account.) [A]nd all knowledge of phase differences is destroyed[.] (Dirac and Heisenberg have pointed out, that if beam of atoms are illuminated long time or short time intermittently, you can never say at a moment an atom is in a stationary state. By the experiments the knowledge of the past is destroyed.)

All this is general and by considering wave [ ] we get over all troubles.

If we send α-particles through an atom, that is very difficult to connect with quantum theory because it goes very quick. After collision the atom will be in a stationary state [ ] Classical picture [ ] electron [ ] and the α-particle in such short time agreed about the future. But if we know anything about α-particle it can only be represented by wave train of great extension and the trouble disappears. (How can α-particle agree with electron in stationary state. [ ] trivial. If the α-particle is in the atom at a definite time, its energy is uncertain, connected as before with the time[.])

After some words said by Lorentz, Bohr continues:

BOHR. – I go on now on this common basis using different words perhaps. Hitherto we have spoken of consequences of the insight got by this wonderful new method of last years. But now the problem in quantum theory is this. We have avoided the great difficulty embodied in the word causality[.] If we [ ] the consequences of the wave representation, everything gets statistical to a degree [which] does not correspond to experiments. We have been speaking of question stationary states [ ] quite different sense [ ] some simple ways of speaking about individualities [ ] But if one goes deeper [ ] There is a great difference from our picture in classical theory which we get from the difference between closed and not closed systems. I think this problem is of quite fundamental. The essence of the quantum theory is the discovery any atomic process open to observation involves something symbolised by Planck's constant.

If we consider isolated system then we have no possibility of observing it. An observation involves interaction between system and the tools of observation. Classically we do not [ ] for we [ ] the interferences can be made to tend to zero.

[I]f we will speak of something definite we must have closed system [ ] at the same time we must assume we are speaking of an abstraction[.] But this is not so in quantum theory. The abstraction cannot be observed[.] If observed it interacts with agencies which don't be observed[.] If we ask for [ ] then we must allow interaction [ ] which do not belong to system [ ] means system not well defined [ ] necessity forming statistical considerations.

We say that probability [ ] sort of complementary nature [ ]

We may say [ ] spacetime description and causality are complementary[.] The idea of [ ] theory of a causal description in spacetime is at variance with the very existence of the quantum theory. Conclusion of a very simple kind example: when in matrix theory or Schrödinger $E = h\nu$ is introduced, we deal always with closed systems [ ] avoid irrationality [ ] that can to very large extent be done [ ] exclude prob[ability] of the wave [ ] we speak of close [ ] that simply means that we must try [ ] and therefore I think that in any interpretation of observation [ ] the quantum theory postulate has a place when it explicitly appears [ ] not always been sufficiently emphasised.

GENERAL DISCUSSION CONTINUED ON FRIDAY

BOHR wishes to add something to what he said yesterday.

There is one point more that is essential to any point of view.

We have looked upon two special problems[:]

1) The description of a free particle from point of view of wave theory and the consequences concerning definition of it and experimental interpretation.

2) Question of definition of stationary states. I just tried to show by a few examples that it seems that we can't get proper definition of stationary states without inconsistencies. When we say that an atom enters in a stationary state we can only give meaning to it when we actually have reason to say that experiments [ ] then the theory is such that the phase is undetermined and the past may be ignored [ ] what is to be attributed [ ]

Now the third point: emphasise importance in discussion of observation of the properties of free particles. [ ] to investigate experimentally [ ] depend [ ] interaction free particles. As soon as we don't deal with free particles, we lose possibility of representing statement in space and time[.]

I think that the fundamental importance of the method of Schrödinger is just that it draws attention to this point [ ] matrix [ ] but in wave representation [ ] so many elements [ ] we meet with the feature that these elements are represented [ ] symbolisation [ ] in space and time (represents the waves in higher extension and resign [?] to 3-dimensional space and time)[.] Due to the uncertainty which lies in description of single free particles we cannot define an atom with sufficient exactness to place it in spacetime in 3 dim. [ ] we can't all the same define [ ] potential and kinetic energy [ ] For such visualisation there is no place but still the theory fulfills all requirements.

It seems we may say the whole notion of space and time is connected to free particles[.] In quantum theory free particles just an abstraction [ ] only found by interaction[;] when observed they are not free.

This very point does not come only in the points treated yesterday but comes in every statement. Example of Born [ ] that the Schrödinger function may be used to define probability of the presence of electron at a certain point of atom. What does that mean? It means some statement regarding probability of observation [ ] to that degree electron may be considered as being free, probability of position is given by [ ] If we use some agency which acts so short a time that forces between particles and atom may be neglected, we will have a possibility of finding electrons [ ] that is what happens when we illuminate very briefly by γ-rays, or a very short collision with swift particles [ ] but as soon as we limit time[,] definition of agency is lost.

This is directly connected with a problem discussed in the past days, the problem of width of Compton scattering from bound electron. If we think that we illuminate an atom very briefly we may look apart from interaction between electron and nucleus[.] In that case during the short time we may consider electrons as free, we may say that the Schrödinger function may be resolved in terms of de Broglie waves, and during this time the electron will behave as very short time. That means time must be very short. But short time means bad definition of frequency of light used[:] $\Delta\nu\Delta t \sim 1$.

But if we ask for Doppler effect of electron, it is given by [ ] $\nu\frac{v}{c}$ [ ]. For uncertainty only Δt appear [ ] by making ν high enough we can always make this higher than [ ] $\Delta\nu\Delta t$ [ ] this means that frequency can be made as small as possible so that $\frac{v}{c}$ can be found.

It was just to call attention to certain sides of the problem [ ] define any properties of observation [ ] only [ ] of free particles.

Einstein having exposed his Standpunkt, Bohr says:

BOHR. – I feel myself in a very difficult position because I don't understand what precisely is the point which Einstein wants to [ ]. No doubt it is my fault.

[ ] few remarks about [ ] This use of more-dimensional space as described by Einstein is hardly in conformity with use I meant to mention before. I meant it as a device to avoid difficulties of a technical character to represent interaction between individuals in spacetime, but I think the difficulty is of fundamental character and this method is [ ] many-dimensional spaces are still [ ] real [?] that [ ] spacetime. In an example discussed by Born in a very beautiful way [ ] can be taken too seriously [ ] because we can't get that from that [ ] but the waves [go] out in all directions, as Einstein admits they may.

The use of these spaces is meant to specify [ ] in order to describe interaction with free particles [ ] we can describe free particles as wave packets. But when we want to describe interaction in a proper way so as to calculate statistical laws [ ] no other way [ ] use th[is] more complicated way. We take a particle and call it a wave and then cannot put it in spacetime, [ ] always fall in abstractions [ ] we cannot represent the particle in space and time and this difficulty necessitate[s] us to use more complex mathematical treatment of statistical laws.

That is first point. I do not understand what Einstein's objections are for [ ] and I do not think that to that extent the many-dimensional spaces are a purely mathematical device, but the more-dimensional is not the only place where the difficulties occur.

I would only just emphasise difficulty space and time [ ] mathematical analysis use more than 3 dimensions [ ]

As regards general problem I feel its difficulties. I would put problem in other way. I do not know what quantum mechanics is. I think we are dealing with some mathematical methods which are adequate for description of our experiments. Using a rigorous wave theory we are claiming something which the theory cannot possibly give. [ ] that we are away from that state where we could hope of describing things on classical theories [ ]. Understand same view is held by Born and Heisenberg. I think that we actually just try to meet, as in all other theories, some requirements of nature. [ ] but difficulty is that we must use words which remind of older theories.

The whole foundation for causal spacetime description is taken away by quantum theory, for it is based on assumption of observation without interference. [A]nd that is not the actual state of affairs [ ] excluding interference means exclusion of experiment and the whole meaning of space and time observation [ ] we lose all possibility [ ] because we take interaction [ ] and thereby we put us on a quite different standpoint than we thought we could take in classical theories. If we speak of observations we play with a statistical problem. There are certain features complementary to the wave pictures (existence of individuals). Just this complementary character of definitions and observation so unsatisfactory that we cannot in the description of space and time account for all features [ ] are complementary [ ] but I think, repeating what I said before, that the new

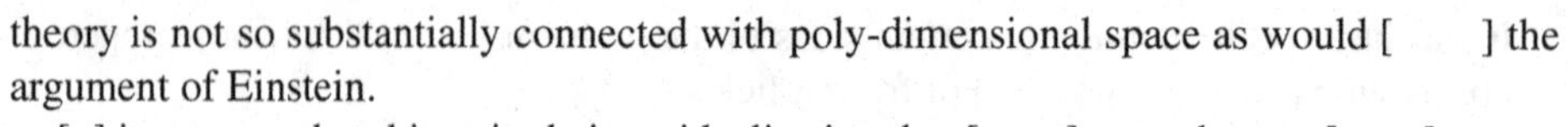

theory is not so substantially connected with poly-dimensional space as would [ ] the argument of Einstein.

[ ] just to say that this attitude is an idealisation that [ ] very abstract [ ]

I think [ ] difficulty of accounting for the interaction of individua[ls] in space and time [ ]

The saying that spacetime is an abstraction might seem a philosophical triviality, but nature reminds us that we are dealing with something of practical interest. Depends on how I consider theory. I may not have understood, but I think the whole thing lies [ ] theory is nothing else as a tool for meeting our requirements and I think it does.

Dirac having exposed his views:

BOHR quite sees that one must go into details of pictures, if one wants to control or illustrate general statements. I think still that you may simpler put it in my way. Just this distinction between observation and definition allows to let the quantum mechanics appear as generalisation. What does mean: get records which do not allow to work backwards[?] Even if we took all molecules in photographic plate one would have closed system. If we tell of a record we give up definition of plate. Whole point lies in that by observation we introduce something which does not allow to go on [ ]

One given [?] experiment to consider and at that moment we introduce [ ] by observation we actually mean trying to stop before [ ] we bring ourselves free will [ ]. How does one know that the waves are not going [ ] put a photographic plate in [ ] and try to [ ] then we exclude what we call observation [ ] and the rest lies in the techni[que]s of the quantum theory.

GENERAL DISCUSSION CONTINUED ON SATURDAY

Nachdem Born hat gesagt: ‘Es ist nicht richtig, dass man von der Energie eines Teilchens reden kann, nur von Energie eines Systems’, sagt

BOHR. – Es ist nur eine scheinbare Schwierigkeit. Bin derselben Meinung wie Born, muss es aber anders formulieren.[a] Man kann das Elektron weit vom Kern nur nachweisen, wenn man Energie zuführt, z. B. man konzentriert Strahlung in ein kleines Gebiet. Die Frequenz muss sehr gross sein, und das bedeutet grosse Zufuhr von Energie. Rotes Licht, das nicht aufs Elektron wirkt, kann eben nicht benutzt werden.

(L’onde de Schrödinger va à l’infini. On ne peut pas prouver par une expérience que l’électron est dans un certain volume. Pour le prouver il faudrait un choc contre autre chose ou bien en concentrant la lumière sur un point de l’espace où on veut observer. Si on est loin du noyeau on pourra définir avec moins de précision le volume [ ] employer lumière rouge (pour éviter effet Compton qui change énergie) alors il faut observer pendant longtemps pour définir la lumière rouge. Je ne peux pas savoir à un instant donné où est l’électron, je pourrai seulement observer que les formules ordinaires de dispersion [ ].[)]

[a] Figure in the margin clearly related to Richardson’s Fig. D (p. 467) from the general discussion (*eds.*).

Verschaffelt's working notes for the discussions[a]

DISCUSSION DU RAPPORT DE M. COMPTON

1. LORENTZ. – On the ether.
2. COMPTON agrees with Lorentz that photons are not enough. Waves don't carry the energy which must [be] carried by photons.
3. BRAGG. – About Williams experiments (already received).
4. WILSON agrees that the forward momentum is considerably in excess of $\frac{h\nu}{c}$. Bragg 0.5–0.7 Å in O_2 and N_2. Auger about the same[.]
5. BORN. – What do we expect in this case? Why 1.0 $\frac{h\nu}{c}$ to be expected?
7. COMPTON. – About same question (already asked).
8. DEBYE. – What is the reason why you think rest of atom does not . . .
9. COMPTON. – Experiment.
13. DIRAC worked out motion of compelled electron in arbitrary field of force and incident radiation. Then the forward momentum increases at a rate (increase of energy) at each moment.
14. LORENTZ. – But that applies to force electron[.]
15. DIRAC. – Not too bound also[.]
16. BOHR. – What is a bound electron?
17. DIRAC. – There is a principle . . .
18. BORN. – Wentzel takes uniform light wave on Schrödinger atom and he says that for very short waves $\frac{h\nu}{c}$ is the forward momentum. He assumes that diffraction in optics / Schrödinger atom taking account electric force / resonance point are not [ ] he says that only short wavelengths / in any other case resonance and so may be explained in this way [ ] we should then find other effect.
19. BOHR asks for what range of wavelengths Williams experiments were made.
20. BRAGG. – Only 0.5–0.7 Å units. We may not speak of variation of this with wavel[ength]. [O]ne can hardly say there is variation.
21. DIRAC. – About classical transfer of momentum compared to energy transfer. At each instant the [ ] rate of change of energy [ ] orbit that electron describes.
22. BOHR. – We get some dissymmetry[.]
23. EHRENFEST. – Example: if you have a box with a small hole / reflection diffused in some way [ ] box explodes / part flows afterward.

You speak of field of force.

24. BRAGG. – Kickback will just compensate.
26. BOHR speaks of elementary points raised by Compton (already asked).
29. EINSTEIN demande à Bohr s'il ne pourrait pas exprimer sa pensée en paroles ordinaires sans recourir à des formules mathématiques difficiles à interpréter. Si l'on pouvait représenter les faits d'une manière tant soit peu enfantine, la question deviendrait peut-être plus claire.[b]

[a] Some obvious amendments have been made tacitly, and some of the spelling, capitalisation and punctuation have been amended. Small amendments have been made also to formulas, where the context made it clear that, e.g. a typed v should be ν, or in order to at least partially uniformise the notation with that in the main text. Other amendments are suggested in square brackets. The sign [ ] denotes a clear blank space in the typescript, the sign [ ] denotes a small one or an uncertain one. The sign / appears in the original (*eds.*).

[b] This contribution is crossed out in Verschaffelt's typescript (*eds.*).

30. BRILLOUIN. – Fr. Perrin et Auger (déjà reçu).
31. LORENTZ. – Il s'agit bien des photo-électrons? Comment passe-t-on des grands cercles au petit cercle?
32. BRILLOUIN. – Par superposition des deux.
33. EHRENFEST. – Two different planes, polarisation plane and wave plane[.]
35. LANGEVIN. – ?
36. BRILLOUIN. – On suppose qu'il y a symétrie autour du vecteur électrique.
37. COMPTON. – Experiments of Bubb and Kirchner (already asked).
38. BRILLOUIN. – Wentzel gives same expression.
39. COMPTON. – As 37.
40. LORENTZ. – Remark on basis of old theory. If you had nucleus and electrons and beam polarised light; the[i]r [?] force direction... and would impart angular momentum [ ] then after all you would find any momentum to be conserved[.]
41. COMPTON. – But in case complete absorption of radiation the angular momentum of system must be conserved.
42. LORENTZ. – When only... then no angular mom[entum]. [Y]ou are right.
43. RICHARDSON. – On the equations of momentum and energy (already asked).
44. LORENTZ. – That term will be extremely small. MV^2 will be much smaller than mv^2.
45. RICHARDSON. – One part in 1900, not to be neglected.
46. KRAMERS. – Question on $\mu = \tau + \sigma$ (already received).
47. COMPTON answers (already asked).
48. BRAGG. – Interesting to see how inaccurate X-ray μ-measurements are. This is very weak experimentally.
49. EHRENFEST. – Are the modified σ and not modified by Compton effect measured together? Monochromatic?
50. COMPTON answers (already asked).
51. PAULI. – How great is the broadening of the modified lines?
52. COMPTON explains (already asked).
53. PAULI macht eine Bemerkung über Verbreiterung der Linien (schon gefragt).
55. DEBYE. – Ein Ding habe ich nicht verstanden, nämlich wie die Endlichkeit des Atoms eine Änderung der λ gibt.
56. PAULI. – Oszillator [ ] man braucht eine [ ] dann kommt die Verschiebung heraus. Diese Sache entspricht einem Gitter.
59. EHRENFEST. – What is precise difference between Pauli and Bohr?
60. BOHR. – If you do use [ ] you disturb spectrum [ ] that falls on new grating / that grating gives rise to not sharp lines.
61. EHRENFEST. – Because the grating you use in atom moves and it moves because it is [ ] shot [short?].
62. BOHR. – No, because it. [ ] you get a moving grating in a free space.
63. EHRENFEST. – Warum läuft dieser fiktive Gitter [ ] sondern läuft es, dann kriegen wir unendlich lange [ ] de Broglie aber das Laufen schadet nicht [ ] Atom / das Gitter ist kurz und es läuft.
64. BOHR. – [ ] diese Verteilung.
65. EHRENFEST. – Und dieser Schwebungseffekt zwischen... [ ] durch die Gruppe [ ] und das schlägt die Brücke zwischen 2 Sprachweisen.

66. BORN erklärt Gruppengeschwindigkeit beim Elektron als verschieden von den ρ-Schwebungen, die den Compton-Effekt geben.
67. COMPTON. – Jauncey has calculated broadening of modified lines by first of Pauli's methods using v given by Bohr's theory. He obtains greater width than observed.
68. BOHR. – Pauli showed merely the connection between two points of view, not the connection between the old theory and the new. It was so to say a) a corpuscular description [b]) a wave theory description. Pauli was not referring to [ ] but to the two new points of view: electrons and waves. Pauli's velocity meant a velocity in the new matrix (?) sense and can be related to [de] Broglie waves and with the [ ] ? [ ] for Compton effect that Pauli pointed out.
71. LORENTZ. – Je n'avais pas bien compris M. Pauli. J'avais cru qu'il s'agissait d'un calcul d'après la théorie ancienne et d'un calcul d'après la théorie ondulatoire.

Il s'agit en réalité de deux calculs d'après la nouvelle mécanique.
72. CURIE expose quelques questions.
76. EINSTEIN. – Ich möchte wissen, wie man findet... [ ] wie begrenzte und [ ] miteinander in Interferenz kommen.
77. PAULI antwortet darauf (schon gefragt).
79. LORENTZ. – Man hat am Ende mit 4 Wellenzügen zu machen 1 Elektron im Anfangs-, 1 Quantum im Anfangs-, 2 im Endzustande.
80. PAULI. – ?

Pendant que Schrödinger dessine au tableau son système d'ondes.
84. EINSTEIN. – Kleine Frage bezüglich Compton-Effekt [ ] Störungsproblem. Es wundert mich, dass hier was Endliches kommt.
87. SCHRÖDINGER explique son dessin.
88. LORENTZ. – Do the two sets of waves coexist allthrough, or does one set of waves spring suddenly into existence at the instant of collision?
89. BOHR. – Klein has tried to give a procedure which allows you to trace connection with classical theory. It is not inconsistent to take the two waves but has classical analogy.

These principles are formal representation of classical theory. We cannot trace too close an analogy.
94. BORN says a word about that question (schon gefragt).
95. LORENTZ. – Obvious on general grounds problem undeterminate / there must be a third parameter / case of two small spheres. We had to know angle of contact[.]
96. BORN. – Quantum mechanics makes no use of microscopic coordinates[.] We only use [ ] ? [ ] we can [ ] ?
100. LANGEVIN. – La mécanique des quanta permet de déterminer la probabilité des déviations de diverses grandeurs. On peut faire la comparaison avec la mécanique classique où l'on détermine la probabilité des déviations après le choc. La nouvelle mécanique doit donner les angles et la probabilité...

DISCUSSION DU RAPPORT DE BORN-HEISENBERG

1. DIRAC attire l'attention sur l'étroite analogie entre l'ancienne et la nouvelle mécanique (already asked).
5. LORENTZ exprime son étonnement de voir que les matrices satisfont à des équations de mouvement (déjà demandé).

6. EHRENFEST. – Potential, for instance.
7. LORENTZ. – On peut changer les phases sans que les équations de mouvement cessent d'être satisfaites (déjà demandé).
8. BORN. – Es ist so etwas wie mit komplexen Zahlen (schrecklich komplex!), die auch den gewöhnlichen Zahlen entsprechen [ ]. Difficulté à former les f. H. (?) [fonctions Hamiltoniennes] des matrices par suite de non-permutabilité.

Aus klassischen Gesetzen niemals exakt schliessen.

9. LORENTZ. – Auf klassischen Gesetzen schützen [stützen] / wunderbar, dass das geht.
10. HEISENBERG répond à la remarque de Lorentz: c'est qu'on a voulu formuler la correspondance.
11. DIRAC. – À propos de la 2me remarque de Lorentz: quand on exprime les p et q en fonction des variables angulaires élémentaires il reste pour chacune une phase indéterminée. C'est le $\alpha_n\alpha_m$ de Lorentz. The arbitrariness in φ's corresponds exactly to arbitrariness in classical theory in w.
12. LORENTZ. – You do not feel what I felt, that this is an induction of [there] being behind all these things something hidden.
13. BORN. – Ces $\alpha_n\alpha_m$ sont les phases relatives à chaque niveau d'énergie [ ] ou en explication ondulatoire les phases des ondulations Schrödinger de chaque niveau.

Die β sind zugeordnete Zustände. Aber ich glaube nicht, dass dahinten noch etwas ist.

14. LORENTZ. – Sie sind weniger vorsichtig als...
15. HEISENBERG. – ?
16. LORENTZ. – Ich wollte es allgemein ausdrücken.
17. BOHR. – I understand quite Lorentz's point, but I think that it has partly the reason stressed by Dirac. It has not very much to do with the possibility of defining the things we are dealing with and it is really a point for the general discussion.
18. EHRENFEST. – Es ist doch gewiss ein Glück, dass nicht genau gemessen werden kann, denn sonst könnte man die Resultate nicht wiedergeben mittels der Quantenmechanik.
19. BOHR. – Es ist gut, dass man die Ungenauigkeit betont.
20. EHRENFEST treibt es auf die Spitze, indem er sagt, dass...
21. DEBYE. – ?
22. LORENTZ. – Bohr behauptet, dass in keiner künftigen Theorie p und q genau gemessen werden können, [ ] dass p und q scharf widersprechen würde die Begriffe, mit denen man die Messungen interpretiert.
23. PLANCK. – In der Relativitätstheorie könnte man Gleichzeitigkeit doch genau messen. Also die Analogie ist falsch.
24. HEISENBERG. – Die Analogie kommt bei anderer Betonung zum Vorschein.
25. WHO? beschreibt durch Kennards Rechnung, dass die Sache sich ausbreitet wie Schrot.

DISCUSSION DU RAPPORT DE SCHRÖDINGER

1A. EHRENFEST ET SCHRÖDINGER donnent des explications sur le sens de la photo instantanée de Schrödinger (zu entwickeln).
1B. BORN fragt, wie man eindeutig mit $\psi\psi^* d\tau$ im dreidimensionalen Raum eine Ladung definiert.
1C. BOHR. – Is it your view that that takes much away of your...

Do you mean that you can still represent densities?
2. LORENTZ. – Bohr's idea is: we can really explain the intensities / very good idea of radiation if you take...
4. BOHR. – I do not think that we can get the intensities in the first way.

You say that by this distribution $\psi\psi^* d\tau$ and classical laws get the frequency and intensity of the light, but your later remarks on difficulties, do they indicate that what you get above is not exact? The difficulties are so large that you may...
5. SCHRÖDINGER. – I meant that the difficulties are purely logical. If you form $\psi\psi^*$ you get $c_k c_k e^{c(\nu_k - \nu_k)}$ and the intensity is $(c_k)^2(c_{k'})^2$ when perhaps it ought to be $(c_k)^2$. In certain cases it seems that this radiation...

It is a logical difficulty, but as a matter of fact we can work out intensities and frequencies accurately quantitatively.
6. BOHR thinks that latter remark about radiation is in contrast with his former description of radiation. I think Dirac has got over the difficulty. I think there are serious difficulties in the first way of explaining intensities, but in the other way Dirac has succeeded. The difference between the two methods explains this.

Do difficulties in principle of which you have spoken prevent your working out intensities and frequencies of lines?
7. SCHRÖDINGER. – I can't believe that these difficulties are such that the proposed densities are meaningless.
8. BRILLOUIN. – Il faut mettre dans ψ toutes les énergies inférieures[.]
9. BORN. – Ist es richtig zu schreiben [ ]...? Il critique la manière dont Schrödinger calcule la densité électrique en 3 dimensions[.] [L]ui ne pense qu'en $3n$ dimensions (schon gefragt).
10. SCHRÖDINGER behauptet, es ist dasselbe.
11. LORENTZ. – Cette question n'est pas si facile à trancher. Nous y reviendrons.
12A. FOWLER demande des explications sur la méthode de résoudre les équations dans le cas des problèmes à plusieurs électrons.
12B. DE DONDER. – L'équation (24) peut être étendue au cas où les n particules électrisées sont différentes etc... (déjà reçu).
13. LORENTZ. – Vous considérez un cas plus général, mais les résultats s'accordent.
14. DE DONDER. – Quand il y a interaction, le problème est beaucoup plus compliqué. Même en négligeant les interactions la solution est encore très grossière.
17. LORENTZ. – On néglige les actions électromagnétiques[.] Pour cette raison pas de potentiel vecteur.
18. BORN a fait des calculs numériques des intégrales; calculs approchés et difficiles (schon gefragt).
20. LORENTZ. – Do you consider results of this long work as satisfactory?
21. BORN. – Experiment is better.
22. LORENTZ. – You get right results in concentration of helium.
23. BORN. – For this approximation 7%, that is the disagreement with experimental number.
24. LORENTZ. – But you hope it will become better.
25. BORN. – Convergence very slow.
26. HEISENBERG. – [Does?] [t]his approximation give hope we can get theory in 3 and many dimensions (schon empfangen).

27. SCHRÖDINGER. – I don't specialise my hope of 3-dimensional calculation, but I hope to look through [understand] better my own equations, so that it can be formulated 3-dimensional.
28. HEISENBERG. – Is not your hope already fulfilled by matrix mechanics?
29. SCHRÖDINGER. – I don't look for a 3-dimensional differential equation precisely.
30. HEISENBERG. – Then why aren't you contented?
31. SCHRÖDINGER. – I am not content because of the [ ]? of the electrostatic potential. I hope there will be possibility of putting the equation in a simpler form.
32. HEISENBERG. – This is a pure mathematical problem / of principal axes / of quadratic forms and we can get the number of variables reduced.
33. SCHRÖDINGER. – ?
34. HEISENBERG. – You seem to suggest that the equations you have given have led to something quite new, but I do not see this.
35. SCHRÖDINGER thought there might be some deeper interest. Unsöld uses it. Fowler told us there is another method of approximation: method of Hartree. This method, apart from the neglect of exchange terms, is still more 3-dimensional / it might be referred to as a really 3-dimensional method.

It has objections, of course, but I mention it as showing that some modifications of it, or extension of it might lead to a satisfactory 3-dimensional model.

If it were correct I should like it better.

36. HEISENBERG. – So should I, but I don't see any argument in that against an infinite number of dimensions. That does not show that your exact calculations may lead to a satisfactory 3-dimensional model.

It cannot be done. Certain problems can only be described by n variables. Approximation from classical model.

38. BORN. – Consider an atom with 3 electrons etc. (schon gefragt).
39. BOHR. – I do not see that you can ever explain.

The 3d electron is subject to the rule of Pauli and therefore the first of electrons don't merely act as a charge [ ] out [?] you can in 3 dimensions to explain how the periodic table is built up.

Another point. Can you say anything about the observation of Wood on resonance radiation?

40. ? – Answer?

DISCUSSION GÉNÉRALE (JEUDI)

2. LORENTZ montre les difficultés qu'on rencontre dans la théorie ancienne.
3. BOHR will try to explain as simply as he can his attitude to problems of quantum theory.
4. BRILLOUIN. – Cette incertitude des mesures est très vieille, elle est l'illustration nette du sens de la notion des cellules d'extension en phase de grandeur h, introduites depuis longtemps par Planck.
5. BOHR. – Yes, through Planck's discovery $E = h\nu$ which connects things of different kinds. All our measures of h such as Planck's earliest are immediately bound up with these uncertainties. / use new constants which have dimension [ ] problem is – because are [u]ndefined.

6. KRAMERS. – Why do you stress the point of the uncertainties / equation of conservation of momentum before the collision.
7. BOHR. – It is just to remark that... [ ] absolute new thing in theory [ ] inconsistencies [ ] these two things go together [ ] it is consistent so far that [ ] then we [ ] have to give up the idea [ ] I meant only [ ] impossibility of starting from an initial state determined by x and p_x.
8. DE DONDER fait un rapprochement entre ce qu'a dit Bohr et la relativité (déjà reçu).
9. FOWLER asks about uncertainty $\mathcal{I}$-vector. The point is the two types vector...

In the case of Compton effect, Bohr says that all we know of the relation of momentum is that uncertainty is the same before and after.
10. BOHR. – I only wanted to point out a trivial thing, which one ough[t] to have understood always. Already in this [ ] we must [ ] certain action [ ] but just as De Donder says it is very [ ] that just according [ ] all these notions fulfill requirements of relativity / they do not contain mass [ ] we feel in quantum theory everywhere [ ] I feel also we have had a great deal of luck in quantum theory [ ] wonderfully connected up.
11. LORENTZ. – But that is due to the fact that after all the theory is a very good one; the good luck is due to the strong foundation of it.
12. BOHR continue l'exposé de son point de vue (demandé).
13. LORENTZ pose une question relative aux matrices. That has nothing to do with this... [ ] virtual vibrations. I think of stationary states. In matrix theory there are no stationary states.
14. BOHR. – The connection is not exact... [ ] used in a way which allows of [ ].
15. BORN. – It is the same [ ] two positions [ ] and opposite one [ ] to describe by matrix elements in row by a matrix [ ]. The phases which occur there are really meaningless.
16. BOHR. – In the calculation of a way [path] we must have idea of phase but this idea comes never in the experiments and therefore there is a possibility of individuals. In the idea of a wave we must have a phase. In a stationary state we cannot have a phase [ ] these arguments allow us just the [ ] to make one [ ] feature of the two.
17. LORENTZ. – When you want to have definite symbols [ ] I see that these phases [ ] in all experiments can never show themselves [ ] I understand that.
18. BORN. – ?
19. LORENTZ. – Question. Let us consider motion of an electron in Stern–Gerlach's experiment but simply [ ] how is that accounted for in wave theory? The ψ-waves ought to be deflected [ ] probability waves also [ ] and now can we really introduce terms which give that deflection? [ ] Can one get this deflection out of the wave equation [ ] vector potential as Schrödinger [ ] can you consider limited beams of these probability waves?
20. BOHR. – Yes, this has been discussed by Darwin in a forthcoming paper [ ]. Just one point is simple. It looked for a time as if we had a difficulty here in the wave theory [ ] stream of atoms [ ] but comes so simple. [ ] The Verzweigung was difficult. [ ] If we take the most general solution we can begin to [ ] (as it is a sum of simple solutions) that it will all [ ] up in the wave required. We can by superposition just represent the thing. [ ] Schrödinger [ ] stationary states [ ] wave calculation / we believe in atoms / only in one of these beams.

21. LORENTZ. – In old days when was question of cathode rays [ ] Lenard [ ] (Jaumann?) suggested idea of longitudinal waves and explained the bending of the rays in a magnetic field.
22. BOHR. – You cannot say atom is in stationary states[.]
23. LORENTZ. – Because you can never know it.
24. BOHR. – Particles sent through atom, only to be represented by wave train so that its energy is uncertain.
25. LORENTZ. – That would be maintained in theory as exposed.
26. BOHR. – There was no difference anywhere in the two views[.]
27. LORENTZ. – Much a question I spoke of [ ] I should say that the ψ-waves of the α-particle are given and we explain the result by probability assumptions, so that this is no contradiction. I feared it very much...
28. BOHR. – On causality (demandé).

DISCUSSION GÉNÉRALE (SUITE VENDREDI)

1. BOHR adds something to what he said yesterday (demandé).
2. BORN. – Bemerkung über Frage [der] Wahrscheinlichkeitspakete in mehreren Dimensionen (demandé).
3.–5. EINSTEIN expose son point de vue dans diverses questions (demandé).
6. EHRENFEST. – In welcher statistischen Theorie ist es notwendig, zwei Partikel als eine einzige...
7.–9. EINSTEIN. – Antwort? Lässt sehen, wie es geht, wenn man nicht mehrdimensionell arbeitet. Die Theorie stellt nun dar, wie gross die Zeit ist, um nach 1 und 2 zu kommen. Dann nach Bohr-Kramers-Slater könnte sowohl 1 wie 2 beinflusst werden.
11. BOHR précise son point de vue (demandé).
12. LORENTZ fait une remarque à propos du mouvement d'un système de points matériels (demandé).
13. EINSTEIN. – Doch besteht, was ich meinte, denn [ ] zu formulieren [ ]. Es ist nicht die Sache [ ] aber [ ] die Schwierigkeit wird nicht ihrem Wesen nach beseitigt.
14. BOHR fasst seine Rede zusammen. Dieser Kunstgriff des mehrdimensionellen Raumes...

Ich glaube nicht... [ ] Man braucht diesen Gesichtspunkt, um gewisse Sachen quantitativ zu beschreiben. Aber es ist nicht so gemeint, dass es derart...

Ich glaube gar nicht [ ] Erfahrung [ ] wenn man genau beschreiben will Anschluss an 3 Dimensionen [ ] bedeutet dass [das?] [ ] die quantentheoretische Schwierigkeit liegt viel tiefer.

Alles, was anschaulich ist, liegt in den freien Teilchen.

15. BORN will populär ausdrücken, was Bohr meint. Wir sind gewohnt, mit anschaulichen Bildern in Raum and Zeit zu operieren. Bohr glaubt nun, $E = h\nu$ genügt uns... [ ] in der Natur möglich ist... ν ist nur definierbar, wenn über ganzen Raum erstreckt ist [ ]. Abstraktion ist... [ ]. Bohr meint nun rein technische [ ] nicht so [ ] schreiben zu können [ ] abstrakt mathematisch [ ] und jedesmal, wenn wir auf Zeit und Raum müssen wir überlegen [ ].
16. EHRENFEST. – I regret that Bohr has not said more precisely one thing: in 3 dimensions we cannot describe interaction of 2 particles because... [ ] kinetic and potential energy of the 2 particles.

17. BOHR. – Alle Punkte hängen so zusammen... [ ] die Worte, die man darüber benutzt... [ ] um Richtung anzugeben... [ ] Eine Weise einsehen [ ] wie Kunstgriff [ ] ob Beschriebung Sinn hat [ ] aber dass es so einfach damit zusammenhängt, dass wir nicht in Raum und Zeit überhaupt die Wechselwirkung zweier Partikelchen [ ] überhaupt nicht möglich, die gesamte Energie anzugeben... [ ] Ein Partikelchen kann ich sagen, wo es ist..., aber wenn ich 2 habe, dann muss ich abstrahieren und kin[etische] Energie...
18. EHRENFEST. – Zwei Wellenpakete [ ] wo sie sind und wie sie laufen [ ] aber wenn ich sie scharf mache... [ ] wenn ich sie aber eine Geschwindigkeit... [ ] unscharf. / Kinetic energy comes not in the differential equ[ation]... [ ] At the first view it looks simple... but we [ ] because we cannot define kinetic and potential energy at the same time... [ ] but in *n* dimensions the difficulty does not exist.
19. BOHR. – Wir bedürfen Abstraktion [ ] dass lässt unsere Mittel nicht zu [das lassen unsere Mittel nicht zu,] Platz in Raum und Zeit anzugeben [ ]. Möglichkeit, kausale Beschreibung zu machen, haben wir die Mittel dazu.
20. PAULI ist vollkommen derselben Meinung wie Bohr und Born.
21. SCHRÖDINGER. – Was ist das für ein Dipol? Warum schwingt es?
22. PAULI erklärt: Er hat es mit Matrizen gemacht, [ ] wie Debye weiss [ ] hat dasselbe vieldimensional gemacht. Nach dieser Methode berechnet kommt etwas heraus, was auch herauskommen würde...
23. LORENTZ. – Kommt das darauf hinaus, dass, wenn... [ ] keine lectr [elektrischen] Momente [ ] keine Wirkung [ ] aber, wenn mehrdimensional, dass dann von selbst [sich] etwas herausstellt, das man mit Momenten vergleichen kann?
24. PAULI. – Ja[.]
25. BORN. – Man kann es auch so auffassen [ ] also durch Störung alle anderen Zustände angeregt sind.
27. DEBYE. – Man kann aber sagen, dass die 2 Atome zusammen in einem Grundzustand sind[.]
28.–30. LORENTZ. – Da trennt man nicht so von vorne herein... [ ] Vieldimensionale Auffassung führt also zu viel besseren Resultaten.

Je propose de ne plus continuer cette discussion; il y a certaines questions sur lesquelles M. Dirac pourrait nous donner des explications.

31. DIRAC gives his ideas thereupon (already received).
32. BORN. – The idea is this one[.] Dirac goes out [begins] from definition of electrical density [ ] seems very interesting and [von] Neumann has sent me a long manuscript in which he constructs the whole theory with common probabilities. All [ ] with diagonal elements of matrix. It seems to me very important because [ ] has shown whole theory may be constructed without using amplitudes.

(Matrix element is... [ ] but [ ] density is not enough for higher... [ ] interaction as Debye's quadrupole attraction in a gas. I see that the important point is that from a definition of electric density of packets in a given volume and not from... This is very important and... [ ] von Neumann has done something of the same kind.)

33. DIRAC answers?
34. BORN. – What you have done is nearly the same.
35. KRAMERS. – I think calculation can be put in easier form, but it is too long to go with now.

36. LORENTZ. – Dirac means improvement you want is just this one.
37. DIRAC continues (already received).
38. BOHR. – I think that Dirac's remarks may be put in simpler words as I tried. The distinction between observation and definition allows us to describe actual state without making a one-sided departure from classical ideas. [What] I meant by emphasising complementary view above was only to try to let the whole view as a rational generalisation of spacetime description. What does it mean that we get a record on a photographic plate? The whole thing that lies behind observation . . . [ ] we would know everything . . . [ ] we had to take all into account[.] (A record which stops the causal . . . [ ] brings in the whole difficulty of whether behind observation . . . / If we know everything about the . . . [ ] photographic plate we should then have a closed system and then the rays might interact back and we should not have made the observation.)
39. LORENTZ. – Do you mean to say that. . .
40. BOHR. – No, we can never be sure that.
41. LORENTZ. – But that action of the reflected rays. . .
42. BOHR. – One gives up to consider and at that moment we introduce. . . by observation we actually mean trying to stop before. . . we bring ourselves [ ] free will [ ]. How does one know that the waves are not going [ ] put a photographic plate in it [ ] and try to [ ] then we exclude what we call observation [ ] and the rest lies in the techni[que]s of the. . .
43. LORENTZ. – Would it be possible in order to give an idea of general question not to consider physical phenomena but. . . [ ] Dirac's view may perhaps be expressed in this way[.]
44. DIRAC. – All numbers have been described by a) action of free will on experiments, b) results afterwards.
45. LORENTZ. – We could think of succession of things A, B, C. . . in that sequel nature makes a choice. According to causality the[re] would be one series. . .

Bohr says it is more or less equivalent.

46. BRAGG. – Bohr's point of view is like killing of any animal in order to examine it. [L]ook at it with a microscope.
47.–49. BOHR. – What does it mean. . . [ ] if x and $\dot{x}$ are undetermined; how can you speak of record? [ ] We never. . . You fix. . . [ ] for the rest of the world, but what do you fix. . . [ ] events connected [ ] then stop.
50. DIRAC.— You cannot stop world going on[.]
51. BOHR. – Nature makes a choice.
52. HEISENBERG. – If . . . [ ] then nature would never make choice. Only if you have an observation at that moment. . .
53. BOHR. – Choice of making it observable.
54. LORENTZ. – There seems to be a fundamental difference of opinion about the meaning of these choices made by nature. I think it means. . . [ ] admitting the possibility of nature making choices it would mean we could never know it beforehand. . . indeterminism. You would admit it as a principle, these are events we cannot predict while we thought till now that may be done.
55. DIRAC doesn't agree with Heisenberg quite.

FIN DE LA DISCUSSION GÉNÉRALE (SAMEDI)

1. LANGEVIN. – Comparaison entre les statistiques ancienne et moderne.
1BIS. HEISENBERG. – You can always have several statistics. Question of interaction in first approximation independent of statistics. In certain way Einstein–Bose better for photons, Fermi–Dirac for particles. This might be the main difference between radiation and matter. However, interaction is not an adequate description; think of magnetic electron for instance. Also neutral molecules will have Bose statistics in first approximation... more argument of feeling... [ ] not quite good[.]
2. BOHR completely agrees with what Heisenberg says. I think it is just as Heisenberg points out, that actually the various statistics correspond to complete quantum mechanical solution, complete multitude of states which can never combine (explanation of this). Most beautifully illustrated by work of Dennison, anomalies of specific heat of H_2 at low temperatures. As Heisenberg says[,] it is not possible to get any possible argument for different statistics. Pauli has strongly emphasised that point... [ ] circumscription of classical theory into quantum... [ ] just these problems deal with questions which cannot be solved by classical analogies[;] not guide in these theories[.]
3. LANGEVIN. – Dans le cas des photons c'est bien la statistique de Bose-Einstein qu'il faut appliquer et aucune autre.
4. BOHR. – Problem is simpler... [ ] we have no sufficient...
5. KRAMERS rappelle les recherches de Dirac. Hohlraum (déjà reçu).
6. LANGEVIN. – On attribue des degrés de liberté aux photons[.]
7. KRAMERS. – Statistique de Fermi-Jordan (déjà reçu).
8. DIRAC. – Cela obligeait à admettre des lois étranges de quantification. Il fallait adopter d'avance des valeurs 0 et 1 pour certains nombres. Jordan a dû changer un signe dans les formules: il y a des erreurs dans ce mémoire.
9. KRAMERS. – N'est-ce pas une question de goût[?]
10. BORN. – On new theory of Dirac the fluctuations of energy come out in the right way, not in the old theory of Debye. Not in conformity with Planck's law but [ ] no difference.
11. EHRENFEST. – I regret Heisenberg did not say some words more on...
12. HEISENBERG. – Proton et électron ont tous deux un spin et une statistique Dirac; ces deux points semblent liés. If we take two particles together... H_2 atom Bose–Einstein statistics is true because... [ ] in other words, Bose–Einstein statistics holds for every neutral molecule... [ ] charge with even number of electrons and Fermi–Dirac for odd number. He nucleus has no spin because contains 6 particles[.] Very easy to say for every set of even number of particles.
13. FOWLER asks if the things fit in with helium bands and intensity changes[.]
14. HEISENBERG. – In He-bands [ ] slightly together... [ ] symmetric and asymmetric [ ] difference.
15. FOWLER. – You cannot tell?
16. HEISENBERG. – No[.]
17. SCHRÖDINGER. – You mentioned an experimental proof for the assumption that the proton has a spin just as the electron and that the protons obey the law of the Fermi–Dirac's statistics. Which proof do you mean?
18. HEISENBERG. – Work of Dennison (déjà reçu).

19. EHRENFEST formulates a question (reçu).
20. HEISENBERG answers (déjà reçu).
21. EHRENFEST. – If it is so complicated, I retract...
22. RICHARDSON. – Evidence of spin much more complete than Heisenberg said. All visible bands of the spectrum of H_2 which have been investigated belong to normal H_2 molecule... [ ] will be explored [ ] confusion... lines mixed up. Quite definite alternation of intensity in successive lines. [ ] McLennan found ratio of intensities equal to 1–3 (5%). Is lot of experimental eviden[ce.]
23. LANGMUIR should like to see a parallel established between electrons and photons. What characterises an electron? A definite charge. What a photon? Velocity perhaps.

What is the analogy, what the difference? Electron: de Broglie waves, photon: electromagnetic waves. In some respect... [ ] perfectly clear what relations are. Cela va-t-il jusqu'au bout? Quelles suggestions d'ordre expérimental?
24. EHRENFEST adds one question. Is astonished whether exists... [ ] Ellipticity of polarised light invariant against Lorentz transformation, but axes are variable; clear relation of analogy and contrast with electronic spin; as much [many] photons as there are elliptical polarisations.
25. COMPTON. – Is ellipticity of light possible when photon has an angular momentum?
26. EHRENFEST. – I should rather say the opposite. Invariance of b/a just made me think that photon could not... [ ] Uhlenbeck–Goudsmit [ ] introduced spin. Can we graphically represent polarisation? I felt intensely hampered to do it because... [ ] running together with electrons... and then spin... [ ] but with photon because not run so fast. Photon has momentum, not mass.

Other point. I was offended by fact... [ ] photon of radio wave in principle not different from photon in Kohlhörster–Millikan wave.
27. LORENTZ. – Can you make them identical by such transformation?
28. EHRENFEST. – ?
29. LORENTZ. – In other language, they lie in the same direction.
30. EHRENFEST. – ?
31. LORENTZ. – If you apply transformation to both of them can you make them to have equal frequency and equal...
32. EHRENFEST. – Yes.
33. LORENTZ. – Then there will always remain difference in direction of motion[.]
34. EHRENFEST. – ... [ ] by classification we have remarked...
35. HEISENBERG. – About ellipticity. Pauli and Jordan ha[ve] tried to obtain resemblance between spin and polarisation of light. Must be difficulty because spin can only have two directions and light waves three $(+\frac{a}{2}, 0, -\frac{h}{2})$ [*sic*].
36. EHRENFEST. – Ellipticity is a scalar!
37. HEISENBERG. – ?
38. EHRENFEST. – Do you speak relativistic or not? Pauli has said he tried not to make his work relativistic... [ ] magneton... [ ] photon runs so hard.
39. HEISENBERG. – ... [ ] some arbitrariness in...
42. PAULI antwortet auf Langmuirs Frage (gefragt).
43. LORENTZ. – Von tiefgehender Bedeutung ist auch: eine haben Ladung, andere nicht[.]
44. PAULI. – Zahl der Lichtquanta nicht konstant, aber diese Differenz nicht wesentlich.
45. LORENTZ. – Ladung wird doch erhalten bleiben.

46. EHRENFEST. – Werden die elektrische Ladung und Masse in die Schrödingers Rechnung hineingesteckt, oder kommen sie heraus?
47. BOHR. – Nur hineingesteckt... [ ] Korrespondenztheorie.
48. COMPTON. – Question of conservation of photons. Work of Lewis.
49. HEISENBERG verbessert seine Antwort 20 (empfangen).
50. BOHR. – What can be regarded as elementary? What you call elementary process is... [ ] because... a superposition of the two polarisations. It is quite right that you can send out elliptical polarised light in electric and magnetic fields, but we cannot be sure that the photons must have a continuous variable in it. The whole thing not so easy... how far... absorbed. We must speak of a continuous...
51. DIRAC. – Difference between photons and electrons very fundamental (déjà développé).
55. LORENTZ. – Pourrions-nous parler de ces photons et de leur mouvement dans un champ de rayonnement?
56. DE BROGLIE n'a pas étudié la question au point de vue électromagnétique[.] [S]e place uniquement au point de vue des ondes ψ.
57. LORENTZ. – Pour photons on aura aussi besoin de ces ondes. Sont-elles de nature différente des ondes lumineuses? Introduire deux ondes me plairait moins.
58. DE BROGLIE. – Quelle est l'onde ψ des photons? Peut-on identifier onde ψ avec l'onde de champ électrique?

En tout cas on peut développer une théorie du photon en l'associant à onde ψ.

59. LORENTZ. – La vitesse de l'onde est égale à celle de la lumière?
60. DE BROGLIE. – Oui, sauf dans un champ d'interférences il faut attribuer à une particule en mouvement une masse propre augmentée d'un $\Box$; alors la vitesse est inférieure à c, mais le photon a une masse $M_0 = \sqrt{m_0 \ldots}$ qui devient grande mais l'énergie reste finie. Dans mouvement rectiligne il ne reste plus que... [ ] mais dans interférences... cesse d'être nul et alors $v < c$... et même être en repos.
61. LORENTZ. – Le $\Box$ est négatif? Autrement la masse deviendrait imaginaire.
62. DE BROGLIE. – Cette variation de la masse donne l'explication de la déviation d'un photon près du bord d'un écran; trajectoire courbe; analogue aux forces autrefois imaginées par Laplace, force provenant du gradient, et l'on retrouve ainsi pour un nuage de photons les tensions de Maxwell complètes en tenant compte de cette variation de masse.
63. LORENTZ. – Cela correspond-il un peu à ce qu'on pensait autrefois?
64. DE BROGLIE. – Oui... [ ] tension, tenseur de tensions... nuage de photons [ ] on trouve 1er terme [formula largely illegible], terme qui disparait lorsque... [ ] est nul[.]
65. LORENTZ répète.
66. PAULI kommt zurück auf Beispiel von Schr. mit de Broglie besprochen.
67. DE BROGLIE. – Pas tout à fait le même que ce que je traitais tout à l'heure[.] [C]'est un problème dans espace ordinaire et espace de configuration; donc un peu artificiel. La difficulté qui se présente est tout à fait la même qu'en optique quand une onde tombe sur un réseau et où il faut aussi des ondes limitées.
68. LORENTZ. – Dans la théorie classique nous sommes contents d'avoir des ondes. La question est de savoir ce qu'une particule devra faire quand elle est plongée dans deux ondes à la fois.

69. DE BROGLIE. – A-t-on le droit de prendre faisceau limité dans espace de configuration? La vitesse de l'électron ne sera définie qu'après que les ondes seront séparées. Que signifie d'ailleurs vitesse dans un espace de configuration? Mes conceptions ne sont valables que dans l'espace réel.
70. LORENTZ. – Pauli dit que pour limiter le faisceau incident il faut aussi admettre une indétermination dans l'état initial du rotateur. On arriverait à la conclusion qu'avant la rencontre le rotateur n'aurait même pas été dans un état stationnaire.
71. BOHR. – Man kann's natürlich nicht wissen, aber kann man jede Möglichkeit vollständig beschreiben im gewöhnlichen Raum? Jede Möglichkeit eines Elektrons beschreiben im gewöhnlichen Raum wäre nach Pauli nicht richtig.
72. LORENTZ. – Cela montre de nouveau la difficulté de décrire tous les phénomènes dans l'espace ordinaire.
73. DE DONDER à propos du tenseur (déjà reçu).
74. LORENTZ. – Ce courant quantique semble donc s'ajouter et se combiner au courant électrique. M. De Donder pourrait-il nous en donner un exemple simple? De même que dans la théorie de Maxwell on introduit un courant de déplacement, M. De Donder introduit une nouvelle notion. Ne pourrait-il nous la rendre familière en donnant un exemple analogue à celui de la décharge d'un condensateur quand il s'agit du déplacement de Maxwell?
75. DE DONDER. – J'ai déjà cherché un exemple avec M. de Broglie; il faudrait résoudre quelques problèmes. Peut-être la discrimination se fera-t-elle sur un exemple particulier; les expériences de Wolfers. Mais je ne vois pour le moment.
76. LORENTZ fait deux remarques (demandé).
76[BIS]. SCHRÖDINGER dit qu'il ne croit pas à l'existence de l'électron dans l'atome.
77. BORN. – Der Satz ist nicht richtig, dass man von der Energie eines Teilchens reden kann, nur Energie eines Systems / Wichtigste Differenz zwischen klassischer und Quantentheorie.
78.–80. BOHR formuliert es anders (gefragt).
81. LORENTZ. – Es handelt sich immer wieder um einerseits die scharfe Vorstellung, die ich liebe, andererseits die Unmöglichkeit... [ ] (je me place toujours au point de vue d'une représentation imagée et toujours on me répond par une impossibilité matérielle)... die Photographie machen... [ ] Vielleicht ist die ganze Auffasung nicht richtig.
82. BOHR. – Si l'électron était si loin l'énergie potentielle serait très petite, donc l'énergie cinétique deviendrait négative.
83. BORN. – Autres conséquences de l'idée que $\psi\psi^*$ est une probabilité. Cela joue un rôle dans la théorie des métaux etc.
85. SCHRÖDINGER. – Missverstanden mit Momentphotographie. Der Begriff, das Elektron befindet sich hier, hat keinen Sinn meiner Meinung nach. (La conception de la position de l'électron dans un atome est à rejeter. On ne peut pas dans un atome définir la position avec cette précision.)
86. DE BROGLIE. – Si l'électron est en mouvement avec la vitesse donnée par mes formules... [ ] alors M_0 [ ] ... [ ] est inférieur à m_0, alors on comprend l'énergie cinétique négative, c'est-à-dire énergie totale inférieure à m_0c^2. L'électron ne peut pas sortir d'une certaine sphère, mais chez moi la masse devient variable et alors l'électron pourrait aller à l'infini. Peut-être faut-il chercher la solution de la difficulté dans le sens...
87. LORENTZ. – Mouvement qui parait ne pas exister.

88. BORN. – Schrödinger sagt, Ort und Bewegung hat überhaupt keinen Sinn. Bohr und ich sind der Meinung, dass es wohl einen Sinn hat, dann, wenn man angeben kann...
[ ] ob man noch über einen Ort sprechen kann [ ] Geschmackssache.
89. PAULI. – Man kann wohl den Ort des Elektrons ausserhalb der Kugel bestimmen, aber nicht ohne zugleich die Energie zu ändern.
90. BOHR. – Wir behaupten nur, den Versuch (siehe 78) kann man machen, und das deutet man so; das Elektron [ ] ist da.
90BIS. KRAMERS. – On ne peut exprimer son résultat par des instruments.
91. EHRENFEST. – If I did receive a letter saying: I have observed... [ ] I would answer: I don't believe either the one or the other[.] The two things are impossible together. So much I have learnt from Heisenberg.
92. LORENTZ. – Remarque au sujet des paquets d'ondes. Tous ces paquets finiront par se fondre (demandé).
93. EHRENFEST. – They may diffuse, but by experiment we find again the electron [ ] very hard γ-rays... new means.
94. LORENTZ continue (demandé).
95. SCHRÖDINGER répond (déjà reçu).
96. BOHR. – Heisenberg hat die Wellenpakete so schön behandelt. Schrödinger sagt, dass es sich nicht um verschiedene Dinge handelt, nur Beschränkung, ganz richtig, dass wir im gewissen Sinne alle Eigenschaften des Atoms kennen müssen. Aber es gibt doch andere Möglichkeiten... [ ] eine Stelle eines Elektrons anweisen... und dann findet man... läuft sehr viele Male umher... aber nach dieser Zeit nicht zu finden, bedeutet, dass sie diesen Versuch machen[,] doch über alle Versuche reden können.
97. SCHRÖDINGER. – Die Bahnen können sie doch nicht beschreiben.
98. BOHR. – Bei hohen Quantenzahlen könnte man durch Superposition benachbarter Zustände das Laufen des Elektrons nachahmen. Sie beschränken sich nur auf eine bestimmte Anzahl von Versuchen.
99. LORENTZ. – Je voudrais me représenter un électron et je pense à ce paquet d'ondes qui l'accompagne. Je voudrais attribuer à ce paquet certaines dimensions. Celles-ci dépendent dans la mesure des... [ ] cela dépendrait donc des moments où il a été observé.
100. BORN. – Das ist doch in der klassischen Theorie auch so... [ ] wenn man weiss, dass bei den ersten Messungen ein kleiner Fehler gemacht ist. [ ] exakt wahrnehmbar... [ ] dass man einen Bereich festlegen kann.
101. HEISENBERG. – Das hängt von der Genauigkeit ab.

Kramers' notes from the general discussion[a]

[9:54] BRAGG. – This morning additional figure for scattering NaCl at different angles. Almost exact agreement with Hartree's curves *if corrected for zero-point energy.*[b]

[9:56] DISCUSSION GÉNÉRALE

LORENTZ a toujours voulu former des images, comme Hertz l'a dit. On les a formées avec [?] notions ordinaires, espace et temps. Cette image est absolument nette. Corpuscule et point lui sont [?] des conceptions toutes nettes. Alors impossible de ne pas penser au[x] trajectoires.

Même si l'on s'écarte des anciennes idées, on a alors les mots anciens. Il n'aime pas de dire qu'on ne peut pas défendre d'avoir l'idéal de décrire tout d'une manière nette.

Introduisons les fonctions ψ en recherchant le choc.

Lorentz wil niet de onzekerheid in de metingen zetten in 't begin van de theorie, maar aan 't eind.

Wil niet doen aan indeterminisme.

BOHR. – Glad explain as simple as I can certain attitude [towards] problems quantum theory. Very little new: only hope to assist in discussion of small points, where we come perhaps to some agreement. Agree heartily with Lorentz's attitude. Very different to speak about ultimate interpretation of words and symbols used in calculation. Must see first what tools are, see what the mathematical theories are in themselves, and then see how they suit requirements for interpreting experiments.

Still like to begin about such points: because in light in free [deleted?] space, and free particles we have already some tools about which we can speak. Therefore consider these tools, and then analyse some problems of bound electrons. Third [?] go a little in details on problem described by Heisenberg. 1) Difficulties in recovering things in space and time. It all lies in wave theory. All theory embodied in $E = h\nu$. Use of word 'energy' connected with 'particle', 'ν' with 'waves'. What can we practically do with waves in order to describe particles? I will write down expression for phase wave used by de Broglie:

$$e^{i2\pi(\tau_x x+\tau_y y+\tau_z z-\nu t)} ,$$

τ_x: number waves unit length in direction x-axis.

Let this one interfere with other waves. Well known from optics how location of wave-field is connected up with analysis in monochromatic parts.

In order to have limited extensions

$$\Delta x \Delta\tau_x \sim 1$$
$$\Delta\tau \Delta\nu \sim 1 .$$

[a] Titled at the top 'Donderdag 27–10–27'. The manuscript includes page numbers, some timings, some line numbers and some numbers corresponding, where possible, with those labelling the discussion contributions in Verschaffelt's typescript reproduced above. We omit all of these, except the timings. Our text tacitly includes any manuscript insertions and corrections, except for a few words that seem to be notes for a translation into French (and are quite possibly in Verschaffelt's hand). Tacit amendments to spelling and punctuation have been made when reasonably obvious. Other interpolations and amendments are denoted by square brackets. We thank Mark van Atten for assistance with the Dutch. (*eds.*).

[b] Sketch of a figure in the margin, which appears similar to Bragg's Fig. 8 (p. 279) (*eds.*).

Now de Broglie fundaments [?] are

$$E = h\nu$$
$$P = h\tau\ .$$

Therefore

$$\Delta x \Delta p_x \sim h$$
$$\Delta t \Delta E \sim h\ .$$

Limitations are the limits of defining E and P in x, y, z, t. So far mathematics. If we really believe in properties of light and material particle these formulae give something connected with measurements.

Think first of Davisson and Germer and second of Thomson. Then of Heisenberg's example of seeing electron in microscope.[a]

If we see light, we know that there was an electron at A. The definition of its position is given by aperture so that $\Delta x \sim \frac{\lambda}{\varepsilon}$. We come to difficulty, that there comes an uncertainty in momentum, due to location.

The fact that the light comes within cone makes that from conservation we do not know exactly from what direction the impulse comes. Consider impulse in x-direction: $\Delta p_x \sim \varepsilon h\tau$, $\Delta x \Delta p_x \sim h$.

This directly connected with other formulae. If we ask for other way of calculating impulse; for instance by Einstein recoil. But then the microscope obeys the laws given. If microscope infinitely heavy, no impulse can be measured. If not, it is represented by de Broglie waves, and its position is unknown.

Other example; how [to] find energy of electrons coming out of hole, and we ask for *time*. The latter being measured by shutting of hole will make the energy *less defined* (if the wave description of the behaviour is right). For instance the energy in photoelectrons cannot be more accurate [?] in the short time.

Again other example. Compton effect. The possibility of getting any consistency in quantum problem [?] due to de Broglie's discovery. We always deal with free particles: which is an abstraction, because we find them due to interaction. Let us therefore consider Compton effect.

We represent the electron and light by packets. I can only point out little point, if I don't go into the details of the theory. We may say the four packets meet at the same time-space point, but this point is not well defined.[b]

The conservation theorem is *expressed* by saying

$$\mathcal{I}'_r + \mathcal{I}'_q = \mathcal{I}''_r + \mathcal{I}''_q\ .$$

But there will be uncertainty in definition of energy-impulse vector, which is the same on both sides: therefore extension the same on both sides. Therefore harmony in description of conservation.

[a] Sketch of the Heisenberg microscope in the margin, with the annotation 'λ and τ are due to the light' (*eds.*).

[b] There is a rough sketch in the margin similar to Richardson's Fig. B (p. 435) in the general discussion (*eds.*).

[10:45] BRILLOUIN. – Cette incertainty is old [*sic*], because it is just Planck's point.

BOHR says: yes, through Planck's discovery $E = h\nu$, which connects things of different kinds.

[10:46] EHRENFEST hopes Stern–Gerlach will be discussed.

KRAMERS asks why necessary.

DE DONDER. – On ne trouve jamais $\Delta t \Delta E$. Relativité 4-dimensionnelle donne quelque chose de plus profond, que M. Bohr étudie.[a]

BOHR. – [Figure of a beam split in two, with annotations: '$T\frac{\delta E}{h} > 1$', '$\frac{\delta E}{h}$ difference of de Broglie frequency between A and B', '$T\delta\nu_H > 1$'.]

Now we cannot [distinguish] experimentally between parts A and B of the beam. $T\delta\nu_H > 1$ says the phase must be lost in the Stern–Gerlach experiments, in order that something can be found.

LORENTZ asks if it has anything to do with what he said yesterday.

BOHR says the connection is not exact.

BORN says it is the same.

BOHR. – The phase comes not in the experiments, and therefore there is a possibility of individuals.

LORENTZ asks how is the wave theory of the Stern–Gerlach experiments.

[11:14] BOHR. – Darwin has discussed it in a forthcoming paper. For a time difficulty: the 'Verzweigung' was difficult. In the *general* solution, we can by superposition just represent the thing.

LORENTZ recalls how in the old days the cathode rays (Lenard) were considered as waves, which were bent.

BOHR. – Since the phase is lost, we can now [say] something [?] stationary states. *Dirac* and *Heisenberg* have pointed out, that [?] if beam of atoms are illuminated long time or short times intermittently. You can never say at a moment atom is in a stationary state. By the experiments the knowledge about the past is destroyed.

Other thing. α-particle through atom: difficulty if it passes quickly. After collision it will be in stationary state. How can α-particle agree with electron in stationary

[a] Here, evidently, a page is missing from Kramers' notes, corresponding to discussion contributions 9, 10 and 11 in Verschaffelt's typescript and the beginning of Bohr's contribution 12 (*eds.*).

state? Now trivial. If the α-particle is in the atom at a definite time, its energy is uncertain.

Causality. Everything seems to become statistical in a way inconsistent with experiments. As a matter of fact the *individuality* does not give these difficulties. Difference with old conception: difference between closed and not-closed systems. I'll use undefinable words [*sic*]. Any atomic process open to experiment involves individuality, governed [?] by Planck's action. Any experiment means interaction between system and tools of measurements. Classically one assumed this interaction can be made as small as possible. In our formulae for closed systems, we speak about things nothing to do with observation. Logically, as soon as interaction, system no longer defined. Therefore uncertainty in system. Have said that causality and spacetime conception are complementary. Classical idea of causal description in space of [?] time at variance with quantum theory. Example: when in matrix or Schrödinger $E = h\nu$ is introduced, we deal always of closed system. In experiments never. We must connect up our usual ideas and the newly discovered ideas. In any interpretation of observations, if done well, there is a place where quantum perturbations come in explicitly. Point which [?] [needs] to be stressed very much.[a]

[a] Sketch at the bottom of the page, which appears again to be of a Stern–Gerlach experiment with two beams A and B (*eds.*).

Bibliography

Aharonov, Y., and Vaidman, L. (1996). About position measurements which do not show the Bohmian particle position. In Cushing, Fine and Goldstein (1996), pp. 141–54.

Bacciagaluppi, G. (2005). The role of decoherence in quantum theory. In *The Stanford Encyclopedia of Philosophy*, Summer 2005 edn., ed. E. N. Zalta. http://plato.stanford.edu/archives/sum2005/entries/qm-decoherence.

Bacciagaluppi, G. (2008a). The statistical interpretation according to Born and Heisenberg, In *HQ-1: Conference on the History of Quantum Physics (Vols. I & II)*, eds. C. Joas, C. Lehner and J. Renn. MPIWG preprint series, vol. 350. Berlin: MPIWG, pp. 269–288.

Bacciagaluppi, G. (2008b). Original typescript of Heisenberg's uncertainty paper discovered. *AIP History Newsletter*, **XL** (Spring 2008), 6.

Ballentine, L. E. (1970). The statistical interpretation of quantum mechanics. *Reviews of Modern Physics*, **42**, 358–81.

Ballentine, L. E. (1972). Einstein's interpretation of quantum mechanics. *American Journal of Physics*, **40**, 1763–71.

Ballentine, L. E. (1986). Probability theory in quantum mechanics. *American Journal of Physics*, **54**, 883–9.

Ballentine, L. E. (1987). Resource letter IQM-2: Foundations of quantum mechanics since the Bell inequalities. *American Journal of Physics*, **55**, 785–92.

Ballentine, L. E. (2003). The classical limit of quantum mechanics and its implications for the foundations of quantum mechanics. In *Quantum Theory: Reconsideration of Foundations – 2*, ed. A. Khrennikov, Mathematical modelling in physics, engineering and cognitive science, vol. 10. Växjö: Växjö University Press, pp. 71–82.

Barbour, J. B. (1994a). The timelessness of quantum gravity, I. The evidence from the classical theory. *Classical and Quantum Gravity*, **11**, 2853–73.

Barbour, J. B. (1994b). The timelessness of quantum gravity, II. The appearance of dynamics in static configurations. *Classical and Quantum Gravity*, **11**, 2875–97.

Bell, J. S. (1964). On the Einstein-Podolsky-Rosen paradox. *Physics*, **1**, 195–200. Reprinted in Bell (1987), pp. 14–21.

Bell, J. S. (1966). On the problem of hidden variables in quantum mechanics. *Reviews of Modern Physics*, **38**, 447–52. Reprinted in Bell (1987), pp. 1–13.

Bell, J. S. (1980). De Broglie-Bohm, delayed-choice double-slit experiment, and density matrix. *International Journal of Quantum Chemistry*, **14**, 155–9. Reprinted in Bell (1987), pp. 111–16.

Bell, J. S. (1984). Beables for quantum field theory. CERN-TH. 4035/84. Reprinted in Bell (1987), pp. 173–80.

Bell, J. S. (1986). Interview. In *The Ghost in the Atom*, eds. P. C. W. Davies and J. R. Brown. Cambridge: Cambridge University Press, pp. 45–57.

Bell, J. S. (1987). *Speakable and Unspeakable in Quantum Mechanics*. Cambridge: Cambridge University Press.

Bell, J. S. (1990). Against 'measurement'. In Miller (1990), pp. 17–31.

Beller, M. (1999). *Quantum Dialogue: The Making of a Revolution*. Chicago: University of Chicago Press.

Belousek, D. W. (1996). Einstein's 1927 unpublished hidden-variable theory: Its background, context and significance. *Studies in History and Philosophy of Modern Physics*, **27**, 437–61.

Berndl, K., Dürr, D., Goldstein, S., and Zanghì, N. (1996). Nonlocality, Lorentz invariance, and Bohmian quantum theory. *Physical Review A*, **53**, 2062–73.

Berry, M. (1997). Slippery as an eel. *Physics World*, **10** (December 1997), 41–2.

Bjorken, J. D., and Drell, S. D. (1964). *Relativistic Quantum Mechanics*. New York: McGraw-Hill.

Böhme, K. (ed.) (1975). *Aufrufe und Reden deutscher Professoren im ersten Weltkrieg*. Stuttgart: Reclam.

Bohm, D. (1951). *Quantum Theory*. New Jersey: Prentice-Hall.

Bohm, D. (1952a). A suggested interpretation of the quantum theory in terms of 'hidden' variables, I. *Physical Review*, **85**, 166–79.

Bohm, D. (1952b). A suggested interpretation of the quantum theory in terms of 'hidden' variables, II. *Physical Review*, **85**, 180–93.

Bohm, D. (1953). Proof that probability density approaches $|\psi|^2$ in causal interpretation of the quantum theory. *Physical Review*, **89**, 458–66.

Bohm, D., Dewdney, C., and Hiley, B. J. (1985). A quantum potential approach to the Wheeler delayed-choice experiment. *Nature*, **315**, 294–7.

Bohm, D., and Hiley, B. J. (1993). *The Undivided Universe: An Ontological Interpretation of Quantum Theory*. London: Routledge.

Bohm, D., and Vigier, J. P. (1954). Model of the causal interpretation of quantum theory in terms of a fluid with irregular fluctuations. *Physical Review*, **96**, 208–16.

Bohr, N. (1925). Über die Wirkung von Atomen bei Stößen. *Zeitschrift für Physik*, **34**, 142–57.

Bohr, N. (1928). Das Quantenpostulat und die neuere Entwicklung der Atomistik. *Die Naturwissenschaften*, **16**, 245–57.

Bohr, N. (1949). Discussion with Einstein on epistemological problems in atomic physics. In Schilpp (1949), pp. 201–41.

Bohr, N. (1963). *Essays 1958–1962 on Atomic Physics and Human Knowledge*, New York: Interscience.

Bohr, N. (1984). *Niels Bohr: Collected Works*, vol. 5, ed. K. Stolzenburg. Amsterdam: North-Holland.

Bohr, N. (1985). *Niels Bohr: Collected Works*, vol. 6, ed. J. Kalckar. Amsterdam: North-Holland.

Bohr, N., Kramers, H. A., and Slater, J. C. (1924a). The quantum theory of radiation. *Philosophical Magazine*, **47**, 785–802. Reprinted in *Sources of Quantum Mechanics*, ed. B. L. van der Waerden. Amsterdam: North-Holland, 1967, pp. 159–76.

Bohr, N., Kramers, H. A., and Slater, J. C. (1924b). Über die Quantentheorie der Strahlung. *Zeitschrift für Physik*, **24**, 69–87.

Bonk, T. (1994). Why has de Broglie's theory been rejected? *Studies in the History and Philosophy of Science*, **25**, 191–209.

Born, M. (1924). Über Quantenmechanik. *Zeitschrift für Physik*, **26**, 379–95.

Born, M. (1926a). Zur Quantenmechanik der Stossvorgänge. *Zeitschrift für Physik*, **37**, 863–7.

Born, M. (1926b). Quantenmechanik der Stossvorgänge. *Zeitschrift für Physik*, **38**, 803–27.

Born, M. (1926c). Das Adiabatenprinzip in der Quantenmechanik. *Zeitschrift für Physik*, **40**, 167–92.

Born, M. (1926d). *Problems of Atomic Dynamics*. Cambridge, Mass.: Massachusetts Institute of Technology. Reprinted Cambridge, Mass.: MIT Press, 1970.

Born, M. (1926e). *Probleme der Atomdynamik*. Berlin: Springer.

Born, M. (1927). Physical aspects of quantum mechanics. *Nature*, **119**, 354–7.

Born, M. (1969). *Atomic Physics*, 8th edn. Glasgow: Blackie and Son. Reprinted New York: Dover, 1989.

Born, M., and Jordan, P. (1925). Zur Quantenmechanik, I. *Zeitschrift für Physik*, **34**, 858–88.

Born, M., Heisenberg, W., and Jordan, P. (1926). Zur Quantenmechanik, II. *Zeitschrift für Physik*, **35**, 557–615.

Born, M., and Wiener, N. (1926a). Eine neue Formulierung der Quantengesetze für periodische und nichtperiodische Vorgänge. *Zeitschrift für Physik*, **36**, 174–87.

Born, M., and Wiener, N. (1926b). A new formulation of the laws of quantization of periodic and aperiodic phenomena. *Journal of Mathematics and Physics M.I.T.*, **5**, 84–98.

Bose, S. N. (1924) Wärmegleichgewicht im Strahlungsfeld bei Anwesenheit von Materie. *Zeitschrift für Physik*, **27**, 384–92.

Bothe, W., and Geiger, H. (1924). Ein Weg zur experimentellen Nachprüfung der Theorie von Bohr, Kramers und Slater. *Zeitschrift für Physik*, **26**, 44.

Bothe, W., and Geiger, H. (1925a). Experimentelles zur Theorie von Bohr, Kramers und Slater. *Die Naturwissenschaften*, **13**, 440–1.

Bothe, W., and Geiger, H. (1925b). Über das Wesen des Comptoneffekts: ein experimenteller Beitrag zur Theorie der Strahlung. *Zeitschrift für Physik*, **32**, 639–63.

Brillouin, L. (1922). Diffusion de la lumière et des rayons X par un corps transparent homogène; influence de l'agitation thermique. *Annales de Physique*, **17**, 88–122.

Brown, L. M. (ed.) (2005). *Feynman's Thesis: A New Approach to Quantum Theory*. New Jersey: World Scientific.

Camilleri, K. (2006). Heisenberg and the wave-particle duality. *Studies in History and Philosophy of Modern Physics*, **37**, 298–315.

Campbell, N. R. (1921). Atomic structure. *Nature*, **107**, 170.

Campbell, N. R. (1926). Time and chance. *Philosophical Magazine* (7), **1**, 1106–17.

Chevalley, C. (1988). Physical reality and closed theories in Werner Heisenberg's early papers. In *Theory and Experiment*, eds. D. Batens and J. P. van Bendegem. Dordrecht: Reidel.

Clifton, R., Bub, J., and Halvorson, H. (2003). Characterizing quantum theory in terms of information-theoretic constraints. *Foundations of Physics*, **33**, 1561–91.

Compton, A. H. (1928). Some experimental difficulties with the electromagnetic theory of radiation. *Journal of the Franklin Institute*, **205**, 155–78.

Compton, A. H., and Simon, A. W. (1925). Directed quanta of scattered X-rays. *Physical Review*, **26**, 289–99.

Cushing, J. T. (1994). *Quantum Mechanics: Historical Contingency and the Copenhagen Hegemony*. Chicago: University of Chicago Press.

Cushing, J. T. (1996). The causal quantum theory program. In Cushing, Fine and Goldstein (1996), pp. 1–19.

Cushing, J. T., Fine, A., and Goldstein, S. (eds.) (1996). *Bohmian Mechanics and Quantum Theory: An Appraisal*. Dordrecht: Kluwer.

Darrigol, O. (1992). *From c-numbers to q-numbers: The Classical Analogy in the History of Quantum Theory*. Berkeley: University of California Press.

Darrigol, O. (1993). Strangeness and soundness in Louis de Broglie's early works. *Physis*, **30**, 303–72. Reprinted with typographical corrections, 1994.

Davisson, C., and Germer, L. (1927). The scattering of electrons by a single crystal of Nickel. *Nature*, **119**, 558–60.

de Broglie, L. (1922). Rayonnement noir et quanta de lumière. *Le Journal de Physique et le Radium* (6), **3**, 422–8.

de Broglie, L. (1923a). Ondes et quanta. *Comptes Rendus Hebdomadaires des Séances de l'Académie des Sciences (Paris)*, **177**, 507–10.

de Broglie, L. (1923b). Quanta de lumière, diffraction et interférences. *Comptes Rendus Hebdomadaires des Séances de l'Académie des Sciences (Paris)*, **177**, 548–50.

de Broglie, L. (1923c). Les quanta, la théorie cinétique des gaz et le principe de Fermat. *Comptes Rendus Hebdomadaires des Séances de l'Académie des Sciences (Paris)*, **177**, 630–2.

de Broglie, L. (1923d). Waves and quanta. *Nature*, **112**, 540.

de Broglie, L. (1924a). A tentative theory of light quanta. *Philosophical Magazine* (6), **47**, 446–58.

de Broglie, L. (1924b). Sur la définition générale de la correspondance entre onde et mouvement. *Comptes Rendus Hebdomadaires des Séances de l'Académie des Sciences (Paris)*, **179**, 39–40.

de Broglie, L. (1924c). Sur un théorème de M. Bohr. *Comptes Rendus Hebdomadaires des Séances de l'Académie des Sciences (Paris)*, **179**, 676–7.

de Broglie, L. (1924d). Sur la dynamique du quantum de lumière et les interférences. *Comptes Rendus Hebdomadaires des Séances de l'Académie des Sciences (Paris)*, **179**, 1039–41.

de Broglie, L. (1924e). Recherches sur la théorie des quanta. Ph.D. Thesis, University of Paris.

de Broglie, L. (1925). Recherches sur la théorie des quanta. *Annales de Physique* (10), **3**, 22–128.

de Broglie, L. (1926). Sur la possibilité de relier les phénomènes d'interférences et de diffraction à la théorie des quanta de lumière. *Comptes Rendus Hebdomadaires des Séances de l'Académie des Sciences (Paris)*, **183**, 447–8.

de Broglie, L. (1927a). La structure atomique de la matière et du rayonnement et la mécanique ondulatoire. *Comptes Rendus Hebdomadaires des Séances de l'Académie des Sciences (Paris)*, **184**, 273–4.

de Broglie, L. (1927b). La mécanique ondulatoire et la structure atomique de la matière et du rayonnement. *Le Journal de Physique et le Radium* (6), **8**, 225–41.

de Broglie, L. (1928). La nouvelle dynamique des quanta. In *Electrons et Photons: Rapports et discussions du Cinquième Conseil de Physique*. Paris: Gauthier-Villars, pp. 105–32.

de Broglie, L. (1930). *An Introduction to the Study of Wave Mechanics*. New York: E. P. Dutton and Company.

de Broglie, L. (1955). *Physics and Microphysics*. New York: Pantheon Books.

de Broglie, L. (1956). *Une tentative d'interprétation causale et non linéaire de la mécanique ondulatoire (la théorie de la double solution)*. Paris: Gauthier-Villars.

de Broglie, L. (1974). Beginnings of wave mechanics. In *Wave Mechanics: The First Fifty Years*, eds. W. C. Price, S. S. Chissick and T. Ravensdale. London: Butterworths.

de Broglie, L. (1999). The wave nature of the electron. Nobel lecture, 12 December 1929. In *Nobel Lectures in Physics (1901–1995)*, CD-ROM edn. Singapore: World Scientific.

de Broglie, L., and Brillouin, L. (1928). *Selected Papers on Wave Mechanics*. London: Blackie and Son.

Debye, P. (1909). Das Verhalten von Lichtwellen in der Nähe eines Brennpunktes oder einer Brennlinie. *Annalen der Physik*, **30**, 755–76.

de Regt, H. (1997). Erwin Schrödinger, Anschaulichkeit, and quantum theory. *Studies in History and Philosophy of Modern Physics*, **28B**, 461–81.

de Regt, H. (2001). Space-time visualisation and the intelligibility of physical theories. *Studies in History and Philosophy of Modern Physics*, **32B**, 243–65.

Despy-Meyer, A., and Devriese, D. (1997). *Ernest Solvay et son temps*. Bruxelles: Archives de l'ULB.

Deutsch, D. (1999). Quantum theory of probability and decisions. *Proceedings of the Royal Society of London A*, **455**, 3129–37.

Dewdney, C., Hardy, L., and Squires, E. J. (1993). How late measurements of quantum trajectories can fool a detector. *Physics Letters A*, **184**, 6–11.

DeWitt, B. S. (1967). Quantum theory of gravity, I. The canonical theory. *Physical Review*, **160**, 1113–48.

DeWitt, B. S., and Graham, N. (1973). *The Many-Worlds Interpretation of Quantum Mechanics*. Princeton: Princeton University Press.

Dirac, P. A. M. (1926a). Quantum mechanics and a preliminary investigation of the hydrogen atom. *Proceedings of the Royal Society A*, **110**, 561–79.

Dirac, P. A. M. (1926b). Relativity quantum mechanics with an application to Compton scattering. *Proceedings of the Royal Society A*, **111**, 405–23.

Dirac, P. A. M. (1926c). On the theory of quantum mechanics. *Proceedings of the Royal Society A*, **112**, 661–77.

Dirac, P. A. M. (1927a). The physical interpretation of quantum dynamics. *Proceedings of the Royal Society A*, **113**, 621–41.

Dirac, P. A. M. (1927b). The quantum theory of emission and absorption of radiation. *Proceedings of the Royal Society A*, **114**, 243–65.

Dirac, P. A. M. (1927c). The quantum theory of dispersion. *Proceedings of the Royal Society A*, **114**, 710–28.

Dirac, P. A. M. (1933). The Lagrangian in quantum mechanics. *Physikalische Zeitschrift der Sowjetunion*, **3**, 64–72.

Dürr, D., Fusseder, W., Goldstein, S., and Zanghì, N. (1993). Comment on 'Surrealistic Bohm trajectories'. *Zeitschrift für Naturforschung*, **48a**, 1261–2.

Dürr, D., Goldstein, S., and Zanghì, N. (1992). Quantum equilibrium and the origin of absolute uncertainty. *Journal of Statistical Physics*, **67**, 843–907.

Dymond, E. G. (1927). On electron scattering in Helium. *Physical Review*, **29**, 433–41.

Eckart, C. (1926). Operator calculus and the solution of the equations of quantum dynamics. *Physical Review*, **28**, 711–26.

Ehrenfest, P. (1917). Adiabatic invariants and the theory of quanta. *Philosophical Magazine*, **33**, 500–13.

Einstein, A. (1909). Über die Entwickelung unserer Anschauungen über das Wesen und die Konstitution der Strahlung. *Physikalische Zeitschrift*, **10**, 817–26. English translation in *The Collected Papers of Albert Einstein*, vol. 2, eds. A. Beck and P. Havas. Princeton: Princeton University Press, 1989, pp. 379–98.

Einstein, A. (1924). Quantentheorie des einatomigen idealen Gases. *Sitzungsberichte der Preussischen Akademie der Wissenschaften, Physikalisch-mathematische Klasse* (1924), 261–7.

Einstein, A. (1925a). Quantentheorie des einatomigen idealen Gases, 2. Abhandlung. *Sitzungsberichte der Preussischen Akademie der Wissenschaften, Physikalisch-mathematische Klasse* (1925), 3–14.

Einstein, A. (1925b). Quantentheorie des idealen Gases. *Sitzungsberichte der Preussischen Akademie der Wissenschaften, Physikalisch-mathematische Klasse* (1925), 18–25.

Einstein, A. (1949). Remarks concerning the essays brought together in this co-operative volume. In Schilpp (1949), pp. 665–88.

Einstein, A., and Besso, M. (1972). *Correspondance: 1903–1955*, ed. P. Speziali. Paris: Hermann.

Einstein, A., Podolsky, B., and Rosen, N. (1935). Can quantum-mechanical description of physical reality be considered complete? *Physical Review*, **47**, 777–80.

Elsasser, W. (1925). Bemerkungen zur Quantenmechanik freier Elektronen. *Die Naturwissenschaften*, **13**, 711.

Englert, B.-G., Scully, M. O., Süssmann, G., and Walther, H. (1992). Surrealistic Bohm trajectories. *Zeitschrift für Naturforschung*, **47a**, 1175–86.

Everett, H. (1957). 'Relative state' formulation of quantum mechanics. *Reviews of Modern Physics*, **29**, 454–62.

Fényes, I. (1952). Eine wahrscheinlichkeitstheoretische Begründung und Interpretation der Quantenmechanik. *Zeitschrift für Physik*, **132**, 81–106.

Fermi, E. (1926). Zur Wellenmechanik des Stoßvorganges. *Zeitschrift für Physik*, **40**, 399–402.

Fermi, E. (1927). Sul meccanismo dell'emissione nella meccanica ondulatoria. *Rendiconti Lincei*, **5**, 795–800.

Feuer, L. S. (1974). *Einstein and the Generations of Science*. New York: Basic Books.

Feynman, R. P. (1942). The principle of least action in quantum mechanics. Ph.D. Thesis, Princeton University.

Feynman, R. P. (1948). Space-time approach to non-relativistic quantum mechanics. *Reviews of Modern Physics*, **20**, 367–87.

Feynman, R. P., Leighton, R. B., and Sands, M. (1965). *The Feynman Lectures on Physics, vol. III: Quantum Mechanics*. Reading, Mass.: Addison-Wesley.

Fine, A. (1986). *The Shaky Game: Einstein, Realism and the Quantum Theory*. Chicago: University of Chicago Press.

Fine, A. (1999). Locality and the Hardy theorem. In *From Physics to Philosophy*, eds. J. Butterfield and C. Pagonis. Cambridge: Cambridge University Press, pp. 1–11.

Fuchs, C. A. (2002). Quantum mechanics as quantum information (and only a little more), quant-ph/0205039.

Gell-Mann, M. and Hartle, J. B. (1990). Quantum mechanics in the light of quantum cosmology. In *Complexity, Entropy, and the Physics of Information*, ed. W. H. Zurek. Reading, Mass.: Addison-Wesley, pp. 425–58.

Ghirardi, G.C., Rimini, A., and Weber, T. (1986). Unified dynamics for microscopic and macroscopic systems. *Physical Review D*, **34**, 470–91.

Goldstein, H. (1980). *Classical Mechanics*. Reading, Mass.: Addison-Wesley.

Gordon, W. (1926). Der Comptoneffekt nach der Schrödingerschen Theorie. *Zeitschrift für Physik*, **40**, 117–33.

Greene, B. (2000). *The Elegant Universe: Superstrings, Hidden Dimensions, and the Quest for the Ultimate Theory*. London: Vintage.

Greene, B. (2005). *The Fabric of the Cosmos: Space, Time, and the Texture of Reality*. New York: Vintage.

Griffiths, R. B. (1984). Consistent histories and the interpretation of quantum mechanics. *Journal of Statistical Physics*, **36**, 219–72.

Griffiths, R. B. (2002). *Consistent Quantum Theory*. Cambridge: Cambridge University Press.

Guth, A. H. (1981). Inflationary universe: A possible solution to the horizon and flatness problems, *Physical Review D*, **23**, 347–56.

Hardy, L. (1995). The EPR argument and nonlocality without inequalities for a single photon. In *Fundamental Problems in Quantum Theory*, eds. D. M. Greenberger and A. Zeilinger. New York: New York Academy of Sciences, pp. 600–15.

Hardy, L. (2001). Quantum theory from five reasonable axioms. quant-ph/0101012.

Hardy, L. (2002). Why quantum theory? In *Non-locality and Modality*, eds. T. Placek and J. Butterfield. Dordrecht: Kluwer, pp. 61–73.

Hardy, L. (2005). Probability theories with dynamic causal structure: A new framework for quantum gravity. gr-qc/0509120.

Hartle, J. B. (1995). Spacetime quantum mechanics and the quantum mechanics of spacetime. In *Gravitation and Quantizations: Proceedings of the 1992 Les Houches Summer School*, eds. B. Julia and J. Zinn-Justin. Amsterdam: Elsevier, pp. 285–480.

Heisenberg, W. (1925a). Über eine Anwendung des Korrespondenzprinzips auf die Frage nach der Polarisation des Fluoreszenzlichtes. *Zeitschrift für Physik*, **31**, 617–28.

Heisenberg, W. (1925b). Über quantentheoretische Umdeutung kinematischer und mechanischer Beziehungen. *Zeitschrift für Physik*, **33**, 879–93.

Heisenberg, W. (1926a). Schwankungserscheinungen und Quantenmechanik. *Zeitschrift für Physik*, **40**, 501–6.

Heisenberg, W. (1926b). Mehrkörperproblem und Resonanz in der Quantenmechanik, I. *Zeitschrift für Physik*, **38**, 411–26.

Heisenberg, W. (1926c). Quantenmechanik. *Die Naturwissenschaften*, **14**, 989–94.

Heisenberg, W. (1927). Über den anschaulichen Inhalt der quantentheoretischen Kinematik und Mechanik. *Zeitschrift für Physik*, **43**, 172–98.

Heisenberg, W. (1929). Die Entwicklung der Quantentheorie 1918–1928. *Die Naturwissenschaften*, **14**, 490–6.

Heisenberg, W. (1930a). *Die physikalischen Prinzipien der Quantentheorie*. Leipzig: Hirzel. Reprinted Mannheim: BI Wissenschaftsverlag, 1991.

Heisenberg, W. (1930b). *The Physical Principles of the Quantum Theory*, translated by C. Eckart and F. C. Hoyt. Chicago: University of Chicago Press. Reprinted New York: Dover, 1949.

Heisenberg, W. (1946). Der unanschauliche Quantensprung. *Neue physikalische Blätter*, **2**, 4–6.

Heisenberg, W. (1948). Der Begriff der 'Abgeschlossenen Theorie' in der modernen Naturwissenschaft. *Dialectica*, **2**, 331–6.

Heisenberg, W. (1962). *Physics and Philosophy: The Revolution in Modern Science*. New York: Harper and Row.

Heisenberg, W. (1967). Quantum theory and its interpretation. In *Niels Bohr: His Life and Work as Seen by his Friends and Colleagues*, ed. S. Rozental. New York: John Wiley and Sons, pp. 94–108.

Heisenberg, W. (1969). *Der Teil und das Ganze*. München: Piper.

Heisenberg, W. (1984). *Gesammelte Werke*, eds. W. Blum, H.-P. Dürr and H. Rechenberg. Berlin: Springer.

Hendry, J. (1986). *James Clerk Maxwell and the Theory of the Electromagnetic Field.* Bristol: Adam Hilger.

Hilbert, D., von Neumann, J., and Nordheim, L. (1928). Über die Grundlagen der Quantenmechanik. *Mathematische Annalen*, **98**, 1–30.

Holland, P. R. (1993). *The Quantum Theory of Motion: An Account of the de Broglie-Bohm Causal Interpretation of Quantum Mechanics.* Cambridge: Cambridge University Press.

Holland, P. R. (2005). What's wrong with Einstein's 1927 hidden-variable interpretation of quantum mechanics? *Foundations of Physics*, **35**, 177–96.

Howard, D. (1990). 'Nicht sein kann was nicht sein darf', or the prehistory of EPR, 1909–1935: Einstein's early worries about the quantum mechanics of composite systems. In Miller (1990), pp. 61–111.

Hylleraas, E. A. (1928). Über den Grundzustand des Heliumatoms. *Zeitschrift für Physik*, **48**, 469–94.

Jammer, M. (1966). *The Conceptual Development of Quantum Mechanics.* New York: McGraw-Hill.

Jammer, M. (1974). *The Philosophy of Quantum Mechanics: The Interpretations of Quantum Mechanics in Historical Perspective.* New York: John Wiley and Sons.

Jordan, P. (1926a). Über kanonische Transformationen in der Quantenmechanik. *Zeitschrift für Physik*, **37**, 383–6.

Jordan, P. (1926b). Über kanonische Transformationen in der Quantenmechanik, II. *Zeitschrift für Physik*, **38**, 513–17.

Jordan, P. (1927a). Über quantenmechanische Darstellung von Quantensprüngen. *Zeitschrift für Physik*, **40**, 661–6.

Jordan, P. (1927b). Über eine neue Begründung der Quantenmechanik, I. *Zeitschrift für Physik*, **40**, 809–38.

Jordan, P. (1927c). Über eine neue Begründung der Quantenmechanik, II. *Zeitschrift für Physik*, **44**, 1–25.

Jordan, P. (1927d). Zur Quantenmechanik der Gasentartung. *Zeitschrift für Physik*, **44**, 473–80.

Jordan, P. (1927e). Die Entwicklung der neuen Quantenmechanik. *Die Naturwissenschaften*, **15**, 614–23, 636–49.

Jordan, P. (1927f). Kausalität und Statistik in der modernen Physik. *Die Naturwissenschaften*, **15**, 105–10.

Jordan, P. (1927g). Philosophical foundations of quantum theory. *Nature*, **119**, 566–9.

Kiefer, C., Polarski, D., and Starobinsky, A. A. (1998). Quantum-to-classical transition for fluctuations in the early universe. *International Journal of Modern Physics D*, **7**, 455–62.

Kirsten, C., and Treder, H. J. (1979). *Albert Einstein in Berlin 1913–1933.* Berlin: Akademie-Verlag.

Klein, O. (1926). Quantentheorie und fünfdimensionale Relativitätstheorie. *Zeitschrift für Physik*, **37**, 895–906.

Koopman, B. O. (1955). Quantum theory and the foundations of probability. In *Applied Probability*, ed. L. A. MacColl. New York: McGraw-Hill, pp. 97–102.

Kramers, H. A. (1924). The quantum theory of dispersion. *Nature*, **114**, 310–11.

Kramers, H. A., and Heisenberg, W. (1925). Über die Streuung von Strahlung durch Atome. *Zeitschrift für Physik*, **31**, 681–708.

Kragh, H. (1990). *Dirac: A Scientific Biography.* Cambridge: Cambridge University Press.

Kuhn, W. (1925). Über die Gesamtstärke der von einem Zustande ausgehenden Absorptionslinien. *Zeitschrift für Physik*, **33**, 408–12.

Lacki, J. (2004). The puzzle of canonical transformations in early quantum mechanics. *Studies in History and Philosophy of Modern Physics*, **35**, 317–44.

Lanczos, K. (1926). Über eine feldmässige Darstellung der neuen Quantenmechanik. *Zeitschrift für Physik*, **35**, 812–30.

Liddle, A. R. and Lyth, D. H. (2000). *Cosmological Inflation and Large-Scale Structure*. Cambridge: Cambridge University Press.

Lochak, G. (1992). *Louis de Broglie*. Paris: Flammarion.

London, F. (1926a). Über die Jacobischen Transformationen der Quantenmechanik. *Zeitschrift für Physik*, **37**, 915–25.

London, F. (1926b). Winkelvariable und kanonische Transformationen in der Undulationsmechanik. *Zeitschrift für Physik*, **40**, 193–210.

Ludwig, G. (1968). *Wave Mechanics*. Oxford: Pergamon Press.

Madelung, E. (1926a). Eine anschauliche Deutung der Gleichung von Schrödinger. *Die Naturwissenschaften*, **14**, 1004.

Madelung, E. (1926b). Quantentheorie in hydrodynamischer Form. *Zeitschrift für Physik*, **40**, 322–6.

Margenau, H. (1939). Van der Waals forces. *Reviews of Modern Physics*, **11**, 1–35.

Mehra, J. (1975). *The Solvay Conferences on Physics: Aspects of the Development of Physics since 1911*. Dordrecht and Boston: Reidel.

Mehra, J., and Rechenberg, H. (1982a). *The Historical Development of Quantum Theory*, vol. 1, part 2. New York: Springer.

Mehra, J., and Rechenberg, H. (1982b). *The Historical Development of Quantum Theory*, vol. 2. New York: Springer.

Mehra, J., and Rechenberg, H. (1982c). *The Historical Development of Quantum Theory*, vol. 3. New York: Springer.

Mehra, J., and Rechenberg, H. (1987). *The Historical Development of Quantum Theory*, vol. 5, part 2. New York: Springer.

Mehra, J., and Rechenberg, H. (2000). *The Historical Development of Quantum Theory*, vol. 6, part 1. New York: Springer.

Miller, A. I. (ed.) (1990). *Sixty-two Years of Uncertainty*. New York: Plenum Press.

Millikan, R. A. (1916). A direct photoelectric determination of Planck's "*h*". *Physical Review*, **7**, 355–88.

Moore, W. (1989). *Schrödinger: Life and Thought*. Cambridge: Cambridge University Press.

Muller, F. (1997). The Equivalence Myth of Quantum Mechanics, parts I and II and addendum. *Studies in History and Philosophy of Modern Physics*, **28B**, 35–61 and 219–47; *ibid.*, **30B**, 543–5.

Nelson, E. (1966). Derivation of the Schrödinger equation from Newtonian mechanics. *Physical Review*, **150**, 1079–85.

Newton, I. (1730). *Opticks: Or, a Treatise of the Reflections, Refractions, Inflections and Colours of Light*, 4th edn. London. Reprinted as I. Newton, *Opticks*. New York: Dover, 1979.

Nicolai, G. F. (1917). *Die Biologie des Krieges*. Zürich: Orell Füssli.

Norsen, T. (2005). Einstein's boxes. *American Journal of Physics*, **73**, 164–76.

Nye, M. J. (1997). Aristocratic culture and the pursuit of science: The de Broglies in modern France. *Isis*, **88**, 397–421.

Omnès, R. (1992). Consistent interpretations of quantum mechanics. *Reviews of Modern Physics*, **64**, 339–82.

Omnès, R. (1994). *The Interpretation of Quantum Mechanics*. Princeton: Princeton University Press.

Oseen, C. W. (1999). Presentation speech, 12 December 1929. In *Nobel Lectures in Physics (1901–1995)*, CD-Rom edn. Singapore: World Scientific.

Padmanabhan, T. (1993). *Structure Formation in the Universe*. Cambridge: Cambridge University Press.

Pais, A. (1982). *Subtle is the Lord: The Science and the Life of Albert Einstein*. Oxford: Oxford University Press.

Patterson, A. L. (1927). The scattering of electrons from single crystals of nickel. *Nature*, **120**, 46–7.

Pauli, W. (1925). Über die Intensitäten der im elektrischen Feld erscheinenden Kombinationslinien. *Matematisk-fysiske Meddelser*, **7**, 3–20.

Pauli, W. (1927). Über Gasentartung und Paramagnetismus. *Zeitschrift für Physik*, **41**, 81–102.

Pauli, W. (1979). *Wissenschaftlicher Briefwechsel mit Bohr, Einstein, Heisenberg u.a., Teil I: 1919-1929*, eds. A. Hermann, K. v. Meyenn and V. F. Weisskopf. Berlin and Heidelberg: Springer.

Pearle, P. (1976). Reduction of the state vector by a nonlinear Schrödinger equation. *Physical Review D*, **13**, 857–68.

Pearle, P. (1979). Toward explaining why events occur. *International Journal of Theoretical Physics*, **18**, 489–518.

Pearle, P. (1989). Combining stochastic dynamical state-vector reduction with spontaneous localization. *Physical Review A*, **39**, 2277–89.

Pearle, P., and Valentini, A. (2006). Quantum mechanics: Generalizations. In *Encyclopedia of Mathematical Physics*, eds. J.-P. Françoise, G. Naber and T. S. Tsun. Amsterdam: Elsevier, pp. 265–76.

Perez, A., Sahlmann, H., and Sudarsky, D. (2006). On the quantum origin of the seeds of cosmic structure. *Classical and Quantum Gravity*, **23**, 2317–54.

Perovic, S. (2006). Schrödinger's interpretation of quantum mechanics and the relevance of Bohr's experimental critique. *Studies in History and Philosophy of Modern Physics*, **37**, 275–97.

Philippidis, C., Dewdney, C., and Hiley, B. J. (1979). Quantum interference and the quantum potential. *Nuovo Cimento B*, **52**, 15–28.

Polkinghorne, J. (2002). *Quantum Theory: A Very Short Introduction*. Oxford: Oxford University Press.

Popper, K. R. (1982). *Quantum Theory and the Schism in Physics*. London: Unwin Hyman.

Przibram, K. (ed.) (1967). *Letters on Wave Mechanics*. New York: Philosophical Library. Originally published as *Briefe zur Wellenmechanik*. Wien: Springer, 1963.

Rovelli, C. (2004). *Quantum Gravity*. Cambridge: Cambridge University Press.

Saunders, S. (1995). Time, decoherence and quantum mechanics. *Synthese*, **102**, 235–66.

Saunders, S. (1998). Time, quantum mechanics, and probability. *Synthese*, **114**, 373–404.

Scheibe, E. (1993). Heisenbergs Begriff der abgeschlossenen Theorie. In *Werner Heisenberg: Physiker und Philosoph*, eds. B. Geyer, H. Herwig and H. Rechenberg. Heidelberg: Spektrum, pp. 251–7.

Schiff, L. I. (1955). *Quantum Mechanics*. New York: McGraw-Hill.

Schilpp, P. A. (ed.) (1949). *Albert Einstein: Philosopher-Scientist*. The Library of Living Philosophers, vol. VII. La Salle, Ill.: Open Court.

Schrödinger, E. (1922). Dopplerprinzip und Bohrsche Frequenzbedingung. *Physikalische Zeitschrift*, **23**, 301–3.

Schrödinger, E. (1924a). Über das thermische Gleichgewicht zwischen Licht- uns Schallstrahlen. *Physikalische Zeitschrift*, **25**, 89–94.

Schrödinger, E. (1924b). Bohrs neue Strahlungshypothese und der Energiesatz. *Die Naturwissenschaften*, **12**, 720–4.

Schrödinger, E. (1926a). Zur Einsteinschen Gastheorie. *Physikalische Zeitschrift*, **27**, 95–101.

Schrödinger, E. (1926b). Quantisierung als Eigenwertproblem (erste Mitteilung). *Annalen der Physik*, **79**, 361–76.

Schrödinger, E. (1926c). Quantisierung als Eigenwertproblem (zweite Mitteilung). *Annalen der Physik*, **79**, 489–527.

Schrödinger, E. (1926d). Über das Verhältnis der Heisenberg-Born-Jordanschen Quantenmechanik zu der meinen. *Annalen der Physik*, **79**, 734–56.

Schrödinger, E. (1926). Der stetige Übergang von der Mikro- zur Makromechanik. *Die Naturwissenschaften*, **14**, 664–6.

Schrödinger, E. (1926f). Quantisierung als Eigenwertproblem (dritte Mitteilung: Störungstheorie, mit Anwendung auf den Starkeffekt der Balmerlinien). *Annalen der Physik*, **80**, 437–90.

Schrödinger, E. (1926g). Quantisierung als Eigenwertproblem (vierte Mitteilung). *Annalen der Physik*, **81**, 109–139.

Schrödinger, E. (1926h). An undulatory theory of the mechanics of atoms and molecules. *Physical Review*, **28**, 1049–1070.

Schrödinger, E. (1926i). Vorwort zur ersten Auflage. In Schrödinger (1928), p. iv.

Schrödinger, E. (1927a). Über den Comptoneffekt. *Annalen der Physik*, **82**, 257–64.

Schrödinger, E. (1927b). Der Energieimpulssatz der Materiewellen. *Annalen der Physik*, **82**, 265–72.

Schrödinger, E. (1927c). Energieaustausch nach der Wellenmechanik. *Annalen der Physik*, **83**, 956–68.

Schrödinger, E. (1928). *Abhandlungen zur Wellenmechanik*, 2nd, enlarged edn. Leipzig: Barth. Translated as *Collected Papers on Wave Mechanics*. London and Glasgow: Blackie and Son, 1928.

Schrödinger, E. (1929a). Was ist ein Naturgesetz? *Die Naturwissenschaften*, **17**, 9–11.

Schrödinger, E. (1929b). Die Erfassung der Quantengesetze durch kontinuierliche Funktionen. *Die Naturwissenschaften*, **13**, 486–9.

Schrödinger, E. (1935). Die gegenwärtige Situation in der Quantenmechanik. *Die Naturwissenschaften*, **23**, 807–12, 823–8, 844–9.

Schrödinger, E. (1936). Probability relations between separated systems. *Proceedings of the Cambridge Philosophical Society*, **32**, 446–52.

Schrödinger, E. (1984). *Gesammelte Abhandlungen*, eds. Österreichische Akademie der Wissenschaften. Braunschweig: Vieweg.

Schrödinger, E. (1995). *The Interpretation of Quantum Mechanics*, ed. and with introduction by M. Bitbol. Woodbridge, Conn.: Ox Bow Press.

Schulman, L. S. (1997). *Time's Arrows and Quantum Measurement*. Cambridge: Cambridge University Press.

Senftleben, H. A. (1923). Zur Grundlegung der 'Quantentheorie', I. *Zeitschrift für Physik*, **22**, 127–56.

Shankar, R. (1994). *Principles of Quantum Mechanics*, 2nd edn. New York: Springer-Verlag.

Shimony, A. (2005). Comment on Norsen's defense of Einstein's 'box argument'. *American Journal of Physics*, **73**, 177–8.

Slater, J. C. (1924). Radiation and atoms. *Nature*, **113**, 307–8.

Smekal, A. (1923). Zur Quantentheorie der Dispersion. *Die Naturwissenschaften*, **11**, 873–5.

Taylor, G. I. (1909). Interference fringes with feeble light. *Proceedings of the Cambridge Philosophical Society*, **15**, 114–15.

Thomas, W. (1925). Über die Zahl der Dispersionselektronen, die einem stationären Zustande zugeordnet sind (vorläufige Mitteilung). *Die Naturwissenschaften*, **13**, 627.

Unruh, W. G. and Wald, R. M. (1989). Time and the interpretation of canonical quantum gravity. *Physical Review D*, **40**, 2598–2614.

Valentini, A. (1991a). Signal-locality, uncertainty, and the subquantum H-theorem, I. *Physics Letters A*, **156**, 5–11.

Valentini, A. (1991b). Signal-locality, uncertainty, and the subquantum H-theorem, II. *Physics Letters A*, **158**, 1–8.

Valentini, A. (1992). On the pilot-wave theory of classical, quantum and subquantum physics. Ph.D. Thesis, International School for Advanced Studies, Trieste, Italy. http://www.sissa.it/ap/PhD/Theses/valentini.pdf.

Valentini, A. (1997). On Galilean and Lorentz invariance in pilot-wave dynamics. *Physics Letters A*, **228**, 215–22.

Valentini, A. (2001). Hidden variables, statistical mechanics and the early universe. In *Chance in Physics: Foundations and Perspectives*, eds. J. Bricmont, D. Dürr, M. C. Galavotti, G. Ghirardi, F. Petruccione and N. Zanghì. Berlin: Springer-Verlag, pp. 165–81.

Valentini, A. (2002a). Signal-locality in hidden-variables theories. *Physics Letters A*, **297**, 273–8.

Valentini, A. (2002b). Subquantum information and computation. *Pramana – Journal of Physics*, **59**, 269–77.

Valentini, A. (2008a). Cambridge: Cambridge University Press. Forthcoming.

Valentini, A. (2008b). Inflationary cosmology as a probe of primordial quantum mechanics. arXiv:0805.0163 [hep-th].

Valentini, A. and Westman, H. (2005). Dynamical origin of quantum probabilities. *Proceedings of the Royal Society of London A*, **461**, 253–72.

von Laue, M. (1914). Die Freiheitsgrade von Strahlenbündeln. *Annalen der Physik*, **44**, 1197–1212.

von Neumann, J. (1927). Mathematische Begründung der Quantenmechanik. *Nachrichten der Akademie der Wissenschaften in Göttingen. II., Mathematisch-Physikalische Klasse*, 20 May 1927, 1–57.

Wallace, D. (2003a). Everett and structure. *Studies in History and Philosophy of Modern Physics*, **34**, 87–105.

Wallace, D. (2003b). Everettian rationality: Defending Deutsch's approach to probability in the Everett interpretation. *Studies in History and Philosophy of Modern Physics*, **34**, 415–39.

Wallstrom, T. C. (1994). Inequivalence between the Schrödinger equation and the Madelung hydrodynamic equations. *Physical Review A*, **49**, 1613–17.

Wentzel, G. (1926). Zur Theorie des photoelektrischen Effekts. *Zeitschrift für Physik*, **40**, 574–89.

Wentzel, G. (1927). Über die Richtungsverteilung der Photoelektronen. *Zeitschrift für Physik*, **41**, 828–32.

Wheeler, J. A. (1978). The 'past' and the 'delayed-choice' double-slit experiment. In *Mathematical Foundations of Quantum Mechanics*, ed. A. R. Marlow. New York: Academic, pp. 9–48. Reprinted in Wheeler and Zurek (1983), pp. 182–213.

Wheeler, J. A. (1986). Interview. In *The Ghost in the Atom*, eds. P. C. W. Davies and J. R. Brown, Cambridge: Cambridge University Press, pp. 58–69.

Wheeler, J. A., and Zurek, W. H. (eds.) (1983). *Quantum Theory and Measurement*. Princeton: Princeton University Press.

Wien, W. (1923). *Kanalstrahlen*, 2nd edn. Vol. IV.1 of *Handbuch der Radiologie*, ed. E. Marx. Leipzig: Akademische Verlagsgesellschaft.

Wigner, E. P. (1961). Remarks on the mind-body question. In *The Scientist Speculates*, ed. I. J. Good. London: Heinemann, pp. 284–302. Reprinted in E. P. Wigner *Symmetries and Reflections*. Bloomington: Indiana University Press, 1967, pp. 171–84. Also reprinted in Wheeler and Zurek (1983), pp. 168–81.

Wigner, E. P. (1980). Thirty years of knowing Einstein. In *Some Strangeness in the Proportion: A Centennial Symposium to Celebrate the Achievements of Albert Einstein*, ed. H. Woolf. Reading, Mass.: Addison-Wesley, pp. 461–8.

Wootters, W. K., and Zurek, W. H. (1979). Complementarity in the double-slit experiment: quantum nonseparability and a quantitative statement of Bohr's principle. *Physical Review D*, **19**, 473–84.

Zurek, W. H. (1991). Decoherence and the transition from quantum to classical. *Physics Today*, **44**, n. 10 (October 1991), pp. 36–44.

Index

Printed in the United States
By Bookmasters